Lecture Notes in Computer Science

Lecture Notes in Artificial Intelligence 14981
Founding Editor

Jörg Siekmann

Series Editors

Randy Goebel, *University of Alberta, Edmonton, Canada*
Wolfgang Wahlster, *DFKI, Berlin, Germany*
Zhi-Hua Zhou, *Nanjing University, Nanjing, China*

The series Lecture Notes in Artificial Intelligence (LNAI) was established in 1988 as a topical subseries of LNCS devoted to artificial intelligence.

The series publishes state-of-the-art research results at a high level. As with the LNCS mother series, the mission of the series is to serve the international R & D community by providing an invaluable service, mainly focused on the publication of conference and workshop proceedings and postproceedings.

Jens Lemanski · Mikkel Willum Johansen ·
Emmanuel Manalo · Petrucio Viana ·
Reetu Bhattacharjee · Richard Burns
Editors

Diagrammatic Representation and Inference

14th International Conference, Diagrams 2024
Münster, Germany, September 27 – October 1, 2024
Proceedings

Editors
Jens Lemanski
University of Münster
Münster, Germany

Emmanuel Manalo
Kyoto University
Kyoto, Japan

Reetu Bhattacharjee
University of Münster
Münster, Germany

Mikkel Willum Johansen
University of Copenhagen
Copenhagen, Denmark

Petrucio Viana
Universidade Federal Fluminense
Niterói, Brazil

Richard Burns
West Chester University of Pennsylvania
West Chester, PA, USA

ISSN 0302-9743 ISSN 1611-3349 (electronic)
Lecture Notes in Artificial Intelligence
ISBN 978-3-031-71290-6 ISBN 978-3-031-71291-3 (eBook)
https://doi.org/10.1007/978-3-031-71291-3

LNCS Sublibrary: SL7 – Artificial Intelligence

© The Editor(s) (if applicable) and The Author(s), under exclusive license
to Springer Nature Switzerland AG 2024, corrected publication 2024
Chapters 4, 5, 13, 19, 22, 32, and 33 are licensed under the terms of the Creative Commons Attribution 4.0 International License (http://creativecommons.org/licenses/by/4.0/). For further details see license information in the chapters.

This work is subject to copyright. All rights are solely and exclusively licensed by the Publisher, whether the whole or part of the material is concerned, specifically the rights of translation, reprinting, reuse of illustrations, recitation, broadcasting, reproduction on microfilms or in any other physical way, and transmission or information storage and retrieval, electronic adaptation, computer software, or by similar or dissimilar methodology now known or hereafter developed.
The use of general descriptive names, registered names, trademarks, service marks, etc. in this publication does not imply, even in the absence of a specific statement, that such names are exempt from the relevant protective laws and regulations and therefore free for general use.
The publisher, the authors and the editors are safe to assume that the advice and information in this book are believed to be true and accurate at the date of publication. Neither the publisher nor the authors or the editors give a warranty, expressed or implied, with respect to the material contained herein or for any errors or omissions that may have been made. The publisher remains neutral with regard to jurisdictional claims in published maps and institutional affiliations.

This Springer imprint is published by the registered company Springer Nature Switzerland AG
The registered company address is: Gewerbestrasse 11, 6330 Cham, Switzerland

If disposing of this product, please recycle the paper.

Preface

The 14th International Conference on the Theory and Application of Diagrams (DIAGRAMS 2024) was held at the Fürstbischöfliches Schloss of the University of Münster between 27 September and 1 October 2024. The conference thus directly followed on from the Congress of the German Society of Philosophy (Deutsche Gesellschaft für Philosophie, DGPhil), which concluded on 26 September in Münster, with the intention of establishing a closer connection between the topic of visual representation and reasoning and the philosophical community.

Submissions to DIAGRAMS 2024 were sought in the form of long papers, short papers, posters and non-archival abstracts. Each submission was subjected to at least three peer reviews using a single-blind methodology. The authors were given the opportunity to respond to the reviews in the form of a rebuttal. Subsequently, for each paper both the reviews and the rebuttal were taken into account in an animated debate among the Program Committee, before a final decision was reached. This rigorous process ensured that only the most meritorious papers were accepted for presentation at the conference. We extend our gratitude to all members of the Program Committee for their dedication and commitment to the discussion.

One and a half years after the World Health Organization declared the global coronavirus pandemic to have reached a conclusion, a modest upward trend is beginning to emerge in the number of submissions to conferences. A total of 80 submissions were received. Of these submissions, 17 were accepted for presentation as long papers. A further 19 submissions were accepted as short papers, eight as abstracts, and 11 for poster presentation. Of the 80 submissions, 41 were submitted in the main track, 24 in the Philosophy track and 15 in the Psychology and Education track. This observation suggests that there is an under-representation of thematic tracks. This can be attributed, however, to several factors. For instance, certain organizers who were also authors opted to submit to the main track due to the peer review process.

Alongside the regular program, DIAGRAMS 2024 included a Graduate Symposium, and the conference was very successful in attracting workshop and tutorial submissions. A total of ten tutorials and three workshops were accepted for the conference. The workshop entitled Peirce's Philosophy of Notation included nine presentations, while the sessions on Diagrams and Mathematical Practice and The Diagrammatic Turn in Quantum Physics each included seven presentations.

This year, for the first time, three keynote speakers, corresponding to each of the tracks, were invited to present. Mateja Jamnik was selected for the main track, Catarina Dutilh Novaes for the philosophy track, and Barbara Tversky represented the psychology and education track. Nevertheless, their exceptional contributions extend far beyond the confines of the respective subject areas they represent.

There are, of course, many people to whom we are indebted for their considerable assistance in making DIAGRAMS 2024 a success. We thank Silvia de Toffoli for her work as Workshop Chair, Andrea Reichenberger for her role as Tutorial Chair, Giulia

Ferrari for her role as Publicity Chair, and Leonie Bosveld for her role as Graduate Symposium Chair. Richard Burns took on the demanding role of Proceedings Chair, for which we are immensely grateful. The local organization was conducted by Jens Lemanski and Reetu Bhattacharjee, who would not have been able to complete this task without the guidance and assistance of Niko Strobach.

Many thanks go to the Zentrum für Wissenschaftstheorie (ZfW) at the University of Münster, which actively supported the conference. We are also grateful to the NSF for providing significant funding, which supported graduate students from US-based institutions with their participation. Jens Lemanski was further supported by the Thyssen project *History of Logic Diagrams in Kantianism*, and Reetu Bhattacharjee by the DFG project *Gestures and Diagrams in Visual-Spatial Communication* in the priority program *ViCom – Visual Communication*. Finally, we would like to express our gratitude to the DIAGRAMS Steering Committee for their unwavering support, guidance, and motivation.

July 2024

Jens Lemanski
Mikkel Willum Johansen
Emmanuel Manalo
Petrucio Viana
Reetu Bhattacharjee
Richard Burns

Organization

Executive Committee

General Chair

Jens Lemanski — University of Münster, Germany

Senior Program Chair

Mikkel Willum Johansen — University of Copenhagen, Denmark

Main Track Chair

Jorge Petrucio Viana — Federal Fluminense University, Brazil

Philosophy Track Chair

Reetu Bhattacharjee — University of Münster, Germany

Psychology and Education Track Chair

Emmanuel Manalo — Kyoto University, Japan

Proceedings Chair

Richard Burns — West Chester University, USA

Local Chairs

Niko Strobach — University of Münster, Germany
Reetu Bhattacharjee — University of Münster, Germany

Graduate Symposium Chair

Leonie Bosveld — University of Groningen, Netherlands

Workshop Chair

Silvia De Toffoli — IUSS Pavia, Italy

Tutorial Chair

Andrea Reichenberger — TUM Munich, Germany

Publicity Chair

Giulia Ferrari — University of Torino, Italy

Program Committee

Andrew Aberdein	Florida Institute of Technology, USA
Mohanad Alqadah	Umm Al-Qura University, Saudi Arabia
Amrita Basu	Jadavpur University, India
David Beisecker	University of Nevada, USA
Francesco Bellucci	University of Bologna, Italy
Jean Paul Van Bendegem	Vrije Universiteit Brussel, Belgium
Reetu Bhattacharjee	University of Münster, Germany
Andrew Blake	University of Brighton, UK
Viktor Blasjo	Utrecht University, Netherlands
Ben Blumson	National University of Singapore, Singapore
Leonie Bosveld-De Smet	University of Groningen, Netherlands
Jean-Michel Boucheix	University of Burgundy, France
Richard Burns	West Chester University, USA
Mihir Chakraborty	Jadavpur University, India
Peter Chapman	Edinburgh Napier University, UK
Peter Cheng	University of Sussex, UK
Daniele Chiffi	Polytechnic University of Milan, Italy
James Corter	Columbia University, USA
Gennaro Costagliola	University of Salerno, Italy

Lorenz Demey	KU Leuven, Belgium
Jacques Fleuriot	University of Edinburgh, UK
Amy Fox	University of California San Diego, USA
Reinhard von Hanxleden	Christian-Albrecht University of Kiel, Germany
Nathan Haydon	Tallinn Institute of Technology, Estonia
Mikkel Willum Johansen	University of Copenhagen, Denmark
Yasuhiro Katagiri	National Institute of Advanced Industrial Science and Technology, Japan
Michael Klasen	University of Münster, Germany
Janina Krawitz	University of Cologne, Germany
Timm Lampert	Humboldt University of Berlin, Germany
Brendan Larvor	University of Hertfordshire, UK
Catherine Legg	Deakin University, Australia
Javier Legris	University of Buenos Aires, Argentina
Jens Lemanski	University of Münster, Germany
Emmanuel Manalo	Kyoto University, Japan
Kim Marriott	Monash University, Australia
Ingolf Max	Leipzig University, Germany
Mark Minas	University of the Bundeswehr Munich, Germany
Amirouche Moktefi	Tallinn University of Technology, Estonia
Laura Ohmes	Carl von Ossietzky University, Germany
Arnold Oostra	University of Tolima, Colombia
Volker Peckhaus	Paderborn University, Germany
Niki Pfeifer	University of Regensburg, Germany
Mario Piazza	Scuola Normale Superiore in Pisa, Italy
Margit Pohl	Vienna University of Technology, Austria
Henri Prade	IRIT – CNRS, France
Uta Priss	Ostfalia University, Germany
Daniel Raggi	University of Cambridge, UK
Andrea Reichenberger	TUM Munich, Germany
Peter Rodgers	University of Kent, UK
Yuri Sato	Ochanomizu University, Japan
Frank Thomas Sautter	Federal University of Santa Maria, Brazil
Fabien Schang	HSE University, Russia
Dirk Schlimm	McGill University, Canada
Stanislaw Schukajlow	University of Münster, Germany
Christoph Daniel Schulze	Christian-Albrecht University of Kiel, Germany
Andrew Schumann	UITM in Rzeszow, Poland
Stephanie Schwartz	Millersville University, USA
Marco Segala	University of L'Aquila, Italy
Atsushi Shimojima	Doshisha University, Japan
Hans Smessaert	KU Leuven, Belgium

Pawel Sobocinski   Tallinn University of Technology, Estonia
Michael Stoeltzner   University of South Carolina, USA
Takeshi Sugio   Doshisha University, Japan
Matthias Thimm   FernUniversität in Hagen, Germany
Silvia De Toffoli   IUSS Pavia, Italy
Yuri Uesaka   University of Tokyo, Japan
Ioannis Vandoulakis   Hellenic Open University, Greece
Ivan Varzinczak   Artois University, France
Petrucio Viana   Federal Fluminense University, Brazil
Erica de Vries   Grenoble Alpes University, France
Leo Wennmann   Maastricht University, Netherlands
Julia Xing   Tarleton State University, USA

Additional Reviewers

Malte Clement   Alfonso Piscitelli
Mattia De Rosa   Niklas Rentz
Nathaniel Gan   Kai Sauerwald
Cem Gündogan   Alexander Schulz-Rosengarten
Maximilian Kasperowski   Jannis Schopp
Kim Mönch   Parinaz Tabari
Jette Petzold   Peter Weinert

Contents

Analysis of Diagrams

Diagrams and Their Role in Economics as Problem-Solving Devices
and Theory-Improving Tools. The Case of the Phillip Machine 3
  Giulia Miotti

Why Feynman Diagrams Are Worth 10,000 Formulae: A Representational
Epistemological Analysis 11
  Peter C-H. Cheng and Timon G. Boehm

A Building-Block Approach to the Diversity of Visualization Types – Each
Type Expressed Visually, and as a Systematically Generated Sentence 28
  Yuri Engelhardt and Clive Richards

Domain-Specific Rules Override Aesthetic Graph Drawing Criteria:
An Exploration of User-Generated Diagrams 44
  Stefan Helmke, Kerem Doğan, Robert Scheffler, and Gregor Wrobel

Generating Qualitative Descriptions of Diagrams with a Transformer-Based
Language Model 61
  Marco Schorlemmer, Mohamad Ballout, and Kai-Uwe Kühnberger

Diagram Control and Model Order for Sugiyama Layouts 76
  Sören Domrös and Reinhard von Hanxleden

B_42: The Geometry of 4-Valued Contradiction 84
  Alessio Moretti

A Way Diagrams Explain: Analysis Based on Consequence Matching 101
  Atsushi Shimojima and Dave Barker-Plummer

Euler Diagrams, Aristotelian Diagrams and Syllogistics 111
  Lorenz Demey and Hans Smessaert

What Does It Mean that Diagrams Represent Constructions? 129
  Piotr Kozak

The Topology of Assertion: A Diagrammatic Rationale for Our Enduring
Love of Truth 137
  Dave Beisecker

Schopenhauer's Sorites Diagram 145
    Christina Kittsteiner

Category Theory for Aristotelian Diagrams: The Debate on Singular
Propositions 153
    Alexander De Klerck, Leander Vignero, and Lorenz Demey

Euler and Venn Diagrams

Rectangular Euler Diagrams and Order Theory 165
    Uta Priss and Dominik Dürrschnabel

Reference by Occurrence 182
    Francesco Bellucci

EulerMerge: Simplifying Euler Diagrams Through Set Merges 190
    Xinyuan Yan, Peter Rodgers, Peter Rottmann, Daniel Archambault,
    Jan-Henrik Haunert, and Bei Wang

Representing Uncertainty with Expanded Ueberweg Diagrams 207
    Amirouche Moktefi, Reetu Bhattacharjee, and Jens Lemanski

Indeterminate Set Space Diagrams 215
    Björn Gottfried

Can Euler Diagrams Improve Syllogistic Reasoning in Large Language
Models? 232
    Risako Ando, Kentaro Ozeki, Takanobu Morishita, Hirohiko Abe,
    Koji Mineshima, and Mitsuhiro Okada

Diagrams in Logic

Mozi's Square of Opposition and Logemes as New Logical Approach 251
    Andrew Schumann

Implicational Existential Graphs 267
    Arnold Oostra

Aristotelian Diagrams as Logic Diagrams 275
    Stef Frijters and Atahan Erbas

Sentence Negation and Term Negation as Syntactic Operations in Diagram
Logic 284
    Sohail Hossain and Mihir Kumar Chakrobarty

Playing Games with Diagrams: Truth Diagrams and Game Semantics 300
Can Başkent

Peirce's Extended Euler Diagrams and the System Atl Based
on Ladd-Franklin's Exclusion Relations 316
Fangzhou Xu and Ahti-Veikko Pietarinen

Diagrams and Applications

Anxiety Moderates the Effects of Drawing Support on Drawing Accuracy
in Mathematical Modeling 327
Johanna Schoenherr and Richard E. Mayer

Learning Magnitudes of Energy Consumption with Symbolic or Iconic
Representations 335
Erica de Vries, Neil Schwartz, and Martin Galilée

Designing a Mind-Mapping-Assisted Comparative Literature Course
in Chinese Academic Settings 350
Binfeng Chen, Jing Zhao, and Lin He

Integration of Learning Through the Use of Self-constructed Diagrams:
Opportunities and Challenges 358
Emmanuel Manalo and Mari Fukuda

Chinese Children' Drawing in Science Class 366
Ran Lu and Emmanuel Manalo

Diagram Tools

Hoop Diagrams: A Set Visualization Method 377
*Peter Rodgers, Peter Chapman, Andrew Blake, Martin Nöllenburg,
Markus Wallinger, and Alexander Dobler*

Building a Large Dataset of Human-Generated Captions for Science
Diagrams 393
Yuri Sato, Ayaka Suzuki, and Koji Mineshima

KIELER: A Text-First Framework for Automatic Diagramming
of Complex Systems 402
*Maximilian Kasperowski, Niklas Rentz, Sören Domrös,
and Reinhard von Hanxleden*

Historical Aspects of Diagrams

Drawing Technology: Sketches of Isambard Kingdom Brunel 421
  Guy Clarke Marshall

On the Expressivity of Byzantine Diagrams in Logic 429
  Jens Lemanski and Reetu Bhattacharjee

Posters

An Innovative Approach to Diagrams Representation: The Marlo
Diagrams Web Page 449
  Fernando Soler Toscano and Marcos Bautista López Aznar

Codifying Visual Representations 454
  *Wode Ni, Sam Estep, Hwei-Shin Harriman, Jiří Minarčík,
  and Joshua Sunshine*

A Diagram Helping the Mathematical Problem Solving Procedure 458
  Tullio Aebischer

Collaborative Graph-Document Composition is Efficient and Enhances
Critical-Thinking Skills Without Extra Cost 462
  *Kôiti Hasida, Zilian Zhang, Zifan Yao, Vili Valtteri Karilas,
  Shitao Fang, Kuanghuan Tan, Kenichi Shibata, and Yusuke Matsubara*

An Eye-Tracking Study on the Effects of Using Highlighted Multi-attribute
Tables: A Preliminary Report 467
  *Masahiro Morii, Takashi Ideno, Yuki Tamari, Kazuhisa Takemura,
  and Mitsuhiro Okada*

On the Formal Cause of Diagrams: Mimesis and Phenomenology 472
  Noah Greenstein

The Region Connection Calculus, Euler Diagrams and Aristotelian
Diagrams 476
  Claudia Anger and Lorenz Demey

Between Pro/Con-Lists and Argument Graphs Finding the Right Level
of Complexity in Argumentation Representation 480
  Joannes B. Campell and Michael A. Müller

Diagrammatic Analogical Reasoning 485
  Henri Prade and Gilles Richard

Correction to: A Building-Block Approach to the Diversity of Visualization Types – Each Type Expressed Visually, and as a Systematically Generated Sentence C1
  Yuri Engelhardt and Clive Richards

Author Index 491

Analysis of Diagrams

Diagrams and Their Role in Economics as Problem-Solving Devices and Theory-Improving Tools. The Case of the Phillip Machine

Giulia Miotti(✉)

Department of Philosophy and Education, Università di Torino, Via S.Ottavio, 20, 10124 Torino, Italy
giulia.miotti@unito.it

Abstract. In my proposal, I advance a description of diagrams, as an instance of visualization within theory-building and theory- enhancing processes, as effective problem-solving devices and as didactical tools with powerful didactical properties. The ability of diagrams and diagrammatic reasoning to work as effective theory-enhancing devices has been described in many contexts, from mathematics to chemistry, in my proposal I claim that such ability is even more clear within the context of economics and economic theory-developing processes. Through the presentation of the case-study of the Philip machine, I will argue that these properties turned out to be of fundamental importance in the understanding, circulation and finally practice of the economic theory they "depicted".

Keywords: Diagrammatic Reasoning · Economic Theories · Theory Improvement

1 Diagrams and Visualization as Knowledge-Enhancing Tools

In the last few decades the study of visualization and diagrams' role within knowledge enhancement has strongly developed. It has been argued that visualization and the use of diagrams in particular show interesting epistemic properties, according to (Morgan 2020) scientist do not merely use diagrams to visualize objects and relations among objects in a given scientific field, but they use diagrams to reason with them as proper tools of their disciplines.

An interesting appraisal of diagrams' characteristics in terms of informational effective tools is provided in (Larkin and Simon 1987). In order to determine when, and whether, diagrams are more effective than sentences they are to be evaluated in terms of information-processing systems. In this perspective, "sentential" and "diagrammatic" representations are alternative information-processing systems and insofar they are informationally equivalent, they differ merely in the information-processing operators that act on them. Such operators vary in terms of their capabilities for recognizing patterns, in the inferences they can carry out directly, and in their control strategies. Diagrams and sentence differ under each of such characteristics, but the most significant difference lies in the efficiency of search for information and in the explicitness of information. In diagrammatic representations information is organized by location, which means that much of the information needed to make an inference is present and explicit at a single locations and hints for the next inferential steps are present at an adjacent location. As Larkin and Simon claim, in diagrammatic representations "solving-problems can proceed through a smooth traversal of the diagram, and may require very little search or computation of elements that had been implicit".

The appraisal of diagrammatic reasoning and visualization as effective reasoning tools has been widely discussed within philosophy of mathematics during the last decades, with different degrees of endorsement and different evaluations concerning the actual possibilities of diagrams. An instance of full endorsement can be found in (Grosholz 2013), where she concentrates her analysis on diagrams' epistemic role within analytic geometry. In this specific context, diagrammatic reasoning seems particularly effective since it allows for an "enrichment" of the discourse due to the introduction of visual elements, thus bringing two different discourses together (in the context of analytic geometry, these would be the algebraic-analytical discourse on the one side and the geometric-visual on the other) and also providing a different perspective on the analysis as, for example, complex numbers are easier to visualise when the one-dimensional number line becomes a part of a two-dimensional plane.

A more cautious approach is provided by (Giaquinto 2008) who points out two critical constraints to which diagrammatic reasoning might be subject, which are the possible relevant mismatch between diagrams and captions and the possible, unwarranted generalization from diagrams. These two constraints may prevent diagrammatic reasoning from acting as a means to a proof or discovery. As Giaquinto claims, verbal descriptions can be discrete, i.e. they do not provide more information than needed. Visual representations, on the other hand, are usually indiscrete, as for many properties of, say, kinds F, a visual representation cannot represent something as being F in a particular way. Nonetheless, diagrammatic reasoning proves useful in augmenting understanding in mathematics and geometry: diagrammatic reasoning provides the ability to rearrange the symbol arrays (both in imagination or on paper) and thus allows for symbol manipulation. In Giaquinto's account these characteristics make visualizing and diagrammatic reasoning a non-superfluous part of thinking through a proof, which in some cases can be irreplaceable, meaning that one could not solve the same proof through a process of thought that does not entail visual thinking.

1.1 Non-triviality and Independence as Two Main Elements in Diagrammatic Reasoning

I argue that diagrams and diagrammatic reasoning show at least two features when employed as proper inquiry tools in scientific thinking: non-triviality and independence.

Non-triviality. By non-triviality, I mean the ability of diagrammatic reasoning to provide a radically different strategy for proving theorems. Any diagram working as a problem-solving device is non-trivial if it provides a reliable and rational visual array that leads to the understanding of theorems in a series of steps that are cognitively "more economical" than those needed to follow the whole formal development. More precisely: there are three important elements that need to be taken into account if we describe a diagram as non-trivial with respect to its formal version: the ability of non- trivial diagrammatic to guarantee an "economical" access to theorems entails, the fact that they exhibit a rational array of visual elements and finally the relation between picture-proofs and canonical proofs. As for the economical access: moving from the theorem and the non-iconic proof to the picture-proof and vice versa, the latter is clearly able to provide an easier and quicker way to access the content of the theorem and, consequently, to prove it. Although "easier", the access to the content of a given theorem by means of diagrams is not to be consider as a mere spatial reproduction: the easier access is granted by the nature of the visual display and by the cognitive engagement it requires. A genuinely informative picture-proof consists in a "reasoned" visual display and array of elements which, in order to be understood and properly read, requires a rational selection of the elements to be taken in consideration and of the order in which they are to be read. Thus in picture-proofs a higher level of abstraction than that usually required in mere physiological visualization is required; and it is precisely this higher level of abstraction that allows diagrams to prove theorems even in highly abstract domains where pictures are often supposed to have little explanatory or proving power.

Independence. On the other hand, by the characteristic of independence, I refer to the ability of diagrams to build a proof autonomously, i.e., to build a proof-search for a given theorem within the diagram itself. At this point, a difference seems to emerge between the way a non-trivial picture proof proves a theorem and the way an independent visual array of figures proves a theorem, and I defend the claim by distinguishing the relation that diagrams have (in these two different cases) from the formal proofs of the theorems they are supposed to prove.

In the case of non-trivial picture-proofs, the epistemic standards the visual proof has to respect are, in a sense, articulated in accord with those set by the formal proof, and the outcome of the picture proof cannot exceed the outcome of the formal proof: even if they are useful and accord with reliable epistemic standards, non-trivial picture-proofs are "ancillary" to formal proofs. By contrast, in the case of independent visual arrays of figures, the visual construction itself sets the standards for the proof, since the proof is built "inside" the visual array. I refer here to the analysis provided by (Macbeth 2014) concerning Euclidean diagrams. According to Macbeth, Euclidean diagrams allow reasoning to develop in the diagrammatic display rather than on it. There are three different aspects of diagrams which make them able to display contents in an independent way. The first aspect consists in the ability of diagrammatic elements to assume a non-natural meaning, the second one consists in the particular relation between parts and whole, and finally the third aspect consists in the fact that in order to understand a diagram, the reader needs to run through three different levels of analysis.

Let me explain a little further each of these aspects. In this account, since diagrams behave as non-natural objects, they escape the limit that seems to be inherent to many visual representations of being individual instances of objects and therefore unable to convey general meaning: on this account, diagrams convey a non-natural meaning insofar as they behave as general icons (for instance, as a geometer draws some lines to form a rectilinear angle to show that an angle can be bisected, he or she does not mean to draw angle that is obtuse or acute, but simply to draw *an angle*). Furthermore, diagrams exhibit a particular relation among their primary elements (their parts) and the diagram as a whole. For example, Euclidean diagrams are made up of primitive elements such as points, lines, segments with a clear theoretical independence, which, within a diagrammatic construction, can combine with one another and therefore they allow more complex structures to emerge within which they behave as parts of a whole. In these constructions, primary elements actually build new elements (different from and more complex than their constituent parts) but nonetheless, they remain "detectable" inside the construction. This latter aspect paves the way to the third one, i.e. the possibility of reasoning in the diagram in a complex sequence which involves three levels: at the first, lowest level of articulation, primitive parts are found; at the second level, the concepts of different objects are found (these could be understood as specified by some kind of "instructions"); finally, at the third level, the whole diagram is found. At this third level, we find the diagram as a whole, within which various second-level objects can be found depending on the specific "instructions" followed in the configuration of the drawn lines within the diagram itself. If we take the diagram as a whole of (intermediate) parts, we may even obtain a diagram working as a discovery tool; because of this specific structure, one may conceive parts of the intermediate wholes as new wholes and thus discover something new.

2 Diagrams in Economic Theories

Economic theories can be described as complex tools composed of different items. The first one can be described as a "narrative part", which usually contains specific assumptions concerning human behavior and economic agents' interactions, the economic environment in which they move and the constraints to which it might be subject (both within and outside its boundaries, e.g. ecological constraints) and how economic agents react, adjust and also shape such environment. The second item consists in a specific mathematical framework, displayed in terms of different arrays of mathematical instruments such as equations or functions which should express in quantitative terms the contents of the assumptions previously proposed in a qualitative form. Finally, the third item consists in a diagrammatic part which in most theories sometimes act as a "synthesis".

Nonetheless, not all theories in the history of economics present such ingredients and within each theory a different degree of importance or refinement is given to each of them. For instance, if we take a look at the first, most famous economic theories and treatises we find that both the narrative and the diagrammatic parts are given a much higher importance than the pure mathematical framework. An example in this sense can be found in the funding document of the French Physiocratic school, the *Tableau économique* devised by the French economist François Quesnay in 1758. As the name itself suggests, the core of Quesnay's work is represented by a diagrammatic representation of flows and incomes in a rural economy by means of a zig-zag diagram, the first one to appear in economics (Morgan 2012).

On the contrary, modern and contemporary economics seems less interested in developing a rich narrative part and seems more inclined to rely on a highly refined mathematical framework. The growing complexity of recent economic theories in terms of the number of variables and relations used brought economists (macroeconomists in particular) to expand their models in terms of systems of equations rather than figures. The epistemic role ascribed to diagrams within economic theories, furthermore, seems to differ according to the different theories and approaches. As I claimed above, sometimes diagrams might act as a specific kind of summary; notwithstanding this, it is more usually the case for economists to recognize diagrams as epistemic devices necessary for a thorough and effective understanding of economic theories (Blaug and Lloyd 2010). The advantage they represent consists in the particular way in which diagrams convey information, bringing together and juxtaposing information contained in two or more curves that would not be equally easily comprehended in equations or even in words. In these instances, diagrams in economics seem to show the same characteristics I outlined in the previous paragraph concerning diagrams as problem- solving and knowledge-improving devices in reasoning. As (Marshall 1949) claims "it happens with a few unimportant exceptions: all the results which have been obtained by the application of mathematical methods to pure economic theory can be obtained independently by the method of diagrams." In Marshall's view this is possible since diagrams are able to display most of the relevant information at a glance, thus suggesting results which may have been not so clear if only methods of mathematical analysis were used.

Therefore, in order to actually work in this way, diagrams must convey information which is not trivial and must also display a certain degree of independence with respect to the formal description.

3 Case study, the Phillip Machine

I propose the Phillip machine (Fig. 1) as an interesting case-study to show how diagrams work within economic theories. The analysis I propose here is built on its diagrammatic two-dimension version, and not on its three-dimensional one, which displays further interesting aspects and abilities (Morgan and Boumans 2004).

The Phillip Machine was built by the engineer and economist A.W.H. Phillip to visually reason on the Keynesian IS-LM model. Keynes' model was one of the first economic models to rely more on its mathematical framework and was perceived by Keynes' contemporaries as highly mathematized. The algebra he used, though, did not seem to guarantee enough clarity to the understanding of his concepts, and many economists tried to devise small algebraic models to understand the meaning of the Keynesian system, but did not manage to represent adequately the ideas at the core of the IS-LM model.

The Machine, instead, managed to surpass this impasse (Ongan 2009). Moving form a diagrammatic display, Phillips visualizes the effects of changes in various macroeconomic parameters such as consumption (C), savings (S), investments (I), government expenditures (G), taxes (T), exports (X), and imports (M) among themselves and on the total income (Y) with respect to the flow of *money-income* represented by water. In other words, using *liquid data* and a hydraulic presentation (the arrows in the diagram show the direction in which fluids are pumped), Phillips has outlined the macroeconomic dynamic modeling of an economy that works through a circular flow of the *water-money-income* in an economy.

In line with the variables listed above, the machine visually illustrates the following *Keynesian equilibrium condition*:

$$Y = AE = C + I + G + (X - M)$$

In the mechanism described by the equation above and Fig. 1 below, total income (Y) enters the system from the top; after the taxes are paid, there remains disposable income; when savings exit from the middle column, consumption spending is left behind. When investments (I) from the right and government spending (G) from the left are added to the consumption spending (C), domestic spending (C + I + G) minus imports (M) and plus exports (X) yields total expenditures [C + I + G + (X − M)]and the machine establishes the equilibrium [Y = AE = C + I + G + (X − M)] above.

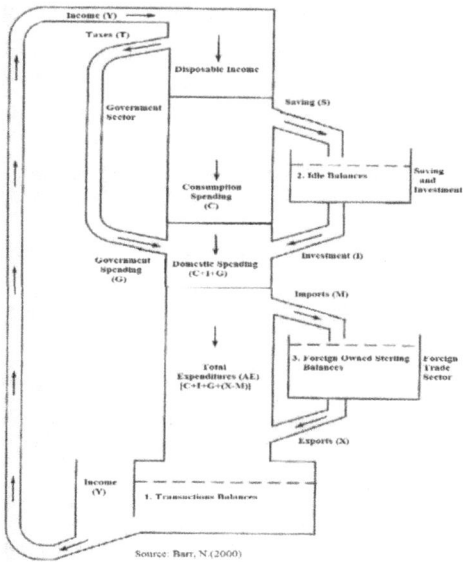

Fig. 1. A simplified version of the Phillips Machine.

Through its diagrammatic arrays, the Phillip Machine manages to clarifies the contents of the IS-LM model, it does not provide a mere synthesis but it also provided efficient proof and explanation to settle disputes concerning the actual content of the Keynesian theory (Barr 2000). The Philip Machine works as a suitable diagram with respect to the IS-LM model since it provides a 2-dimensional description of the economic world analysed in the model. Keynes' IS-LM model does not "photograph" single aspects of the economic context but it rather provides a highly complex description of it, bringing the real-economy sector (IS) and the monetary sector (LM) together. It describes the production of wealth as a flow in time and, it could be argue, in the economic space: in this particular respect the Philip Machine as a 2-dimensional diagram provides a spatialisation of the model's assumptions which would not be allowed by a more classical diagram. The Philip Machine is independent (according to the definition of "independence" I proposed) of the formal modal since it allows reasoning in a complex sequence of levels: it is composed of elementary parts (pipes, valves, liquids and tanks) which are accurately labelled and arranged (each tank represents a specific actor or entity of the economic context and the fluids, which represent the flow of resources, can run only from a specific tank to another in a specific quantity, which may vary from passage to passage) so as to provide "instructions" to read the diagram. Bringing these different parts and instructions together, the diagram-machine can be read as a whole. As for the non-trivial characteristic of the Machine: the Philip Machine provides a spatiotemporal lecture of the IS-LM model. This spatiotemporal dimension allows for an easier access to the content of the Keynesian model since it "lays-out" the theoretical conception of the economy and wealth circulation at a glance, providing an immediate contextual framework which would otherwise be rather complex to grasp with the aid of the mathematical framework alone.

Following the description of diagrams as problem-solving and knowledge-improving devices, I claim that the Philips Machine actually works as a good instance of such a diagrammatic tool. In particular, it shows four elements: it bring different languages together (the visual language of the hydraulic machine and the terms from the Keynesian theoretical outline); it provides "at a glance" necessary, relevant information on how an open economy works; it is made up of different elements, each of which may work and be analysed separately and then be built back together; finally, in order to be meaningfully read, it must be seen according to specific instructions and through different, logical steps. This last characteristic, though, partly represents a limitation: as already stated, the Philip Machine diagram must be read according to specific "instructions" or constraints exemplified above all by the flow direction of the liquids. Such constraints are directly connected to the theoretical assumptions of the model and in this respect the diagram strictly obeys such assumptions and, given its structure, it cannot let new information in. This aspect prevents the diagram to show a heuristic or knowledge-advancing activity but it nonetheless confirms its role as a strong knowledge-aiding device.

Disclosure of Interests. The author has no competing interests to declare that are relevant to the content of this article.

References

1. Blaug, M., Lloyd, P.: Famous Figures and Diagrams in Economics, 1st edn. Edward Elgar Publishing, Cheltenham (2010)
2. Barr, N.: Development between Phillips and the London school of economics. In: Leeson, R. (ed.) A.W.H. Phillips: The Collected Works in Contemporary Perspective. Cambridge University Press, Cambridge (2000)
3. Giaquinto, M.: Visualizing in mathematics. In: Mancosu, P. (ed.) The Philosophy of Mathematical Practice, pp. 22–42. Oxford University Press, Oxford (2008)
4. Grosholz, E.: Teaching the complex numbers: what history and philosophy of mathematics suggest. J. Humanist. Math. **3**(1), 62–73 (2013)
5. Larkin, J.H., Simon, H.A.: Why a diagram is (sometimes) worth ten thousand words. Cogn. Sci. **11**, 65–99 (1987)

6. Macbeth, D.: Realizing Reason: A Narrative of Truth and Knowing, 1st edn. Oxford University Press, Oxford (2014)
7. Marshall, A.: Pure Theory of Foreign Trade: The Pure Theory of Domestic Values. Series of Reprints of Scarce Tracts in Economic and Political Science, (1). London School of Economics and Political Science, London (1949)
8. Morgan, M.S., Boumans, M.: Secrets hidden by two-dimensionality: the economy as a hydraulic machine. In: de Chadarevian, S., Hopwood, N. (eds.) Models, the Third Dimension of Science, pp. 369–401. Stanford University Press, Stanford (2004)
9. Morgan, M.S.: The World in the Model: How Economists Work and Think, 1st edn. Cambridge University Press, Cambridge (2012)
10. Morgan, M.S.: Inducing visibility and visual deduction. East Asian Sci. Technol. Soc.: Int. J. **14**(2), 225–252 (2020)
11. Ongan, S.: The economy machine. J. Am. Acad. Bus. **11**(1), 1–16 (2009)

Why Feynman Diagrams Are Worth 10,000 Formulae: A Representational Epistemological Analysis

Peter C-H. Cheng[1](✉) and Timon G. Boehm[2]

[1] University of Sussex, Brighton, UK
p.c.h.cheng@sussex.ac.uk
[2] Stuttgart Research Centre for Text Studies, Stuttgart, Germany
philosophie@timonboehm.ch

Abstract. Feynman Diagrams are an indispensable and powerful tool in particle physics. They effectively support comprehension and computation in quantum electrodynamics (QED), and are generalisable to other quantum field theories. They are a paradigmatic example of how diagrams enable understanding and problem solving in complex knowledge domains. This paper examines why they are so effective from a cognitive perspective by adopting a *representational epistemological* approach that shows the completeness and coherence of the epistemic functions of Feynman Diagrams when encoding the concepts of QED.

Keywords: Physics diagrams · Feynman diagrams · Quantum electrodynamics · Representational epistemology · Cognitive efficacy

1 Introduction

Quantum electrodynamics (QED) is the theory of the interactions of electrically charged particles with other such particles or with light in a relativistic and quantum mechanical framework. It explains on a microscopic level macroscopic phenomena such as diffraction, interference, polarisation, absorption, and scattering of light. Together with the quantum field theories of strong and weak interactions it covers all phenomena of the physical world except those involving gravitation. One major tool in QED are Feynman Diagrams (FDs), a few examples of which are shown in Fig. 1. They are much appreciated by physicists because of their enormous comprehensional and computational benefit. For example, the publisher's description of David Kaiser's book [7] on the postwar dispersion of FDs throughout physics boldly states: "Feynman diagrams have revolutionized nearly every aspect of theoretical physics since the middle of the twentieth century."

Richard P. Feynman first introduced his diagrams in two papers in 1949 accompanying his theoretical analysis of QED processes [8,9] that would lead him to win the Nobel Prize for Physics in 1965. For a lay reader's introduction written

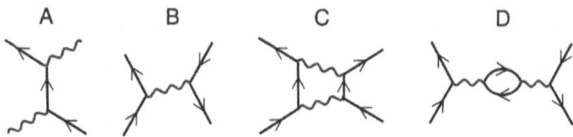

Fig. 1. Some Feynman Diagrams displaying the interactions of electrons (solid lines) and photons (wavy lines). A and B are first order, C and D are second order in the electromagnetic coupling constant.

by Feynman himself see [16]. The diagrams are fundamental to his approach and are specifically mentioned in the commendation statement for his award of the prize (https://www.nobelprize.org/prizes/physics/1965/feynman/facts/).

Historically, there were other approaches to the study of particle physics processes, including the methods of Feynman's fellow Nobel laureates, Julian Schwinger and Sin-Itiro Tomonaga, see [11-14]. However, only Feynman's diagrams, with the additional systematisation of Dyson, have become the state of the art in quantum field theory (of which QED is a restricted case for electromagnetic interactions). So an obvious question is: what makes FDs so successful that they are now the favoured diagrammatic representations in this area of theoretical physics?

The research on FDs so far is mostly done from a historical perspective, see e.g. [15], or from particle physicists themselves. However, FDs have not been studied from a *cognitive science* perspective, even though they are paradigmatic examples of what diagrammatic representations can do for human information processing. They are graphical representations for the rich, highly abstract and conceptually demanding knowledge that constitutes QED.

The effectiveness of notational systems can be examined from many perspectives [1]. For our analysis we adopt a *representational epistemological* (REEP) approach [2,3]. It considers the efficacy of representations in terms of how readily accessible they make the full range of concepts of their target domain to the reader. In the next section the REEP approach is introduced by comparing a novel diagrammatic representation of circuit electricity with the conventional approach that deploys circuit diagrams and algebraic formulae. Then, the approach is applied to the analysis of FDs across Sect. 3, 4, and 5. A concluding discussion highlights the implications for research on diagrams.

2 Representational Epistemological Analysis

Representational epistemology (REEP) is an approach for the analysis of diagrams, notations and graphical user interfaces [2,3,6]. It claims that effective representations make the essential conceptual contents of the represented domains easily accessible by providing representational schemes that are *semantically transparent* and *syntactically plastic* [2,3,6]. Users can find concepts they need and malleably transform expressions to make inferences with little cognitive effort. REEP has been used to reveal the limitations of conventional representations for many topics and applied to the design of novel graphical representations for those topics including, among others: circuit electricity [2]; set and probability theory [3]; Newtonian particle collisions [1]; propositional logic [4]; event timetabling [6]; production planning and scheduling [5]. Of the empirical studies that have been conducted, all have shown that the novel representations were superior for problem solving and learning, e.g. [2,3,6].

To illustrate REEP, consider a case study that contrasts the conventional representations for electric circuit theory (circuit diagrams and formulae) with a novel diagrammatic alternative - *AVOW diagrams* (Amps, Volts, Ohms, Watts) [2]. Empirical studies have shown that students using AVOW diagrams not only learn more quickly, but that their understanding is deeper and that they are able to solve problems that students using the conventional representations are barely able to start [2]. This REEP analysis will explain the relative efficacy of alternative representations of a knowledge rich domain by examining the *epistemic functions* of those representations.

What is an epistemic function? An epistemic function of a representational system is a representational device that encodes a concept from the target domain (e.g. circuit electricity) in some graphical object. The concepts may be abstract or concrete. Graphic objects are perceived configurations of ink or pixels in a display, and the properties of those configurations, such as alphanumeric strings, glyphs, shapes, spatial or geometric properties, formats, and colour. A representation typically embodies many and diverse epistemic functions. REEP claims that the cognitive efficacy of a representation for a knowledge rich domain is largely determined by the *completeness* and *coherence* of its epistemic functions: in other words, how well does a representation make the full conceptual panoply of the domain accessible to the user? A REEP analysis comprises three steps. **(Step 1)** Enumerate all the concepts needed for a good comprehension of the domain. The first column of Table 1 lists the concepts for the circuit electricity domain. **(Step 2)** Assess the extent to which a representation of interest provides epistemic functions for the concepts. The second, third, and fourth column of Table 1 list the representational devices that conventional representations and AVOW diagrams use to encode the concepts. Each row corresponds to an epistemic function, so an effective representation should have no empty cells. **(Step 3)** Evaluate the coherence of a representation's epistemic functions. In Table 1 this is performed by examining the overall directness and parsimony of the representational devices in the column(s) for each representation.

(Step 1) Electric Circuit Concepts. The various concepts to be addressed are listed in the first column of Table 1. These include (**general concepts** in bold and *electricity concepts* in italics): physical **objects** are *circuit* components (e.g., *resistors* or *light bulbs*); electrical **properties** (e.g., *voltage V, current I, resistance r*, and *power P*); **laws** that interrelate the properties (e.g., $V = Ir$, $P = VI$); laws that capture overall **invariants** of any *network* or *sub-networks*, such as the **conservation** of *current* and the *distribution of voltage*; **system boundary** that demarks the network of interest from the rest of the circuit;

Table 1. Epistemic functions of the conventional representation (circuit diagram and formulae) and AVOW diagrams

Concepts	Circuit diagram	Formulae	AVOW Diagrams
Entities	Icons (rectangles, lines)	Subscript labels, e.g., A,B,C	Boxes
Properties	—	Letters as variables	Box properties
Relations	—	Formulae	Box geometry
Values	Labels	Assign values to variables	Box dimensions
Network configuration	Lines between rectangles	Embedded resistance formulae	Box arrangements
Invariant conservation laws:	—	Embedded in formulae	No overlaps or gaps
System boundary	—	—	Overall box
System inputs	—	V_{total}, I_{total}	Overall box height & width
System outputs	—	P_{total}	Overall box area
Extreme cases	—	$r = 0, r = \infty$	Vertical/horizontal lines
2nd order models	Add component	Complex formulae	Edges of boxes, Fig. 3 left.
3rd order models	—	Complex formulae	Zoom in, Fig. 3 right
Hydraulic interpretation	—	—	Boxes as fluid flows
Electron path interpretation	—	—	Path downwards line; Probability aspect ratio

inputs and **outputs** to the network of interest (e.g., *current* and *voltage in* versus *power produced*); **topological arrangements** of components (e.g., *series* and *parallel networks* of *resistors*); **relations** that specify the properties of such networks (e.g., *sum of resistances for series loads*); **extreme cases** in which some property tends to zero or infinity, as with *insulators* and *conductors*. Typical analysis assumes **ideal** insulators and conductors. But when **high precision** is required, **second order** analysis takes non-infinite or non-zero *resistances* into account. **Third order** analysis may consider the magnitudes of the trickle of current running through the insulation of covering resistor and conductor to be equal. Further, different **models of operation** or **perspectives** of circuits can be adopted. In the *causal hydraulic* analogy, current is like fluid flow rate, voltage like pressure, and resistance like the friction within pipes. The *stochastic path* model considers alternative *paths* that *electrons* might traverse, with resistance interpreted in terms of the *probability* of taking a path. Clearly, there are many concepts that a representational system for electricity should encode.

(Step 2) Completeness of Epistemic Functions. Fig. 2 shows a typical circuit diagram and just a few of the equations that characterise it. (Graphical

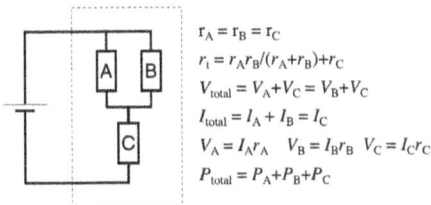

Fig. 2. Circuit diagram and some electrical circuit formulae

components of representations are underlined). The rectangles are the loads, lines are conductors, and the overall topology of the circuit is given by the "wiring" of the diagram. The formulae in this particular case indicate that all the resistors are identical and the sum of the current in A and B equals that in C, for example. The conventional representations do not fully capture the concepts. To emphasise the point, in Fig. 2 we have added an explicit dashed rectangle to show the system boundary, which the reader would typically be expected to imagine for themselves. The middle two columns of Table 1 are incomplete for the circuit diagram or formulae alone, but more importantly the coverage is incomplete even for them taken together. If both cells in a row are empty, we must resort to natural language to cover the missing content, or supplement the canonical representations with additional representational devices.

In contrast, AVOW diagrams have epistemic functions for *all* the concepts. Figure 3A shows an AVOW diagram for the same circuit. The completely filled fourth column of Table 1 lists explicit representational devices exploited by AVOW diagrams. Each rectangular *AVOW box* stands for one resistor. The overall AVOW box (A+B+C) is the whole network. The geometric properties of an AVOW box represent electrical properties, see Fig. 3A, so the total voltage, current, resistance, and power of the network are the overall height, width, gradient of the diagonal, and area. The stacking and horizontal adjacency of AVOW boxes captures series and parallel sub-networks. A syntactically valid AVOW diagram must have no gaps between or overlaps of AVOW boxes. This encodes *Kirchhoff's laws* about conservation of current and distribution of voltage. To model second order circuits we draw tall-thin (red) and squat-wide (green) AVOW boxes for the insulators and conductors as in Fig. 3B. To model third order effects, we would expand the relevant parts of the AVOW diagram;

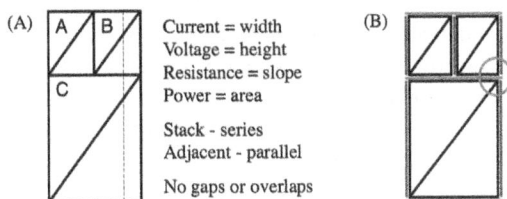

Fig. 3. AVOW diagrams: (A) model of the circuit in Fig. 2; (B) model with insulators and conductors.

zooming into the blue circle in Fig. 3B would allow us to model tiny currents through the insulators of resistors and wires. The alternative perspectives are accessed by attending to different graphical objects or their features. For example, the electron path interpretation requires one to envisage lines running down an AVOW diagram, such as the dashed line in Fig. 3.

(Step 3) Coherence of Epistemic Functions. AVOW diagrams are superior in this regard. First, AVOW diagrams are a single representational system whereas the conventional approach distributes epistemic functions between two distinct (sub)representations (or not all), which each has its own syntax and semantics. Second, all the representational devices of AVOW diagrams are based on properties of, or relations among, AVOW boxes, including the higher order models and alternative interpretations. In many cases, alternative concepts can be accessed merely by shifting our perception to different aspects of the same AVOW diagram, rather than needing to generate new formulae, which means those aspects may mutually serve as contexts to support each other's interpretations.

See [2] for a full REEP analysis of these representations and [1] for principles for judging the coherence of epistemic functions. We now apply the three steps of this approach to FDs.

3 QED Concepts

Since the REEP methodology asks "What are the concepts?" and "How well are they encoded?", we start by providing an overview of the conceptual landscape of QED (**Step 1**). Once this landscape has been mapped, the epistemic functions can be enumerated (**Step 2**), and thence the efficacy of Feynman's representation examined (**Step 3**). For readers unfamiliar with QED we refer to textbooks such as [17] or [18]. As previously, **general concepts** appear in boldface, more *specific physics concepts* in italics.

The elementary **particles** are *electrons*, *positrons*, and *photons*. They have certain fixed **properties**, for instance, *mass*, *charge*, and *spin*. Whereas mass and spin are the same for electrons and positrons, their electric charge is exactly opposite. Therefore, electrons belong to *matter*, positrons to *anti-matter*. Photons have zero (rest)mass, are electrically neutral and so are their own anti-particles.

The simplest case is a particle propagating freely from an **initial** to a **final state**. But as in classical electrodynamics, a charged particle is the source of a *potential* whose spatial gradient is the *field*. This field will now be quantised. (To be more precise: the gradient of the scalar potential is the electric field, the curl of the vector potential is the magnetic field.) If another particle is electrically charged too, it is able to 'feel' the presence of that field. Therefore the charge plays the role of a *coupling*. Photons having zero charge cannot couple to the field, rather they are the *quanta* of that field.

Initial and final states can be determined and fixed in an experimental setup where, for instance, a source is emitting electrons and a detector is receiving them. Such a setup defines a **system** with a **boundary** in a region of *space-time*. Note that the term does not conflict with the fact that ingoing and outgoing particles are 'unbound', i. e., asymptotically free, but simply delimits a region of space-time where the **processes of interactions** may happen. In particular, in quantum mechanics (QM) the processes inside the system are not observable, but only theoretically conceivable. This is precisely what FDs do. Accordingly, a system boundary gives rise to a distinction that is not intrinsic to photons. There are *real photons* as the particles of light (or more general of electromagnetic radiation) that are observable or measurable in *reflection, refraction, emission*, or *absorption*. In contrast, *virtual photons* are not observable, but are theoretical constructs to model the mediation of the electromagnetic field (as we shall explain in more detail in the next section). Notice, that in a relativistic theory such as QED, *space* and *time* cannot be separated. Together they form a four-dimensional manifold on which the interactions take place.

A property of particles that is not fixed is their *momentum*. It is specified with the initial and final state. When talking about momentum, one it tempted to imagine the movement of electrons, positrons, or photons along *paths* or *trajectories* much like classical particles in Newton's mechanics or light rays in geometrical optics. But in QM, according to Heisenberg's uncertainty principle, the momentum and the position of a particle cannot be determined simultaneously, so the notion of a path is awkward. Instead, a quantum mechanical particle has an *amplitude* indicating the probability to move from an initial to a final state. Between these states any – truly any – path can be adopted and its possible occurrence has to be summed up to yield the total amplitude. The notion of an amplitude reveals the *dual particle and wave-like nature* of photons and notably also of electrons or positrons. FDs will combine both aspects by representing the propagation of particles with lines that originally could stand for trajectories and wave-vectors. "It would seem that only someone as bold as Feynman would have dared to apply this pre-quantum theoretical representation to quantum theoretical problems, and be rewarded for his boldness with finding a one-to-one correspondence to field theoretical expressions." ([15], p. 30f.)

For physicists much of their understanding resides in the equations that encode the laws of the domain. So we now turn to the concepts, some rather abstract, that are captured by the equations that are central to QED. Many of these concepts will reappear in FDs. In QED, electrons, positrons, and photons are described mathematically by *wave functions* related to the above mentioned amplitudes. Wave functions are the solutions of the equations of motion making explicit the physical laws. In non-relativistic QM the equation of motion is the *Schrödinger equation* which relates the time evolution of a wave function to the *Hamiltonian* H by the formula $i\frac{\partial}{\partial t}\psi = H\psi$. The Hamiltonian is an operator describing all contributions to the total energy of a system such as kinetic and potential energy. Its eigenvectors ψ are the wave functions and its eigenvalues are the energy levels. In relativistic QM, instead, the equation of motion for electrons and positrons is the *Dirac equation*. Its solution is no longer a simple wave function, but a *spinor*, a vector with four components, because the Dirac equation is a system of four first order differential equations. The puzzling innovation here is the prediction of solutions which correspond to negative energy states. These solutions were interpreted as representing an electron with positive energy (the first two components) and an electron with negative energy or, equivalently, a positron. The first FDs actually stem from Feynman's efforts to give a physical interpretation of the Dirac equation.

The laws for a specific domain have to respect more general principles demanding, for instance, the continuity and constancy of properties within a system in the form of **conservation laws**, which here amount to the *conservation of energy-momentum*, *spin*, or the *Fermi principle* claiming that no two electrons can be in exactly the same state. Also note that some laws are universally applicable across **varying spatial scales** while others apply at just one scale such as Snell's law on a macroscopic scale claiming that light incident on a mirror is reflected symmetrically.

QED adopts all of these concepts stemming from a variety of physical theories: geometrical optics and wave optics, Newtonian mechanics and classical electrodynamics as well as their relativistic extensions, and of course QM and its relativistic extensions. In an ingenious way, Feynman combined elements of these into a single theory that was able to take account of two supposedly mutually exclusive requirements: maximum range and maximum precision. These QED concepts capture the essence of particle physics processes and are necessary to calculate, for instance, the probability for them to happen. In general, without FDs, these calculations are extremely tedious and complicated to manage. In the next section we link these various concepts to graphical elements in order to explain the epistemic functions of FDs.

4 Epistemic Functions of Feynman Diagrams

4.1 Representing Space, Time, and Free Particles

Examples of FDs were given above in Fig. 1. But these are just a few simple cases. Even for the same initial and final states there would be infinitely many diagrams refining the internal structure of the process. In this section, we explain what they express and how they are able to do what is seemingly impossible. These are the prerequisites for **Step 2**, i.e. the identification of the extent to which a representation provides epistemic functions for QED concepts.

A first virtue of FDs is that they break down the seemingly hopeless complexity of QED into manageable pieces. How do they do this? To deal with the quantum mechanical difficulties mentioned in the previous section and to compute the amplitude or probability for particle processes to occur, FDs make a few radical abstractions. First of all, they reduce space-time from $3+1$ dimensions (which could not be drawn on a sheet of paper) to $1+1$ dimensions, suppressing two spatial dimensions. Time is the vertical axis, space the horizontal axis. This immediately raises the question of whether certain processes are represented incorrectly. For example, one can imagine that two one-dimensional paths do not intersect in three-dimensional space, but do intersect when projected onto

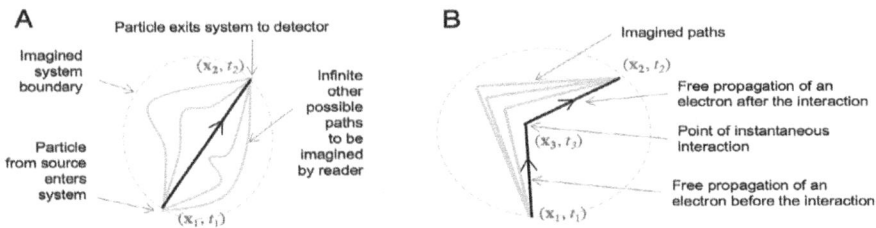

Fig. 4. A 'pre-Feynman-diagrammatic' analysis of (A) free propagation and (B) an interaction (see Sect. 4.2). Only the paths in black appear in actual FDs and they represent the sum of possible amplitudes. The grey graphic elements, indicating a few of all possible paths, do not appear in FDs. The dashed ovals represent system boundaries that the reader imagines, ($\mathbf{x}$, t) are space-time coordinates.

a two-dimensional surface. In a FD, however, this is not a problem because there is no need to represent concrete trajectories. What is relevant and what is represented is only the fact whether there is an intersection or not.

Let us have a closer look at this. When a quantum mechanical particle 'travels', it must be considered as simultaneously taking all possible paths between its start and end points; or, more figuratively speaking, all shapes of paths are possible: straight, curved, curly, backward (see Fig. 4A) and they may even reach out to infinity. Thus, a straight line in a FD does not represent a concrete trajectory, but an *abstract path* composed of many possible paths. The fact that the line is straight simply means that external influences, such as electromagnetic fields, are absent. What a line represents is the *free propagation* of a particle. Like any other in QM, this process has an amplitude or probability to occur. Explicitly, Feynman writes: "Although I will represent [...] a straight line between two points, we can think of it as the sum of many amplitudes" ([16], p. 92). As a convention one uses solid lines for electrons and positrons and wavy lines for photons. The precise slope of the line does not matter, but the overall direction, either upwards or downwards, indicates whether a particles is propagating forward or backward in time, respectively. This has deep implications that we discuss below.

4.2 Representing Interactions and Their Transmission

To deal with interactions, Feynman made further approximations and abstractions: (1) The electromagnetic potential is decomposable into point-like sources. (2) The potential is effective only during a very short time (as if it were switched on and off again). (3) The transmission of the potential is conceptualised by particles called virtual photons. (4) Anti-particles are automatically included via the slope of paths. (5) Particles may be created and annihilated at any time in space. In this section, we discuss each of these steps and how they are assembled to give the complete picture.

(1) Feynman started from the observation that the value of the wave function at a point in space-time $(\mathbf{x_2}, t_2)$, i.e. $\psi(\mathbf{x_2}, t_2)$, can be found by the convolution with a Green's function K, namely $\psi(\mathbf{x_2}, t_2) = \int K(\mathbf{x_2}, t_2, \mathbf{x_1}, t_1)\psi(\mathbf{x_1}, t_1) d^3\mathbf{x_1}$. Intuitively, this integral collects the influences from all points around $\mathbf{x_2}$ (see Fig. 4A). Here, we adopt Feynman's original notation in [8] with some abbreviations of his written out here for convenience (see also textbooks [17,18]). The Green's function K is at the same time a solution of the Schrödinger equation, where the potential is assumed to be point-like and symbolised by a Delta-function δ. The Schrödinger equation then reads: $(i\partial/\partial t_2 - H_2) K(\mathbf{x_2}, t_2, \mathbf{x_1}, t_1) = i\delta(t_2 - t_1)\delta(\mathbf{x_2} - \mathbf{x_1})$. The subscript on H_2 denotes that the Hamiltonian contains derivatives with respect to $\mathbf{x_2}$ only. The Delta-function specifies that the contribution of the point-like potential is non-zero only for $(\mathbf{x_2}, t_2) = (\mathbf{x_1}, t_1)$.

The use of Green's functions is a technique well known in classical electrodynamics. It is the solution of an equation of motion for a charged particle in a point-like potential. Quantum mechanically, $K(\mathbf{x_2}, t_2, \mathbf{x_1}, t_1)$ can be interpreted as the amplitude for a particle to travel from $\mathbf{x_1}$ at t_1 to $\mathbf{x_2}$ at t_2 including all possible paths in space-time in between.

(2) Assume that a particle is 'travelling on a path'. In classical mechanics or electrodynamics the effect of a potential would be a continuous deflection of that path. If the potential is only switched on and off briefly, this deflection occurs abruptly and instead of a continuous curve there will be a kink. Before and after the particle is free and the path is straight. In QED, since the concept of a well defined trajectory is not applicable, the analogous situation can be represented as well by a straight line, a kink, and another straight line (see Fig. 4B). But now, as explained above, the straight line represents many possible ways of free propagation which, what we don't know and cannot know, could themselves be straight, curved, curly, backward etc.

To calculate the amplitude K (what a physicist is ultimately interested in) one has to find the wave function ψ by solving the above Schrödinger equation. Since this is difficult in general, Feynman applied another technique well-known in classical mechanics and electrodynamics: perturbation theory. Thereby, the amplitude is expanded as:

$$K(\mathbf{x_2}, t_2, \mathbf{x_1}, t_1) = K^0(\mathbf{x_2}, t_2, \mathbf{x_1}, t_1) + K^1(\mathbf{x_2}, t_2, \mathbf{x_1}, t_1) + \ldots, \quad (1)$$

where the zero order term $K^0(\mathbf{x_2}, t_2, \mathbf{x_1}, t_1)$ is the amplitude for a particle to move freely from $\mathbf{x_1}$ at t_1 to $\mathbf{x_2}$ at t_2 and the first order term $K^1(\mathbf{x_2}, t_2, \mathbf{x_1}, t_1)$ is the contribution due to a single interaction between those initial and final points. If the potential (call it U) is weak, one may hope to capture the main contribution already by the zero and first order terms, whereas higher orders are negligible. That this is a valid approximation in QED, makes it such a powerful theory.

Feynman calculated the first order term to be (see [8], p. 751, Eq. (9)):

$$K^1(\mathbf{x_2}, t_2, \mathbf{x_1}, t_1) =$$
$$-i \int \int \int \int K^0(\mathbf{x_2}, t_2, \mathbf{x_3}, t_3) U(\mathbf{x_3}, t_3) K^0(\mathbf{x_3}, t_3, \mathbf{x_1}, t_1) d^3\mathbf{x_3} dt_3. \quad (2)$$

This seemingly complicated formula with a fourfold integral has quite a simple interpretation that can be visualised by a FD as in Fig. 4B. Note that K^1 is a function of K^0. If one reads the factors in the integrand from the right to the left, this interpretation is: The term $K^0(\mathbf{x_3}, t_3, \mathbf{x_1}, t_1)$ is the amplitude for free propagation of the particle from $\mathbf{x_1}$ at t_1 to $\mathbf{x_3}$ at t_3. Next, $U(\mathbf{x_3}, t_3)$ is the value of the (scalar) potential that acts upon the particle at the point of interaction. Finally, $K^0(\mathbf{x_2}, t_2, \mathbf{x_3}, t_3)$ is a free propagation to the point $(\mathbf{x_2}, t_2)$ where the wave function has to be evaluated. There is an implicit time ordering $t_1 < t_3 < t_2$. To make this plausible, Feynman resorted to basic probability theory: "if the event occurs as a *succession of steps* – or depends on a number of things happening 'concomitantly' (independently) – then we *multiply* the probabilities of each of these steps" ([16] p. 78).

Since the actual point of interaction $(\mathbf{x_3}, t_3)$ is not known, one has, in addition, to sum over all possible locations of it in space-time. As this point is given by four coordinates, the integral is fourfold. Again, the idea stems from basic probability theory: "if something can happen in *alternative ways*, we *add* the probabilities for each of the different ways" ([16] p. 78). Hence, "by means of the Green's functions and the established method of perturbative expansion, he [Feynman] was able to represent the interaction of an electron in an electromagnetic field as a sequence of free propagations that are interrupted by infinitesimal amounts of disturbance by the external potential. In order to determine the total effect of the potential, Feynman integrated over all the space-time points where an infinitesimal interaction with the potential occurs" ([15], p. 183). To each mathematical factor in Eq. 2 corresponds a graphical element in a FD. All complicated maths and physics just described is thus encoded in a simple arrangement of line segments as shown in Fig. 4B. The epistemic function of a FD is to capture the involved structure of the process. In more complicated cases, such as higher order interactions, this representational epistemological (REEP) virtue will prove indispensable.

(3) "Feynman's first published graphical representations [Fig. 1 in [8], p. 751] were in the tradition of quantum theoretical wave mechanics: an incoming probability wave is scattered off a potential and produces a modified outgoing wave" ([15], p. 183). In his subsequent paper [9], p. 772, one finds in Fig. 1 what can be considered as the first proper FD, which is reproduced here in Fig. 5.

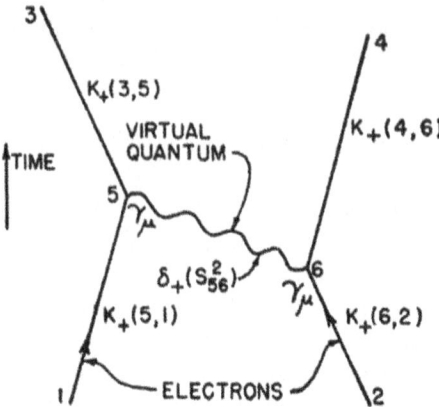

Fig. 5. The first proper FD showing the scattering of two electrons (solid lines) by exchange of one virtual photon (wavy line). The labels K_+ denote Green's functions corresponding to positive energy solutions. Graphics from [9], p. 772.

As another innovation, Feynman has thus "introduced a new graphical element: a wavy line representing the propagation of the photon that brings about the interaction." ([15], p. 137) The idea that the field associated with the potential is made up of such 'quanta' is essential to quantum field theory (and the reason why it is so named). The function of these quanta, in the case of electromagnetism *virtual photons*, is to mediate the field. Notice that the concept was already known before as 'second quantisation', but Feynman put the emphasis on paths: The virtual photons trace out an abstract path in space-time transmitting the potential from its origin (say electron 2 in Fig. 5) to the target (electron 1). An analogy in classical electrodynamics are retarded potentials, but in the quantum mechanical case, again, there is no question of a real trajectory nor needs the virtual photon to travel at the speed of light. It even does not matter whether the emission at point 6 in Fig. 5 is earlier than the absorption at point 5. FDs are topological and combinatorial devices. All that matters is the existence of such a type of event. And, as before, one has to sum over all possible places in space-time where this could happen.

But could a mere kink without photons not be sufficient to represent the interaction? The virtue of the virtual photons is to make the **cause** explicit. If they were absent, the deflection would appear ghostlike. In more complicated cases involving many particles and/or higher orders the virtual photon lines help in addition to sort out the combinatorial structure of the process. Finally, the three-vertex established this way (two solid lines and one wavy line) is the only type of event in QED. Everything else is made from these building blocks. From the REEP perspective this is an example of how FDs are complete (i. e. a graphical element is in one-to-one correspondence to a concept) and coherent (i. e. there is no overlap or contradiction in the representation) as we shall discuss in Sect. 5.

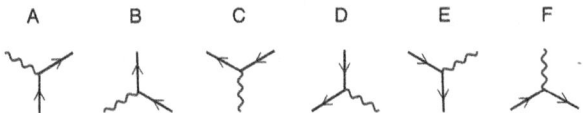

Fig. 6. Six elementary FD vertices for the six basic interactions. (A) A photon emitted by an electron. (B) A photon absorbed by an electron. (C) An electron position pair production from a photon. (D) A photon absorbed by a positron. (E) A photon emitted by positron. (F) Annihilation of an electron-positron pair producing a photon.

(4) Since the lines in FDs are abstract paths, their concrete slope (as depicted in Figs. 4 and 5) does not matter. However, if the slope turns negative, it represents a particle travelling backward in time. If this particle is an electron, it means that negative charge is flowing backward in time. Physically, this is equivalent to positive charge flowing forward in time. Although this equivalence was proposed already before Feynman (e.g. by Stückelberg), this case is now automatically included in his diagrams. Strictly speaking, positrons arise only in relativistic QM as a solution of the Dirac equation. This solution is not a (scalar) wave function, but a four-component vector called spinor, the first two components of which are interpreted as electrons, the last two as positrons (see Sect. 3). In the non-relativistic case described by the Schrödinger equation, future events simply follow from past events and there is no question of a particle moving backward in time.

(5) Due to the equivalence of energy and mass, $E = mc^2$, new particles can be produced anywhere at any time. QED is a genuine multi-particle theory, where the particle number is not conserved. The basic possibilities are the creation of an electron-positron pair from a photon and its inverse, the annihilation of an electron and a positron into a photon.

All considerations in (1)-(5) can be captured elegantly by only six types of FDs shown in Fig. 6. As every interaction is a vertex comprising a photon and an electron path with a kink, there are by symmetry six syntactically valid vertices. Notice that the vertices have the same form but are merely rotated. Each of these processes happens with same probability given by the *coupling constant*.

4.3 Encoding Higher Order Processes and Obtaining Correct Values of Charge and Mass

So far we have only discussed free propagation and the types of interaction shown in Fig. 6, i.e. zero and first order terms in the perturbative approach. This gives already a fairly good approximation of the total amplitude. For more accuracy, higher order terms in Eq. 1 must be taken into account. With the same basic assumptions (1)-(5) one can then construct a more finely subdivided structure of the process. Higher order contributions include loops, examples of which are given in Fig. 1C and D. Still, all interactions can be composed of the six types of vertices. For instance, Fig. 1A and Fig. 5 are composites of Fig. 6A and Fig. 6B. In the total amplitude, each higher order correction appears with a higher power of the coupling constant which is the electric charge. Because its value is smaller than one (in suitable units), these corrections contribute less and less to the perturbation series.

The miracle of FDs is that they allow, like this, to figure out all possible types of processes by assembling the six basic building blocks in a syntactically correct way. Syntactically correct means that each vertex must contain two electron/positron lines and one photon line. Thus FDs are combinatorial devices that indicate the physically admissible processes in a necessary and sufficient way. Notice again, that each diagram is only a type and not a token of a process or, in other words, stands for a class of topologically equivalent events. This goes as far as the "diagrams articulate the unobservable but nevertheless theoretically important distinction (because of the exclusion principle) of the processes, which differ only in an interchange of the particles." ([15], p. 125)

Finally, we address the issue that one has to distinguish between a 'bare charge' and 'physical charge'. The bare charge is a theoretical coupling that would lead to infinite amplitudes in the calculations. The physical charge, in contrast, is what is measured in an experiment. To arrive at the correct value of the physical charge (by a technique called renormalisation), one must consider all processes a particle is subject to, such as the creation of electron-positron pairs and the self-energy of the electron. It is another astonishing feature of FDs that, by naturally including those processes, they account for the real charge of electrons. Similar considerations are valid for the mass of electrons.

5 Completeness and Coherence of FD Epistemic Functions

We are now in a position to evaluate the completeness and coherence of the epistemic functions of FDs (REEP **Step 3**) on the basis of the QED concepts presented in Sect. 3 (**Step 1**), listed in the left column of Table 2, and the epistemic functions in Sect. 4 (**Step 2**), which are the rows of Table 2.

Regarding completeness Table 2 shows that there is a one-to-one correspondence, where each QED concept standing in the left hand column has a graphical device on the right. Some of these concepts and devices are simple, such as particles and lines, or space as the 1D horizontal axis. Others are more complex and sophisticated, such as the modelling of particular phenomena by the specific configuration of vertices. Most correspondences are explicit. Others are implicit, but they are not difficult for the user to imagine, for instance the system boundary or possible paths as in Fig. 4. Notably, imagination is not a technical term here, but simply denotes the act of envisaging (in the reader's mind) graphical elements in diagrams that are not explicitly present (and need not be present in view of completeness). The completeness of the epistemic functions of FDs means that when they are drawn "[t]he diagram thus obtained can then be translated, element by element, into a complex mathematical expression, the evaluation of which yields an observable quantity" [15]. Each symbol (line, vertex) along a

Table 2. Epistemic functions of Feynman Diagrams

QED concepts	Feynman Diagram representation
3D space	1D horizontal axis
Time	Vertical axis (up forwards)
Amplitude for free propagation	Line without vertices
Particles (electron & positron, photon)	Line format (straight, wavy)
Particle type (electron, positron)	Line slope (up, down)
Non-relativist/relativist interaction	Vertices with electron/positron lines
System boundary	Closed curve intersecting free ends of lines
Initial and final states	Lines from bottom, lines to top
Interaction	Vertex
Interaction types	Vertex orientation
EM potential/field	Virtual photons
Conservation laws (momentum, energy, charge, spin)	Continuity of lines/vertices
Modelling phenomena	Configurations of vertices
Renormalisation, Accuracy	Higher order vertices

path maps to a multiplicative factor and each path maps to a summative term in the equations (as with Eq. 2).

The epistemic functions of FDs have a high degree of coherence — read down the contents of the right hand column of Table 2. At a basic level, there are no contradictions, overlaps or redundancies among the ways in which the epistemic functions encode the many QED concepts. Strikingly, this coherence arises from the remarkably parsimonious set of graphic objects found in FDs. The exploitation of different facets of that small set of graphic objects in different representational devices for related concepts drives the systematicity of the epistemic functions. At a high level FDs deploy three closely related representational devices to integrate three sets of core concepts. (1) A line segment represents all possible paths of free propagation from its beginning to its end (not a concrete particle trajectory). (2) One type of vertex in six alternative orientations models all types of interactions (Fig. 6). Those six types form topological equivalence classes (they do not indicate concrete locations of interactions in space-time). (3) The composition of those types of vertices is sufficient to model all possible configurations of interactions. It is the sum of these configurations that miraculously captures the reality of physical processes.

The REEP analysis of FDs parallels that of AVOW diagrams (Table 1). However, there are differences. First, the range and esoteric nature of QED concepts is certainly greater than that of circuit electricity. The greater extent of concepts encoded by FDs emphasises the noteworthiness of the completeness and coherence that Feynman achieved in his selection of the epistemic functions of FDs. Second, unlike AVOW diagrams, the metric properties of symbols in FDs do not stand for specific quantities; the actual length and slope of a line is arbitrary, magnitudes of momentum or energy cannot be read off the diagrams. For quantitative results one must switch to the formulae and solve equations, but it may be argued that this is a small compromise given all the other representational epistemic advantages of FDs.

6 Conclusion

To answer the title question concerning why FDs are seen as so special, we have applied the three steps of REEP analysis. In sum, FDs deploy parsimonious representational devices providing a system of epistemic functions to encode the concepts of QED completely and coherently, despite the diverse and abstract character of those concepts. Where FDs do not provide simple explicit representations, they nevertheless offer a ready means to conceptualise the ideas without recourse to supplementary notations.

There are wider implications for the study of diagrams from the REEP analysis of FDs. First, the successful analysis adds some weight to Cheng's [2,3,5,6] claim that accounts of the efficacy of representational systems for *knowledge rich domains* should be grounded in a thorough elaboration of the conceptual structure of the topic. Second, the successful application of REEP analysis to FDs stands, in a reflexive fashion, as a test of the validity of the representational epistemological approach to the analysis of representations. REEP has been applied to representational systems for other domains, but its application to FDs for QED was, we consider, a noteworthy challenge.

Physicists often explain the value of the FDs in terms of their pragmatic facility for guiding calculations, via the so-called Feynman rules (see [18], Chap. 4.4). What this paper shows is that there is an alternative richer cognitive explanation of why FDs are special among representational systems. However, to fully answer the question, it will be necessary to compare FDs with REEP analyses of Tomonaga's and Schwinger's approaches to QED — in the way AVOW diagrams were compared to the conventional representations for electricity above. Will the epistemic functions in Tomonaga's and Schwinger's representations of QED be as complete and as coherent as those in FDs?

References

1. Cheng, P.C-H.: What constitutes an effective representation? In: Jamnik, M., Uesaka, Y., Elzer Schwartz, S. (eds.) Diagrams 2016. LNCS (LNAI), vol. 9781, pp. 17–31. Springer, Cham (2016). https://doi.org/10.1007/978-3-319-42333-3_2
2. Cheng, P.C-H.: Electrifying diagrams for learning: principles for effective representational systems. Cogn. Sci. **26**(6), 685–736 (2002)
3. Cheng, P.C-H.: Probably good diagrams for learning: representational epistemic re-codification of probability theory. Top. Cogn. Sci. **3**(3), 475–498 (2011)

4. Cheng, P.C-H.: Truth Diagrams versus extant notations for propositional logic. J. Lang. Log. Inf. **29**, 121–161 (2020)
5. Cheng, P.C-H., Barone, R.: Representing complex problems: a representational epistemic approach. In: Learning to Solve Complex Scientific Problems, pp. 97–130. Lawrence Erlbaum Associates, Mahmah, N. J. (2007)
6. Cheng, P.C-H., Barone, R., Cowling, P.I., Ahmadi, S.: Opening the information bottleneck in complex scheduling problems with a novel representation: STARK diagrams. In: Hegarty, M., Meyer, B., Narayanan, N.H. (eds.) Diagrams 2002. LNCS (LNAI), vol. 2317, pp. 264–278. Springer, Heidelberg (2002). https://doi.org/10.1007/3-540-46037-3_26
7. Kaiser, D.: Drawing Theories Apart. The Dispersion of Feynman Diagrams in Postwar Physics. The University of Chicago Press, Chicago (2005)
8. Feynman, R.P.: The theory of positrons. Phys. Rev. **76**(6), 749–759 (1949)
9. Feynman, R.P.: Space-time approach to quantum electrodynamics. Phys. Rev. **76**(6), 769–789 (1949)
10. Kaiser, D., Ito, K., Hall, K.: Spreading the tools of theory. Soc. Stud. Sci. **34**(6), 879–922 (2004)
11. Koba, Z., Tomonaga, S.: On radiation reactions in collision processes I. Progress Theor. Phys. **III**(3), 290–303 (1948)
12. Koba, Z., Takeda, G.: Radiation reactions in collision processes II. Progress Theor. Phys. **III**(4), 407–421 (1948)
13. Koba, Z., Takeda, G.: Radiation reactions in collision processes III/1. Progress Theor. Phys. **IV**(1), 60–70 (1949)
14. Koba, Z., Takeda, G.: Radiation reactions in collision processes III/2. Progress Theor. Phys. **IV**(2), 130–141 (1949)
15. Wüthrich, A.: The Genesis of Feynman Diagrams. Springer, Dordrecht (2010). https://doi.org/10.1007/978-90-481-9228-1
16. Feynman, R.P.: The Strange Theory of Light and Matter. Princeton University Press, Princeton (2014)
17. Halzen, F., Martin, A.D.: Quarks and Leptons. An Introductory Course in Modern Particle Physics. Wiley, New York (1984)
18. Peskin, M.E., Schrder D.V.: Quarks and Leptons. An Introduction to Quantum Field Theory. Addison-Wesley, Boston (1997)

A Building-Block Approach to the Diversity of Visualization Types – Each Type Expressed Visually, and as a Systematically Generated Sentence

Yuri Engelhardt[1](✉) and Clive Richards[2]

[1] University of Twente, Enschede, The Netherlands
`yuri.engelhardt@utwente.nl`
[2] Birmingham City University, Birmingham, UK
`clive.j.richards@me.com`

Abstract. A building-block approach is presented that shows commonalities and differences within a large corpus of different types of visualization. Refining and testing our earlier work, we have analyzed more than one hundred different types of visualization in terms of 19 fundamental visual encoding techniques, and 13 types of questions that visualizations can answer with those encoding techniques. We characterize each type of visualization accordingly, with a natural language sentence, generated in a systematic way. We also provide guidance on how to choose (combinations of) visual encoding techniques, depending on the questions to be answered by a visualization. A few examples of visual encoding techniques are *grouping by boundary*, *unit-based tallying* (as in Isotype charts), *proportional space-filling*, *extending along a coordinate axis*, and *coupling by adjacency*. Various other building-block approaches to visualization are reviewed.

Keywords: Types of Visualization · Comparing Visualization Types · Choosing Visual Encoding Techniques · Visualization Design Options

1 Background and Purpose

The *language of visualization* framework builds on *Diagrammatics* (Richards 1984), *The language of graphics* (Engelhardt 2002), and our joint publications from 2018 onwards, including 'A framework for analyzing and designing diagrams and graphics' (Engelhardt and Richards 2018).

The work we are presenting here is the result of refining our framework and testing it on a large corpus of different types of visualization. The framework can be used to show commonalities and differences between types, and to characterize each type of

---

Both authors contributed equally to the work.

The original version of the chapter has been revised. Figure 4 was displayed incorrectly. A correction to this chapter can be found at
https://doi.org/10.1007/978-3-031-71291-3_45

visualization with a specific natural language sentence, generated in a systematic way. In addition, the framework enables the exploration of alternative design choices and may therefore help designers to generate visualization options. The framework also offers a tool for research, a basis for formalization, and the potential for computer-based visualization advice. The framework involves a set of conceptual building blocks.

2 The Conceptual Building Blocks of Visualization

The **conceptual building blocks** and core aspects of the framework are the following:

- Visualizations are composed of (one or more of fifteen) **types of visual components** that have been manipulated using (one or more of nineteen) **visual encoding techniques** in order to answer (one or more of thirteen) **types of questions**.
- Some types of visualization are composite visualizations, in which the components are themselves composed of (sub)components that have been manipulated using visual encoding techniques.
- Every type of visualization (more than a hundred of which are generally known by name) can be characterized by a specific natural language sentence, systematically generated, in terms of the above.

The nineteen *visual encoding techniques* are shown in Figs. 1 and 2. This list of visual encoding techniques draws on Bertin (1967/1983), Twyman (1979), and on the Gestalt principles of perception. There are also parallels with the work of Johnson (1987), Lakoff (1987), Tversky (1995), Card and Mackinlay (1997), and Ware (2008). These parallels are set out in Engelhardt and Richards (2018, p. 202, table 1).

Visual encoding techniques can be used to show six types of information which can answer a total of thirteen different *types of questions*. *Which* visual encoding techniques can be used to answer *which* type of question is shown in Fig. 3. This figure thus offers design options that creators of visualizations can combine into a range of different possible types of visualization.

Some of the visual encoding techniques (for example those involving a coordinate axis) can be specified with an *orientation*: horizontal, vertical, third dimension, angular, radial, spiral, and array of axes (the symbols for these can be seen in Figs. 1 and 3). Changing the orientations that are used within a visualization will often result in another visualization type. For example, switching the orientations from horizontal/vertical to angular/radial, will turn a stacked bar into a pie chart, a parallel coordinates chart into a radar chart (items 018 and 019 in Fig. 7b), and an icicle chart into a sunburst chart (items 025 and 026 in Fig. 7c). In addition to the orientations there are a few other possible *qualifiers* for visual encoding techniques, these are shown in Fig. 4.

Visual encoding techniques are applied to the *visual components* that make up a visualization. The types of visual components that are available in the framework are shown in Fig. 3: symbols, bars, segments, connectors, directed connectors, boundaries, bands, blocks, pictures, pictorial components, textual components, line locators, surface locators, and invisible components (this list of types of visual components was introduced in Engelhardt 2002 pp. 74–78, and extended in Richards and Engelhardt 2019). The type of a set of visual components depends on the visual encoding technique that is applied to those components, in other words, our classification of visual components is driven by the semantics of the encoding that is involved. For example, a rectangle may be of any of various types. A rectangle may be a *bar* when *extending, ranging,* or *diverging along a coordinate axis*. A rectangle may be a *segment* when arranged using *proportional space-filling*. A rectangle may be a *boundary* when used for *grouping by boundary*. A rectangle may be a *block* when arranged using *grid-based proportional space-filling*. A rectangle may be a *symbol* or a *surface locator* when positioned using *mapping (of spatial locations)*. A rectangle may even be a *pictorial component*, when arranged using *picturing*, for example to represent the window of a house.

Together, the *types of questions* that can be answered, the *visual encoding techniques* for answering them, along with their possible *orientations* (or other *qualifiers*), and the *types of visual components* to which these visual encoding techniques can be applied, constitute the conceptual building blocks of the language of visualization. Interdependencies between the building blocks are shown in Fig. 4. Using these building blocks, every type of visualization can be characterized by a specific natural language sentence.

3 Sentences and Icons for Describing Types of Visualization

We have created colour-coded icons to represent the different visual encoding techniques. In the late 90s a basic set of such icons was introduced, which were combined into visual analyses of visualization examples (Engelhardt et al. 1996 and Engelhardt 1999). In 2019 this basic set of visual encoding techniques was extended and colour-coded to distinguish different categories of building blocks (Richards and Engelhardt 2019).

Each different type of visualization can be characterized (in terms of its conceptual building blocks) by a specific natural language sentence, generated in a systematic way. See Figs. 5 and 6 for instructions on how to construct these sentences.

A selection of different types of visualization (35 examples), described using the *language of visualization* framework, is set out in Fig. 7a. Each description includes (only) what all specimens of the concerned type of visualization have in common.

4 Other Building-Block Approaches to Visualization

We have reviewed a number of published proposals with some similarities to the building-block approach of the *language of visualization* framework.

In his paper 'A visualization-based taxonomy for information representation: Identifying images with icon schematics' (2008), William Bevington offers a building-block approach for analyzing visualizations. His four main categories are represented by colour-coded icons that can be combined, for a given type of visualization, into a visually represented analysis. The categories are: 'pictorial structures', 'quantitative structures', 'relational structures', and 'symbolic structures'. Bevington's 'symbolic structures' concern mainly the textual layer in a visualization, such as labels. Bevington's 'pictorial structures' correspond to *picturing* and *mapping* in the *language of visualization*, 'quantitative structures' correspond to *positioning along a coordinate axis, proportional space-filling* and *sizing*, and 'relational structures' correspond to *positioning into slots* and *connecting*. While the *language of visualization* framework is much more

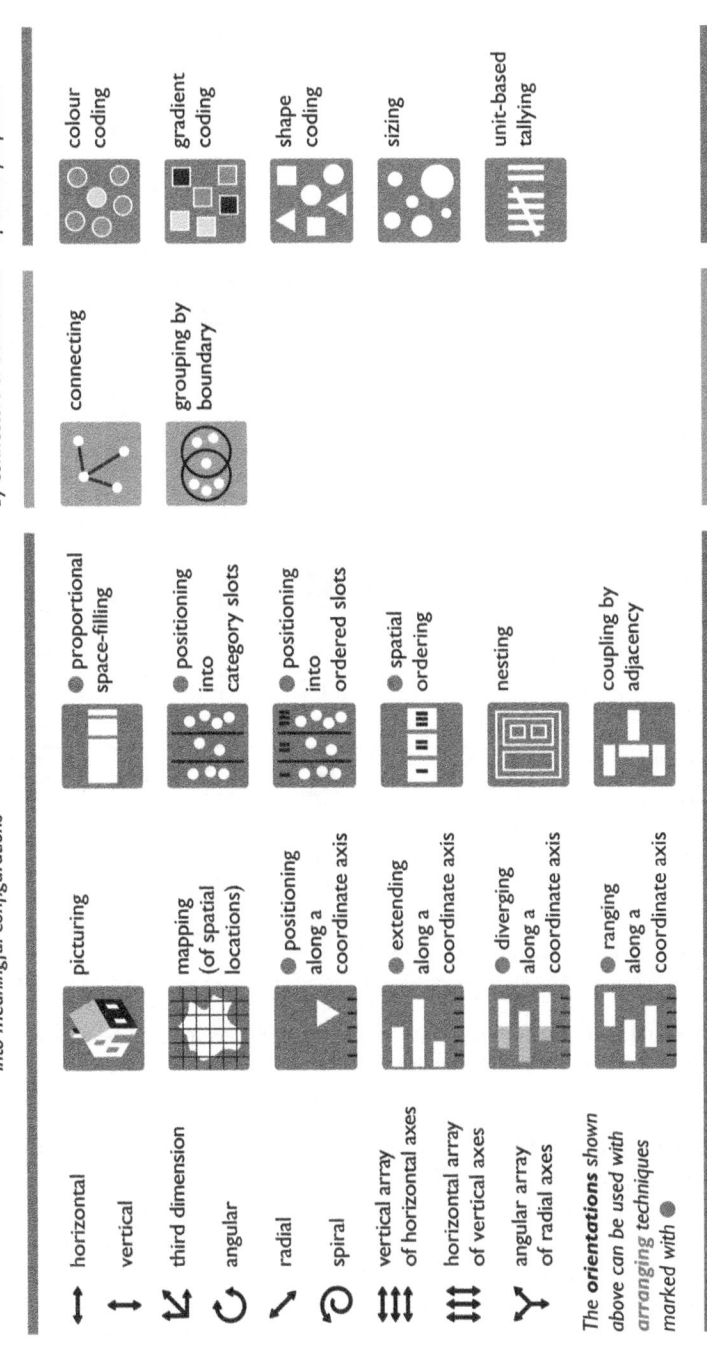

Fig. 1. The 19 visual encoding techniques, categorized into *arranging*, *linking*, and *varying*.

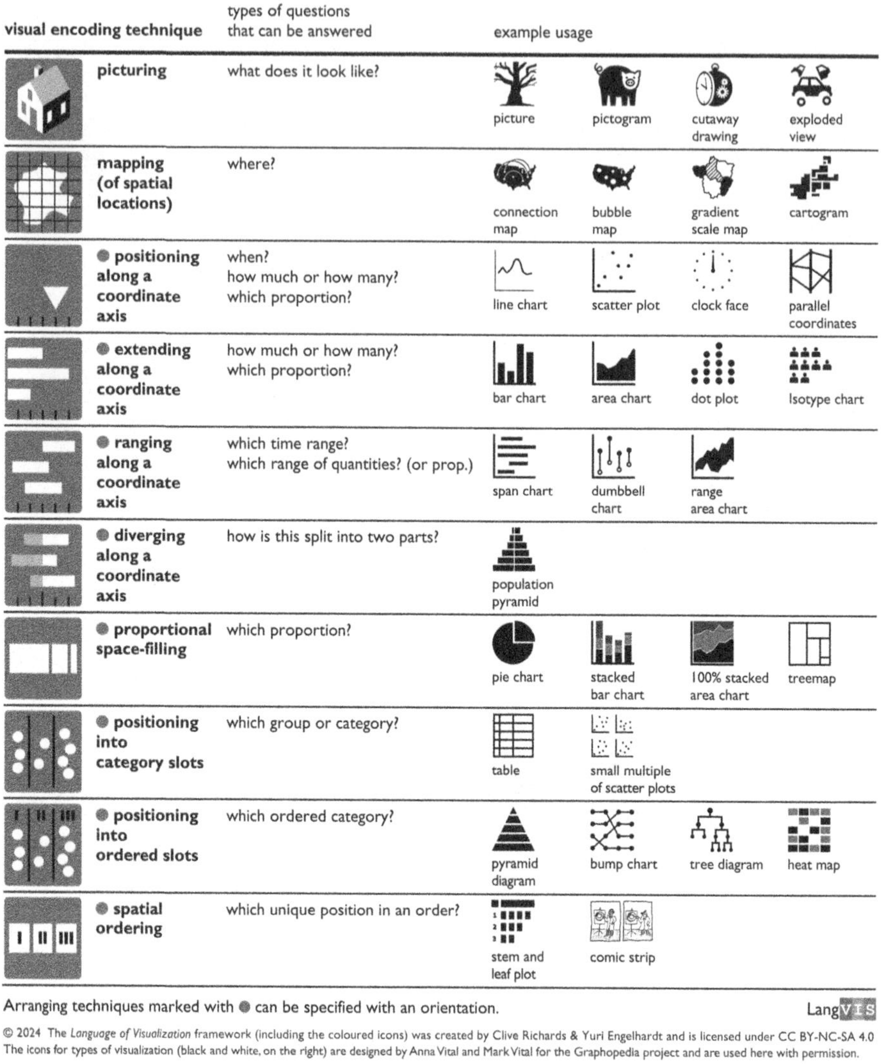

visual encoding technique	types of questions that can be answered	example usage			
picturing	what does it look like?	picture	pictogram	cutaway drawing	exploded view
mapping (of spatial locations)	where?	connection map	bubble map	gradient scale map	cartogram
● positioning along a coordinate axis	when? how much or how many? which proportion?	line chart	scatter plot	clock face	parallel coordinates
● extending along a coordinate axis	how much or how many? which proportion?	bar chart	area chart	dot plot	Isotype chart
● ranging along a coordinate axis	which time range? which range of quantities? (or prop.)	span chart	dumbbell chart	range area chart	
● diverging along a coordinate axis	how is this split into two parts?	population pyramid			
● proportional space-filling	which proportion?	pie chart	stacked bar chart	100% stacked area chart	treemap
● positioning into category slots	which group or category?	table	small multiple of scatter plots		
● positioning into ordered slots	which ordered category?	pyramid diagram	bump chart	tree diagram	heat map
● spatial ordering	which unique position in an order?	stem and leaf plot	comic strip		

Arranging techniques marked with ● can be specified with an orientation.   LangVIS

© 2024 The *Language of Visualization* framework (including the coloured icons) was created by Clive Richards and Yuri Engelhardt and is licensed under CC BY-NC-SA 4.0
The icons for types of visualization (black and white, on the right) are designed by Anna Vital and Mark Vital for the Graphopedia project and are used here with permission.

Fig. 2a. The 19 visual encoding techniques, the types of questions that can be answered with them, and examples of types of visualization using these techniques.

refined and includes more aspects of visualizations, several of the top-level distinctions in it correspond to aspects of Bevington's taxonomy.

Sedig and Parsons (2016) offer 'a pattern-based framework' for designing visualizations: "We have focused on identifying patterns that map information items from an information space to a representation space at a basic level of abstraction". The 14 patterns that they offer are: 'area', 'branch', 'cell', 'coordinate', 'cycle', 'fusion, group', 'hierarchy', 'link', 'list', 'spectrum', 'stack', 'token', and 'track'. Most of these patterns correspond loosely to *visual encoding techniques* in the *language of visualization* (e.g.

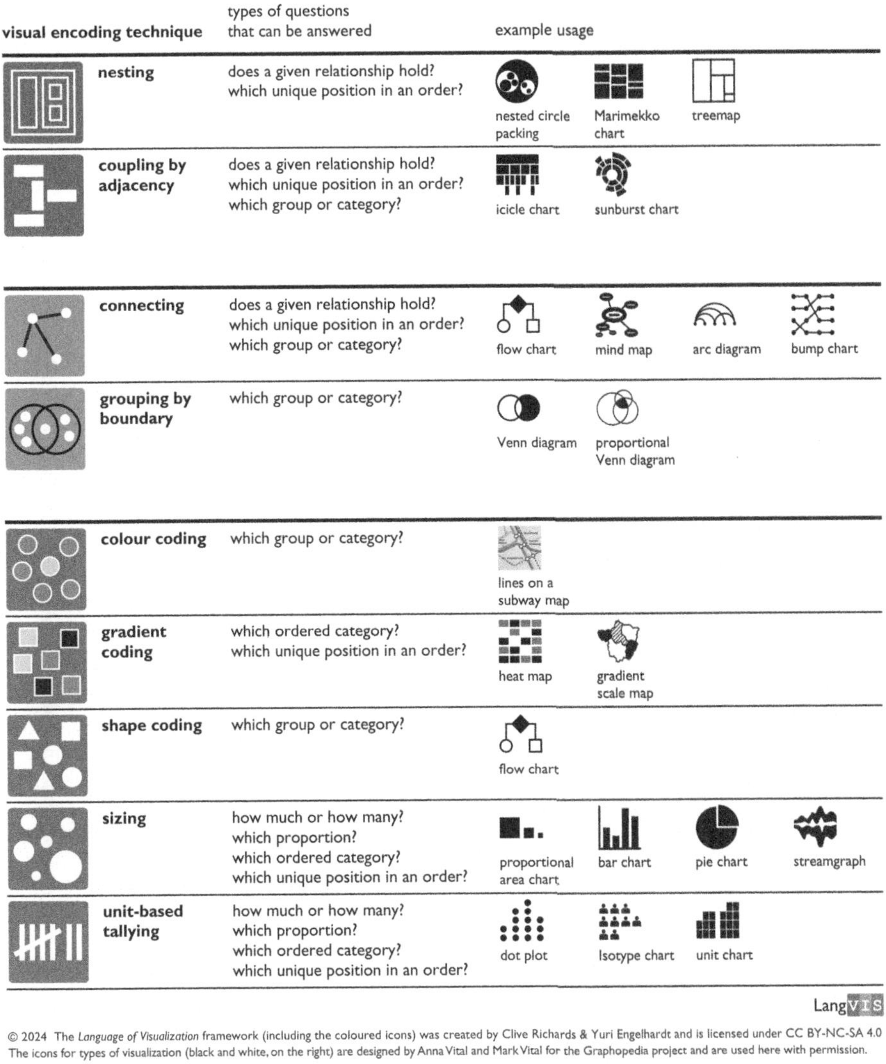

A Building-Block Approach to the Diversity of Visualization Types   33

visual encoding technique		types of questions that can be answered	example usage			
	nesting	does a given relationship hold? which unique position in an order?	nested circle packing	Marimekko chart	treemap	
	coupling by adjacency	does a given relationship hold? which unique position in an order? which group or category?	icicle chart	sunburst chart		
	connecting	does a given relationship hold? which unique position in an order? which group or category?	flow chart	mind map	arc diagram	bump chart
	grouping by boundary	which group or category?	Venn diagram	proportional Venn diagram		
	colour coding	which group or category?	lines on a subway map			
	gradient coding	which ordered category? which unique position in an order?	heat map	gradient scale map		
	shape coding	which group or category?	flow chart			
	sizing	how much or how many? which proportion? which ordered category? which unique position in an order?	proportional area chart	bar chart	pie chart	streamgraph
	unit-based tallying	how much or how many? which proportion? which ordered category? which unique position in an order?	dot plot	Isotype chart	unit chart	

LangVIS

© 2024 The *Language of Visualization* framework (including the coloured icons) was created by Clive Richards & Yuri Engelhardt and is licensed under CC BY-NC-SA 4.0
The icons for types of visualization (black and white, on the right) are designed by Anna Vital and Mark Vital for the Graphopedia project and are used here with permission.

Fig. 2b. Continued from Fig. 2a.

'link' corresponds to *connecting*, and 'spectrum' to *gradient coding*). However, this list of patterns seems incomplete to us, missing, for example, a pattern that would correspond to the fundamental visual encoding technique of *sizing*. We also notice that not all of Sedig and Parsons' patterns seem to be visual encoding techniques. The 'hierarchy' pattern seems to concern the *type of questions* that visualizations can answer (specifically, regarding the presence of hierarchical *relationships*), rather than indicating a specific visual encoding technique for displaying such a hierarchy. And the 'token' pattern corresponds to the use of *visual components* – the very objects to which the other patterns are applied. In comparison, the *language of visualization* framework is more

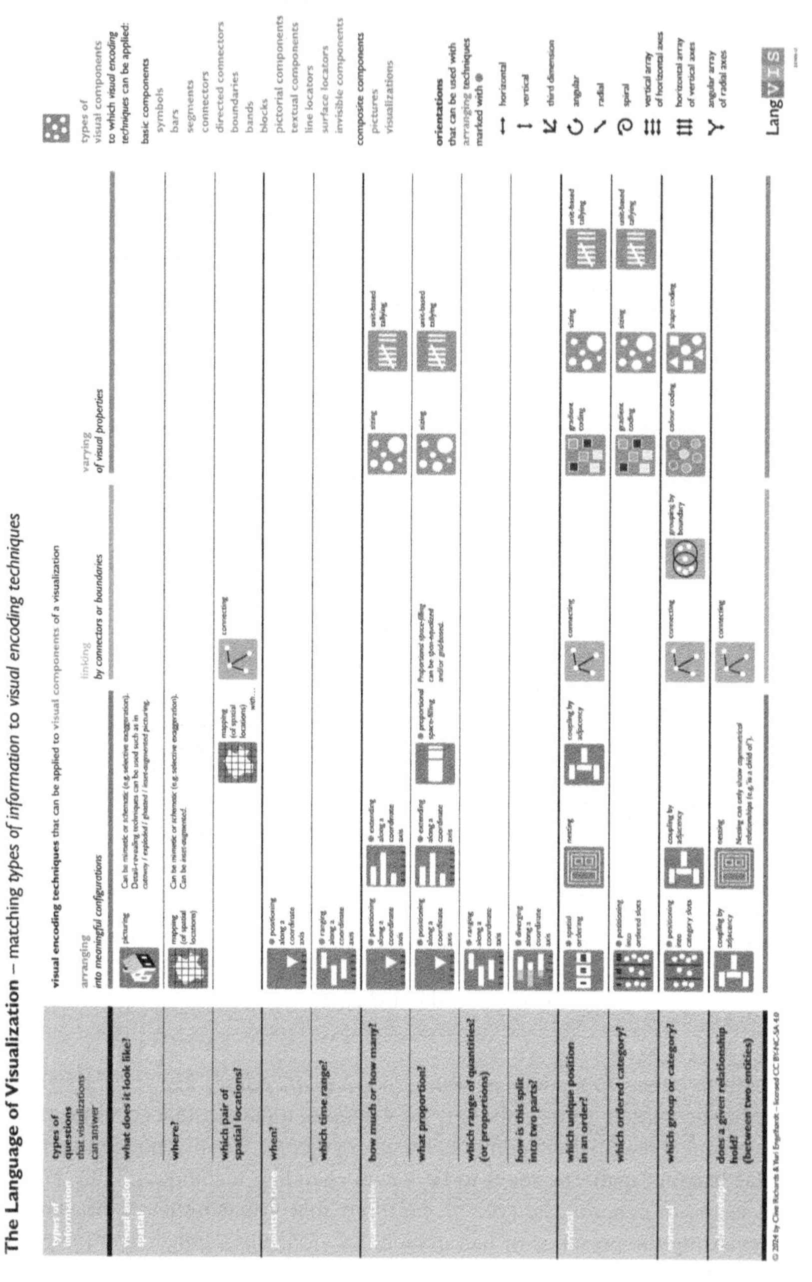

Fig. 3. Matching *types of information/types of questions* to *visual encoding techniques.*

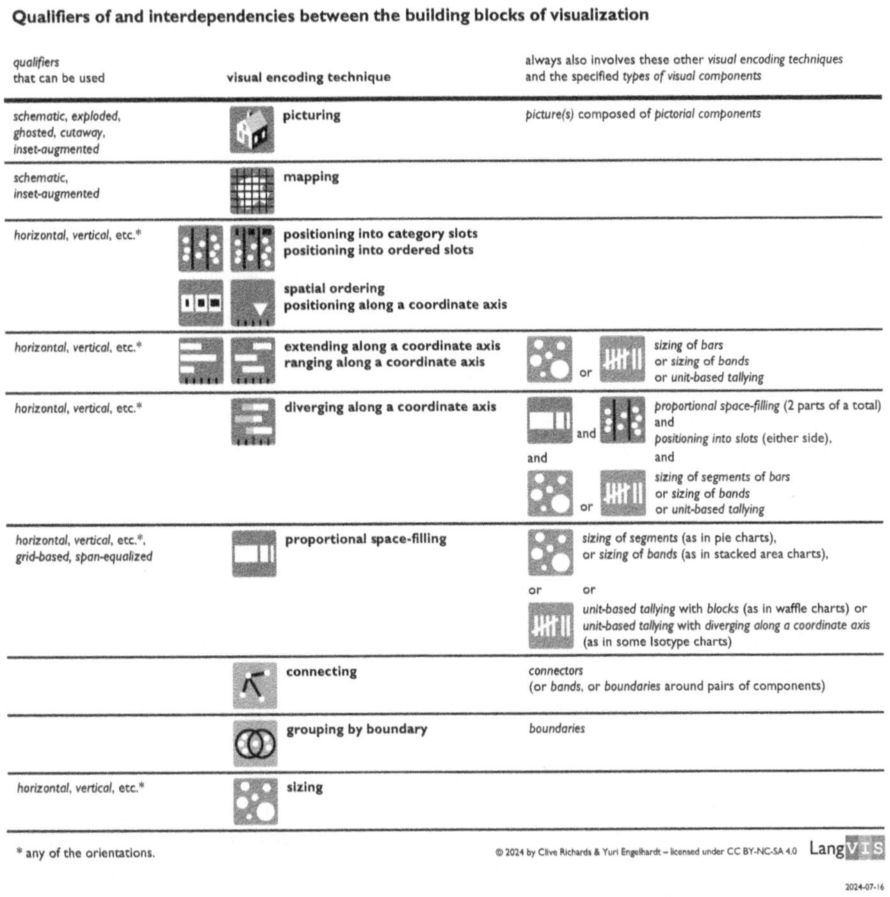

Qualifiers of and interdependencies between the building blocks of visualization

qualifiers that can be used	visual encoding technique	always also involves these other *visual encoding techniques* and the specified types of *visual components*
schematic, exploded, ghosted, cutaway, inset-augmented	picturing	picture(s) composed of *pictorial components*
schematic, inset-augmented	mapping	
horizontal, vertical, etc.*	positioning into category slots / positioning into ordered slots	
	spatial ordering / positioning along a coordinate axis	
horizontal, vertical, etc.*	extending along a coordinate axis / ranging along a coordinate axis	*sizing of bars* or *sizing of bands* or *unit-based tallying*
horizontal, vertical, etc.*	diverging along a coordinate axis	*proportional space-filling* (2 parts of a total) and *positioning into slots* (either side), and *sizing of segments of bars* or *sizing of bands* or *unit-based tallying*
horizontal, vertical, etc.*, grid-based, span-equalized	proportional space-filling	*sizing of segments* (as in pie charts), or *sizing of bands* (as in stacked area charts), or *unit-based tallying with blocks* (as in waffle charts) or *unit-based tallying with diverging along a coordinate axis* (as in some Isotype charts)
	connecting	*connectors* (or *bands*, or *boundaries* around pairs of components)
	grouping by boundary	*boundaries*
horizontal, vertical, etc.*	sizing	

* any of the orientations.

© 2024 by Clive Richards & Yuri Engelhardt – licensed under CC BY-NC-SA 4.0 LangVIS
2024-07-16

Fig. 4. Some visual encoding techniques (middle column in bold type) can be specified by *qualifiers*, such as *orientations* (left column). In addition, interdependencies exist between building blocks – some visual encoding techniques are always used in combination with one or more other visual encoding techniques and/or with a certain type of visual component (right column).

comprehensive, and much more specific regarding the ways in which its building blocks can be combined.

Keck et al. (2020) present 'A didactic methodology for crafting information visualizations'. This work, and Keck's PhD thesis (2019), offer a comprehensive system with many parallels to the *language of visualization*. Keck uses colour-coded categories of building blocks, including 'visual elements' (*visual components* in the *language of visualization*) and 'layout structures' (*arranging* and *linking* in the *language of visualization*). Keck also includes building blocks for interaction (which we do not cover), and the 2020 paper describes a user study. In various other regards however, the *language of visualization* is more encompassing. In 2023, we joined forces with Keck at the Information + conference in Edinburgh. Together we ran a workshop, titled 'Explore the zoo

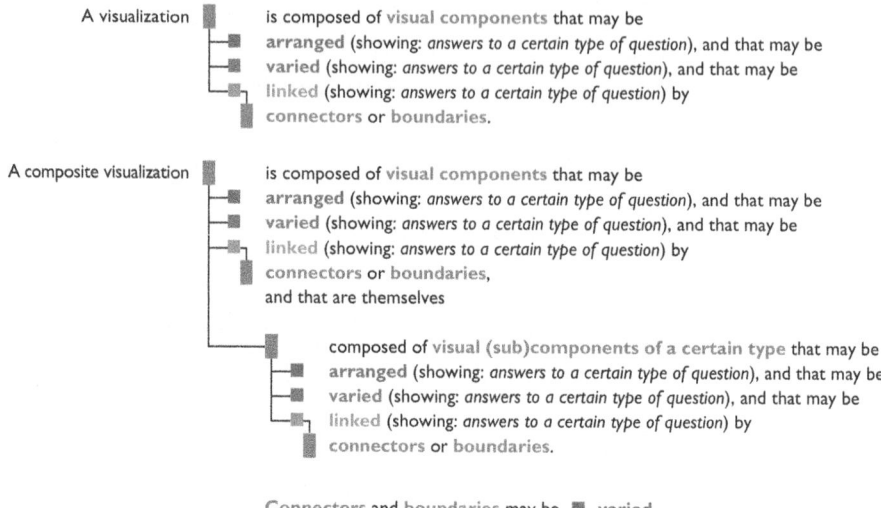

Fig. 5. The *structure* for generating a sentence to describe a type of visualization.

of visualization species', in which participants sketched visualization possibilities using the *language of visualization* framework.

Lachenmeier and Hil (2022) set out another approach – the 'Modular Information Design' (MID) system, which is impressive. Their approach has parallels to the *language of visualization*, notably in terms of its visual appearance. Distributed over four colour-coded 'dimensions', MID has a larger number of building blocks than the *language of visualization*. Many of MID's 80 elements can be expressed as combi-nations of two (more 'atomic') building blocks in the *language of visualization*. While the *language of visualization* uses fewer building blocks than the MID system, it covers a broader range of different types of visualization. This is due to our inclusion of visual encoding techniques such as *coupling by adjacency* (e.g. for icicle charts and sunburst charts), and of qualifiers such as, for example, *span-equalized* (e.g. for 100% stacked bar charts and 100% stacked area charts), and *schematic*, *exploded*, *ghosted*, and *cutaway* (e.g. for technical illustrations). In addition, rather than simply enumerating building blocks, the *language of visualization* offers a grammar and a precise language for describing *how building blocks relate to each other* in any given visualization, expressed visually and in a natural language sentence. Unlike the MID system, the *language of visualization* also offers guidelines for matching *types of questions* that visualizations can answer – i.e. the characteristics of the information to be visualized – to the building blocks for possible visualization designs (Fig. 3).

Pollock et al., in 'Bluefish: A relational grammar of graphics' (2023), note that "Wilkinson's GoG [Grammar of Graphics 2005] and its descendants are largely focused on statistical graphics (i.e., graphic representations that compare quantities …)" and present "a grammar that extends the benefits of compositional specification to data-driven

How to generate a sentence
that describes a type of visualization

Each sentence that describes a type of visualization starts with:

- A [type of visualization] is composed of
  [a type of visual component, e.g. symbols, bars], that are

Followed by one or more visual encoding technique (connected with 'and'):

- arranged using **picturing**
- arranged using **mapping**
- arranged using ● **positioning into category slots**
- arranged using ● **positioning into ordered slots**
- arranged in a ● **spatial order**
- arranged using ● **proportional space-filling**
- **positioned along a** ● **coordinate axis**
- **extending along a** ● **coordinate axis**
- **diverging along a** ● **coordinate axis**
- **ranging along a** ● **coordinate axis**
- **coupled by adjacency**
- **nested**
- **colour-coded**
- **gradient-coded**
- **shape-coded**
- **sized**
- **repeated using unit-based tallying**
- connected
- grouped by boundary

Techniques marked with ● can be specified with an orientation:

> horizontal, vertical, in the third dimension, angular, radial, spiral,
> in a vertical array of horizontal axes where they are [something with ●],
> in a horizontal array of vertical axes where they are [something with ●],
> in an angular array of radial axes where they are [something with ●]

Optionally, each of the visual encoding techniques can be followed by:

> (showing: [the type of question that is answered by using that technique])

*In the case that the components identified in the beginning of the sentence
are composite components, we add:*

- and that are, in turn, composed of [a type of visual component]

*Such (sub)components, as well as connectors and boundaries, can also be specified
by visual encoding techniques. For visualizations with composite components
(i.e. components that are themselves made up of components), we like to use indenting
to indicate the part of the sentence that refers to a further level of decomposition.*

© 2024 by Clive Richards & Yuri Engelhardt – licensed under CC BY-NC-SA 4.0

2024-06-30

Fig. 6. *Instructions* for generating a sentence to describe a type of visualization.

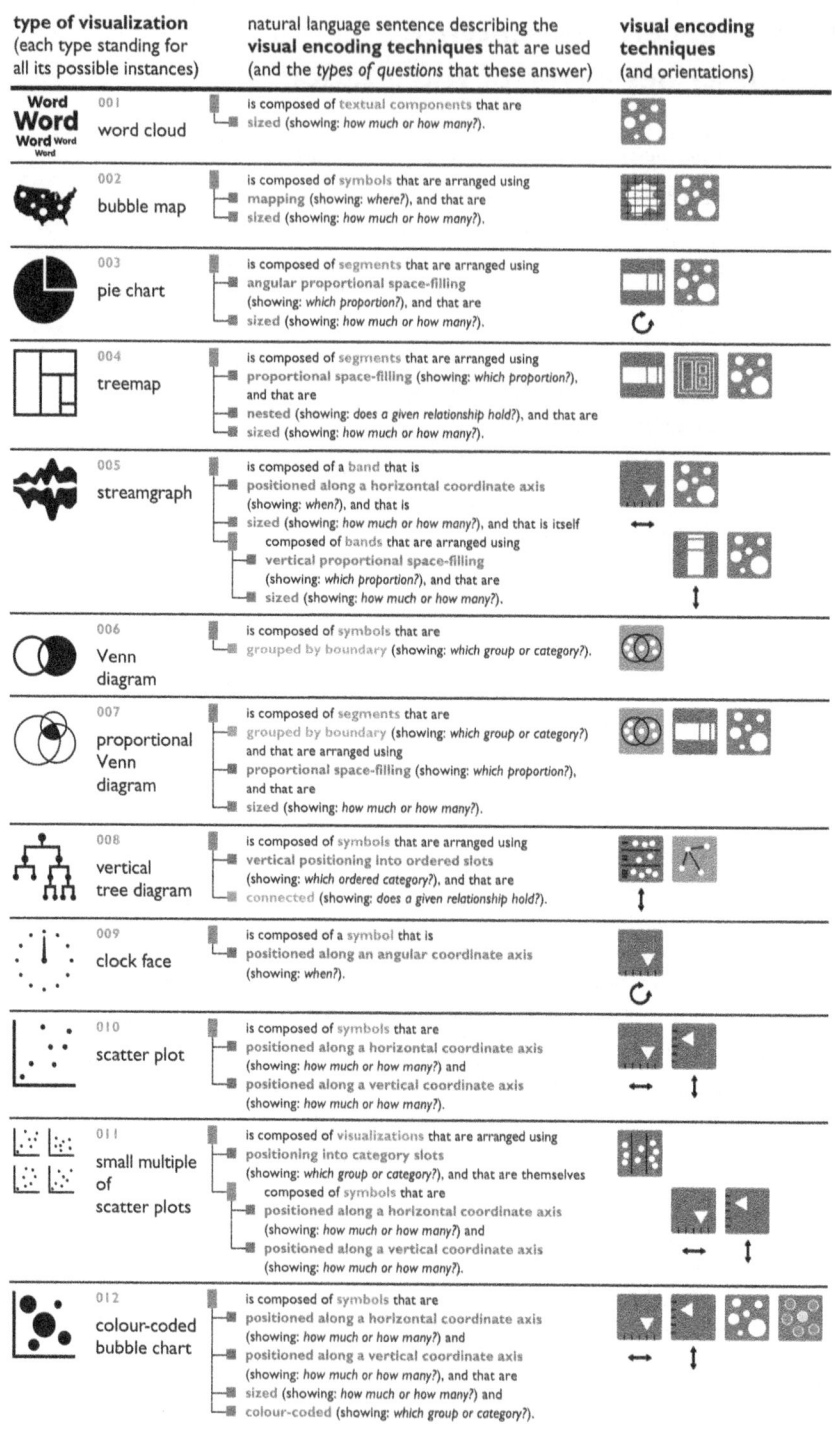

Fig. 7a. Analyses of different types of visualization (continued on the following pages).

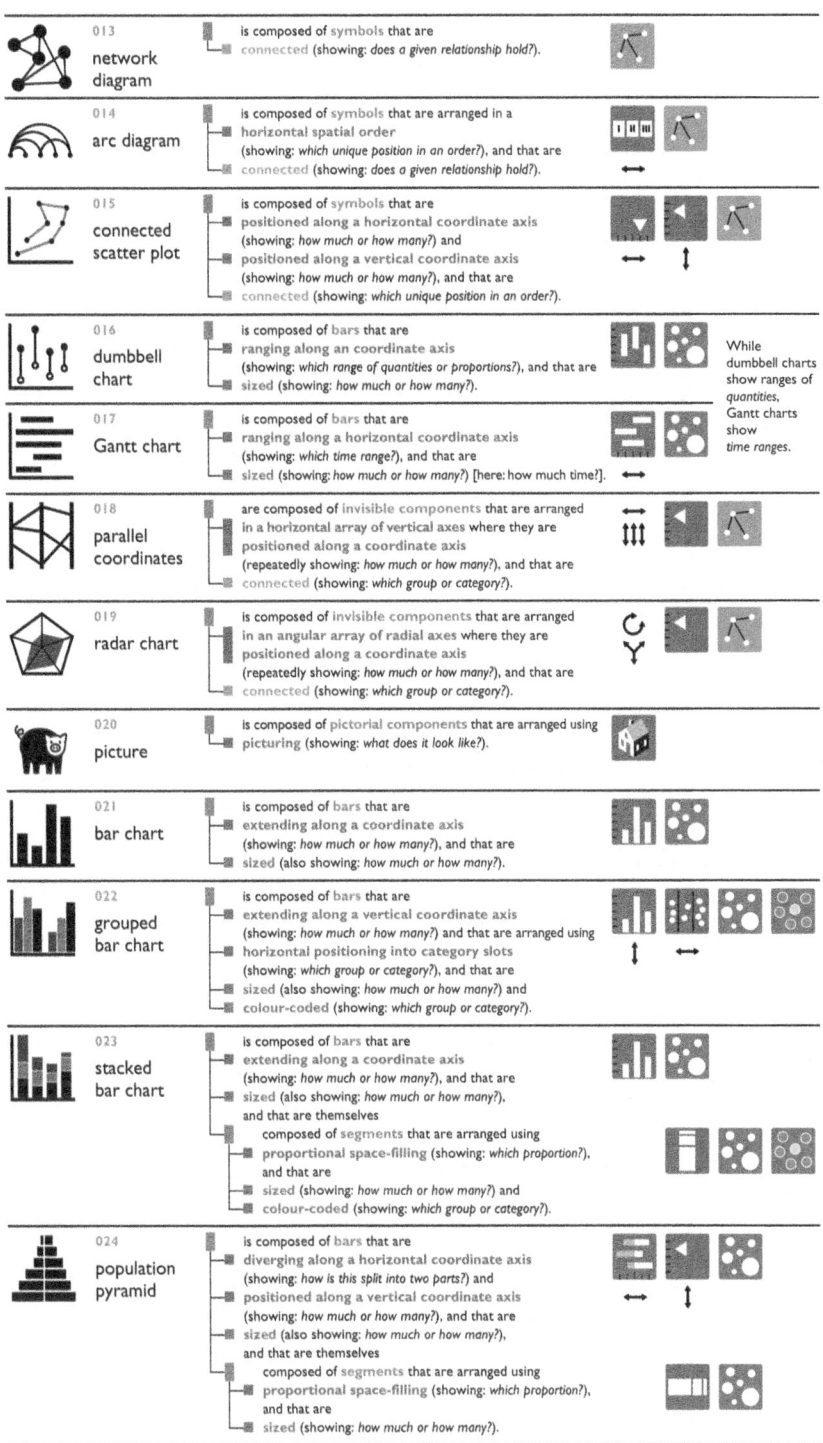

Fig. 7b. Continued from Fig. 7a.

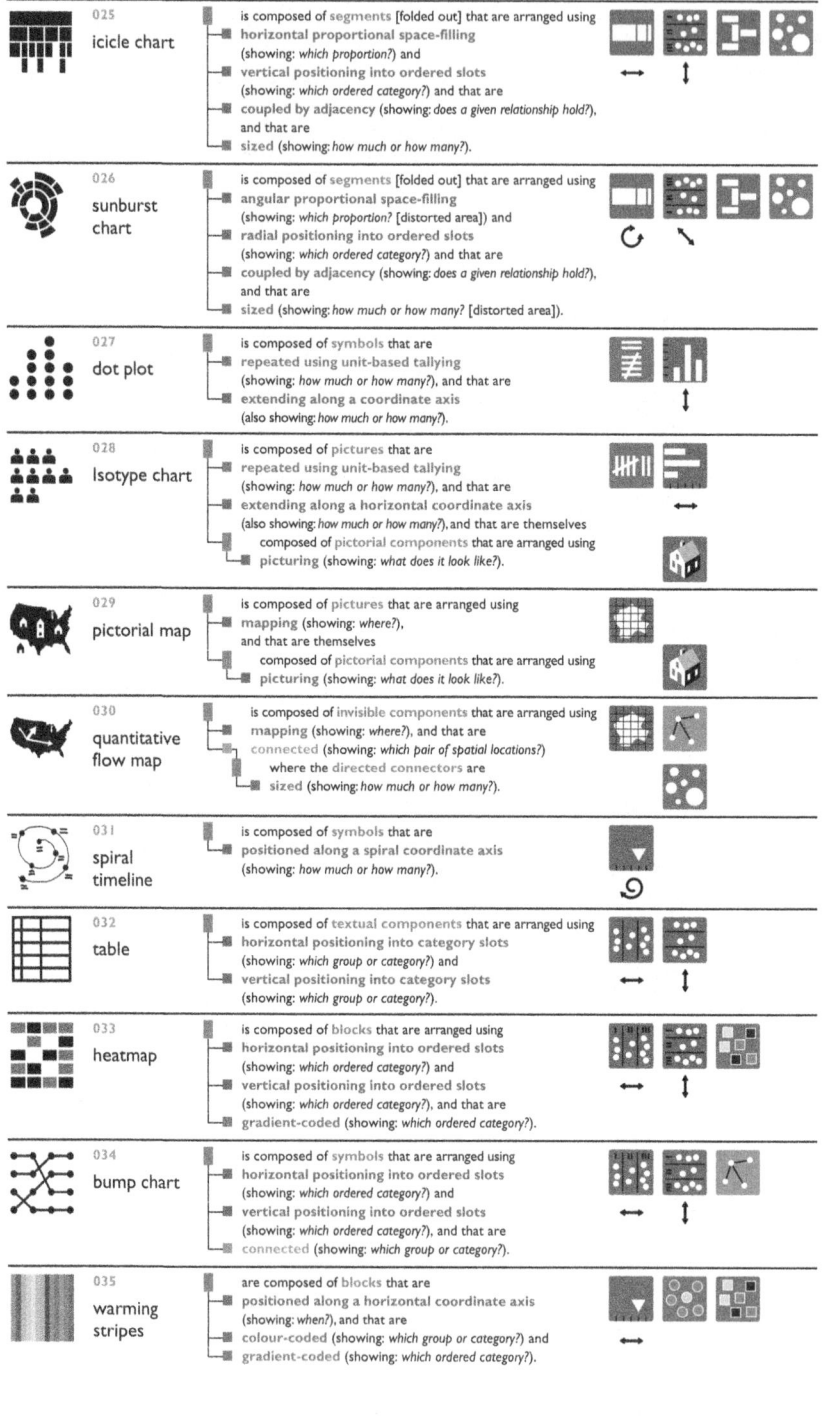

Fig. 7c. Continued from Figs. 7a and 7b.

graphic representations beyond statistical charts." Considering such non-quantitative aspects of visualization, we have pointed out in the past (e.g. in Richards and Engelhardt 2019) that Gestalt principles play an important role as *visual encoding techniques*. Similarly, Pollock et al. point out in their paper that "Perceptual groupings, also known as Gestalt relations, can not only spatially arrange elements but also denote other relationships including containment and connection." They offer five 'perceptual grouping operators': 'element connectedness', 'similar attribute', 'common region', 'spatial proximity' and 'spatial alignment'. All five of these operators have long been at the core of the *language of visualization* – 'element connectedness' as *connecting*, 'similar attribute' as *colour coding* and *shape coding*, and (combinations of) 'common region', 'spatial proximity' and 'spatial alignment' as *grouping by boundary, positioning into slots* and *coupling by adjacency*.

Deng et al., in 'Revisiting the design patterns of composite visualizations' (2023), propose specific 'composition patterns' for combining visualizations into a composite visualization: 'repetition', 'mirror', 'stack', 'co-axis', 'coordinate', 'annotation', 'large panel', and 'nesting'. All of these 'patterns' are covered in the *language of visualization*. For example, Deng et al.'s 'nesting' pattern corresponds to *visual components* that are themselves *composed* of *components* (see Fig. 5 and items 005, 011, 023, 024, 028 and 029 in Fig. 7a). The 'repetition' pattern corresponds to arranging variations of the same visualization into a 'small multiple', using *positioning into category slots* (shown as item 011 in Fig. 7a). In the *language of visualization* such patterns of composition and arrangement can be applied recursively at any level of decomposition.

5 Conclusions

Our review of building-block approaches to visualization has revealed that these approaches use (combinations of) building blocks that correspond to *subsets* of the building blocks of the (more comprehensive and more 'atomic') *language of visualization* framework.

A large number of different types of visualization have been analysed using the conceptual building blocks of the *language of visualization* framework. Through this process it has been demonstrated that the *language of visualization* can be used as an analytical tool for representing the fundamental similarities and differences between different types of visualization. Such similarities and differences can be found by comparing the various analyses in Fig. 7a. The framework seems suitable as a tool for analyzing collections of a wide range of visual representations in terms of their fundamental conceptual building blocks and their combination.

The *language of visualization* is an evolving programme. Because of the flexible structure of the framework, further building blocks can be added in order to accommodate any additional types of visualization that one may want to describe and that cannot be fully characterized using the current scheme. For example, a visual encoding technique *positioning in a ternary plot* can be added, which can answer an additional type of question: *how is this split into three parts (proportions)?* Also, building blocks may be added for animation or for interactivity in visualizations.

The prescription of 'rules for good design' falls outside the aims of the *language of visualization* framework, and the framework will *not* tell a designer which type of visualization is best for any particular task. The framework also does not guarantee that when using a particular type of visualization, the designer will make good implementation choices (e.g. a *colour coding* scheme can be designed in numerous ways).

However, the framework and the associated terminology offer a *language* for discussing design options – ways of discussing possible visual encoding techniques and their combinations for a given visualization challenge. Working with the framework may lead to a better understanding of the structure of the visualization design space.

Future work will explore the framework's potential as an aid for designing visualizations, as an aid for choosing between different types of visualization, and maybe for generating combinations of building blocks that result in novel types of visualization. The framework may provide a basis for formalization, and the potential for computer-based visualization advice.

A more detailed exposition of the *language of visualization* framework will be published in the book *Elements of diagramming* (Richards, forthcoming).

References

Bertin, J.: Semiology of Graphics: Diagrams, Networks, Maps. University of Wisconsin Press, Madison (1967/1983). (original: Sémiologie graphique. Gauthier-Villars)

Bevington, W.: A visualization-based taxonomy for information representation: identifying images with icon schematics. PIIM Paper 01 part 3, Parsons Institute for Information Mapping, New York (2008)

Card, S.K., Mackinlay, J.: The structure of the information visualization design space. In: Proceedings of the 1997 IEEE Symposium on Information Visualization (InfoVis 1997), pp. 92–99 (1997)

Deng, D., et al.: Revisiting the design patterns of composite visualizations. IEEE Trans. Vis. Comput. Graph. **29**(12), 5406–5421 (2023)

Engelhardt, Y., de Bruin, J., Janssen, T., Scha, R.: The visual grammar of information graphics. In: Narayanan, N.H., Damski, J. (eds.) Proceedings of the AID 1996 Workshop on Visual Representation, Reasoning and Interaction in Design, Stanford University (1996)

Engelhardt, Y.: Meaningful space. In: Abrams, J. (ed.) If/then: Play—Design Implications of New Media. Netherlands Design Institute / BIS Publishers (1999)

Engelhardt, Y.: The Language of Graphics. Ph.D. thesis, University of Amsterdam (2002)

Engelhardt, Y., Richards, C.: A framework for analyzing and designing diagrams and graphics. In: Chapman, P., Stapleton, G., Moktefi, A., Perez-Kriz, S., Bellucci, F. (eds.) Diagrams 2018. LNCS (LNAI), vol. 10871, pp. 201–209. Springer, Cham (2018). https://doi.org/10.1007/978-3-319-91376-6_20

Johnson, M.: The Body in the Mind. University of Chicago Press, Chicago (1987)

Keck, M.: Visuelle Exploration multidimensionaler Informationsräume. Ph.D. thesis Dresden University of Technology (2019)

Keck, M., Groh, R., Vosough, Z.: A didactic methodology for crafting information visualizations. In: IEEE Visualization Conference (VIS), pp. 186–190 (2020)

Lakoff, G.: Women, Fire, and Dangerous Things. The University of Chicago Press, Chicago (1987)

Lachenmeier, N., Hil, D.: Visualizing Complexity: Modular Information Design Handbook. Birkhäuser, Basel (2022)

Pollock, J., Mei, C., Huang, G., Jackson, D., Satyanarayan, A.: Bluefish: a relational grammar of graphics. arXiv preprint arXiv:2307.00146v1 [cs.GR] (2023)

Richards, C.: Diagrammatics. Ph.D. thesis, Royal College of Art, London (1984). diagrammatics.com

Richards, C., Engelhardt, Y.: The DNA of information design for charts and diagrams. Inf. Des. J. **25**(3), 277–292 (2019)

Richards, C.: Elements of Diagramming. Routledge (forthcoming)

Sedig, K., Parsons, P.: Design of Visualizations for Human-Information Interaction: A Pattern-Based Framework. Morgan & Claypole, Williston (2016)

Tversky, B.: Cognitive origins of graphic productions. In: Marchese, F.T. (ed.) Understanding Images, pp. 29–53. Springer, New York (1995)

Twyman, M.: A schema for the study of graphic language. In: Kolers, P.A., Wrolstad, M.E., Bouma, H. (eds.) Processing of Visible Language, vol. 1, pp. 117–150. Plenum Press, New York (1979)

Ware, C.: Visual Thinking for Design. Morgan Kaufmann Publishers, San Francisco (2008)

Wilkinson, L.: The Grammar of Graphics, 2nd edn. Springer, New York (2005). https://doi.org/10.1007/0-387-28695-0

Domain-Specific Rules Override Aesthetic Graph Drawing Criteria: An Exploration of User-Generated Diagrams

Stefan Helmke[✉], Kerem Doğan, Robert Scheffler, and Gregor Wrobel

Society for the Advancement of Applied Computer Science (GFaI), Volmerstraße 3, 12489 Berlin, Germany
{helmke,dogan,scheffler,wrobel}@gfai.de

Abstract. This paper investigates the impact of domain-specific rules on classical aesthetic graph drawing criteria. In the field of graph drawing, several aesthetic criteria are understood to improve the comprehensibility of graphs. Domain experts consider the technical implementation in the real world resulting in domain-specific rules for graph drawing, and some of these rules contradict general aesthetic criteria. Experts and novices drew graphs from scratch in two controlled experiments. In the first experiment, the participants modeled a fictional social scenario with graphs. In the second experiment, the diagrams represented digital twins of energy systems. The experiment analysis shows that some classical criteria like reducing crossings are less relevant for domain experts while others, like minimizing bends, are still important.

Keywords: User-Generated Diagrams · Domain-Specific Rules · User Studies

1 Introduction

Schematic representations are important artifacts and have many applications. They mainly serve as tools for visualizing complex structures, with the most widespread application being the modeling of objects, systems, and their relations in modeling languages. Model-driven engineering (MDE) [1] uses models as a bridge between drafts and implementations. In software development, problem domains are described in models that are represented by schematic diagrams. These models can be transformed, transferred, and understood by different tools and users. Model-based systems engineering (MBSE) extends this context to the general application of modeling to the full life cycle of any technical systems. These models are typically referred to as digital twins and are used for requirements specification, system overviews, and analysis, or behavior models of single components. Even technical artifacts that predate MBSE advancement like CAE-drawings can be part of a modeling approach.

When models and modeling techniques are used, schematic representations help make these artifacts visually clear and descriptive. This makes schematic drawings an important human-machine-interface in many domains. As an interface, the schematics also allow for interactions: either passive (e.g., zooming, filtering, changing layout aspects) or active (e.g., editing, extending, or otherwise modifying the model). Understanding diagrams as interfaces composed of interactions puts the topic of user experience in the focus. This paper aims to further the understanding of the process of drawing diagrams by real-world users of modeling tools.

In domains where the MDE paradigm has been widely adopted (e.g., automotive, aerospace, or communications), there are several graphical modeling languages like UML and its various specifications through the profile mechanism [2, 3]. In addition, there are modeling languages with a narrower scope like BPMN or Matlab/Simulink [4]. Whereas all these languages were created with a strong domain focus, they are increasingly used universally. We therefore collectively refer to these modeling languages as general-purpose modeling languages (GPML). Domain-specific languages (DSL) [5] are more restricted in their application and usage. Here, the semantics of the language are the focus of language design. The concrete syntax of these languages is a tool to arrive at certain goals [6]. Language artifacts (i.e., models) are used as the base for ongoing work in model transformations, optimizations, or simulations.

An important distinction between GPML and DSL is their user base. Users of DSL are often not modeling experts, so they have little experience in drawing schematic diagrams. They are, however, experts in their respective domains.

The graphic representation of modeling languages is of central importance for the acceptance of the languages by their intended users [7]. That means that the diagrams and visualizations of DSL should be a focus of their design. People using graphic languages should be able to generate new models and understand existing diagrams. In doing so they interact with the static (e.g., symbols and representations) and dynamic (e.g., layout) aspects of model visualizations.

The central question of this paper is how people interact with schematic drawings. With a background in the development of graphical editors for various graphical, domain-specific languages (GDSL), we have found this to be an important aspect of our work. In order to implement a successful graphical language, the requirements and usage patterns of the users must be taken into account. In this paper, we investigate the practical considerations concerning the generation of schematic diagrams. What do users consider when they draw models, and which criteria do they pose to their work?

The answers to these questions provide a basis for deriving requirements for the development of graphical editors. Specialized layout algorithms should support criteria for schematic drawings that are important to users, and interaction implementations should take them into account.

Graphic representations used by modeling languages and tools often follow aesthetic criteria set in the discipline of graph drawing (see Sect. 2) [8]. These criteria can be contraindicated by domain-specific requirements, and it is not directly clear how users solve this conflict. The importance of aesthetic criteria for the readability of diagrams is well established, but we focus on their effect on the generation of domain-specific schematics by users.

Another consideration when designing graphical languages and tools is the intended user group. Therefore, we consider the knowledge level of different users and how this affects their work on schematic diagrams. Knowledge or experience can be interpreted in two different ways: The first differentiation is made between modeling experts and novices (modeling experience), and the second differentiation is made between domain experts and novices (domain experience).

We conducted two experiments with diverse participants to answer several questions on their interactions with schematic drawings. This study aims to further the understanding of tool support needed for drawing technical models. This should allow us to develop better graphical languages and editors and increase the acceptance of graphical modeling as a tool in technical, e.g., engineering, domains.

We present the relevant research that forms the base for our work in Sect. 2. In Sect. 3, we set out our research questions and hypotheses for the study. Following this, we describe in Sect. 4 the performance of the experiments. The experiment design consists of several modeling tasks, some of which are carried out in a GPML and some in a specific GDSL. In Sect. 5, we present and statistically analyze the results of the experiments. The results are discussed thoroughly in Sect. 6. There, we also answer the research questions and interpret some important aspects of the evaluation.

2 Related Work

In this section, we present common graph drawing criteria and discuss several studies investigating the effect of aesthetic criteria on human understanding. We do this for automatically generated diagrams and for user-generated diagrams. We also consider studies with the focus on the particular domain UML.

The field of graph drawing deals with the design, analysis, implementation, and evaluation of algorithms for drawing graphs. The goal is a clear drawing (layout) of a graph. Such a visualization can be realized by the placement of vertices and routing of edges following aesthetic criteria and need to be embedded in its specific problem domain to acquire good readability. Aesthetic criteria to enhance graph layout include the reduction of crossings, bends, edge length, and the occupied area of drawing [8]. Formal metrics to measure these criteria were introduced by Purchase [9].

Several studies have been conducted to check aesthetic criteria in automatically generated drawings. The minimization of bends and crossings leads to a better comprehensibility of graphs [10, 11], and maximizing symmetry also has a positive effect on the readability of diagrams [11].

Besides the study of automatically generated diagrams, there are also several studies concerning user-generated layouts. The manipulation of automatically generated diagrams indicates that symmetry and minimizing crossings are important for users, but uniform edge length is not relevant [12, 13]. An experiment where participants drew graphs from scratch showed that the minimization of edge crossings and the alignment of vertices and edges to an underlying grid is essential for users [14].

There are also studies considering aesthetic criteria in a particular domain, e.g., in UML. These studies claim that the most relevant criterion for the automatic layout of UML class diagrams is the minimization of crossings and bends, while other criteria like edge length or symmetry are not important [15, 16].

3 Research Questions and Hypotheses

In this section, we define our research questions (Q) and hypotheses (H). In the GPML experiment, we wanted to find out what influence the modeling experience of participants has on the quality of diagrams. Of course, we expected the quality of diagrams from modeling experts to be better (concerning the criteria number of errors, bends, branches, crossings, the total edge length, the size of the layout area) than that from people with less experience.

- Q1: *Do modeling experts create better diagrams than people with less experience?*
- H1: *Modeling experts create better diagrams than people with less experience.*

In the GDSL experiment, we wanted to figure out whether the particular domain has an impact on the layout of the diagrams. We expected that domain experts do not create diagrams of lower quality than people with less domain experience and that classical aesthetic criteria are not as important for domain experts as for modeling experts in the GPML experiment.

- Q2.1: *Do domain experts create better diagrams than people with less experience?*
- H2.1: *Domain experts do not create worse diagrams than people with less experience.*
- Q2.2: *Are aesthetic criteria less important for domain experts?*
- H2.2: *Aesthetic criteria are less important for domain experts.*

The third goal concerned both experiments. We wanted to know whether modeling and domain experts require less time or are more efficient than people with less experience. We expected that modeling and domain experts are more efficient than people with less experience, i.e., they are faster with the creation of diagrams, but invest more time in the quality of diagrams. In other words: they create diagrams of higher quality in the same or less time as people with less experience.

- Q3: *Are experts more efficient than people with less experience?*
- H3: *Experts are more efficient than people with less experience.*

4 The Experiment: Experts vs. Novices

To test our hypotheses, we carried out two experiments: a GPML experiment and a GDSL experiment. The first one used a general diagram editor. The second one used the software TOP-Energy®. The diagram editor and TOP-Energy® were installed on computers with a similar hardware configuration. Both experiments were conducted using a mouse and keyboard. We invited the participants to the experiment in groups one after the other. The seating arrangement was chosen in such a way that mutual interference between the subjects could be ruled out.

4.1 Experimental Procedure and Tasks

Before the first experiment started, the participants got an information sheet and the declaration of consent. The researcher verbally informed the participants about the procedure and scientific benefits of the experiment and gave them the opportunity to ask questions. The participants and the researcher signed the declaration of consent. The procedures of both experiments were identical and included several rest breaks.

In all tasks, participants had to place vertices and connect them starting from a basic layout (see Fig. 1). Hyperedges connected the vertices, i.e., one edge could connect more than two vertices. The participants got tables in which all hyperedges were listed with their vertices. They could choose the order in which to connect the vertices (i.e., the order of the hyperedges) at will. The researcher did not make the participants aware of typical aesthetic criteria such as minimizing crossings in graph drawing. Besides the tables, there was also a textual description to increase understanding of the task. The participants had as much time as they wanted for each task and could also cancel tasks if they felt that they could not solve them with the given tools. The instruction was the following.

– *Create diagrams that correspond to your individual ideas and that you think look good and neat. Create diagrams that you can easily explain to other people and that you or other people could reuse.*

The results of the tasks, i.e., the created diagrams, were recorded. All tasks were also screen-recorded. For each task, the required time, the number of errors, bends, branches, crossings, the total length of all hyperedges, and the layout area used were collected.

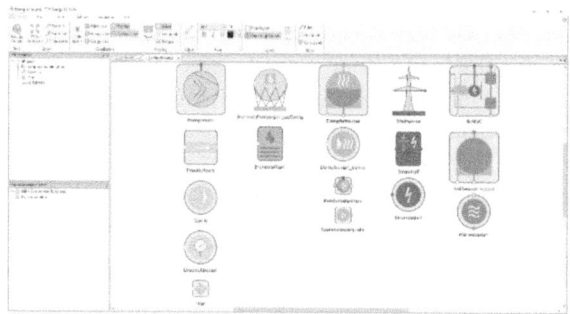

Fig. 1. Software TOP-Energy® used in the GDSL experiment: The figure shows task 3 before the vertices were connected.

At the beginning of an experiment, a tutorial task demonstrated all functionalities that could be used. Participants got a document with a descriptive explanation of the tutorial task and the functionalities they could use in the general diagram editor and TOP-Energy®. It was an opportunity for the participants to get familiar with the tasks. The diagram of the tutorial task was of minimal size so as not to influence participants in later tasks. At this time, participants could ask questions.

In both experiments, the participants solved four tasks. The tasks were available both as files on the computers and as printed documents at the workstations. The first and the second tasks were practice tasks to minimize the learning effect and acclimate the participants to the tasks. The participants were not aware that these tasks were not part of the experiment. The third and the fourth tasks were the tasks of the experiment. During the practice and experiment tasks, participants were able to ask individual questions to the researcher.

At the end of the last experiment, participants answered a questionnaire with questions regarding prior knowledge, test procedure, and suggestions for improvement.

In the first experiment, the graphs were based on an imaginary scenario where vertices were fictional students. All vertices had the same rectangular shape, but included the different names of the students. All names were short and easy. The task was to connect students if they shared a common attribute. These common attributes were birthplace (tutorial task), university attended (first task), age (second task), a movie watched together (third task), and a concert attended together (fourth task), see Fig. 2.

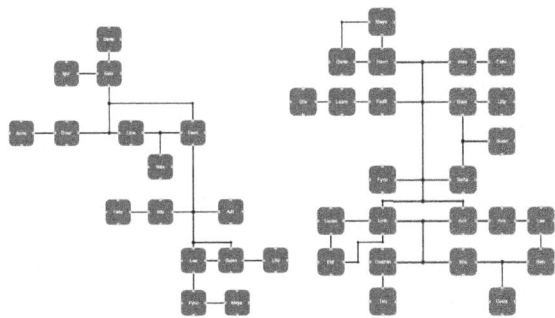

Fig. 2. GPML experiment: diagram from task 3 (left), diagram from task 4 (right).

In the second experiment, diagrams represented digital twins of energy systems. The tutorial task and the first task represented simple power supply systems. The second task represented a simple cooling supply system. The third task represented a combined steam, heat, power, and compressed air supply system; while the fourth task represented a complex coupled heat, power, and steam supply system (see Fig. 3).

The size of the hypergraphs and the number of tasks were determined by pilot tests[1] as follows.

- GPML, third task: 16 vertices, 11 hyperedges
- GPML, fourth task: 24 vertices, 18 hyperedges
- GDSL, third task: 17 vertices, 13 hyperedges
- GDSL, fourth task: 27 vertices, 24 hyperedges

For the second experiment in particular, the size of the diagrams had to be carefully weighed up. On the one hand, they had to represent energy systems that were as realistic as possible. On the other hand, the structure of the diagrams should not overburden users.

---

[1] Pilot tests were carried out separately before the experiments with experts who did not take part in the actual experiments. Based on their results, we determined the size of the hypergraphs and the number of tasks in both experiments.

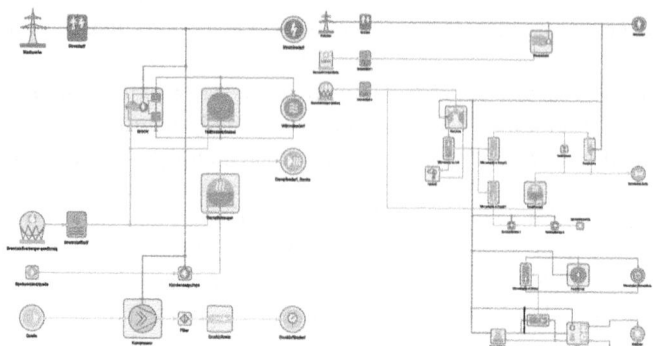

Fig. 3. GDSL experiment: diagram from task 3 (left), diagram from task 4 (right).

4.2 Possible User Interactions

Participants could use a range of interactions to solve the tasks. They could select single vertices or lines by mouse click and multiple vertices or lines by mouse click while holding down the shift key. Alternatively, multiple vertices could be selected by a selection rectangle. The participants could move vertices or lines via drag and drop. Further, participants could move vertices with the arrow keys on the keyboard. Vertices could be mirrored or rotated using a context menu and aligned using a button. An edge between two vertices could be created by dragging and dropping from one vertex to another vertex. An edge or hyperedge could be expanded by connecting a vertex with the edge or hyperedge via drag and drop. Edges or parts of hyperedges could be deleted using a context menu or the delete key. Zoom options were available, and interactions could be reversed.

4.3 The Layout Algorithm

In both experiments, the modification of vertices (i.e., moving, mirroring, rotating) triggered an automatic, orthogonal routing (OrthoRoute) of hyperedges that are connected to these vertices [17]. This algorithm does not strive for a minimization of length or crossings. It takes advantage of large free spaces, so the lines do not run closely along the shapes of the vertices. The main idea of the OrthoRoute algorithm is the construction of horizontal or vertical buses. In large diagrams, the buses help users to identify the single hyperedges. This is an advantage over grid-oriented and length-minimizing methods. The OrthoRoute algorithm only routes hyperedges and does not place vertices automatically.

Besides the modification of vertices, participants were also able to modify the geometry of hyperedges directly. They were allowed to move horizontal lines vertically and vertical lines horizontally. Then, the OrthoRoute algorithm considers such hyperedge modifications [18].

4.4 Participants

We recruited 20 participants for both experiments. Among the 20 participants were 10 employees of the GFaI, six students of the Berliner Hochschule für Technik (BHT, degree programs: Facility Management, Media Informatics), two students of the Technical University of Applied Sciences Wildau (TH Wildau, degree program: Telematics), one student of the Hochschule für Technik und Wirtschaft Berlin (HTW Berlin, degree program: Computer Engineering), and one student of the Technische Universität Berlin (TU Berlin, degree program: Mathematics).

Five GFaI employees took part in the GPML experiment and had modeling experience with our diagram editors, while the other five GFaI employees took part in the GDSL experiment and had a lot of domain and modeling experience with energy systems and TOP-Energy®. All 10 students took part in both experiments. Five students had a rough domain experience with energy systems but no modeling experience with TOP-Energy® and the general diagram editor. Two students had no modeling experience with TOP-Energy®, but had modeling experience with our diagram editors. The remaining students had no background domain or modeling knowledge of energy systems, TOP-Energy®, or our diagram editors. For each experiment, we formed groups divided by experience levels (see Tables 1 and 2).

Table 1. Groups in the GPML experiment.

Group	Modeling experience with diagram editors	Participants
Expert	Yes	7
Novice	No	8

Table 2. Groups in the GDSL experiment.

Group	Domain experience with energy systems	Modeling experience with TOP-Energy®	Participants
Expert	Yes	Yes	5
Competent	Yes	No	5
Novice	No	No	5

5 Results

In this section, we investigate the influence of modeling and domain experience with diagram editors and energy systems on user-generated diagrams. Section 5.1 shows the results of the GPML experiment, while Sect. 5.2 shows the results of the GDSL experiment. We first checked the data for variance homogeneity and normal distribution to determine which test was appropriate. After that, we ran tests to check for significant differences between the groups with different experience. In all tests, we used a significance level of $\alpha = 0.05$.

In both experiments, the dependent variables are the required time and the number of errors plus the following aesthetic criteria: the number of bends, branches, crossings, the total length of all hyperedges, and the layout area used. We normalized the data for each dependent variable and each task, so that the two tasks in each experiment became comparable. The participant with the greatest value got the normalized value 1, and the participant with the smallest value got the normalized value 0. In addition, we set the number of errors and the aesthetic criteria in relation to the time required, i.e., we tested the efficiency concerning the number of errors and the aesthetic criteria. We defined the quality of a diagram as the difference between 1 and the normalized value. The greater the value of a dependent variable, the smaller the quality, and vice versa. We defined the efficiency as the ratio of quality to time. The efficiency is high when the quality of a diagram is high and the required time is short, and vice versa.

Figure 4 shows the diagrams of task 3 in the GDSL experiment, sorted by groups.

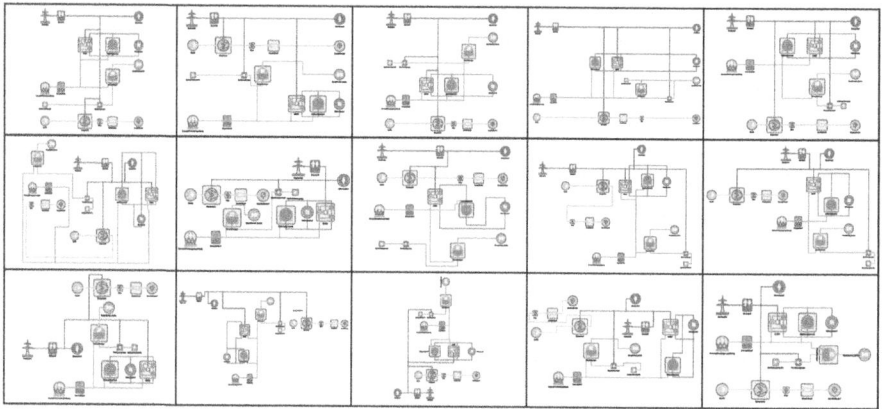

Fig. 4. GDSL experiment, task 3: expert (top), competent (middle), novice (bottom).

5.1 GPML Experiment

In the first experiment, the independent variable is the modeling experience with diagram editors. For each dependent variable, we investigated whether significant differences between the two groups *novice* and *expert* occur (Table 3).

Test for Homogeneity of Variance and Normal Distribution. We used Levene's test to determine whether the variances of both groups were equal and the Shapiro-Wilk test to determine whether the data within both groups were normally distributed. The variances of both groups were not equal regarding the number of crossings and the branches efficiency, whereas the variances of both groups for all other dependent variables were equal. The data of the dependent variables time, errors, bends, branches, crossings, edge

Table 3. Average data from the GPML experiment.

Group (Task)	Time	Errors	Bends	Branches	Crossings	Total length	Layout area
Expert (3)	12:12	0.14	3.9	4.9	0.4	508	26,782
Novice (3)	11:03	0.38	12.5	5.8	2.4	719	32,291
Expert (4)	23:42	0.71	8.0	8.7	0.4	1,001	45,515
Novice (4)	18:59	0.25	21.0	9.5	3.4	1,483	54,354

length, and branches efficiency were not normally distributed, while the data of all other dependent variables were normally distributed (see Table 4).

Table 4. The table shows the p-values of Levene's test and the Shapiro-Wilk test for each dependent variable in the first experiment. Homogeneity of variance and normal distribution were not assumed if $p < 0.05$. The last column shows the p-values of the t-test or the exact p-values of the Mann-Whitney U-Test. Significant differences were assumed if $p < 0.05$.

Dependent variable	Levene's test	Shapiro-Wilk test		p-value
		Expert	Novice	
Time	0.260	0.258	0.014	0.224
Errors	0.758	<0.001	<0.001	0.918
Bends	0.133	0.002	0.364	**<0.001**
Branches	0.081	0.032	<0.001	**0.019**
Crossings	0.001	<0.001	<0.001	0.142
Edge length	0.521	0.002	0.005	**<0.001**
Layout area	0.355	0.206	0.196	0.106
Errors efficiency	0.982	0.900	0.701	0.237
Bends efficiency	0.450	0.161	0.168	**0.006**
Branches efficiency	0.009	0.046	<0.001	**0.025**
Crossings efficiency	0.385	0.152	0.146	0.732
Edge length efficiency	0.518	0.496	0.223	0.057
Layout area efficiency	0.837	0.140	0.440	0.266

Test for Significant Differences. We used the t-test if the variances of both groups were equal and the data in both groups were normally distributed. Otherwise we used the Mann-Whitney U test. In the case of the Mann-Whitney U test, we used the exact p-value due to the small sample size. We found significant differences in the total hyperedge length and the number of bends and branches. Experts used significantly less bends and branches. They created significantly faster diagrams with fewer bends and branches and with significantly shorter hyperedges (see Table 4 and Fig. 5).

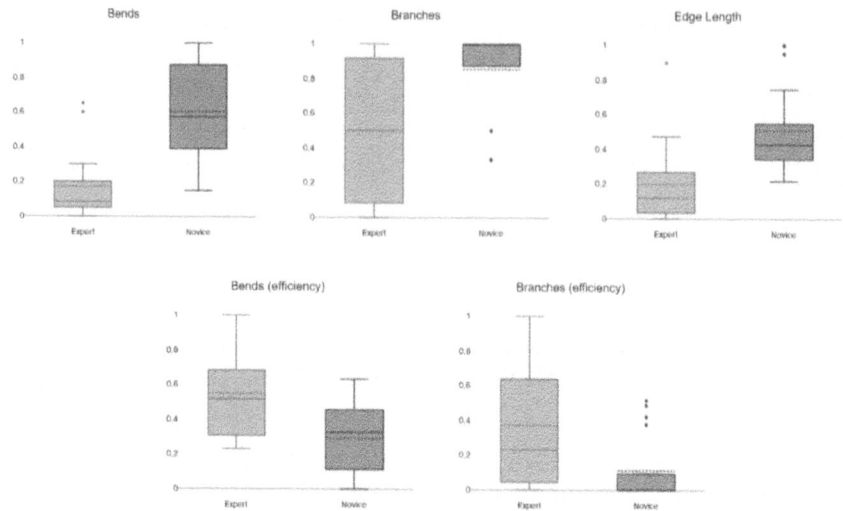

Fig. 5. Distribution of the number of bends and branches, total hyperedge length, and efficiency concerning the number of bends and branches.

Data Separated by Tasks. We separated the data by the tasks but found no further significant differences.

5.2 GDSL Experiment

In the second experiment, the independent variable is the modeling and domain experience with TOP-Energy® and energy systems. For each dependent variable, we investigated whether significant differences between the three groups *novice*, *competent*, and *expert* occur (Table 5).

Table 5. Average data from the GDSL experiment.

Group (Task)	Time	Errors	Bends	Branches	Crossings	Total length	Layout area
Expert (3)	7:33	0.8	9.2	7.2	6.6	2,030	92,155
Competent (3)	12:35	0.6	19.4	5.8	3.0	2,109	92,353
Novice (3)	15:05	0.0	18.4	6.8	2.4	1,522	87,955
Expert (4)	23:03	3.4	33.2	16.6	12.0	4,852	244,690
Competent (4)	26:49	2.6	63.0	14.6	27.6	5,505	234,826
Novice (4)	34:51	1.2	47.4	14.6	11.2	4,475	216,364

Test for Homogeneity of Variance and Normal Distribution. We used Levene's test to determine whether the variances of the three groups were equal and the Shapiro-Wilk test to determine whether the data within the three groups were normally distributed. The variances of the three groups were not equal regarding the number of errors, bends, crossings, and the bends, edge length, and layout area efficiency; while the variances of the three groups regarding all other dependent variables were equal. The data of the dependent variables number of errors, branches, total hyperedge length, layout area, and branches efficiency were not normally distributed, while the data of the other dependent variables were normally distributed (see Table 6).

Test for Significant Differences. We used the analysis of variance (ANOVA), if the variances in the three groups were equal and the data were normally distributed. Otherwise we used the Kruskal-Wallis test. We found significant differences in the required time, the number of bends, branches, and the bends efficiency (see Table 6).

Table 6. The table shows the p-values of Levene's test and the Shapiro-Wilk test for each dependent variable in the second experiment. Homogeneity of variance and normal distribution were not assumed if $p < 0.05$. The last column shows the p-values of the ANOVA or the Kruskal-Wallis test. Significant differences were assumed if $p < 0.05$.

Dependent variable	Levene's test	Shapiro-Wilk test			p-value
		Expert	Competent	Novice	
Time	0.957	0.089	0.250	0.624	**<0.001**
Errors	0.003	0.014	0.043	<0.001	0.163
Bends	0.002	0.504	0.172	0.200	**<0.001**
Branches	0.547	0.033	0.488	0.007	**0.029**
Crossings	0.009	0.057	0.202	0.063	0.130
Edge length	0.654	0.002	0.173	0.123	0.162
Layout area	0.395	0.006	0.310	0.107	0.917
Errors efficiency	0.316	0.242	0.751	0.077	0.945
Bends efficiency	0.044	0.080	0.550	0.731	**<0.001**
Branches efficiency	< 0.001	0.011	0.061	0.766	0.449
Crossings efficiency	0.935	0.373	0.789	0.557	0.774
Edge length efficiency	0.015	0.502	0.358	0.684	0.099
Layout area efficiency	0.025	0.499	0.958	0.614	0.298

Post-hoc Analysis. We examined significant differences between two groups. We used the Bonferroni post-hoc test if the variances in the three groups were equal and the data were normally distributed. Otherwise we used the Dunn-Bonferroni post-hoc test. The significance level was adjusted with Bonferroni correction.

Experts needed significantly less time than novices. On the one hand, experts used significantly less bends than competent participants and novices. On the other hand, experts used significantly more branches than competent participants. In addition, experts created significantly faster diagrams with less bends than competent participants and novices (see Table 7 and Fig. 6).

Table 7. The table shows the p-values (adjusted with Bonferroni correction) of the Bonferroni test or the Dunn-Bonferroni test. Significant differences were assumed if $p < 0.05$.

Dependent variable	Group 1	Group 2	Adjusted p-value
Time	Expert	Competent	0.062
	Expert	Novice	**<0.001**
	Competent	Novice	0.111
Bends	Expert	Competent	**<0.001**
	Expert	Novice	**0.014**
	Competent	Novice	0.806
Branches	Expert	Competent	**0.026**
	Expert	Novice	0.278
	Competent	Novice	1.000
Bends efficiency	Expert	Competent	**<0.001**
	Expert	Novice	**0.001**
	Competent	Novice	1.000

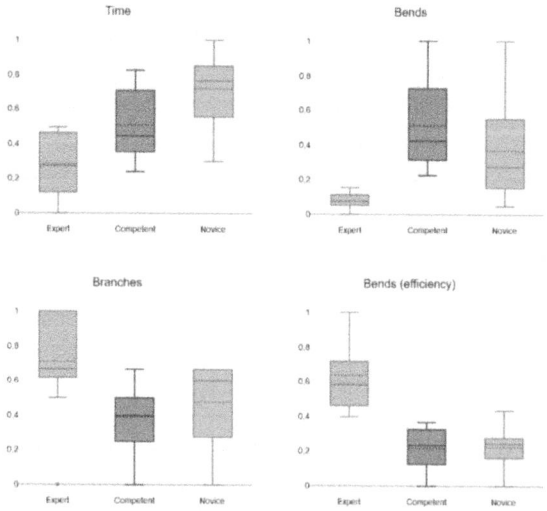

Fig. 6. Distribution of required time, number of bends and branches, and efficiency concerning the number of bends.

Data Separated by Tasks. We separated the data by the tasks and found that experts used significant more crossings than competent participants (adjusted p-value: 0.007) and novices (adjusted p-value: 0.002) in task 3. We found no further significant differences.

6 Discussion and Conclusion

This section discusses the hypotheses from Sect. 3 based on the results from Sect. 5. We observed five aesthetic criteria to measure the quality of diagrams: the number of bends, branches, crossings, the total length of all hyperedges, and the layout area used (see Sect. 5). The overall quality of a diagram is better if it is better in at least one criterion and not worse in any other criteria while the number of errors is not greater.

The first hypothesis H1 relates to the GPML experiment. We claimed that the quality of diagrams created by modeling experts is higher than that of novices. The results of Sect. 5.1 indicate that modeling experts were better in three criteria (number of bends, branches, and total length of all hyperedges) and not worse in any other criteria. We could confirm H1: Modeling experts created better diagrams than novices.

This result is not surprising: Modeling experts pay more attention to aesthetic criteria than novices. Former studies showed that the reduction of bends is an important factor [10, 11]. We assume that branches of hyperedges play a similar role to bends. At first glance, it may seem surprising that modeling experts did not use fewer crossings than novices. This contradicts other studies [10–14]. One possible explanation for this is that the layout algorithm used (see Sect. 4.3) does not strive for a minimization of crossings [17].

The hypotheses H2.1 and H2.2 relate to the GDSL experiment. H2.1 claims that domain experts do not create worse diagrams than competent participants or novices. We saw in Sect. 5.2 that domain experts were better in one criterion (number of bends) compared to competent participants and novices, while they were worse in one criterion (number of branches) compared to competent participants. In task 3, domain experts were worse in one criterion (number of crossings) compared to competent participants and novices. We could confirm H2.1: Domain experts did not create worse diagrams than competent participants or novices. H2.2 claimed that aesthetic criteria are less important for domain experts compared to the GPML experiment. In the GPML experiment, modeling experts were better in three criteria and not worse in any criteria. In the GDSL experiment, domain experts were better in one criterion, but worse in two criteria. We could confirm H2.2: Aesthetic criteria are less important for domain experts. We discuss these results below.

Besides aesthetic graph drawing criteria, domain experts consider also domain-specific rules that override classical criteria. The diagrams are digital twins of real world energy systems and consider the technical implementation in the real world [7]. For example, power grids often take the form of cable routes, from which power grids branch off at distribution points. This often leads to branches in the diagrams. Water pipes for heat supply are circuits in the real world. These power and heating domain-specific rules demand a routing with few bends and can be found in the diagrams of the domain experts. Another domain-specific rule can be found: Energy sources generate energy that is converted and supplied to consumers. Domain experts placed vertices representing energy sources left (e.g. power/gas source) and vertices representing consumers right (e.g. power/gas/cooling consumption). Groups of vertices of one model (e.g., power, heat, cooling) with sources and consumers were placed in a row (see task 3, Figs. 3 and 4). Vertices of models without sources were placed near consumers (see task 4, Fig. 3). Sources and consumers were placed one below the other and centered horizontally. This leads to longer hyperedges and corresponds to the answers from the questionnaire. The domain experts had an arrangement in mind before they created the diagrams.

The syntax of the language defines that the energy models have different colors (e.g., power is red). The vertices related to power are placed at the top of the diagram. The layout algorithm connects distant vertices of any energy model via long horizontal busses (see Sect. 4.3 and [17]). The line-by-line placement and the long horizontal hyperedges lead to crossings with hyperedges that connect vertices from different rows. In the created diagrams, these crossings are always crossings between different models and therefore lines with different colors. Competent participants and novices avoid these crossings, but domain experts do not due to the real world relation. This corresponds to the answers from the questionnaire: Domain experts found crossings less disturbing than competent participants or novices.

We found that competent participants and novices did not consider the above-mentioned domain-specific rules. Besides that, no further differences between these groups were found in the diagrams.

The result of our observation is the following: Domain-specific rules stand in contrast to aesthetic graph drawing criteria. In the particular domain of the GDSL experiment, branches or crossings are not necessarily bad. The most important aesthetic criteria that are in line with domain-specific rules are the alignment of vertices and the minimization of bends.

H3 remains to be verified: Modeling and domain experts are more efficient than people with less experience. This hypothesis relates to both experiments. Section 5.1 shows that modeling experts in the first experiment did not need less time than novices. But they created diagrams with less bends and branches in the same time. We could confirm H3 for the GPML experiment: Modeling experts are more efficient than novices. Section 5.2 shows that domain experts needed less time than competent participants and novices. They also created diagrams with less bends in less time. We could confirm H3 for the GDSL experiment: Domain experts are more efficient than competent participants and novices.

This result was expected. Modeling and domain experts knew both the diagram editors and the layout algorithm. They had fewer problems using the software and were able to anticipate how the layout algorithm would react to their interactions. Thanks to their experience, they already had certain ideas about what a diagram should look like and pursued certain, aesthetic criteria consequently.

All results of this work should be treated with caution due to the small sample size.

Acknowledgments. We thank all participants of the experiments for their motivation and effort. This R&D project of the Society for the Advancement of Applied Computer Science (GFaI) is supported by the funding program Innovation Competence (INNO-KOM) of the German Federal Ministry for Economic Affairs and Climate Action (BMWK), based on a resolution of the Ger-man Parliament (project number: 49VF220032).

Disclosure of Interests. The authors declare no competing interests.

References

1. Brambilla, M., Cabot, J., Wimmer, M.: Model-Driven Software Engineering in Practice. Synthesis Lectures on Software Engineering, vol. 1, pp. 1–182 (2012). https://doi.org/10.2200/S00441ED1V01Y201208SWE001
2. OMG UML 2.5.1: Unified Modeling Language, v2.5.1 (2017). https://www.omg.org/spec/UML/2.5.1/PDF
3. OMG: UML Profile for MARTE: Modeling and Analysis of Real-Time Embedded Systems, v1.3 (2023). https://www.omg.org/spec/MARTE/1.3/PDF
4. Simulink - Simulation und Model-Based Design (2021). https://de.mathworks.com/products/simulink.html
5. Fowler, M., Parsons, R.: Domain-specific languages. Addison-Wesley, Upper Saddle River/Boston/Indianapolis/San Francisco/New York/Toronto/Montreal/London/Munich/Paris/Madrid/Sydney/Tokyo/Singapore/Mexico City (2011)
6. Gupta, R., Jansen, N., Regnat, N., Rumpe, B.: Design Guidelines for Improving User Experience in Industrial Domain-Specific Modelling Languages (2022)
7. Wrobel, G., Scheffler, R., Kehrer, T.: Rethinking the traditional design of meta-models: layout matters for the graphical modeling of technical systems. In: 2021 ACM/IEEE 24th International Conference on Model Driven Engineering Languages and Systems Companion (MODELS-C), pp. 351–360 (2021). https://doi.org/10.1109/MODELS-C53483.2021.00058
8. Tamassia, R., Di Battista, G., Batini, C.: Automatic graph drawing and readability of diagrams. IEEE Trans. Syst. Man Cybern. **18**, 61–79 (1988). https://doi.org/10.1109/21.87055
9. Purchase, H.C.: Metrics for graph drawing aesthetics. J. Vis. Lang. Comput. **13**, 501–516 (2002). https://doi.org/10.1006/jvlc.2002.0232
10. Purchase, H.C., Cohen, R.F., James, M.: Validating graph drawing aesthetics. In: Brandenburg, F.J. (ed.) GD 1995. LNCS, vol. 1027, pp. 435–446. Springer, Heidelberg (1996). https://doi.org/10.1007/BFb0021827
11. Purchase, H.: Which aesthetic has the greatest effect on human understanding?. In: DiBattista, G. (ed.) GD 1997. LNCS, vol. 1353, pp. 248–261. Springer, Heidelberg (1997). https://doi.org/10.1007/3-540-63938-1_67
12. van Ham, F., Rogowitz, B.E.: Perceptual organization in user-generated graph layouts. IEEE Trans. Visual Comput. Graphics **14**, 1333–1339 (2008). https://doi.org/10.1109/TVCG.2008.155
13. Dwyer, T., et al.: A comparison of user-generated and automatic graph layouts. IEEE Trans. Visual Comput. Graphics **15**, 961–968 (2009). https://doi.org/10.1109/TVCG.2009.109
14. Purchase, H.C., Pilcher, C., Plimmer, B.: Graph drawing aesthetics-created by users, not algorithms. IEEE Trans. Visual Comput. Graphics **18**, 81–92 (2012). https://doi.org/10.1109/TVCG.2010.269
15. Purchase, H.C., McGill, M., Colpoys, L., Carrington, D.: Graph drawing aesthetics and the comprehension of UML class diagrams: an empirical study, Sydney (2001)
16. Purchase, H.C., Allder, JA., Carrington, D.: User preference of graph layout aesthetics: a UML study. In: Marks, J. (ed.) GD 2000. LNCS, vol. 1984, pp. 5–18. Springer, Heidelberg (2001). https://doi.org/10.1007/3-540-44541-2_2
17. Goetze, B.: OrthoRoute - Ein gitterloses Verfahren zur Generierung orthogonaler Verbindungen in Schaltplänen. In: Roller, D., Opletal, S. (eds.) Tagungsband Workshop Elektrotechnik CAD 2007, pp. 13–32. Shaker, Aachen (2007)

18. Helmke, S., Goetze, B., Scheffler, R., Wrobel, G.: Interactive, orthogonal hyperedge routing in schematic diagrams assisted by layout automatisms. In: Basu, A., Stapleton, G., Linker, S., Legg, C., Manalo, E., Viana, P. (eds.) Diagrams 2021. LNCS, vol. 12909, pp. 20–27. Springer, Cham (2021). https://doi.org/10.1007/978-3-030-86062-2_2

Open Access This chapter is licensed under the terms of the Creative Commons Attribution 4.0 International License (http://creativecommons.org/licenses/by/4.0/), which permits use, sharing, adaptation, distribution and reproduction in any medium or format, as long as you give appropriate credit to the original author(s) and the source, provide a link to the Creative Commons license and indicate if changes were made.

The images or other third party material in this chapter are included in the chapter's Creative Commons license, unless indicated otherwise in a credit line to the material. If material is not included in the chapter's Creative Commons license and your intended use is not permitted by statutory regulation or exceeds the permitted use, you will need to obtain permission directly from the copyright holder.

Generating Qualitative Descriptions of Diagrams with a Transformer-Based Language Model

Marco Schorlemmer[1]([✉]), Mohamad Ballout[2], and Kai-Uwe Kühnberger[2]

[1] Artificial Intelligence Research Institute (IIIA), CSIC, Barcelona, Spain
marco@iiia.csic.es
[2] Institute of Cognitive Science, Osnabrück University, Osnabrück, Germany
{mohamad.ballout,kkuehnbe}@uni-osnabrueck.de

Abstract. To address the task of diagram understanding we propose to distinguish between the perception of the geometric configuration of a diagram from the assignment of meaning to the geometric entities and their topological relationships. As a consequence, diagram parsing does not need to assume any particular a priori interpretations of diagrams and their constituents. Focussing on Euler diagrams, we tackle the first of these subtasks—that of identifying the geometric entities that constitute a diagram (i.e., circles, rectangles, lines, arrows, etc.) and their topological relations—as an image captioning task, using a Vision Transformer for image recognition combined with language model GPT-2 to generate qualitative spatial descriptions of Euler diagrams with an encoder-decoder model. Due to the lack of sufficient high-quality data to train the pre-trained language model for this task, we describe how we generated a synthetic dataset of Euler diagrams annotated with qualitative spatial representations based on the Region Connection Calculus (RCC8). Results showed over 95% accuracy of the transformer-based language model in the generation of meaning-carrying RCC8 specifications for given Euler diagrams.

Keywords: diagram understanding · Euler diagram · region connection calculus · transformer-based language model

1 Introduction

In computer science, diagram understanding has been primarily addressed by focussing on the following cognitive actions: first, the action of identifying the kind of diagram (flowchart, entity-relationship model, Hasse diagram, etc.) and its overall structure, i.e., its constituting entities (icons, abstract illustrations, links, textual annotations, etc.) and the meaning-carrying relations between these entities (temporal dependencies, semantic relationships, ordering, etc.); next, the action of generating a formal representation of the diagram employing some suitable abstract mathematical structure, for example, a mathematical graph with its vertices and edges; and, finally, the action of performing query answering on the diagram, processing the formal representation.

With the recent advances in deep-learning approaches based on artificial neural networks (ANNs), and given the success rates of these techniques for image classification and visual query answering, there have been several proposals for tackling the actions underlying diagram understanding with ANNs [15,21,28] (see Sect. 5). However, when diagrams are composed of geometric objects, such as circles, rectangles, and other polygons, the task of diagram understanding via ANNs is more challenging than processing natural images because the geometric shapes by themselves do not have specific semantics, only when attached to particular knowledge concepts, such as when we interpret a diamond in a flowchart as a decision step.

In the proposals mentioned above, diagram parsing is carried out assuming particular interpretations of diagrams and their constituents, either as science diagrams, arrow-connected diagrams, geometry diagrams, flowcharts, binary trees, etc. Consequently, the formal representations for diagrams that are generated are determined by these a priori interpretations. This is analogous to the logic-based approach to formal diagrammatic representation and reasoning that has been carried out since the 1990s [1]. Abstract representations have been proposed for different kinds of mathematical diagrams, assuming particular semantics for diagrammatic entities and their relations (e.g., see [9] for abstract descriptions of Euler diagrams). By adopting such an approach, training an ANN for diagram classification and parsing requires richly annotated dedicated datasets, and thus annotators with sufficient knowledge backgrounds on how particular diagrams are to be understood. Consequently, there is a lack of readily available, high-quality data that can be used for training ANN-based architectures, and thus relevant datasets are still scarce.

We propose to tackle the task of diagram understanding by distinguishing (1) the perception of the geometric configuration of a diagram from (2) the assignment of meaning to the geometric entities and their topological relationships. For the perception of the geometric configuration, we aim to identify the geometric entities that constitute a diagram (circles, rectangles, lines, dots, etc.) and generate a qualitative spatial description of their topology. For the assignment of meaning to the geometric entities and their relations, we adopt the stance that meaning and understanding are ultimately grounded on our humanly embodied interaction with our environment and with other living organisms [13,14]. Consequently, we aim to use the previously obtained qualitative spatial description of a diagram's geometry to identify basic cognitive structures that capture the embodied understanding of the diagram.

Bourou & al. [3–5] have been narrowing down the problem of meaning to that of making sense of a diagrammatic representation, drawing from the theories of conceptual blending [8] and image schemas [10], which capture the logic and dynamics of basic human embodied sensorimotor experiences with the environment. For the second of the two subtasks of diagram understanding mentioned above, we plan to follow Bourou & al. and assume that the embodied act of observing a diagram can be modelled as a network of conceptual blends that combine image schemas grounding our embodied understanding with the perceived geometric entities and their topological relationships that constitute a diagram.

Here, though, we shall focus on the first of the two subtasks of diagram understanding that we have distinguished above, and our objective will be to computationally model, using an ANN, the perception of the geometric configuration of a diagram. We do this by generating a qualitative spatial description of the geometric entities and their topological relations that constitute a diagram. To further narrow down our objective, we decided to focus initially only on one particular kind of diagram—given that diagram classification is not our goal—and preferably a kind of diagram for which we do not require many image schemas to model their understanding. For this reason, we started our endeavour with Euler diagrams, because of the simplicity of their geometric entities—they essentially are formed by labelled circles on a two-dimensional plane—and their straightforward semantics, based on our embodied understanding of inclusion, which is grounded on the so-called CONTAINER image schema [16,17]. In addition, we also fixed initially the number of circles in an Euler diagram to three.

2 Materials and Methods

As mentioned above, for the perception of the geometric configuration of a diagram we aim to generate, with an ANN, a qualitative spatial description of the geometric entities and their relations. Although ANNs have previously been used for investigating their accuracy for syllogistical reasoning with Euler diagrams [27], to the best of our knowledge, there is currently no dataset of sufficiently diverse Euler diagrams that are also annotated with *qualitative descriptions* of their spatial representation, so that it can be used to train an ANN for generating qualitative descriptions out of Euler diagrams. Consequently, we opted to create such a dataset. The approach we followed was to first generate qualitative spatial descriptions of as diverse Euler diagrams as possible and then generate several alternative visual renderings of Euler diagrams for each of the generated qualitative descriptions. These descriptions would then constitute the annotations of the Euler diagrams required for an ANN's training.

Automatically generating Euler diagrams from a given description of its entities and topological relationships is a research problem on its own [20, Section 5]. When generating these diagrams, certain desired properties that make up a well-formed Euler diagram need to be taken into account [25]:

- that at most two curves intersect at one point (i.e., no n-points with $n > 2$);
- that no curve segments are concurrent;
- that no two or more curves have the same label;
- that curves are simple (i.e., they do not self-intersect);
- that zones are not disconnected (i.e., no zone consists of more than one minimal region);
- that there are no brushing points (i.e., points in which two or more curves meet but do not cross).

Most approaches to date for generating Euler diagrams take abstract descriptions of diagrams as input, rather than qualitative descriptions. An abstract description of an Euler diagram is the enumeration of its zones, i.e., regions of the two-dimensional plane that are closed by a contour, and that can be identified by the set of labels of curves in which they are included [9].

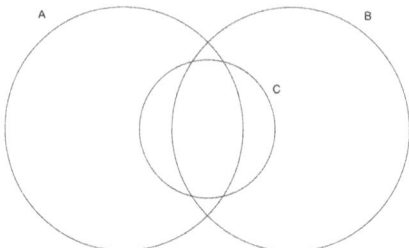

Fig. 1. Euler diagram with disconnected zone: The intersection of A and B that does not intersect C comprises two disconnected minimal regions.

For example, Fig. 1 shows an Euler diagrams that corresponds to the abstract description $\{\emptyset, \{A\}, \{B\}, \{A,B\}, \{A,C\}, \{B,C\}, \{A,B,C\}\}$. Notice that this diagram does not satisfy that all zones are connected, because zone $\{A,B\}$ comprises two disconnected minimal regions. In general, given n curves, there are $2^{(2^n-1)}$ different abstract descriptions of Euler diagrams. Hence, for $n = 3$ the number of different abstract descriptions is 128. But not all these descriptions can be rendered as Euler diagrams that satisfy all well-formedness properties. If the abstract description above is fed, for instance, into the drawing tool that implements the techniques described by Stapleton & al. for automatically drawing Euler diagrams with circles [24], we do not get the diagram of Fig. 1, but instead one that, although does not have disconnected zones, uses repeated labelling, thus not satisfying a different well-formedness property (see Fig. 2).

For creating our dataset we decided thus to focus on well-formed Euler diagrams first. In addition, we needed to associate diagrams with qualitative spatial descriptions of the geometric configuration of the diagram's entities—the circles—describing the topological relations between them, as assumed by Bourou & al. [3–5]. In particular, we needed qualitative descriptions of Euler diagrams that state that curves overlap or not, or are included one within the other.

The only visualisation tool we found that is capable of generating Euler diagrams given a qualitative spatial description of it is the one by Schwarzentruber [23], a web-based application written in JavaScript (see Fig. 3). It is an interactive drawing tool that generates an initial layout for an Euler diagram

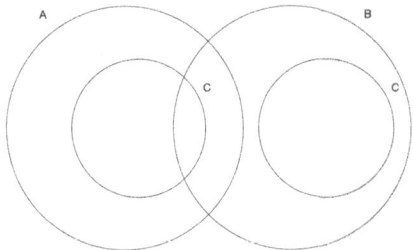

Fig. 2. Euler diagram with repeated label (C).

based on circles out of a given specification written in the Region Connection Calculus (RCC8) [6], which can then be further modified on the fly by mouse control. To generate a diagram, it uses either a local-search-based approach, a process inspired by the gradient method, or a hybrid strategy mixing both. Consequently, we have chosen RCC8 as the formalism to provide qualitative spatial descriptions of Euler diagrams. But, to satisfy all well-formedness properties, we restrict specifications to use only four of the eight available predicates of RCC8:

- $DC(x,y)$, which we use to specify that two circles are drawn without any intersection (x is disconnected from y);
- $PO(x,y)$, which we use to specify that two circles intersect (x partially overlaps y);
- $NTPP(x,y)$, which we use to specify that one circle is drawn entirely within another circle (x is a non-tangential proper part of y);
- $NTPPi(x,y)$, which we use in specifications using its inverse $NTPP(y,x)$.

Hence, we do not use:

- $EQ(x,y)$ as it would yield diagrams with concurrency;
- $EC(x,y)$, $TPP(x,y)$, and $TPPi(x,y)$, as they would yield diagrams with brushing points.

RCC8 is not expressive enough to uniquely determine diagrams with or without n-points ($n > 2$) or disconnected regions, so we cannot provide a univocal RCC8 description of the Euler diagram corresponding to the abstract description $\{\emptyset, \{A\}, \{B\}, \{A,B\}, \{A,C\}, \{B,C\}, \{A,B,C\}\}$ and illustrated in Fig. 1 with a disconnected zone. Consequently, when generating diagrams out of an RCC8 description we will favour the drawings that only have 2-points and connected regions.

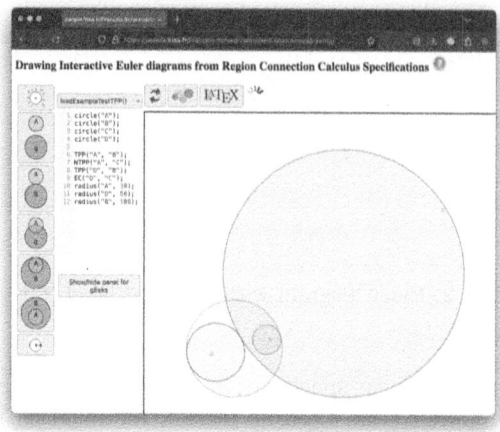

Fig. 3. Schwarzentruber's tool for drawing interactive Euler diagrams from RCC8 specifications [23]

Given the composition table of RCC8 predicates focussing only on DC, PO, NTPP, and NTPPi (Table 1), we can generate 41 different specifications of three-circle Euler diagrams out of the 128 abstract descriptions that exist. Thus, RCC8 is not expressive enough to specify all Euler diagrams, not even those that can be drawn with circles (as the one in Fig. 1). Still, the visualisation tool by Schwarzentruber [23] is the most straightforward option we have found to generate a sufficiently diverse dataset of Euler diagrams annotated with qualitative spatial descriptions to train an ANN.

Table 1. Composition table for the subset of RCC8 predicates DC, PO, NTPP, NTPPi.

∘	DC	PO	NTPP	NTPPi
DC	DC,PO,NTPP,NTPPi	DC,PO,NTPP	DC,PO,NTPP	DC
PO	DC,PO,NTPPi	DC,PO,NTPP,NTPPi	PO,NTPP	DC,PO,NTPPi
NTPP	DC	DC,PO,NTPP	NTPP	DC,PO,NTPP,NTPPi
NTPPi	DC,PO,NTPPi.	PO,NTPPi.	PO,NTPP,NTPPi	NTPPi

2.1 Preparation of the Dataset

We used the visualisation tool developed by Schwarzentruber for drawing Euler diagrams given qualitative spatial specifications in RCC8 [23]. We added to the JavaScript code a loop to generate all 41 different specifications using RCC8 predicates DC, PO, and NTPP with three circles, and for each configuration we generated 1 000 diagrams, getting a total amount of 41 000 three-circle Euler diagrams. To get a certain variety of circle sizes during the generation of the 1 000 diagrams for each RCC8 specification, we randomly specified each of the circles to have either small (with radius randomly set between 20 and 40), medium (with radius randomly set between 50 and 80), or large sizes (with radius randomly set between 90 and 120). Before uploading a specification into Schwarzentruber's drawing tool, we checked that circle sizes were consistent with the RCC8 specification so that the Euler diagram could be drawn. For example, if $NTPP(A, B)$ then A necessarily has to be of smaller size than B.

The tool then generated the Euler diagrams with circles that respected configuration and sizes with labels attached to them, using a gradient method. We allowed at least one second for the algorithm to search for a suitable layout for each specification. For three circles this time interval seemed to be sufficient for the search algorithm to stabilise. Still, for some Euler diagrams the visualisation tool had not enough time to draw circles according to the description or were not entirely within the canvas. We decided not to clean up the dataset and keep these incorrectly or partially drawn diagrams. For each diagram, we used also Schwarzentruber's tool to generate a tikz-based LaTeX code corresponding to the drawn diagram. At the end of the generation loop, we saved the LaTeX file with all 41 000 diagrams and also the text file with the corresponding RCC8 specifications used for the generation.

The LaTeX file and the text file required some post-processing to generate the final dataset. We added colour commands to the LaTeX code to draw the three circles and their corresponding labels in different colours (red, green, and blue), compiled the LaTeX code to generate the PDF file containing all 41 000 diagrams, and subsequently exported each diagram in the PDF file to a separate PNG file. We also replaced the size predicates of quantitative radius dimensions in the specifications with qualitative *small*, *medium*, or *large* predicates, to have entirely qualitative descriptions of the diagrams, and we prepared a CSV spreadsheet with the association of each diagram PNG file with its qualitative description. We run this process three times to get three different sets of 41 000 three-circle Euler diagrams with their corresponding qualitative descriptions. Figure 4 shows one of these annotated Euler diagrams of our dataset.

2.2 Set-up of the Transformer-Based ANN and the Experiments

Based on the research by Ballout & al. using pre-trained language models on non-language tasks [2], we decided to approach the Euler diagram parsing problem as an image captioning problem, to get a qualitative spatial specification in RCC8 from a given Euler diagram. Image captioning is the process of generating textual descriptions of an image, and this task has been successfully tackled by way of deep-learning architectures [12,26]. In our particular case, we decided to use Vision Transformer (ViT) for image recognition [7] combined with language model GPT-2 [18] to generate the text, by way of an encoder-decoder model. ViT extracts features from the images, while GPT-2 generates the captions of the

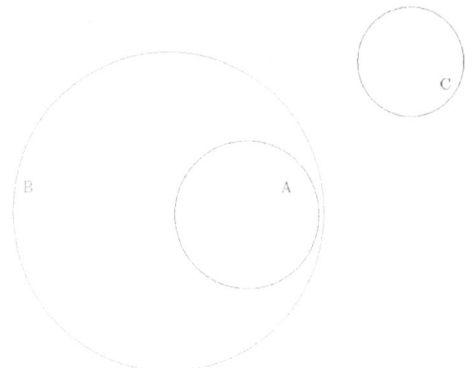

circle("A"); circle("B"); circle("C"); NTPP("A", "B"); DC("B", "C"); DC("A", "C");
small("A"); medium("B"); small("C");

Fig. 4. Generated Euler diagram, annotated with its corresponding qualitative spatial description using RCC8.

images. For our implementation, we resorted to the pre-trained image-captioning model based on ViT-GPT2 from HuggingFace.[1]

In the course of this project, we trained and tested our transformer-based ANN four times with four different datasets. For this, we used the High-Performance Cluster at the Computer Centre (*Rechenzentrum*) of Osnabrück University.[2] For all four datasets, we took 90% of data points for training and 10% for testing.[3] Train and test batch sizes were set to 64, and we trained for 20 epochs. To compute the model's accuracy, only predictions that were identical to the specifications of the test diagrams were deemed to be correct, because our objective was to generate well-formed sentences in RCC8.

The first dataset (ED0) consisted of 41 000 Euler diagrams, annotated with their respective qualitative description in RCC8. Unlike the subsequent datasets, we did not generate coloured drawings of Euler diagrams and stuck to the black-and-white output generated initially when compiling the tikz-based LaTeX code.

---

[1] Available at https://huggingface.co/nlpconnect/vit-gpt2-image-captioning. It consists of 12 layers for the encoder and 12 layers for the decoder model, with approx. 240 million parameters. We chose this particular model because it is well-established and highly used. It is advisable to start with a smaller model and move to larger ones if the results are not satisfactory; there is no necessity to use larger models if a small one is good enough.

[2] This cluster consists of 53 nodes with each node having 128 CPU cores and 1TB Memory. Two of the nodes additionally have 4 NVidia A100 GPUs (6912 CUDA cores, 80GB RAM). The cluster also provides BeeGFS Storage of 1PB with 12GB/s sequential bandwidth (https://www.rz.uni-osnabrueck.de/Dienste/HPC/index.htm).

[3] For large amounts of data, as in our case, we can settle for a small sample of test data, because test variance is low; also, a single validation is sufficient.

This constituted our preliminary attempt to assess the accuracy of the qualitative spatial representation generated by our model. The results and analysis of this preliminary training round set the basis for the subsequent three training rounds:

- ED1: 41 000 coloured three-circle Euler diagrams (red, green, blue), annotated with their respective qualitative descriptions in RCC8.
- ED2: 82 000 coloured three-circle Euler diagrams (consisting of ED1 + 41 000 additional coloured three-circle Euler diagrams), with labels in larger font, and annotated with their respective qualitative descriptions in RCC8.
- ED3: 123 000 coloured three-circle Euler diagrams (consisting of ED2 + 41 000 additional coloured three-circle Euler diagrams), annotated with their respective qualitative descriptions in RCC8.

Each set of 41 000 Euler diagrams consisted of 1 000 variants (in terms of the layout generated by Schwarzentruber's visualisation tool) for each of the 41 different three-circle Euler diagrams that could be specified with RCC8 predicates DC, PO, NTPP, and NTPPi. Although up to 3000 images for the same RCC8 specification were generated (for dataset ED3), we programmed Schwarzentruber's tool to draw circles picking the radius size randomly within nonoverlapping ranges for small, medium, and large circles. In addition, the positions of circles and labels on the canvas were also randomly chosen by the drawing tool and then further adjusted using a gradient method to satisfy the spatial constraints. All in all, this randomness and variability generate a very large variety of different layouts, even for the same RCC8 specification, which makes the test set sufficiently different from the training set.

3 Results

With the preliminary dataset ED0 we obtained an accuracy of around 60%. When comparing the predictions with the test dataset, we realised that most mistakes were in the qualitative descriptions of circle sizes. Because of this preliminary result, we decided to generate coloured Euler diagrams (with red, green, and blue circles and labels), and to distinguish the accuracy with predicting the correct sizes, and without. Circle sizes do not carry any meaning in the semantics of Euler circles, unlike the topological relationship of disjointness, overlap or inclusion. Consequently, we could eventually get by without achieving correct descriptions concerning circle sizes. For datasets ED2 and ED3, we also increased the font size of the labels a bit.

As expected, accuracy improved significantly when ignoring circle sizes, and also with larger datasets, reaching an accuracy over 95% for dataset ED3. Table 2 summarises our results for the training rounds with datasets ED1, ED2, and ED3.

Table 2. Accuracy in predicting the qualitative description specified in RCC8 for the different training rounds using datasets with an increased number of data points.

dataset	diagrams	overall accuracy	accuracy neglecting circle sizes
ED1	41 000	63.1%	87.7 %
ED2	82 000	80.5%	93.4 %
ED3	123 000	87.0%	95.6 %

4 Discussion

The results above show that a sufficiently trained transformer-based ANN architecture is suitable for modelling the perception of the meaning-carrying geometric entities and spatial relations constituting an Euler diagram. This constitutes the first of the two subtasks of diagram understanding that we distinguished in Sect. 1. The qualitative spatial description of an Euler diagram generated by the ANN is thus a symbolic representation of its geometric configuration, which can be subsequently used to tackle the second subtask, that of assigning to this geometric configuration its embodied understanding. Bourou & al. proposed to model this subtask as conceptual blends of the qualitative spatial description of the diagram with relevant image-schematic structure [3]. Both approaches together would constitute a neurosymbolic system for diagram understanding.

In the case of Euler diagrams, it has been claimed that circles are understood in terms of a CONTAINER image schema [16,17]. Image schemas represent patterns derived from recurrent bodily experiences acquired early in life. Repetitive experiences of a similar nature give rise to the formation of these subconscious cognitive structures, which reflect the invariants shared among these experiences. The CONTAINER schema captures our bodily experiences with objects contained in other objects. It is a gestalt structure formed by a boundary, an interior, and an exterior.

Each circle in the Euler diagram could thus be put into correspondence with the boundary of the CONTAINER image schema, thus giving rise to an interior and an exterior for each of the circles of the diagram. The next step would be to develop the computational approaches that establish this correspondence with some formal representation of image schemas to be able to compute the required conceptual blends. Several logic-based specifications of image schemas have been proposed that could be explored for this task [5,11], together with the representation-independent approach to conceptual blending developed by Schorlemmer & Plaza [22].

In this work, we have decided to generate RCC8 specifications of Euler circles, given the availability of the visualisation tool by [23] to create our dataset. However, RCC8 predicates are already interpreted in terms of containment, so blending it with a CONTAINER image schema would seem to be redundant in this case. The NTPP predicate of RCC8, for instance, focuses on the interiors of the circles, and the composition table shown in Table 1 captures the logic of containment. Still, our approach is not limited to RCC8 specifications, and its applicability could be explored for other qualitative spatial representation formalisms such as those used by Bourou & al. [3], provided we can generate diagrams from these qualitative representations to synthesise the training dataset.

5 Related Work

Even though ANN-based diagram understanding has drawn much less attention than natural-image classification or image-caption generation, we have reviewed some of these approaches to see if they could provide some insights that might be relevant to our research objectives. Below we briefly mention three of these works, which are representative of the state-of-the-art about how diagram understanding and reasoning is tackled by way of ANNs.

Kembhavi & al. focussed on the syntactic parsing of grade-school science diagrams to detect constituents and their syntactic relationships, and the semantic interpretation of the constituents of a diagram to answer grade-school science questions [15]. They created a dataset of diagrams (AI2D) with annotations of their constituents and relationships by scraping a Google Image Search based on seed terms derived from the chapter titles in grade 1–6 science textbooks. As a representation to model the structure of diagrams, they defined Diagram Parse Graphs (DPG) and trained a Long Short Term Memory (LSTM) network to generate DPG representations for diagrams. To tackle the subtask of question answering, they set up another neural network architecture that learned to attend to useful relations in a DPG.

Schäfer & al. tackled the identification of node symbols of varying shapes connected by arrows with optional text annotations in handwritten diagrams, to generate fully-structured models [21]. Thus, they focussed on the subtask of detecting the diagram constituents and their localisation, as well as processing the overall diagram structure, building upon the work of [19] with deep convolutional neural networks (CNNs).

Wang & al. addressed the problem of processing computer-science diagrams (binary trees, graphs, flow charts, etc.), by using an ANN to classify the diagram—twelve kinds of computer-science diagrams were considered—and subsequently parsing them to capture the topological structure and text information of the diagrams and further complete also diagram answering-question tasks [28]. For this, they compiled a dataset of computer science diagrams and generated rich annotations (CSDia).

As already mentioned in Sect. 1, in these works—unlike the approach described in this paper—diagram parsing is carried out assuming particular a priori interpretations of diagrams and their constituents. Consequently, training an ANN for diagram classification and parsing requires richly annotated dedicated datasets, and thus annotators with sufficient knowledge backgrounds on how particular diagrams are to be understood. To the best of our knowledge, our work is the first that explores how to tackle the understanding of diagrams by generating qualitative descriptions with a transformer-based language model.

6 Conclusion and Future Work

We have been able to show that, with a transformer-based ANN architecture using the ViT-GPT2 encoder-decoder model we were capable of generating qualitative spatial representations of three-circle Euler diagrams describing their meaning-carrying geometric entities and topological structure, with an accuracy of over 95%. The lack of datasets to train the language model for our purposes required the additional task of exploring automatic Euler diagram generation tools, choosing the right specification formalisms, and generating our own datasets for training. The initial difficulties in obtaining high-quality results after training our pre-trained language model also led to carrying out an analysis of the reasons for this and generating several post-processed datasets of increasing size. We suspect that the problem with learning circle sizes reported in Sect. 3 is due to the way we generated Euler circles with Schwarzentruber's tool. Although, in most cases, one second was enough for the algorithm to find a suitable layout satisfying each combination of RCC8 predicates, the tool did not settle for drawing circles satisfying the qualitative specification of sizes. We think a longer stabilisation interval has to be used, which we will explore in our future work.
As the output of the work presented in this paper, we deliver:

1. A modification of Schwarzentruber's visualisation tool to generate datasets of Euler diagrams from RCC8 specifications.
2. A collection of datasets of Euler diagrams annotated with RCC8 specifications as qualitative descriptions.
3. A a script for training a transformer-based ANN architecture using the ViT-GPT2 encoder-decoder model on our datasets.

Code and data produced in the context of the work presented in this paper is available at:

https://saco.csic.es/index.php/s/raQ8rTn8pezn9XW

In future work, we would like to further investigate the applicability of our approach to Euler diagrams with varying numbers of circles and evaluate the trained ANN architecture on Euler diagrams that were not generated with our modification of Schwarzentruber's visualisation tool. We would also like to explore different kinds of diagrams to evaluate the accuracy of the generation of qualitative spatial descriptions when dealing with multiple sorts of diagrammatic notations. We believe that this line of research might be fruitful for producing diagram understanding systems that can be deployed to tackle different diagrammatic representations.

Acknowledgements. This work has been carried out during a research stay by Marco Schorlemmer at the Institute of Cognitive Science of Osnabrück University, funded with a scholarship granted by the *Deutscher Akademischer Austauschdiest* (DAAD) for Research Stays for University Academics and Scientists (no. 91881609). The cluster used to train the models was funded by the *Deutsche Forschungsgemeinschaft*

(DFG, no. 456666331). The preliminary work that led to this research stay was supported by project CORPORIS (no. PID2019-109677RB-I00) funded by Spain's *Agencia Estatal de Investigación* (AEI) and by a grant for *Grups de recerca consolidats* (no. 2021 SGR 00754) funded by Catalonia's *Agència de Gestió d'Ajuts Universitaris i de Recerca* (AGAUR).

Disclosure of Interests. The authors have no competing interests to declare that are relevant to the content of this article.

References

1. Allwein, G., Barwise, J. (eds.): Logical Reasoning with Diagrams. Oxford University Press, Oxford (1996)
2. Ballout, M., Krumnack, U., Heidemann, G., Kühnberger, K.: Investigating pre-trained language models on cross-domain datasets, a step closer to general AI. In: Jayne, C., et al. (eds.) International Neural Network Society Workshop on Deep Learning Innovations and Applications, INNS DLIA@IJCNN 2023, Gold Coast, Australia, 23 June 2023. Procedia Computer Science, vol. 222, pp. 94–103. Elsevier (2023)
3. Bourou, D., Schorlemmer, M., Plaza, E.: Image schemas and conceptual blending in diagrammatic reasoning: the case of hasse diagrams. In: Basu, A., Stapleton, G., Linker, S., Legg, C., Manalo, E., Viana, P. (eds.) Diagrams 2021. LNCS (LNAI), vol. 12909, pp. 297–314. Springer, Cham (2021). https://doi.org/10.1007/978-3-030-86062-2_31
4. Bourou, D., Schorlemmer, M., Plaza, E.: Modelling the sense-making of diagrams using image schemas. In: Proceedings of the Annual Meeting of the Cognitive Science Society (CogSci 2021), pp. 1105–1111 (2021)
5. Bourou, D., Schorlemmer, M., Plaza, E.: Euler vs hasse diagrams for reasoning about sets: a cognitive approach. In: Giardino, V., Linker, S., Burns, R., Bellucci, F., Boucheix, JM., Viana, P. (eds.) Diagrammatic Representation and Inference. Diagrams 2022. LNCS, vol. 13462, pp. 151–167. Springer, Cham (2022). https://doi.org/10.1007/978-3-031-15146-0_13
6. Cohn, A.G., Bennett, B., Gooday, J., Gotts, N.M.: Qualitative spatial representation and reasoning with the region connection calculus. GeoInformatica **1**(3), 275–316 (1997)
7. Dosovitskiy, A., et al.: An image is worth 16 × 16 words: transformers for image recognition at scale. In: 9th International Conference on Learning Representations, ICLR 2021, Virtual Event, Austria, 3–7 May 2021. OpenReview.net (2021)
8. Fauconnier, G., Turner, M.: Conceptual integration networks. Cogn. Sci. **22**(2), 133–187 (1998)
9. Fish, A., Flower, J.: Abstractions of Euler diagrams. Electron. Notes Theor. Comput. Sci. **134**, 77–101 (2005)
10. Hampe, B. (ed.): From Perception to Meaning. Image Schemas in Cognitive Linguistics. De Gruyter Mouton, Berlin, New York (2005)
11. Hedblom, M.M., Kutz, O., Mossakowski, T., Neuhaus, F.: Between contact and support: introducing a logic for image schemas and directed movement. In: AI*IA 2017 Advances in Artificial Intelligence - XVIth International Conference of the Italian Association for Artificial Intelligence, Bari, Italy, 14–17 November 2017, Proceedings, pp. 256–268 (2017)

12. Hossain, M.Z., Sohel, F., Shiratuddin, M.F., Laga, H.: A comprehensive survey of deep learning for image captioning. ACM Comput. Surv. **51**(6), 118:1–118:36 (2019)
13. Johnson, M.: Embodied understanding. Front. Psychol. 875 (2015)
14. Johnson, M.: The Meaning of the Body. The University of Chicago Press, Chicago (2007)
15. Kembhavi, A., Salvato, M., Kolve, E., Seo, M., Hajishirzi, H., Farhadi, A.: A diagram is worth a dozen images. In: Leibe, B., Matas, J., Sebe, N., Welling, M. (eds.) ECCV 2016. LNCS, vol. 9908, pp. 235–251. Springer, Cham (2016). https://doi.org/10.1007/978-3-319-46493-0_15
16. Lakoff, G.: Women, Fire, and Dangerous Things. University of Chicago Press, Chicago (1987)
17. Lakoff, G., Nuñez, R.E.: Where Mathematics Comes From. Basic Books, New York (2000)
18. Radford, A., Wu, J., Child, R., Luan, D., Amodei, D., Sutskever, I.: Language models are unsupervised multitask learners. OpenAI blog **1**(8), 9 (2019)
19. Ren, S., He, K., Girshick, R.B., Sun, J.: Faster R-CNN: towards real-time object detection with region proposal networks. IEEE Trans. Pattern Anal. Mach. Intell. **39**(6), 1137–1149 (2017)
20. Rodgers, P.: A survey of Euler diagrams. J. Vis. Lang. Comput. **25**(3), 134–155 (2014)
21. Schäfer, B., Keuper, M., Stuckenschmidt, H.: Arrow R-CNN for handwritten diagram recognition. Int. J. Doc. Anal. Recognit. **24**(1), 3–17 (2021)
22. Schorlemmer, M., Plaza, E.: A uniform model of computational conceptual blending. Cogn. Syst. Res. **65**, 118–137 (2021)
23. Schwarzentruber, F.: Drawing interactive Euler diagrams from region connection calculus specifications. J. Log. Lang. Inform. **24**(4), 375–408 (2015)
24. Stapleton, G., Flower, J., Rodgers, P.J., Howse, J.: Automatically drawing Euler diagrams with circles. J. Vis. Lang. Comput. **23**(3), 163–193 (2012)
25. Stapleton, G., Rodgers, P., Howse, J., Taylor, J.: Properties of Euler diagrams. Electron. Commun. Eur. Assoc. Softw. **7** (2007)
26. Stefanini, M., Cornia, M., Baraldi, L., Cascianelli, S., Fiameni, G., Cucchiara, R.: From show to tell: a survey on deep learning-based image captioning. IEEE Trans. Pattern Anal. Mach. Intell. **45**(1), 539–559 (2023)
27. Wang, D., Jamnik, M., Liò, P.: Investigating diagrammatic reasoning with deep neural networks. In: Chapman, P., Stapleton, G., Moktefi, A., Perez-Kriz, S., Bellucci, F. (eds.) Diagrams 2018. LNCS (LNAI), vol. 10871, pp. 390–398. Springer, Cham (2018). https://doi.org/10.1007/978-3-319-91376-6_36
28. Wang, S., et al.: Computer science diagram understanding with topology parsing. ACM Trans. Knowl. Discov. Data **16**(6), 114:1–114:20 (2022)

Open Access This chapter is licensed under the terms of the Creative Commons Attribution 4.0 International License (http://creativecommons.org/licenses/by/4.0/), which permits use, sharing, adaptation, distribution and reproduction in any medium or format, as long as you give appropriate credit to the original author(s) and the source, provide a link to the Creative Commons license and indicate if changes were made.

The images or other third party material in this chapter are included in the chapter's Creative Commons license, unless indicated otherwise in a credit line to the material. If material is not included in the chapter's Creative Commons license and your intended use is not permitted by statutory regulation or exceeds the permitted use, you will need to obtain permission directly from the copyright holder.

Diagram Control and Model Order for Sugiyama Layouts

Sören Domrös[(✉)] and Reinhard von Hanxleden

Department of Computer Science, Kiel University, Kiel, Germany
{sdo,rvh}@informatik.uni-kiel.de

Abstract. Graphical WYSIWYG editors for programming languages are popular since they allow to *control* the diagram layout to express *intention* via *secondary notation* such as proximity and topology. However, such editors typically require users to do manual layout. Conversely, automatic layout of diagrams typically fails to capture intention because graphs are usually considered to not contain any order. *Model order* can combine the desire for control of secondary notation with automatic layout, without additional overhead, since the textual model already employs secondary notation. We illustrate how model order can exert control on the example of the programming languages SCCharts and Lingua Franca and collect developer feedback to validate our findings.

Keywords: Automatic Layout · Model Order · User Intentions

1 Introduction

Automatic layout rises in popularity, as seen in the example of elkjs[1] with, as of this writing, more than 700.000 weekly downloads. Even though automatic layout improved over time and gets more widely used, WYSIWYG editors that rarely employ automatic layout are still very common. WYSIWYG editors place the burden of layout on the user, which can be a severe impediment to productivity [10]. However, WYSIWYG editors have the advantage that they allow controlling secondary notation [10] by creating order, grouping, or alignment in a very direct way and on a graphical level, which is desirable. As Taylor reported for WYSIWYG type-setting [16]: "People like having feedback and control."

One approach to augment automatic layout of diagrams with control is to let the user formulate explicit layout constraints, e.g., through (textual) model annotations or via some WYSIWYG-like graphical interaction [5,11]. Constraints have the advantage to be integrated into layout algorithms. Hence, layout does not need to be done on a pixel granularity, which when done manually often has inconsistencies [14]. Instead, they focus on topology, alignment, or proximity of nodes. Constraints, however, require additional effort beyond creating a textual

---

[1] https://npmtrends.com/elkjs.

model, and sometimes require knowledge about the underlying layout algorithm to be used effectively, as reported by users and developers of ELK [4].

In this paper, we investigate the research question how to exert control by using *model order* [2] for *Sugiyama or layered layouts* [15] whenever the textual model in a tool that employs modeling pragmatics [7] by using a textual model and a graphical model side-by-side expresses secondary notation (R1). We also explore how model order integrates with the mental map, layout stability, aesthetic criteria, and secondary notation in text and diagram (R2).

The Sugiyama algorithm consisting of the phases cycle breaking, layer assignment, crossing minimization, node placement, and edge routing may produce drawings as the one depicted in Fig. 1. Here, nodes are assigned to (horizontal) layers such that nodes in the same layer are not connected by an edge. The cycle breaking step determines which edges should go against the layout direction and hence the order of Send and Receive. Blindly optimizing aesthetic criteria disregards the intention of the model and leaves no leverage to control the layout. The two versions are semantically identical, however, the vertical ordering in Fig. 1b suggests that Send happens before Receive while Fig. 1d suggests the inverse. Figure 1b was created using the textual model in Fig. 1a, while Fig. 1d was created using Fig. 1c. Hence, the textual order should be considered, since the mental map, the inner representation of the model, and secondary notation of the textual model should not diverge from their representation in the diagram.

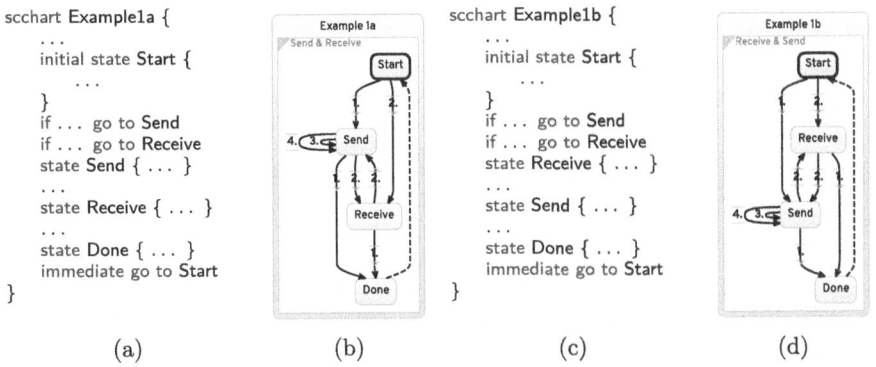

Fig. 1. Two semantically identical variants of an obfuscated SCCharts model created by and used with permission of Scheidt & Bachmann System Technik GmbH.

To answer the research question R1 and R2 stated above, we contribute

- an investigation how controlling the diagram via model order influences the mental map, layout stability, and secondary notation in Sect. 3 and
- an analysis of SCCharts and Lingua Franca (LK) in the context of model order and control in Section 3.1 and 3.2.

Section 4 summarizes and generalizes ours insights and suggests future research.

A long version of this paper [3] contains additional examples and insights for model order and its interaction with the mental map, secondary notation, stability, aesthetic criteria, and control, more detailed evaluation results, and a guidebook how to get model order information for a given language.

2 Related Work

Purchase [13] investigated what aesthetic criteria humans adhere to when drawing graphs given by a textual description. As stated by Purchase, other studies [8,12] focus on reduction of edge crossings, symmetry, placement of important nodes at the top, large angles between incident edges, and average edge length. Purchase investigated whether there are additional criteria people favor when drawing graphs, by analyzing the final drawings and the intermediate steps of two graphs created by the participants. The study revealed that people prefer to place nodes on a grid and that they initially use aesthetic criteria such as only vertical or horizontal edges or nodes ordered lexicographically or by their occurrence in the graph representation. Moreover, the study revealed that the participants worked through the graph node by node or edge by edge depending on the graph representation. They intuitively used the given model order of the graph representation and revised this order partially if the result created undesired crossings or clutter. To evaluate model order, we hence need real models and preferably the developers that built them to investigate what their intention in specific placements was. Moreover, evaluation of model order configurations should be done in an interactive tool to considering the different creation steps.

3 Control and Model Order

How model roder relates to common aesthetic criteria was already covered by previous work [1,2]. Here, we investigate how model order relates to mental map, stability, secondary notation, and control on a meta level (R2) by focusing on tools for model-driven-engineering that use text and diagram side-by-side utilizing the concept of modeling pragmatics [7].

The *mental map* describes the mental image one has of a graphical model [9]. Misue et al. define preserving the mental map as preserving orthogonal ordering and proximity. Preserving the mental map and with it the *stability* of the drawing is especially important when working with models for real use-cases since they are typically large, hierarchical, and do not fit on a single screen[2]. Developers already create a mental map while creating the textual model and not only by looking at the accompanying diagram. Hence, we might compromise the mental map if the textual model does not match the graphical model, as it would be the case if Fig. 1a would create the diagram in Fig. 1d. Since the textual ordering

---

[2] This can be explored interactively in the LF playground https://github.com/lf-lang/playground-lingua-franca by opening the cdn_cache_demo model created by Magnition taken from https://github.com/MagnitionIO/LF_Collaboration.

only has one dimension but the drawing has two to express order, we have to further determine which dimension in the diagram corresponds which textual ordering for a given language.

Similarly, *secondary notation* exists in textual models. E. g., developers begin to write the textual model with an initial state at the top, final states are typically at the bottom, and nodes that should be next to each other in the drawing are typically also placed next to each other in the textual model, as seen in Fig. 1. Since intentional secondary notation exists in the textual source, we should control the layout using the textual model order to bring secondary notation from the text into the diagram.

Control is a very desirable aspect of layout. This might be the reason WYSI-WYG editors are still implemented even though moving boxes around is a tedious process [10]. Moving boxes to desired positions directly creates secondary notation. This level of control can also be achieved using interactive constraint frameworks [5,11] with the advantage of automatic layout. However, using model order inside a layout algorithm can similarly exert control by treating the textual ordering as a constraint. Moreover, model order can also be used as a tie-breaker together with common aesthetic criteria creating different levels of control [2].

To understand how a language can be controlled using model order, we have to identify what part of the textual model is intended secondary notation and how this should be transferred to the two-dimensional diagram during the cycle breaking and crossing minimization steps of the Sugiyama algorithm.

3.1 Controlling SCCharts Layout via Model Order

SCCharts [6] is a sequentially constructive statechart dialect that models control-flow. Figure 1 depicts an SCChart drawing with the corresponding textual model. An SCChart consists of a declaration of inputs, outputs, constants, variables and actions, which are here filtered out of the diagram. Moreover, it has concurrent regions (the white box called Send & Receive in Fig. 1b) with states and transitions between them. The textual syntax only allows to define outgoing transitions of a state directly under the state declaration, which constraints the global edge order. E. g., the state Start has two outgoing transitions defined below it. States might have internal behavior including everything an SCChart may consist of.

The developer may reorder states without changing the semantics of the model. However, the initial state, i.e., the state Start, is usually defined at the top of a textual model. Moreover, the textual SCCharts model employs secondary notation such that the node model order indicates the control-flow direction. E. g., the textual model for Fig. 2 defines the states Ok and Active that belong to the same control-flow branch above the state Failed, Disconnected defined first, and the states Inactive and Disconnect at the end.

Only transitions with mutually exclusive transition guards can be freely reordered without changing the behavior since reordering changes their priority, which SCCharts users want to visualize as secondary notation (R2). Hence, if the textual ordering of model elements has semantics, they cannot be freely reordered to exert control (R1).

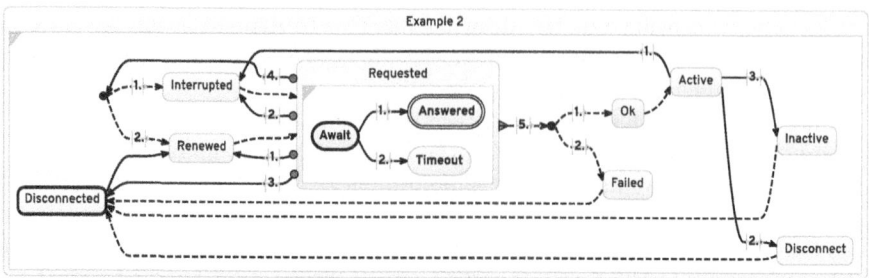

Fig. 2. An obfuscated SCCharts model created by and used with permission of Scheidt & Bachmann System Technik GmbH. It uses model order to constrain the flow and as tie-breaker for the (vertical) order inside a layer.

Previous work on model order for SCCharts [1,2] resulted in the following model order configurations. Strategy 1 uses the strict model order cycle breaker that enforces the node model order along edges since the node model order employs intended secondary notation. Strategy 1 uses the edge model order to pre-order node and edges before the crossing minimization step, which may revise the pre-order to reduce edge crossings. Additionally, Strategy 1 uses the node and edge model order as secondary criterion for crossing minimization such that the solution with minimal crossings that respects most of the model order will be chosen. Strategy 1 was already field-tested while teaching two students courses about Embedded System and Synchronous Languages and is currently also employed by Scheidt & Bachmann System Technik GmbH (S&B). Strategy 2 has a different use-case and fully controls the layout by model order to create layouts for documentation [2].

Additionally to our own experience with SCCharts, we interviewed five S&B developers about 36 SCCharts models created for non-safety critical projects in the railway domain. 30 models fully conformed to the model order not counting models where one edge needed reordering to achieve the desired result.

On no occasion, developers or students reported anything out of the ordinary or problems as a result of the cycle breaking step constrained by model order, e.g., backward edges that should not be there. However, S&B developers did not notice that the node model order constrained the flow of the diagram before pointing it out via Fig. 2. Here, the connector state (black dot) above the initial state Disconnected is a dangling node—a node with only edges to the right—even though it is not a source node. Moving this node somewhere else may, however, hide the symmetry of the Interrupted and Renewed states.

The strict model order cycle breaking has been active as the default layout for about a year, which allowed us to gather feedback on it. During this time this strategy did not produce Obviously Non-Optimal (ONO) or left developers confused. Hence, model order should per default control the flow (R1).

Strategy 2 cannot be the default option for SCCharts, since it may produce ONO layouts. However, if a controlling, more strict, strategy controls most parts of the layout, a bad layout often points to a bad textual model. How this may affect developers needs further evaluation.

3.2 Controlling Lingua Franca Layout via Model Order

Lingua Franca [7] is a polyglot coordination language for reactive real-time systems that models data-flow. An example LF model can be seen in Fig. 3.

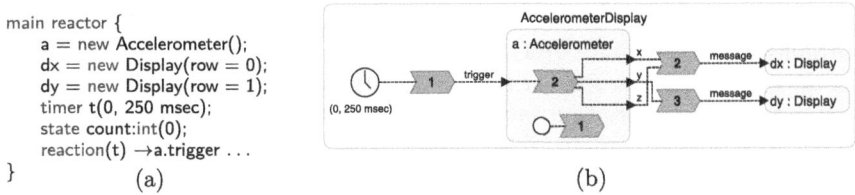

Fig. 3. The AccelerometerDisplay LF example model with abbreviated textual model and graphical representation.

Figure 3a depicts the main reactor of the LF model and Fig. 3b depicts the corresponding diagram with a timer (clock), reactions (numbered arrows shapes), and reactors (gray boxes). Users may define states, actions, timers, reactors, edges between reactors, and reactions and often define the components in the mentioned order forming *ordering groups*. E. g., users define reactors in a separate group above the ordering group of reactions. A reactor may again consist of everything the main reactor has to offer but may define inputs and outputs. Edges between reactors and other elements are created implicitly based on their interfaces, i.e., their declared inputs and outputs, which corresponds to the port model order. The reactor instantiation order can be freely changed as it is the case for all elements despite reactions and typically expresses desired secondary notation that could be controlled via model order (R1). The order of reactions defines their scheduling order. Hence, we cannot freely reorder them to control the layout (R1). However, this scheduling order is secondary notation that developers want to represent in their drawing.

Interviews and presentations as part of the weekly LF team meeting, interviews with Magnition that want to use LF to model cache structures, and one-on-one interviews resulted in the following statements.

1. Edge-crossings should be reduced.
2. Stability is important for large hierarchical models.
3. The ports of reactors define their interface and should hence be respected.
4. The reactor-instantiation-order could be used to control their ordering.
5. Actions should be placed such that the flow indicates their activation.

Note that statement 1, 2, and 3 came from Magnition developers and 1, 3, 4 from the person in the LF community that mainly interacted with us during the presentations and discussions. All the statements above were additionally verified by one-on-one interviews with a small group of developers regarding the layout of LF models.

We see that LF developers value common aesthetic criteria (statement 1) but prefer stability create by model order for large models (statement 2). Reactor order and port order are the primary elements that can be used to control the layout (statement 3 and 4). However, different model elements are not directly comparable, even though there might be relative orderings between the different ordering group (statement 5).

We conclude that we should not use model order strategies that completely constrain the layout for LF (R1). Layout algorithms for LF should use model order only as a tie-breaker as the default strategy and may only constrain nodes of the same ordering group to avoid misleading secondary notation. For cycle breaking the depth-first strategy that uses model order to determine the visiting order proved to be good to handle the often intertwined action and reaction networks quite well. Since most models are small, finding and analyzing all cycle might be a promising strategy. Additionally, big models might want to constrain the port or node order by model order to increase stability and preserve the mental map (R2).

4 Conclusion

How SCCharts and LF layout and secondary notation can be controlled by model order can be generalized for similar languages, as SCCharts are just state-machines and LF is only a special kind of actor oriented data-flow language. Hence, constraining the flow of all state-machines by the state model order creates intentional backward edges for all state-machine dialects. However, the model order should only be used to control the layout and used as a constraint if the underlying textual ordering matches the user intention and can be controlled by the user. Model elements that are ordered by convention, restrictions in the grammar, or are only programmatically created should only control the diagram if their order expresses intention.

While flow of information can often be constrained to create secondary notation that visualizes how data or control may go backwards, constraining crossing minimization by model order should not be a default strategy for automatic layout. Even if model order is only considered as a tie-breaker it always increases stability, which helps to maintain the mental map, and may be one option to control the layout. The control given by strict model order strategies is, however, always desired to create specific layouts for documentation or presentations.

For languages with cross-hierarchical edges and expandable and collapsible hierarchical nodes, stability is a very important aspect. Considering node or port order to constrain the layout can solve this problem without an editing overhead.

References

1. Domrös, S., von Hanxleden, R.: Preserving order during crossing minimization in Sugiyama layouts. In: Proceedings of VISIGRAPP 2022 - Volume 3: IVAPP, pp. 156–163. INSTICC, SciTePress (2022). https://doi.org/10.5220/0010833800003124
2. Domrös., S., Riepe., M., von Hanxleden., R.: Model order in Sugiyama layouts. In: Proceedings of VISIGRAPP 2023 - Volume 3: IVAPP, pp. 77–88. INSTICC, SciTePress (2023). https://doi.org/10.5220/0011656700003417
3. Domrös, S., von Hanxleden, R.: Diagram control and model order for Sugiyama layouts (2024). https://doi.org/10.48550/arXiv.2406.11393
4. Domrös, S., von Hanxleden, R., Spönemann, M., Rüegg, U., Schulze, C.D.: The Eclipse Layout Kernel (2023). https://doi.org/10.48550/arXiv.2311.00533
5. Dwyer, T., Marriott, K., Wybrow, M.: Dunnart: a constraint-based network diagram authoring tool. In: Revised Papers of GD '08. LNCS, vol. 5417, pp. 420–431. Springer, Berlin, Heidelberg (2009). https://doi.org/10.1007/978-3-642-00219-9
6. von Hanxleden, R., et al.: SCCharts: sequentially constructive statecharts for safety-critical applications. In: Proceedings of the ACM SIGPLAN PLDI '14, pp. 372–383. ACM, Edinburgh, UK, June 2014. https://doi.org/10.1145/2594291.2594310
7. von Hanxleden, R., et al.: Pragmatics twelve years later: a report on Lingua Franca. In: Margaria, T., Steffen, B. (eds.) Leveraging Applications of Formal Methods, Verification and Validation. Software Engineering. ISoLA 2022. LNCS, vol. 13702, pp. 60–89. Springer, Cham (2022). https://doi.org/10.1007/978-3-031-19756-7_5
8. Huang, W., Eades, P., Hong, S.H., Lin, C.C.: Improving multiple aesthetics produces better graph drawings. JVLC **24**(4), 262–272 (2013). https://doi.org/10.1016/j.jvlc.2011.12.002
9. Misue, K., Eades, P., Lai, W., Sugiyama, K.: Layout adjustment and the mental map. JVLC **6**(2), 183–210 (1995). https://doi.org/10.1006/jvlc.1995.1010
10. Petre, M.: Why looking isn't always seeing: readership skills and graphical programming. Commun. ACM **38**(6), 33–44 (1995). https://doi.org/10.1145/203241.203251
11. Petzold, J., Domrös, S., Schönberner, C., von Hanxleden, R.: An interactive graph layout constraint framework. In: Proceedings of VISIGRAPP 2023 - Volume 3: IVAPP, pp. 240–247. INSTICC, SciTePress (2023). https://doi.org/10.5220/0011803000003417
12. Purchase, H.: Which aesthetic has the greatest effect on human understanding? In: DiBattista, G. (eds.) Graph Drawing. GD 1997. LNCS, vol. 1353, pp. 248–261. Springer, Berlin, Heidelberg (1997). https://doi.org/10.1007/3-540-63938-1_67
13. Purchase, H.C.: A healthy critical attitude: revisiting the results of a graph drawing study. J. Gr. Algorithms Appl. **18**(2), 281–311 (2014). https://doi.org/10.7155/jgaa.00323
14. Purchase, H.C., Archambault, D., Kobourov, S., Nöllenburg, M., Pupyrev, S., Wu, H.-Y.: The Turing test for graph drawing algorithms. In: GD 2020. LNCS, vol. 12590, pp. 466–481. Springer, Cham (2020). https://doi.org/10.1007/978-3-030-68766-3_36
15. Sugiyama, K., Tagawa, S., Toda, M.: Methods for visual understanding of hierarchical system structures. IEEE Trans. Syst. Man Cybern. **11**(2), 109–125 (1981). https://doi.org/10.1109/TSMC.1981.4308636
16. Taylor, C.: What has WYSIWYG done to us? Seybold Rep. Publ. Syst. **26**(2) (1996)

B_42: The Geometry of 4-Valued Contradiction

Alessio Moretti

Università Telematica eCampus, Via Isimbardi 10, 22060 Novedrate, CO, Italy
thalnalessio@gmail.com

Abstract. We study the "quadri-segment" B_42, a point in the space B_MN of the "poly-simplexes" of "oppositional geometry". This is the mathematical 4-valued counterpart of the expression, inside the classical 2-valued "bi-simplicial" space B_2N, of the segment of 2-valued contradiction B_22 (e.g. any of the 3 diagonals in a "logical hexagon" B_23, cf. [7]). We rely on a method of poly-simplicial construction put forward in our study [10] of the "tri-segment" B_32 (i.e. 3-valued contradiction), based on sheaf-theory (Angot-Pellissier, [2]) and on the "Pascalian simplexes" P_M [3]. We show that the quadri-segment B_42 is composed of 8 hexagonal tri-segments B_32, 4 of which are "strong" (B_32*) while the other 4 are "weak" ($B_32\#$). These $4 + 4$ hexagonal tri-segments are "glued" (each strong to its correlated weak) by another sub-structure of the B_42, made of 3 squares. The quadri-segment B_42, with its 12 vertices, has as "geometric attractor" a 3D cuboctahedron. It will be recalled how "logical geometry" (Smessaert, Demey [12, 13]) is a fragment of oppositional geometry, where the 2 Smessaertian sub-geometries ("oppositive" and "implicative") undergo, in each poly-simplex, "Aristotelian fusion" into a unique geometry. The quadri-segment B_42 will be useful, in the future, for exploring a higher poly-simplex, the quadri-triangle B_43, i.e. the 4-valued counterpart of the classical (bi-simplicial) "logical hexagon" B_23.

Keywords: Oppositional Geometry · Many-valued logic · Pascalian Simplexes

1 Bi-Simplexes, Pascalian Simplexes, Poly-simplexes and B_32

The concept of "opposition" is captured by "bi-simplexes", when 2-valued, and by "poly-simplexes", when m-valued [5, 6]. The mathematical identity of each lays in their "closures". In [10] we showed that the "closures" B_2N [11] of the bi-simplexes of opposition A_2N [4] are encrypted in the rows of Pascal's triangle P_2 (see Fig. 1).

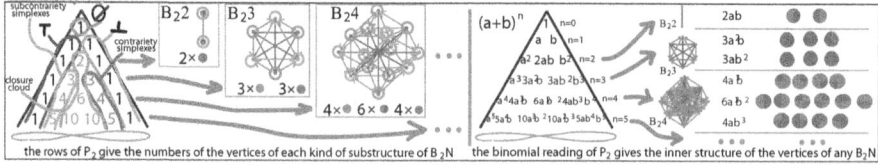

Fig. 1. Each closure B_2n of any A_2n is fully expressed by one row of "Pascal's triangle" P_2.

© The Author(s), under exclusive license to Springer Nature Switzerland AG 2024
J. Lemanski et al. (Eds.): Diagrams 2024, LNAI 14981, pp. 84–100, 2024.
https://doi.org/10.1007/978-3-031-71291-3_7

In [10] we also showed that the poly-simplexes, accessible *via* Angot-Pellissier's "sheaf-theoretic technique" [2], are encrypted in the "horizontal sections" of the "Pascalian simplexes" P_M (a generalization of Pascal's triangle P_2 [3]) (see Fig. 2).

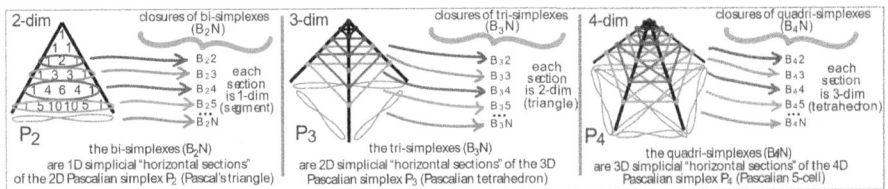

Fig. 2. The "Pascalian simplexes" P_M: their "horizontal sections" capture the closures $B_M N$.

In the discrete space $B_M N$, where "M" is the parameter for "poly-" (bi-, tri-, quadri-, ...) and "N" is that for "simplex" (-segment, -triangle, -tetrahedron, ...), each point $B_m n$ is an oppositional structure (generalizing that of the "logical hexagon" $B_2 3$ [7]). In this space, seen either as a 2D matrix or as a numerical triangle, each $B_m n$ can be represented by the number $V_m n = m^n - m$ of its vertices. Conjecturally, the geometric attractor of each $B_m n$ is n-dimensional iff it lays in the n-th row of the numerical triangle (see Fig. 3).

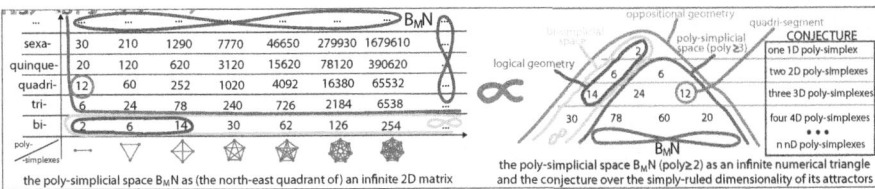

Fig. 3. The space $B_M N$ of the poly-simplexes (poly ≥ 2), seen either as a matrix, or as a triangle

We studied in [10] the first poly-simplex, the "tri-segment" $B_3 2$, using sheaf-theory [2] and a new "Pascalian technique". Its "geometrical attractor" is a new kind of hexagon, and it diffracts classical contradiction $B_2 2$ into 3, by adding a paracomplete (i.e. intuitionist) and a paraconsistent (i.e. co-intuitionist) contradiction segment (see Fig. 4).

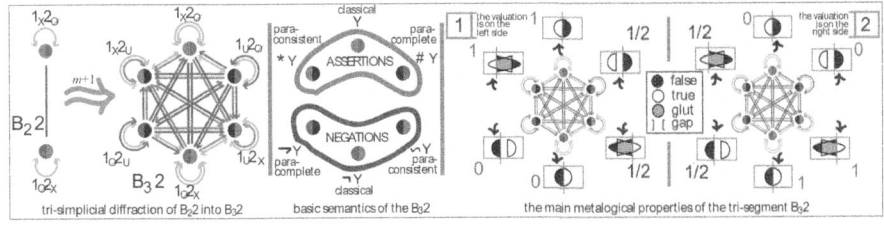

Fig. 4. The "tri-segment" $B_3 2$ expresses the oppositional geometry of 3-valued "contradiction".

86 A. Moretti

In this paper we produce, by a similar technique, the first study of the quadri-segment B_42: its main structures, its 3D attractor, its semantics, its valuations and 2 applications.

2 Quadri-Segment B_42: The First 5 Steps of Its Construction

2.1 First Step: "Sheaf-Theoretical Analysis" of the Quadri-Segment B_42

In the case of the n-oppositions (i.e. the geometry of 2-valued oppositions) the "closures" B_2N of the bi-simplexes A_2N are reached through a "set-theoretic technique" [11]. The poly-simplexes (i.e. p-valued oppositions, with $p \geq 3$) can be studied with an equivalent, but slightly more complex, sheaf-theoretic method [2] which, however, does not rule out trivial elements (which hinder the emergence of important geometrical properties) and therefore does not give yet clear geometric hints for reaching an "attractor" (this is done more clearly – as we will see – by our "Pascalian method", see [10]). The aim of this Sect. 2.1 is to start determining all the vertices of the quadri-segment, including the trivial ones (we will be able to exclude them "à la Pascal" in Sect. 2.2).

So, following Angot-Pellissier's method [2], let us construct a "topos" for 4-valued logic, upon which it will be possible to calibrate an "enumerating and naming method" for 4-valued oppositional geometrical vertices (for any simplex "n" of a quadri-simplex B_4N). Upon this 4-valued topos (whose 4 fundamental elements are X, V, U, Ø, so strictly ordered) let us construct a "numerical sheaf" for the concept of oppositional quadri-segment. This is the workable image of "truth", or "total oppositional reality" (which can be split into a complete finite set of alternative possibilities – potential vertices of an oppositional geometrical structure), such that by partitioning it (sheaf-theoretically, i.e. not only horizontally, but also vertically) in all the possible ways one will be able to obtain all possible "opposites", whatever their shape, for that universe.

The first parameter (horizontal) to be dealt with to do that is the "length" of the numerical sheaf: for a quadri-*segment* (a 4-valued *contradiction*) this length is 2 (for a bi-*triangle* of 3-*contrariety*, or for any poly-*triangle*, it would have been 3, etc.). The second parameter (vertical) is the "depth" of the numerical sheaf (measuring and ordering its truth-value levels): for the *quadri*-segment (as for any *quadri*-simplex) this depth is 4 (for any *quinque*-simplex it would have been 5, etc.). From this total numerical sheaf $1_X 2_X$ one goes (looking for the oppositional closure of the quadri-segment B_42), downwards through all its "sub-sheaves" (i.e. its possible weakenings, leading from the maximum $1_X 2_X$ to the minimum $1_\emptyset 2_\emptyset$): here there are exactly 16 of them (see Fig. 5).

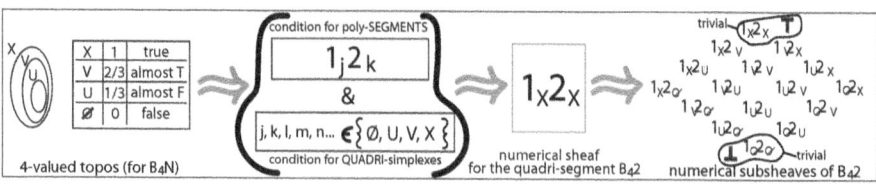

Fig. 5. The total "numerical sheaf" for the quadri-segment B_42 and all its possible sub-sheaves

Thus, up to now we have 16 "numerical sub-sheaves", from which we should obtain, once excluded the trivial sub-sheaves (cf. next step), the "non-trivial vertices" (i.e. the real vertices) of the oppositional-geometrical solid expressing the quadri-segment B_42.

2.2 Second Step: "Pascalian-Simplicial Analysis" of the Quadri-Segment B_42

A warning: our "Pascalian method" [10] aims at providing deep geometrical, qualitative insight on the studied poly-simplex, but it becomes less intuitive in higher-dimensional spaces (i.e. from 4D upwards). We are producing the first "Pascalian-simplicial" study of a quadri-simplex B_4N (a B_42), and this is less easy than what seen for tri-simplexes B_3N (like in the study of B_32, cf. Sect. 1, Fig. 4): we have now to consider the 3D "horizontal (tetrahedric) sections" (i.e. tetrahedron) of a 4D Pascalian simplex P_4.

As we established in [10], and as recalled (cf. Sect. 1, Fig. 2 *supra*), poly-simplexes $B_m n$ are finite "horizontal sections" of infinite "Pascalian simplexes" P_M: the horizontal sections of Pascal's infinite 2D triangle (finite 1D segments) give the bi-simplexes, the horizontal sections of the Pascalian infinite 3D tetrahedron (finite 2D triangles) give the tri-simplexes, the horizontal sections of the Pascalian infinite 4D 5-cell (finite 3D tetrahedra) give the quadri-simplexes, and so on. The quadri-segment B_42, which we now study, is a quadri-simplex (the one whose simplex is the segment). Focussing on the series of the 3D tetrahedric "horizontal sections" of the "Pascalian simplex" of dimension 4 (which are relative to the quadri-simplexes B_4N), the third downwards is the one relative to the quadri-segment B_42 (in the figure, right side, we also show the fourth horizontal tetrahedric section, relative to the quadri-triangle B_43) (see Fig. 6).

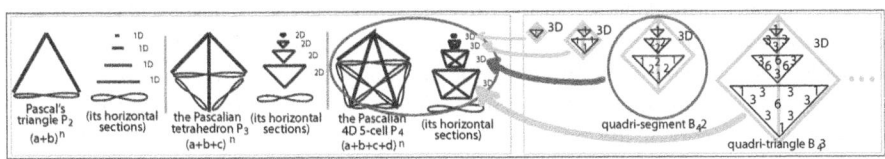

Fig. 6. From P_2 to the Pascalian P_4 for the B_4N: its third horizontal section corresponds to B_42.

If we now focus on this top-down third 3D tetrahedric "horizontal section", namely the one proper to the quadri-segment B_42, we can study it also by comparing it to the lower-dimensional "horizontal sections" of the already known lower poly-segments (poly ≥ 2), namely the (bi-) segment B_22 and the tri-segment B_32. This tetrahedric 3D

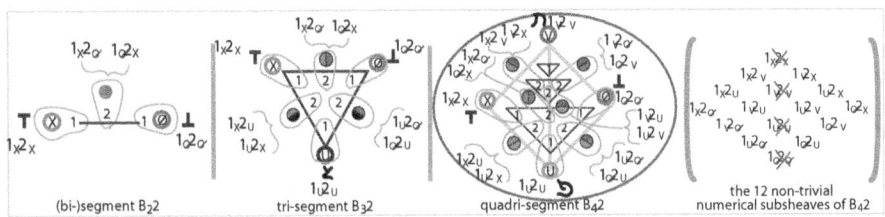

Fig. 7. "Pascalian charts" of B_22, B_32 and B_42 and, for this last, the 4 trivial vertices it discards

horizontal section provides several informations, both on the fundamental geometry and on the main sub-structures of the quadri-segment B_42, by means of a "Pascalian chart", and it highlights four "trivial vertices", which must be discarded (see the right side of Fig. 7).

We will come back to the Pascalian chart of the quadri-segment B_42 (in Sects. 3.1 and 3.3) and we will see, then, how powerful and useful this tool is. By now, having examined the "Pascalian side" of the quadri-segment B_42, and having thus established that its trivial vertices are not 2 (i.e. T and ⊥), but 4 (i.e. T, ⊥, ח and פ), and that therefore its non-trivial vertices are 12 and not 14, let us study, in the next step, the relations holding between any pair of non-trivial vertices of B_42: its edges or (when internal) its lines.

2.3 Third Step: The "Oppositive Sub-geometry" of the Quadri-Segment B_42

The main (and almost only) real discovery of "logical geometry" [12, 13] is the truly important fact that, when speaking of the structures of opposition, the "geometries" at stake are, in some sense, not one, but two. Smessaert added to what he renamed "opposition geometry" (a slightly modified version of the "geometry of oppositions") an "implication geometry". In our study of the tri-segment B_32 (which is a poly-simplex, poly ≥ 3 – and "logical geometry" never took poly-simplexes in consideration) we demonstrated that this Smessaertian property of admitting two (sub-)geometries holds also there: there are an "oppositive geometry" and an "implicative geometry" (with slightly modified terminology) of the hexagonal tri-segment B_32. In this section we will show that, similarly, there is an "oppositive sub-geometry" of the quadri-segment B_42. In order to unfold it one has to apply, almost unchanged, Angot-Pellissier's sheaf-theoretic method: the only difference is that the possible game-theoretical answer "[1|1]" ("yes, yes") must be read not as "subalternation" (as used to do [2, 5, 6]), but à la Smessaert, i.e. meaning rather "non-contradiction" (for simplicity we keep however the label 'I' to express this "[1|1]" possible answer). The resulting outcomes of the Angot-Pellissierian calculation (cf. *infra*) of the relations between pairs of non-trivial vertices give the possible "oppositive relations" of the quadri-segment B_42 (see Fig. 8).

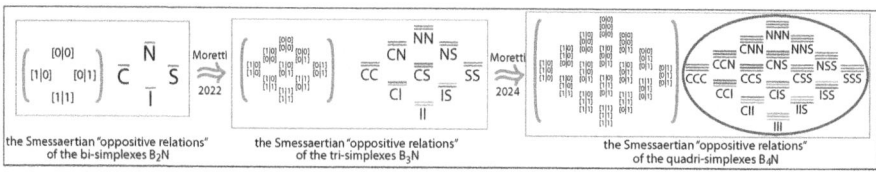

Fig. 8. The 4, 9 and 16 kinds of the "oppositive relations" of the B_2N, B_3N & B_4N respectively

Let us recall, through four examples of suited Angot-Pellissierian calculations, how proceeds the determination of the "oppositive quality" of any of the possible pairs of numerical sub-sheaves of B_42. Let us consider first the pair 1_X2_U–$1_\emptyset 2_V$. We build a left column of answers A_1 and a right column of answers A_2 (each 3-layered) by answering the questions Q_1 and Q_2, for each level X, V and U. The Q_1 is: "Can the (inclusive) disjunction of the two subsheaves be false at level X (respectively at level V, or U)?".

The correlated Q_2 is: "Can the conjunction of the two subsheaves be true at level X (respectively at level V, or U)?". Let us examine the answer A_1. At level X the disjunction of the indexes "2" can be false, but at level V and at level U it can't, so the global answer A_1 is: "1, 0, 0". Similarly, for A_2, at levels X and V the conjunction cannot be true (neither in index "1", nor in index "2"), but it can at level U (for the index "2"), so the global answer A_2 is: "0, 0, 1". Putting these two results, as vertical columns, side by side, we get as vertical 3-layered answer (top-down): "[1|0], [0|0], [0|1]" (to be seen as a vertical 3-layered column), which gives: "CNS" (to be read vertically and top-down). A similar calculation is done for any other pair of vertices (see Fig. 9).

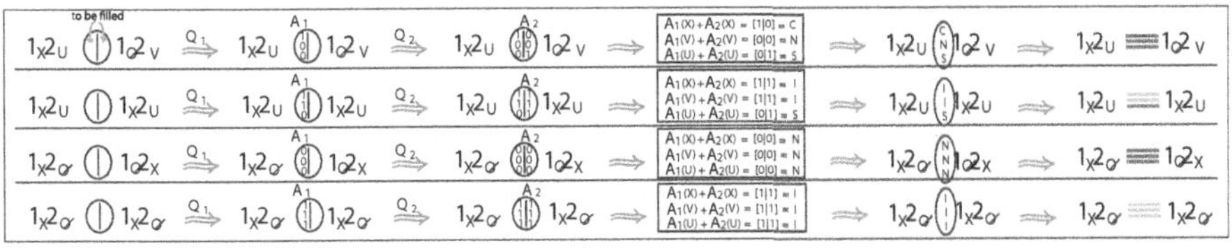

Fig. 9. B_42: four examples (over the possible 78) of calculations of the "oppositive relations"

We will give soon, as total result of the 78 such calculations among all possible pairs of non-trivial vertices, a global distribution of "oppositive qualities" (cf. left side of Fig. 12 *infra*). We will not yet represent the "final" geometry of this, since to depict the Smessaertian "oppositive half" of the quadri-segment B_42 we would need to rely on a "geometrical attractor" (cf. Sect. 3.2) which we do not have yet. But having calculated the "oppositive sub-geometry" of the quadri-segment B_42, we can now turn, in the next step, to the second Smessaertian sub-geometry, the "implicative" one.

2.4 Fourth Step: The "Implicative Sub-geometry" of the Quadri-Segment B_42

The second Smessaertian sub-geometry differs from the first in so far it examines the compatibilities and incompatibilities between *different* truth-values. Three consequences follow from that: (1) the "implicative kinds" follow a different combinatorial pattern; (2) the algorithm for calculating them can be obtained from the Angot-Pellissierian sheaf-theoretic method, but it needs a suited adaptation (like in the case of the tri-segment B_32); (3) these relations are order-sensitive: "α—β" is not always the same as "β—α". The adaptation here of the sheaf-theoretic method is the following: (A) one has to ask, for any of the three top-down sheaf-theoretic levels X, V and U, two "meta-questions", namely Q'_1: "Can the first be false and the second true (at level...)?"; and Q'_2: "Can the first be true and the second false (at level...)?"; (B) the possible outcomes, by answering "0" (i.e. "no") or "1" (i.e. "yes") to the two questions, that is [0|0], [0|1], [1|0], [1|1], work as in the "oppositive" case (Sect. 2.3), and give the possible "implicative relations" of the quadri-segment B_42: "B" (bi-arrow), "L" (left-arrow), "R" (right-arrow) and "A" (no-arrow); (C) remark that "R" and "L", being geometrically the same, just considered in two opposite directions ("αRβ" is geometrically equivalent to "βLα", for any vertices "α" and "β"), some combinations of them (i.e. 2- or 3-layered implicative relations) like "AR" and "AL", or like "RAA" and "LAA", or like "LLR" and "RRL", etc., are equivalent; (D) so, the combinatorially possible 4, 16 and 64 implicative kinds (for B_22, B_32 and B_42) can be reduced respectively to 3, 10 and 36 non-redundant implicative kinds (see Fig. 10).

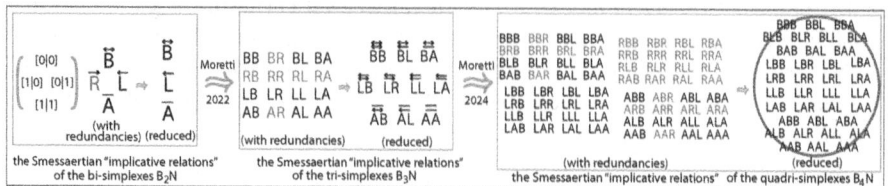

Fig. 10. The 3, 10 and 36 non-redundant kinds of "implicative relations" in B_2N, B_3N and B_4N

According to what said, we propose, here, some examples of the way in which the "implicative quality" of any of the possible pairs of numerical sub-sheaves of B_42 can be obtained through Angot-Pellissierian calculations (see Fig. 11).

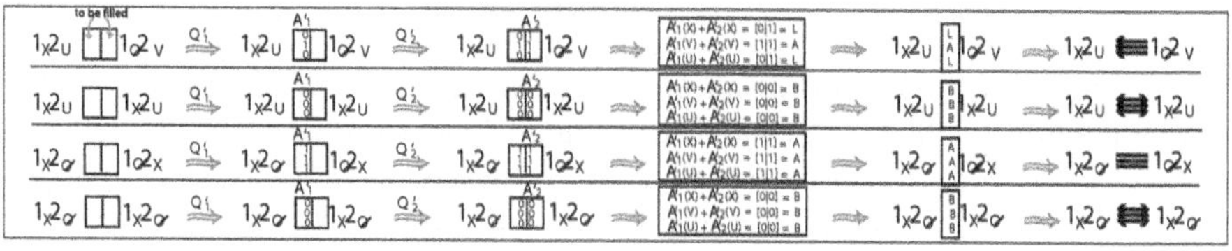

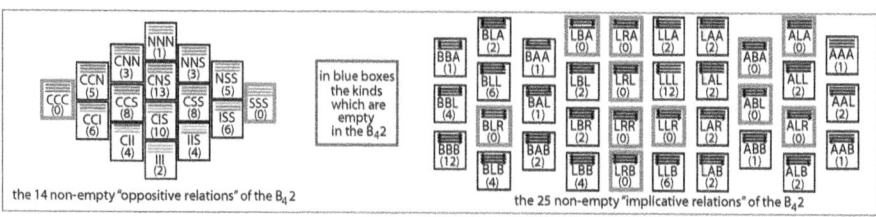

Fig. 11. Four examples (over the 78) of calculations of "implicative relations" of the B_42

The final result of the calculation of the "implicative quality" of each of the 78 possible pairs of numerical sub-sheaves of B_42 gives a distribution of 25 non-null main kinds (right side of next figure). Remark that, among the oppositive relations, "CCC" and "SSS", i.e. contrariety and subcontrariety, are empty in the B_42, which is normal for a "contradiction", as is any poly-segment B_M2; as for the implicative relations, of the 36 non-redundant kinds 11 are empty in B_42 (they are non-empty as soon as one considers the quadri-triangle B_43, as another study will show) (see Fig. 12).

Fig. 12. Number and frequency of "oppositive" (14) and "implicative" (25) relations in the B_42

Here as well, in order to depict the Smessaertian "implicative half" of the quadri-segment B_42 we would need to rely on a geometrical attractor (cf. Sect. 3.2). For now, having calculated the "implicative sub-geometry" of the quadri-segment B_42, we can turn to the fifth and final step of the elementary analysis of the quadri-segment: the production of the "Aristotelian fusion" of its two Smessaertian sub-geometries into a unique one.

2.5 Fifth Step: "Aristotelian Fusion" of the Two Sub-geometries of the B_42

Having obtained the elements (i.e. the relations between any pair of vertices, i.e. the edges – or lines – and curls) of the two Smessaertian sub-geometries of the quadri-segment B_42, the next step consists in showing that the two sub-geometries can (and must) in fact be joined into a unique geometry (oppositional), called "Aristotelian".

Our algorithm for Aristotelian fusion (making of "logical geometry" a strict, incomplete fragment of "oppositional geometry") is in fact rather simple (in [10] we proposed a sub-optimal version of it, seeming to be much more complex), it consists in producing a relation composed of two fusioned parts: a "color" and a "shape". The color is that of the corresponding "Smessaertian "oppositive relation", whereas the shape is that of the corresponding Smessaertian "implicative relation". At this point, to go further, the main distinction to be grasped is that between "simple" (i.e. one-layered) and "complex" (i.e. many-layered) Aristotelian relations. The difficulty lays in the fact of being composed (here) not of one, but of 3 layers (which are the effect of the starting 4-valued topos on the structure, cf. Sect. 2.1, Fig. 5), the simple element is, again, the combination of a simple color and of a simple shape. (see Fig. 13).

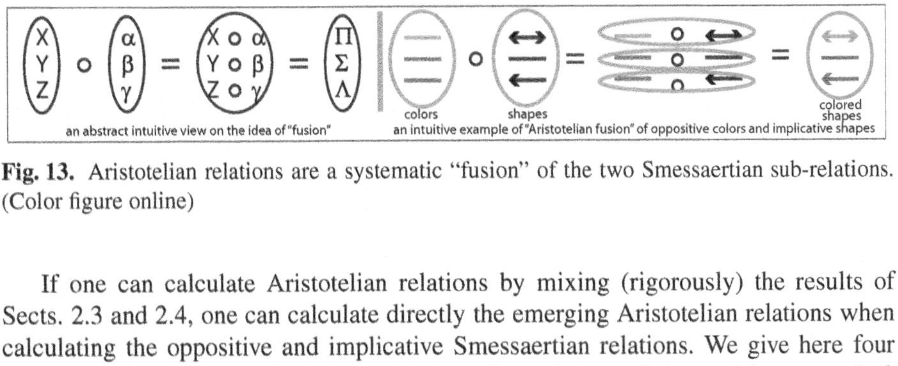

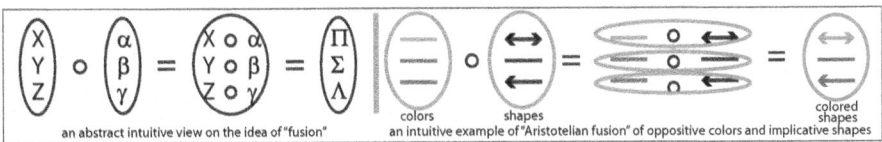

Fig. 13. Aristotelian relations are a systematic "fusion" of the two Smessaertian sub-relations. (Color figure online)

If one can calculate Aristotelian relations by mixing (rigorously) the results of Sects. 2.3 and 2.4, one can calculate directly the emerging Aristotelian relations when calculating the oppositive and implicative Smessaertian relations. We give here four examples of such direct calculations of Aristotelian relations of the quadri-segment B_42 (see Fig. 14).

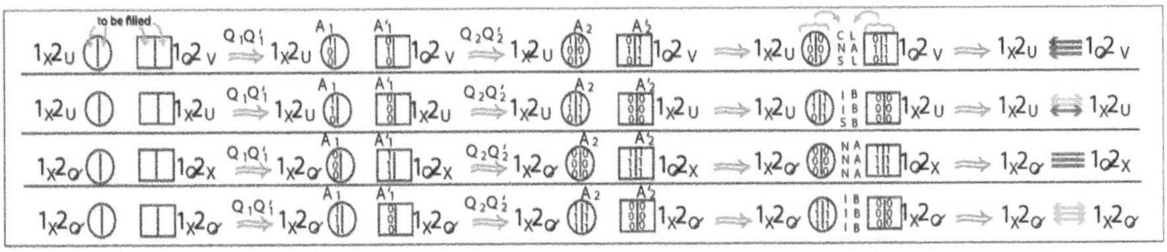

Fig. 14. Four examples (over the 78) of direct calculations of the Aristotelian relations of B_42

The composed (i.e. 3-layered) Aristotelian relations of the B_42 can be displayed in a table (in each box: the graphical relation and the ordered pairs of numerical sub-sheaves satisfying it in the B_42). There are 37 kinds of Aristotelian relations in the B_42 (in the B_43 there are more, as for instance the "CA/CA/CA" and "SA/SA/SA") (see Fig. 15).

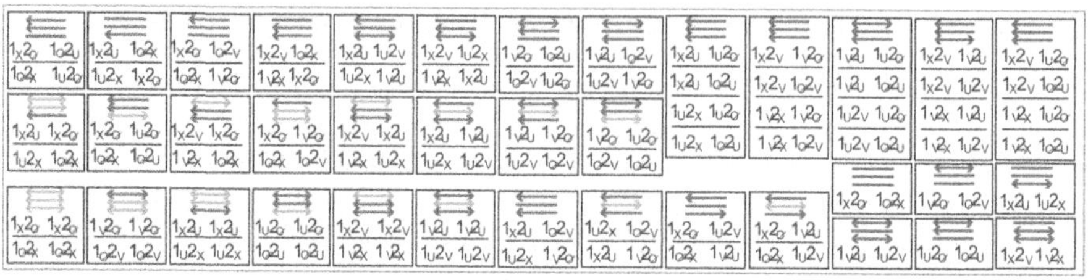

Fig. 15. The 37 kinds of composed Aristotelian relations in the B_42 (its 66 edges and 12 curls)

2.6 The Meaning of the "Aristotelian Relations" Obtained Through "Fusion"

Before better explaining the Aristotelian relations let us summarize the situation. In the bi-simplicial B_2N, and *a fortiori* in the traditional (pre-2004) "Aristotelian fragment" (as "logical geometers" say), the "Aristotelian (Apuleian) relations" were 4: "contradiction", "contrariety", "subcontrariety" and "subalternation". To them Smessaert added (with reason) 4 more ones: 3 missing "implication relations" ("L-arrow", "bi-arrow", "no-arrow") and one missing "opposition relation" ("non-contradiction"). In our study [10] of the first poly-simplex, the tri-segment B_32, six new simple Aristotelian relations have been added: a "blue L-arrow" CL, a "blue R-arrow" CR, a "blue bi-arrow" CB, a "green L-arrow" SL, a "green R-arrow" SR and a "green bi-arrow" SB. In the quadri-segment B_42 we find 2 new simple Aristotelian relations, previously unknown: the "red L-arrow" NL and the "red R-arrow" NR (we leave aside the case of the "impossible" "red bi-arrow" NB, seemingly possible if it stays for $\bot$) (see Fig. 16).

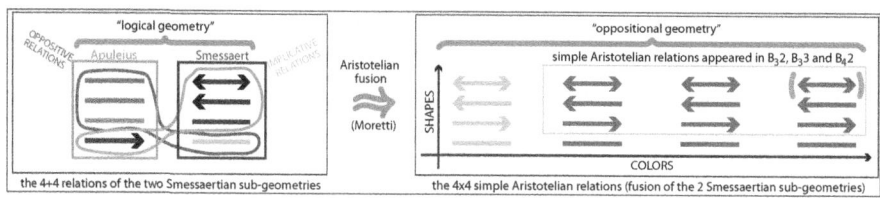

Fig. 16. The 16 simple "Aristotelian relations" of poly-simplicial oppositional geometry (B_MN) (Color figure online)

Consistently with the "Aristotelian fusion" algorithm (cf. Sect. 2.5), the meaning of any Aristotelian relation is determined as follows: (1) it assumes the meaning conditions provided by its color (its Smessaertian "oppositive geometry"); (2) to that it adds ($\cap$) the meaning conditions provided by its shape (its Smessaertian "implicative geometry").

For instance, the meaning of the "red right-arrow" is the intersection of the solutions of both: the truth-conditions for "red" and those for "right-arrow", i.e. "[0|1] or [1|0]" (for "red") and "[0|0] or [0|1] or [1|1]" (for "right-arrow"), that is, as fusion result, "[0|1]" (for "red right-arrow"). A similar reasoning shows that in fact each Aristotelian simple relation has exactly the meaning of a binary connective and that, not surprisingly, there are 16 possible such Aristotelian simple relations (including 2 "trivial" simple Aristotelian relations, T and ⊥, i.e. the "grey non-arrow" and the "red bi-arrow").

The 16 possible relations we thus have (i.e. with the 2 trivial ones, T and ⊥, included), stand in an isomorphic relation to the 16 binary connectives (which, as is known, can be modelled by a 4D hyper-cube, called "Boolean tesseract"). Since the 14 non-trivial connectives can be modelled by a 3D tetrakis-hexahedron B_24 (cf. [5]: deprived of the 2 trivial vertices T and ⊥ the 4D hypercube collapses to the 3D tetrakis-hexahedron B_24), so can the 14 non-trivial Aristotelian relations (see Fig. 17).

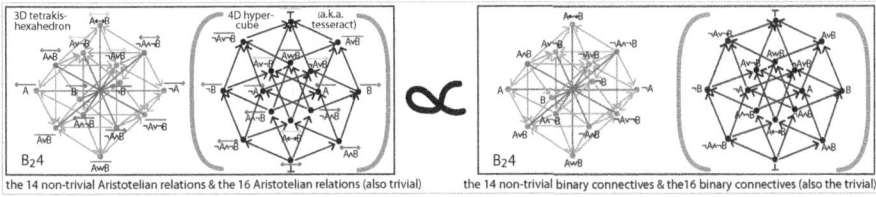

Fig. 17. Complete system of the simple Aristotelian relations of the poly-simplicial space B_MN

We thus reached a complete description of the 16 simple Aristotelian relations of the poly-simplicial space B_MN. As, now, for the complex (i.e. p-layered, with $p \geq 2$) Aristotelian relations (those of the poly-simplexes, with poly ≥ 3), a possible meaningful and clarifying question here seems to be that of knowing whether their larger manyfold (in the case under study of the B_42, we saw that there are 37 such Aristotelian qualities, cf. Sect. 2.5, Fig. 15) can be parted into qualitative kinds. In fact, it seems it can: there seem to be six "macro-kinds" (not excluding other viewpoints). We will be soon in a position of giving meaningful examples of these macro-kinds of composed Aristotelian relations of the quadri-segment B_42. Remark that in each of these six kinds of composed Aristotelian relations their different instances seem to be possibly ordered, so one can speak of "strong" and "weak" instances of each of these kinds (see Fig. 18).

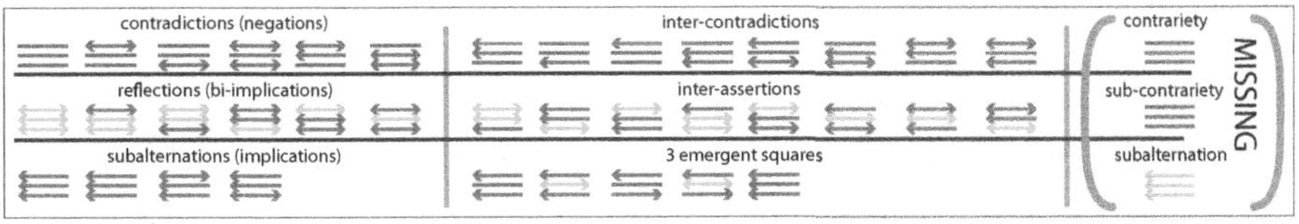

Fig. 18. The 6 macro-kinds over the 37 kinds of composed Aristotelian relations of the B_42

Having determined the 12 vertices of the quadri-segment B_42 (Sects. 2.1 and 2.2) and the 78 Aristotelian relations holding between all the possible pairs of these 12 vertices (including the 12 pairs of identical vertices, which give "reflexive curls"), let us now turn to the problem of determining a "geometrical attractor" for the B_42.

3 The 3D "Geometrical Attractor" of the Quadri-Segment B_42

Having seen the basic elements of the B_42, it is time to try to see its whole shape. We propose to call "geometrical attractor" of a given poly-simplex B_mn the geometrical shape that expresses at best the structural properties of that B_mn. As it happens, in order to find out the attractor of the B_42 (which, conjecturally, is expected to be 3D, cf. Fig. 3 *supra*) it is useful to highlight first one of the quadri-segment's main sub-structures.

3.1 An Important Sub-structure of the B_42: Its 4 Hexagonal Tri-Segments B_32^*

If we go back to the tetrahedric Pascalian chart of the quadri-segment B_42 (cf. Sect. 2.2, Fig. 7), we can see that it contains 4 times the triangular Pascalian map of the tri-segment B_32, which led to the attractor of the latter: a 2D hexagon (cf. Sect. 1, Fig. 4). This strongly suggests that inside the B_42 one finds four instances of hexagonal tri-segment B_32^* – by the asterisk "*" we express that these four hexagons B_32^* are sub-structures of a bigger one (in a similar way, in the bi-simplicial space B_2N the tetrakis-hexahedron B_24 of 4-opposition contains six instances of logical hexagons B_23^*, [11]) (see Fig. 19).

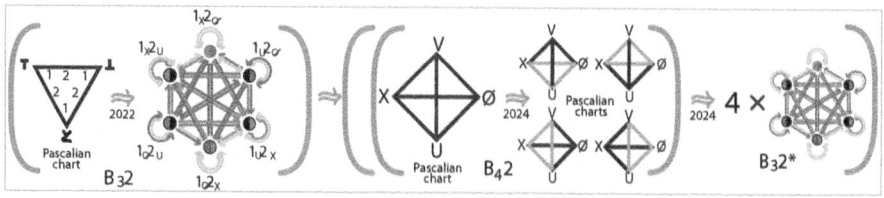

Fig. 19. The quadri-segment B_42 contains as sub-structures four hexagonal tri-segments B_32^*.

In fact, we can draw oppositionally-geometrically these four sub-structures B_32^* of the B_42. Therefore, as it seems, among the main sub-structures of the quadri-segment B_42 we find the following four tri-segments B_32^* (see Fig. 20).

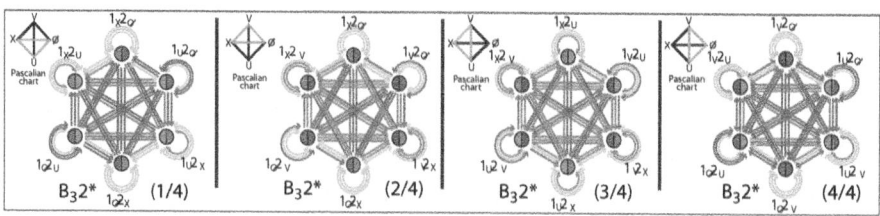

Fig. 20. The four tri-segments B_32^* contained in the quadri-segment B_42 as sub-structures (Color figure online)

Remark that in each of these four B_32^* there is, as expected, a dominant (here: vertical) contradiction segment (among the 3) and a dominant rectangle of subalternation (perpendicular to the dominant contradiction segment, so here horizontal). There also is, as expected, a predominance of paracompletude (i.e. more blue than green) in one of the two remaining diagonals, and the other way round (predominance of paracompleteness) in the other. Another remark, made now possible by the previous figure, is that each of the 12 vertices of the quadri-segment B_42 belongs exactly to two of the four hexagons B_32^*. More precisely: $1_X2_\emptyset$ and its centrally symmetric $1_\emptyset 2_X$ belong to the B_32^* hexagons 1/4 and 2/4, 1_X2_U and 1_U2_X belong to the hexagons 1/4 and 3/4, $1_U2_\emptyset$ and $1_\emptyset 2_U$ belong to the hexagons 1/4 and 4/4, 1_X2_V and 1_V2_X belong to the hexagons 2/4 and 3/4, $1_V2_\emptyset$ and $1_\emptyset 2_V$ belong to the hexagons 2/4 and 4/4, and 1_V2_U and 1_U2_V belong to the hexagons 3/4 and 4/4. This is quite informative for us, since this is then a property that a geometrical attractor of the quadri-segment B_42 needs to be able to express.

3.2 The Geometrical Attractor of the B_42 is a 3D Solid: A "Cuboctahedron"

We are looking now for a geometrical figure, which, following our conjecture (cf. Sect. 1, Fig. 3), is expected to be 3D. We know it should have 12 vertices (Sect. 2.2). It should contain, however shaped (i.e. planar or not), four hexagons (the four inner B_32^* of the B_42), such that each of the 12 vertices belongs to 2 of these 4 hexagons (cf. Sect. 3.1). As it happens, one well-known 3D geometrical figure fulfilling these four *desiderata* does exist: the "cuboctahedron", a rather classical geometrical object, in fact an "Archimedean solid" (here we highlight the main diagonal of each B_32^* in it) (see Fig. 21).

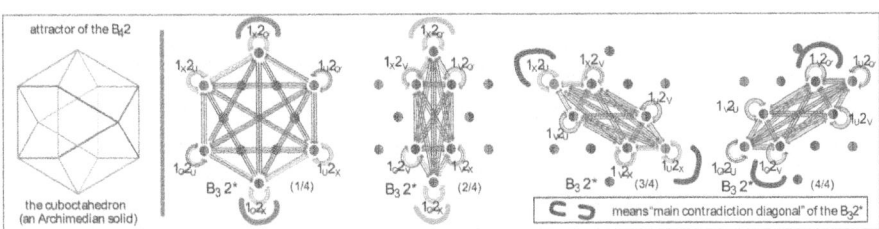

Fig. 21. A good candidate as "attractor" for the B_42, the cuboctahedron, is made of 4 hexagons.

Having reached this important point, let us come back, in this cuboctahedric attractor, to the question of the possible sub-structures of the B_42.

3.3 Two Sub-structures of the Cuboctahedric B_42: Glueing Squares and $B_32\#$

A further emerging sub-structure of the quadri-segment B_42 is related to an important fact, observable in the figure: any of its 12 vertices is related to any of the other 11 vertices by one of the 15 lines of one of the 2 hexagonal tri-segments B_32^* to which it belongs: with the exception of two vertices, to which none of the aforementioned 15 lines of any of the two hexagons B_32^* can lead (any vertex has 2 such "unreachable vertices"). Remark that this behavior is readable in the Pascalian chart of B_42 (cf. Sect. 2.2, Fig. 7):

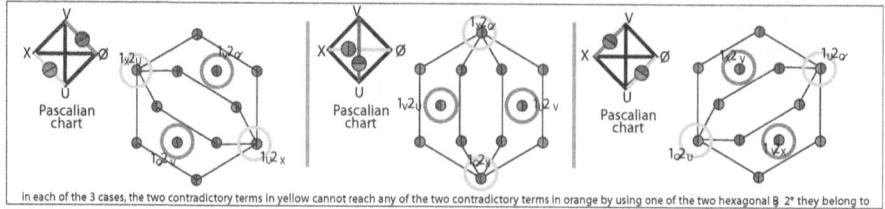

Fig. 22. The Pascalian chart of the B_42 suggests a sub-structure which joins unrelated vertices.

in the fact that any of the 6 pairs of contradictory vertices represented there as "2" is centrally symmetric to another such pair of contradictory vertices (see Fig. 22).

The link between any pair of so disjoined vertices is one of the twelve diagonals of the six "surface squares" of the cuboctahedric attractor of the B_42. These twelve surface diagonals let emerge three mutually disjoined "closed circuits", which are three squares, each cutting the cuboctahedric 3D volume into halves (see Fig. 23).

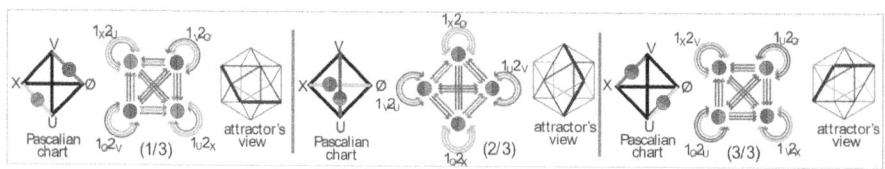

Fig. 23. A sub-structure of the cuboctahedric attractor of the B_42: three "glueing squares"

These three emerging squares have, each, what we defined as a "chromatic signature" [9]. Each will express, as such, a particular "oppositional meaning" (see Figs. 27 and 28 below) and, as we are going to see, they "glue" disjoint sub-structures of the B_42.

Before that, the Pascalian chart delivers a third important sub-structure of the quadri-segment B_42. On the cuboctahedric attractor this can be viewed as a complementarity between any of its 4 hexagons B_32* and the "double crown" hexagon emerging through it from the remaining 6 vertices of the cuboctahedric quadri-segment B_42. Equivalently, maybe more clearly, this can be viewed in the Pascalian chart of the B_42 (Sect. 2.2, Fig. 7), where one can see that, parallel to any triangular face of the tetrahedron, one can find a set of three "2": and these are three diagonals (over the 6 of the B_42) which together determine a hexagonal tri-segment $B_32\#$ (by "#" we mean that this hexagonal tri-segment $B_32\#$ is weaker, i.e. oppositionally less regular than a B_32*, strong) (see Fig. 24).

This third meaningful sub-structure of the B_42 might be called the "internal hexagon" $B_32\#$ of any of the four hexagons B_32* contained in the cuboctahedric attractor of the B_42. These 4 hexagons are "weak tri-segments" $B_32\#$, insofar if they embody the structure "hexagonal tri-segment" B_32, their leading diagonal of contradiction is not the strongest one in them, hence the characterisation as "weak" (with respect to the retrospectively "strong" hexagonal tri-segment B_32* of Sect. 3.1) (see Fig. 25).

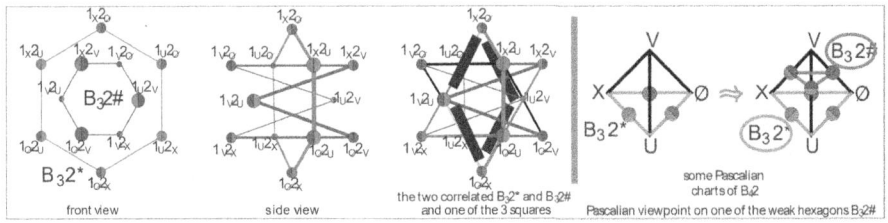

Fig. 24. Another emerging sub-structure of the cuboctahedric B_42: its 4 "weak hexagons $B_32\#$"

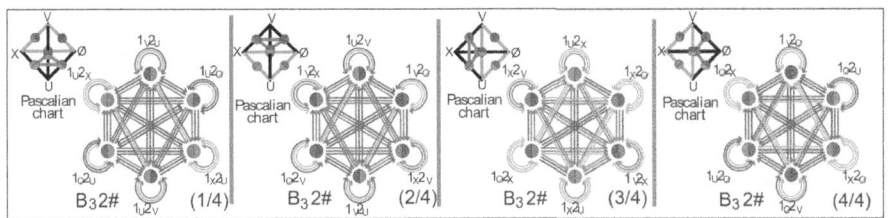

Fig. 25. The 4 "weak hexagons $B_32\#$" are important sub-structures of the quadri-segment B_42.

We stop studying the sub-structures of B_42 (there is a simplex of its 4 trivial sub-sheaves), to have an idea of the dense "inner jungle" of oppositional structures see [9].

4 Semantics, Valuations and Applications of the B_42

4.1 The Semantic Properties and the Valuations of the Cuboctahedric B_42

Semantically speaking (left side of next figure), poly-segments B_M2 do express assertion and negation. This is classically true for the bi-simplicial contradiction segment B_22, and this also true for the tri-segment B_32 (cf. Sect. 1, Fig. 4). In the case of the quadri-segment B_42 this still holds, but it is made more complex by the presence of its inner 4 tri-segments B_32^* (cf. Sects. 3.1 and 3.2). One can visualize assertions by focusing on the "assertive vertex" of the leading contradiction segment in any of the four B_32^* of the B_42. Each of such 4 "leading assertive vertices" determines 2 "subordinate assertive vertices". When one focusses on the superposition of these 4 upper triads of assertions this covers the upper half of the cuboctahedron (5 vertices) plus the vertex 1_V2_U in the middle. Then the 6 vertices so covered have this property of assertiveness in different degrees. $1_X2_\emptyset$ "dominates" twice, 1_X2_U and $1_V2_\emptyset$ dominate once and are each dominated once, 1_X2_V is dominated trice, $1_U2_\emptyset$ is dominated twice, 1_V2_U is dominated once (1_U2_V is left untouched by assertions). A symmetric reasoning can be carried on for the negations (lower half of the cuboctahedric B_42), which leads to covering with negations the lower half of the cuboctahedron (5 vertices) plus one of the two vertices in the middle: 1_U2_V. So, the top 5 vertices plus 1_V2_U express assertion to different degrees, while the bottom 5 vertices plus 1_U2_V express negation to different degrees ($*_1 > *_2$ and $\#_1 > \#_2$, etc.). If we read this in metalogical terms, in each pair of the upper-half 4 vertices dominated by $1_X2_\emptyset$, two (on the right of $1_X2_\emptyset$) express a paracomplete form of assertion, the other (on the left of $1_X2_\emptyset$) a paraconsistent form; of the 2 middle vertices, very symmetric and both logically non-classical but half-way between paracompleteness and paraconsistency (they weaken X and strengthen ∅) and therefore "paranormal", 1_V2_U expresses paranormal assertion, whereas, symmetrically, 1_U2_V expresses paranormal negation. As for the "valuations" of the B_42 (i.e. the global simultaneous possible attribution of truth-values to the 12 vertices of the structure, compatible with the latter), the quadri-segment admits, by construction, 2: one which is ruled by the first of the 2 indices in each vertex, another which is ruled by the other. This, as in all poly-segments B_M2, generates 2 cases (which can be named "1" and "2"), in each of which the 2D surface of the 3D cuboctahedric attractor of the quadri-segment B_42 is partitioned into 4 triangular "patches" of 3 vertices, all with the same truth-value (see Fig. 26).

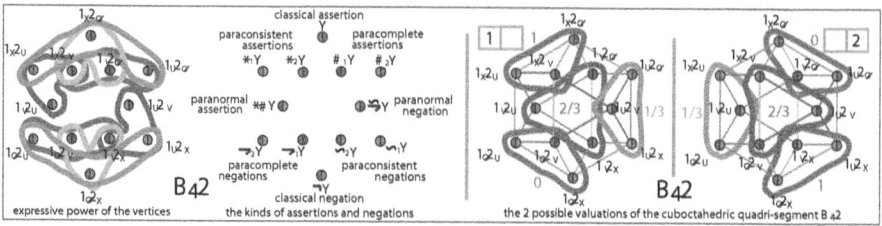

Fig. 26. Assertions and negations (semantics) and the two valuations of the quadri-segment B_42

4.2 Two Applications of the B_42: To a Formal and to an Informal Concept

Let us recall that oppositional structures do apply to any kind of "oppositional phenomena": formal as well as informal [1, 5, 7, 8, 10, 11]. Let us provide an example of each. As for formal concepts, one of the most famous examples on that respect is Sesmat's "hexagon of order" (1951, cf. [7]). Each of its three red diagonals of classical bi-simplicial contradiction B_22 gives now a cuboctahedric quadri-segment B_42. We represent what becomes the middle one, namely the classical red contradiction segment B_22 relating "=" and its classical contradiction "¬=" (alias " ≠"). The figure shows the global B_42 (a cuboctahedron, slightly deformed by us) as well as its decomposition into its main sub-structures (four hexagons B_32^* and the three squares): the main idea is that

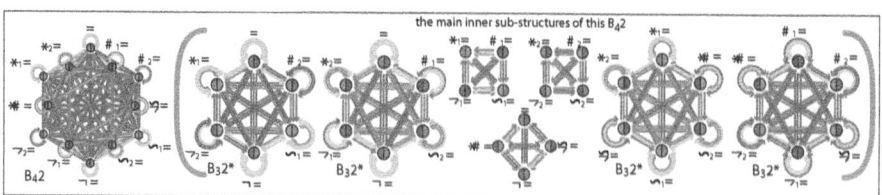

Fig. 27. Application of B_42 to a formal concept: order-theoretic equality ("=") and its negation (Color figure online)

with 4 truth-values one gets 6 varieties of assertion and 6 of negation, the 12 being taken in a complex, but meaningful oppositional-geometrical structure (see Fig. 27).

The same remarks apply to informal concepts, as with "lion" and "¬lion" (see Fig. 28).

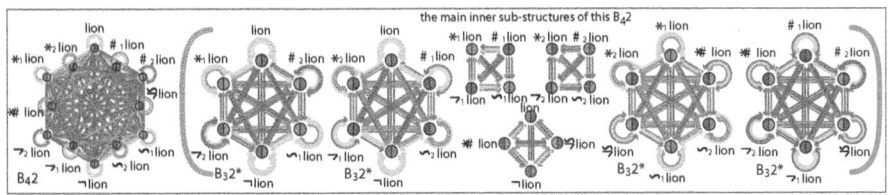

Fig. 28. An application of the B_42 to an informal concept: "lion" and its negation "¬lion"

Disclosure of Interests. The author has no competing interests to declare that are relevant to the content of this article.

References

1. Angot-Pellissier, R.: 2-opposition and the topological hexagon. In: Béziau, J.-Y., Payette, G. (eds.) The Square of Opposition. A General Framework for Cognition, pp. 351–375. Peter Lang, Bern (2012)
2. Angot-Pellissier, R.: Many-valued logical hexagons in a 3-oppositional trisimplex. In: Beziau, J.Y., Vandoulakis, I. (eds.) The Exoteric Square of Opposition. Studies in Universal Logic, pp. 333–345. Birkhauser, Cham (2022). https://doi.org/10.1007/978-3-030-90823-2_15
3. Green, T.M.: The Simplex, the Duplex and Pascal's Triangles. With Excursions into Hyperspace. CreateSpace, Charleston (2015)
4. Moretti, A.: Geometry for Modalities? Yes: Through n.Opposition Theory. In: Béziau, J.-Y., Costa-Leite, A., Facchini, A. (eds.) Aspects of Universal Logic, N. 17 of Travaux de Logique, pp. 102–145. University of Neuchâtel, Neuchâtel (2004)
5. Moretti, A.: The Geometry of Logical Opposition. Ph.D. Thesis, University of Neuchâtel, Neuchâtel, Switzerland (2009)
6. Moretti, A.: The critics of paraconsistency and of many-valuedness and the geometry of oppositions. Logic Log. Philos. **19**(1–2), 63–94 (2010)
7. Moretti, A.: Why the logical hexagon? Log. Univers. **6**(1–2), 69–107 (2012). https://doi.org/10.1007/s11787-012-0045-x
8. Moretti, A.: Was Lewis Carroll an amazing oppositional geometer? Hist. Philos. Logic **2**(5), 383–409 (2014). https://doi.org/10.1080/01445340.2014.981022
9. Moretti, A.: Arrow-hexagons. In: Koslow, A., Buchsbaum, A. (eds.) The Road to Universal Logic. Studies in Universal Logic, pp. 417–487. Birkhäuser, Cham (2015). https://doi.org/10.1007/978-3-319-15368-1_20
10. Moretti, A.: Tri-simplicial contradiction: the "Pascalian 3D Simplex" for the oppositional tri-segment. In: Beziau, J.Y., Vandoulakis, I. (eds.) The Exoteric Square of Opposition. Studies in Universal Logic, pp. 347–479. Birkhäuser, Cham (2022). https://doi.org/10.1007/978-3-030-90823-2_16

11. Pellissier, R.: «Setting» *n*-opposition. Log. Univers. **2**(2), 235–263 (2008). https://doi.org/10.1007/s11787-008-0038-y
12. Smessaert, H., Demey, L.: Logical geometries and information in the square of opposition. J. Logic Lang. Inform. **23**(4), 527–565 (2014). https://doi.org/10.1007/s10849-014-9207-y
13. LOGICAL GEOMETRY. https://logicalgeometry.org. Accessed 03 July 2024

A Way Diagrams Explain: Analysis Based on Consequence Matching

Atsushi Shimojima[1(✉)] and Dave Barker-Plummer[2]

[1] Faculty of Culture and Information Science, Doshisha University,
1-3 Tatara-Miyakodani, Kyotanabe 610-0394, Japan
ashimoji@mail.doshisha.ac.jp
[2] CSLI/Stanford University, Cordura Hall,
210 Panama Street, Stanford, CA 94305, USA
dbp@stanford.edu

Abstract. Diagrams are often used as a means to provide explanations as part of interactions between two agents. While common, this use of diagrams has not been widely examined in the literature. We show that diagrams are particularly useful in these situations because it is possible to observe and reason about an appropriately constructed diagram as a proxy for reasoning about the target domain. This is possible because of "consequence matching" which can occur when the logic of the diagram is designed to match the logic of the target domain in ways that are relevant to the explanation. We illustrate the relevant phenomena by way of examples using a range of common information graphics. We contrast the kind of diagrammatic explanations that we have identified with various kinds of "visual explanations" that have been explored in cognitive science, computer science and information design.

1 Introduction

In this paper, we will show that an important way in which diagrams can help with explanation can be identified when we adopt a particular pragmatic conception of what a good explanation is. We will give some examples where one can observe and reason about an appropriately constructed diagram as a proxy for reasoning about the target domain, providing one with an alternative way to grasp the logical relation between the explicantia and the explanandum. This is possible because of "consequence matching" which can occur when the logic of the diagram is designed to match the logic of the target domain in ways that are relevant to the explanation.

The pragmatic conception of explanatory goodness, mainly due to Achinstein [1], lets us take up certain cases of explanations where the role of diagrams is particular clear. In assessing goodness of explanation, this criterion prioritizes pragmatic factors such as knowledge, interests, and desires of the explainee, over fixed criteria such as the reference to general laws, causal factors, and unificatory potential. As a result, a wide range of discourses come into scope as cases of

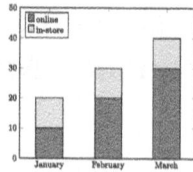

Fig. 1. Bar chart used to explain why March's sales were twice as much as January's.

explanations, making room for cases where the role of diagrams mentioned above can be identified clearly.

The specific pragmatic criterion of explanatory goodness that we adopt is the degree to which the explanation reduces surprise about the explanandum. In a typical case, the explainer (**A**) has described a situation to the explainee (**B**), and the explainee has evinced puzzlement about the situation. The explainer then attempts to provide an explanation to **B** which reduces **B**'s surprise about the situation. The goodness of **A**'s explanation is determined on the basis of how much it actually reduces **B**'s surprise. As we will discuss in Sect. 4, this criterion underlies many theories of explanatory goodness, classical [6,8] or contemporary [4,5,10].

In the following, we start with our main example of diagrammatic explanation where a bar chart is used to illustrate what we call *explanation by consequence matching* (Sect. 2). We then give further examples with connection maps and iconic tables to show the generality of this type of explanation (Sect. 3). We close the paper with a review of related work, to clarify the nature of the particular conception of explanatory goodness that we adopt, as well as the characteristics of our account of the role of diagrams in explanation (Sect. 4).

2 Explanation by Consequence Matching

Consider the following example using a bar chart.

Example 1. A store sells its merchandise at the store and online. The store manager, Sue, asks, "Were March's sales about the same as January's?" The accountant, Hiroko, answers, "Actually, they were twice as much." Sue asks, "How was it so?" In response, Hiroko shows the bar-chart in Fig. 1 to explain. "Sales at the store were 10,000 dollars both in January and March, but online sales tripled from 10,000 dollars, and that made the total sales twice as much."

The fact that March's sales is as twice as January's is a surprise to Sue, since she lacks information about online sales. Hiroko then explains how that is so, by shifting focus from the comparison of overall sales of the two months (the explanandum) to the breakdown of their sales (the explanantia), especially, to the contribution of online sales. Given that the power of an explanation consists in reduction of surprising-ness of the explanandum, an explanation will be helpful if the explicantia are less surprising than the explicandum. Here, the tripling of online sales is assumed to be less surprising than the doubling of overall sales, because Sue has not thought of the online-sale factor and thus has no expectations about its increase or decrease. In this case, the shift of focus enabled by Hiroko's explanation does reduce surprise and therefore has an explanatory force.

Perhaps Sue finds the tripling of online sales as surprising as, or even more surprising than, the doubling of total sales. In this case, Hiroko's reply would fail to be an explanation for Sue, but even then, it serves her by shifting focus to the issue of online sales. Assuming this new factor is an essential component of a complete explanation, Sue's reply constitutes the first link in whatever chain of explanations eventually reduces or eliminate surprise. It is therefore a partial explanation, although lacking an effect of reducing surprise on its own. A further explanation of the situation might involve the fact that there was an online promotion of the product, for example. Here the focus shifts away from the sales figures, and to the reasons behind the sales increase.

Now that we have seen that our sample explanation does at least contribute to an explanation, let us address the particular role the diagram plays. On our analysis, the diagram provides us with the opportunity to perform a spatial deductive inference that tracks a more abstract deductive inference required by the explanation. Figure 2 helps us to see this point. To understand Hiroko's explanation, Sue needs to make an inference along the consequence relation depicted in the lower half of the figure, where the antecedents (premises) consist of the three pieces of information in the curly bracket while the consequent (conclusion) is the piece of information to the right of $\vdash_T$. We call $\vdash_T$ the "target consequence relation" in this explanation. Crucially, there is another consequence relation, depicted in the upper part, that runs in parallel with this target relation. Its antecedents and consequent are not pieces of information about the store's sales situation—not the sales of individual months, nor their breakdown, nor comparisons of sales. Rather, they are properties of the diagram itself—lengths of the bars, lengths of their components, and comparisons of lengths. These properties of the diagram indicate particular pieces of information about the store's sales situation, by the semantic relation ($\wr$) established in the given system of bar charts. As Fig. 2 shows, this semantic relation mediates the exact match between the upper consequence relation and the lower consequence relation. We call $\vdash_S$ the "source consequence relation".

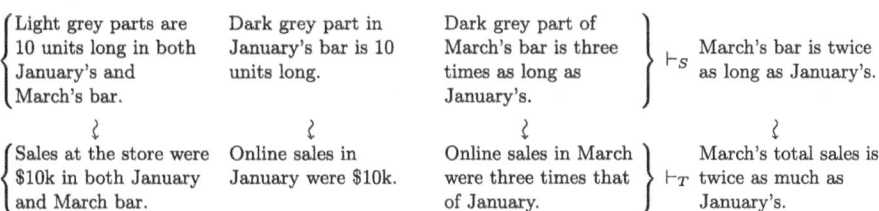

Fig. 2. Match of consequence relations provided by the system of bar charts.

Because of this match between consequence relations, we may perform a deduction along the upper consequence relation and use our grasp of this relation as the scaffolding for our grasp of the lower consequence relation. In fact, we often seem to be satisfied with the grasp of an source consequence relation, without clearly distinguishing it from the target consequence relation This would not be a problem for the purpose of checking the validity of the given explanation, as long as there is a close match between the two consequence relations.

In many cases, inferences along the upper and lower consequence relations are qualitatively different. In our example, an inference along the upper is what may be called spatial deductive inference [7], handling metric and topological relations among bounded regions, points, and lines on the perceptually accessible surface. In contrast, an inference along the lower consequence relation is more purely arithmetical, handling numbers or quantities associated with abstract constructs such as sales in different periods or sales in different outlets. Having the upper inferential route is therefore a significant addition, with great potential to facilitate our understanding of the explanation.

Notice that the upper consequence relation is not a general consequence relation. It holds only because the syntactic conventions of the present system. For example, our system of bar charts require March's and January's bars to consist exclusively of one light grey bar and one dark grey bar directly stacked with no gap. Geometrical and topological constraints together with the syntactic conventions of the bar-chart system create a "proxy logic" that tracks the logic on the represented objects.

3 Extending Examples

To sum up the discussion so far, the following are the characteristics of the class of diagrammatic explanations that we seek to clarify: (1) they explain a surprising state of affairs by describing the circumstances leading to it, thereby shifting focus to an issue or issues that the explainee have not been aware of, and (2) they facilitate the grasp of the consequence relation from the described circumstances to the state of affairs by means of a consequence relation on a diagram itself that closely matches the target consequence relation.

This type of explanation is quite common and not confined to those using bar charts. It can be performed with varieties of diagrammatic systems. In this section, we give examples using iconic tables and connection maps to demonstrate how common they are. Our analysis of additional examples also clarifies factors influencing the efficacy and significance of diagrammatic explanations.

Example 2. Jon, Ken, Bob, and nobody else work on Monday through Friday at a store. The store manager asks, "There are at least two days on which all three people work, right?" Jon replies, "Actually all work only on Fridays." The manager is alarmed, "Why is it that way?" Jon shows the work-shift table in Fig. 3 (a) to explain. "Bob and I work three days with one day off in between, and Ken works only on Thursday and Friday. So, Friday is the only day when all three of us work."

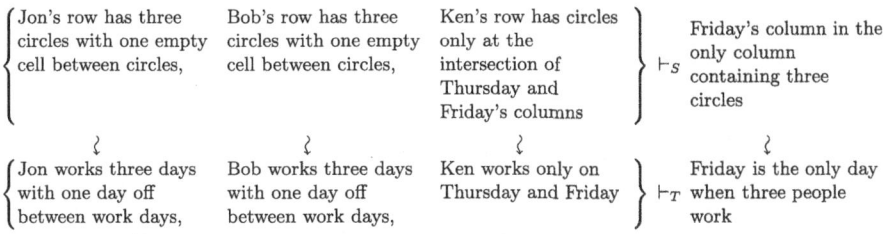

(a) Work-shift table used to explain why Friday is the only day when three people work.

$$\left\{\begin{array}{l}\text{Jon's row has three} \\ \text{circles with one empty} \\ \text{cell between circles,}\end{array}\quad \begin{array}{l}\text{Bob's row has three} \\ \text{circles with one empty} \\ \text{cell between circles,}\end{array}\quad \begin{array}{l}\text{Ken's row has circles} \\ \text{only at the} \\ \text{intersection of} \\ \text{Thursday and} \\ \text{Friday's columns}\end{array}\right\} \vdash_S \begin{array}{l}\text{Friday's column in the} \\ \text{only column} \\ \text{containing three} \\ \text{circles}\end{array}$$

$$\wr \qquad\qquad \wr \qquad\qquad \wr \qquad\qquad\qquad\qquad \wr$$

$$\left\{\begin{array}{l}\text{Jon works three days} \\ \text{with one day off} \\ \text{between work days,}\end{array}\quad \begin{array}{l}\text{Bob works three days} \\ \text{with one day off} \\ \text{between work days,}\end{array}\quad \begin{array}{l}\text{Ken works only on} \\ \text{Thursday and Friday}\end{array}\right\} \vdash_T \begin{array}{l}\text{Friday is the only day} \\ \text{when three people} \\ \text{work}\end{array}$$

(b) Match of consequence relations provided by the system of iconic tables.

Fig. 3. Work shift example

This time, the lower turnstile $\vdash_T$ in Fig. 3b stands for the consequence relation on work days of a set of people and numbers of workers on individual days, whereas the upper turnstile $\vdash_S$ stands for the consequence relation on the positions of circles in the table.

The grasp of the latter is largely a matter of visual routines applied to the circles and cells of the table, [12]. We inspect the particular placement of circles in the cells of the table, observing that the antecedents and the consequent hold in it. To check that the consequent is in fact a consequence of the antecedents, we mentally move the circles to see if there are alternative counterexample configurations. Finding no such ways, we conclude that the consequent is in fact a consequence of the antecedents. We can thus grasp the upper consequence relation.

In contrast, the grasp of the lower consequence relation could be a fairly difficult task if tried only with Jon's verbal explanation, i.e., without the supporting table. How many possible work-shifts there are, where all the antecedents of the lower consequence relation hold? How difficult it would be to correctly determine that there is only one and that the consequent holds in it? At least this is required for you to determine the consequent is in fact a consequence of the antecedents. Such reasoning could be cognitively difficult. The benefit of the diagram here is that we can leverage our spatial abilities to perform similar reasoning about the diagram, and exploit consequence matching to underwrite the inference of interest.

A comparison of the exact cognitive loads involved in these inference tasks is beyond the scope of this paper. Our point is only that the gap between them is an important factor that modulates the efficacy of the diagrammatic explanation, or more specifically, the relevance of the diagram for the explanatory process. Thus, our analysis produces a theoretical prediction about when a diagrammatic explanation based on consequence relation match should be used. It is when one can find or design a diagram for which the visual routines necessary to grasp the upper consequence relation are simple and significantly easier than making an inference along the lower route.

Example 3. The subway in a city has stations A, B, C, D, and E. The passenger asks, "The stations are directly connected in the order of A, B, C, D, and E, and C is in their center. So, C must have the shortest access to other stations, right?" The station employee replies, "Actually, B has a better access to other stations." The passenger wonders, "Why is it so?" The employee shows the connection map in Fig. 4a to explain. "Yes, A, B, C, D, and E are directly connected in this order, but a new direct connection has been built between B and E, so B has a better access to other stations."

Generally, a good explanation has a minimal set of explanantia to entail the explanandum. Every explanans of such an explanation is *relevant*, meaning that the explanandum is entailed only when it is included as an explanans, and if it is removed from the set of explanantia, the explanandum is no longer entailed. For the case of example 4, this means that removing any member of the antecedents in the lower part of Fig. 4(b) results in non-consequence of the consequent. Since the point of explanation 4 is to shift the passenger's focus to a newly constructed connection between stations B and E, it is particularly important to check the relevance of this particular explanans. As it is stated in the second antecedent of the consequence relation shown in the lower part of Fig. 4(b), recognizing the relevance of this factor is to recognize this consequence relation along with the non-consequence relation shown in the lower part of Fig. 4(c).

Now as Figs. 4(b) and 4(c) already indicate, these relations are tracked by particular consequence and non-consequence relations on the part of the diagram. Here we can see another contribution that the connection map is making in this explanation. It helps one to check the *relevance* of an explanans by providing alternative inferential routes of spatial inference, to compare the cases where the explanantia include and do not include the explanans in question. The importance of this contribution is then modulated by the gap of difficulty between when we make inference along these alternative routes and when we take lower routes of non-spatial inference.

Example 3 is chosen for discussion because it is a case where this gap is wide, at least apparently. The visual routines involved in taking the upper routes mainly consist of following different paths on the connection map, and the comparison of the cases where the second antecedent is considered and ignored is done by comparing the total numbers of line segments one encounters when different sets of paths are followed. No mental manipulations on the elements of the diagrams are involved.

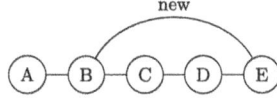

(a) Connection map used to explain why station B has the shortest access to other stations.

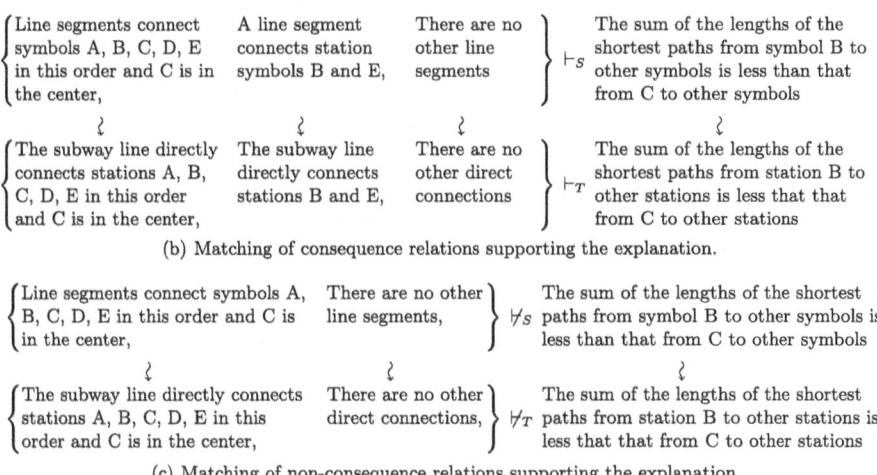

(b) Matching of consequence relations supporting the explanation.

(c) Matching of non-consequence relations supporting the explanation

Fig. 4. Station connections example.

Thus, our analysis based on consequence tracking reveals another factor affecting the efficacy of the explanations of the sort we are concerned with. This time, it is a factor affecting how well the diagram helps us recognize the relevance of explanantia in an explanation, rather than how well it helps us recognize the consequence relation from the explanantia to the explanandum.

4 Related Work

Notion of Explanatory Goodness. First, we contextualize the notion of explanation that we have appealed to in our illustration of diagrammatic explanations. "Reduction of surprise" as a measure of good explanation has been fundamental to many theories of explanation, though not stated using this exact term. Examples include a classical theory of explanation by Hempel and Oppenheim [8], where a good explanation makes an "unexpected", "unpredictable" characteristic of an object predictable by a general laws and antecedent conditions (p. 147). Also, many of the probabilistic approaches to explanatory power [4,6,10] can be considered attempts to quantify the degree to which a hypothesis reduces surprise associated with the explanandum. Glass gives a summary of various formula that has been proposed for this purpose, [5]. In these theories, subjectivism often associated with the notion of surprise is avoided by the assumption that how much an explanandum is surprising is not arbitrarily set by the explainer or the audience, but is objectively determined by comparing it with the theory or knowledge available to these people at the time of inquiry. Theoretically, a certain amount of surprise can be attributed to an explanandum, irrespective of the surprise subjectively felt by the explainer and explainee.

Our choice of examples of explanation (and discussion of them) also draws on the pragmatic conception of explanation put forward by Achinstein [1]. Under this conception, usual criteria of sound explanations such as the introduction of general laws, causal factors, and unification of explananda are " 'prima facie' virtues" (p. 291). They may or may not contribute to the goodness of explanation depending on how well they serve the interlocutors' intellectual necessity determined by their interest, desire, and even values at that occasion.

In fact, all our sample explanations provide only "circumstances leading up to" their explananda, with no general laws explicitly mentioned. The rationale is that the explainee has little or no expectation for the explanation to have a predictive force for future. For example, in Example 1 Sue is only interested in the reason for the jump of sales in a *particular* month of the year. She does not expect or desire any general law to be supplied that may let her predict a similar jump in the future. Thus, Hiroko's reply is just a sufficient explanation for her. Under the pragmatic conception, the reference to general laws and completeness are not absolute requirements for explanation. It puts a large additional field of explanations into our view, which was not possible under a classical, non-pragamatic conception of explanation such as Hempel-Oppenheim's [8]. As our examples suggest, it may be in this field that the contribution of diagrams to explanation can be made most transparent.

Role of Diagrams in Explanation. Existing work in cognitive science, and learning research in particular, describes the use of diagrams as explanations. Building on work describing the advantages of generating self explanations in learning [3], the effect of the modality of those explanations has been further investigated. In one experiment, [2], students were asked to (1) learn material on a new subject, (2) take a test, (3) make a self-explanation of the material, and then (4) take a second test. The condition of the experiment differed on the modality of the explanation given in step 3, either textual, or diagrammatic. An increase in performance between the tests in steps 2 and 4 was observed in both cases, but was larger in the diagrammatic modality.

In [9], an information graphic is used to display information about some domain, and a system can answer questions about the situation. In addition to providing an answer to questions either about the domain or about the diagram, the system offers an explanation of how the answer was obtained. These explanations exclusively make reference to the parts of the graphic that were referenced in the production of the answer. Thus it implicitly relies on the user understanding that the relevant consequence relations match in the way that we have explained.

Tufte cites a wide varieties of graphical representations that can be used for explanatory purposes, [11]. He takes the term "explanation" broadly, as "presenting information about motion, process, mechanism, cause and effect" (p.9).

Thus, not all of his examples are graphics to be used in communications intended to reduce surprise. Yet, many of them seem to be apt for this purpose. His contribution is then to broaden our view of possible explanatory strategies that can be taken with the help of graphics, which almost certainly go beyond the discussion in this paper.

5 Conclusion

In this paper we have analyzed a particular use of diagrams in their role of supporting explanation. We have in minds situations in which one is surprised at a certain state of affairs, and a diagram helps one to grasp not only the content of the explanantia, but their logical relation to the explanandum.

Consequence matching is the fundamental reason why diagrams can play such a role. Consequence matching makes it possible for inferences about the diagram itself to stand in for inferences about the target domain. Systems of diagrams tend to have this property because they evolved among the communities of users precisely to facilitate such consequence tracking. In performing inference about the diagram, we can use our visual-spatial abilities, which can result in the inferences requiring a lower cognitive load relative to inference about the target domain. The critical properties of a useful explanatory diagram, then, are both that it supports consequence tracking, and also that inference about the diagram should involve less effort than inference about the domain.

An additional benefit of diagrams as explanations is that use of the diagram also facilitates determining the relevance of explanatory information, by facilitating the visualization of inferences in which the information is not used. The counterfactual inferences enable explainees to see how this particular part of the explanation contributes to the conclusion and why it is necessary.

A limitation of this paper is that it takes the matching of consequence relation as given. The relation is established and maintained through incremental design or cultural evolution within the community of users. However, in a given diagrammatic system, the logic on the diagram side is almost sound and complete with respect to the logic on the target side. Such extensive matches hold not only within an individual diagrammatic system, but across a variety of diagrammatic systems. Thus, there seem factors other than cultural evolution and incremental design that contribute to the phenomenon. Investigations into these factors are left for future research.

References

1. Achinstein, P.: The pragmatic character of explanation. In: PSA: Proceedings of the Biennial Meeting of the Philosophy of Science Association, pp. 274–292. Cambridge University Press (1984)
2. Bobek, E., Tversky, B.: Creating visual explanations improves learning. Cogn. Res. Principles Implications **1**(1), 27 (2016). https://doi.org/10.1186/s41235-016-0031-6

3. Chi, M.T., De Leeuw, N., Chiu, M.H., Lavancher, C.: Eliciting self-explanations improves understanding. Cogn. Sci. **18**(3), 439–477 (1994). https://doi.org/10.1016/0364-0213(94)90016-7, https://www.sciencedirect.com/science/article/pii/0364021394900167
4. Crupi, V., Tentori, K.: A second look at the logic of explanatory power (with two novel representation theorems). Philos. Sci. **79**(3), 365–385 (2012)
5. Glass, D.H.: How good is an explanation? Synthese **201**(2), 1–26 (2023)
6. Good, I.J.: Corroboration, explanation, evolving probability, simplicity and a sharpened razor. Br. J. Philos. Sci. **19**(2), 123–143 (1968)
7. Hamami, Y., Mumma, J., Amalric, M.: Counterexample search in diagram-based geometric reasoning. Cogn. Sci. **45**(4), e12959 (2021)
8. Hempel, C.G., Oppenheim, P.: Studies in the logic of explanation. Philos. Sci. **15**(2), 135–175 (1948)
9. Kim, D.H., Hoque, E., Agrawala, M.: Answering questions about charts and generating visual explanations. In: Proceedings of the 2020 CHI Conference on Human Factors in Computing Systems, pp. 1–13. CHI 2020, Association for Computing Machinery, New York, NY, USA (2020). https://doi.org/10.1145/3313831.3376467
10. Schupbach, J.N., Sprenger, J.: The logic of explanatory power. Philos. Sci. **78**(1), 105–127 (2011)
11. Tufte, E.R., Robins, D.: Visual Explanations. Graphics Press, Cheshire (1997)
12. Ullman, S.: Visual routines. Cognition **18**, 97–159 (1984)

Euler Diagrams, Aristotelian Diagrams and Syllogistics

Lorenz Demey[1](✉)[iD] and Hans Smessaert[2][iD]

[1] Center for Logic and Philosophy of Science, KU Leuven, Kardinaal Mercierplein 2, 3000 Leuven, Belgium
`lorenz.demey@kuleuven.be`
[2] Department of Linguistics, KU Leuven, Blijde-Inkomststraat 21, 3000 Leuven, Belgium
`hans.smessaert@kuleuven.be`

Abstract. Euler diagrams and Aristotelian diagrams are two of the most important types of diagrams to visualize (relations between) sets. We have previously shown that Euler diagrams for two sets systematically give rise to various Aristotelian diagrams, such as classical squares of opposition. In this paper, we expand this analysis to Euler diagrams for three sets, and show that they give rise to various kinds of hexagons of opposition as well. This move from two to three sets is philosophically well-motivated and technically non-trivial. On the philosophical side, there is a connection with syllogistics, since syllogisms consist of three terms/sets. On the technical side, moving from two to three sets requires us to take the phenomenon of Boolean subtypes into account.

Keywords: Euler diagram · Aristotelian diagram · syllogistics · logical geometry · square of opposition · hexagon of opposition · syllogism

1 Introduction

Throughout history, philosophers, mathematicians and other thinkers have devised various types of diagrams to visualize sets and the various kinds of logical relations that may hold between them. Among the most important such diagram types, there are Euler diagrams,[1] Hasse diagrams and Aristotelian

---

[1] An important subcase of Euler diagrams concerns the Venn diagrams. Given n sets, there are 2^n so-called zones/minimal regions. While an Euler diagram need not show each of these zones (e.g., because some zones are known to be empty anyway), a Venn diagram is required to show *all* zones, and to use shading to explicitly indicate that a given zone is known to be empty (thus also allowing us to express partial knowledge) [23,24]. Venn diagrams can be viewed as a proper subclass of Euler diagrams, i.e., "every Venn diagram is an Euler diagram, but not every Euler diagram is a Venn diagram" [41, p. 134]. Venn diagrams will play a crucial role later in the paper.

The first author holds a Research Professorship (BOFZAP) from KU Leuven. This research was funded through the KU Leuven research project 'BITSHARE: Bitstring Semantics for Human and Artificial Reasoning' (3H190254, 2019–2023).

diagrams.[2] Interestingly, each of these labels turn out to be historically misleading: Euler and Hasse diagrams can already be found much earlier than the works of Leonhard Euler (1707–1783) and Helmut Hasse (1898–1979) [17,30–32,40]; vice versa, although Aristotelian diagrams have their theoretical roots in the logical works of Aristotle (384–322 BCE), the actual diagrams were drawn only in the 2nd century by Apuleius of Madaura [21,34].[3] Each of these diagram types is well-understood on its own, and over the past decade, much research has been done on their various interconnections, i.e., the relation between Aristotelian and Euler diagrams [11,16,33], between Aristotelian and Hasse diagrams [2,13,15,43], and finally, between Hasse diagrams and Euler diagrams [3,4,37,38].

In this paper we will delve deeper into the interconnection between Euler and Aristotelian diagrams. Previous work in this area has been historically motivated [11,33], but also includes a more systematic study [16]. This research has hitherto remained limited to Euler and Aristotelian diagrams for *two* sets (and their complements). The overarching goal of the present paper is to study the interaction between Euler and Aristotelian diagrams for *three* sets (and their complements). This move from two to three sets is well-motivated and non-trivial.

On the one hand, it bears emphasizing that this paper is philosophically *well-motivated*. After all, one might think that this research line is merely cumulative in nature: after previously considering (Euler and Aristotelian) diagrams for two sets, we now move to three sets, and future papers could be dedicated to four sets, five sets, and so on *ad nauseam*... However, the case of three-set diagrams is of particular interest, because these are precisely the diagrams that allow us to draw a connection with the logical system of syllogistics. Indeed, a syllogism is required to consist of precisely three terms/sets (traditionally called the 'major term', the 'minor term' and the 'middle term') [26, p. 143].[4] A typical example is the famous Barbara syllogism: 'all M are P, all S are M, so all S are P'.

On the other hand, the results presented in this paper are technically *non-trivial*. It is well-known in logical geometry that Aristotelian diagrams can have multiple Boolean subtypes [8,14] (this will be explained later in the paper). As long as we restrict ourselves to two sets, the only Aristotelian diagrams we encounter are pairs of contradictories (PCDs), degenerate squares, and classical

---

[2] Aristotelian diagrams are traditionally considered to visualize *propositions* (and the relations between them), rather than *sets* (and the relations between them). However, note that (i) propositions can themselves be viewed as sets, viz., sets of possible worlds [45], and (ii) Aristotelian diagrams are most naturally defined relative to some Boolean algebra, which can consist of sets just as well as of propositions [10,16].

[3] See [5] for a dissenting voice, ascribing the square of opposition directly to Aristotle.

[4] This specific number of terms even lies at the source of one of the traditional fallacies: the *quaternio terminorum*. If one of the terms in a syllogism is ambiguous between two distinct meanings, then we are dealing with a total number of four, rather than three, terms/meanings/sets, which renders the syllogism invalid [6, p. 206ff.].

squares of opposition [16], which all have a unique Boolean complexity.[5] In other words, as long as we restricted ourselves to two-set diagrams, the entire issue of Boolean subtypes simply did not arise. However, once we move to three sets, we will also encounter various hexagons of opposition, such as the Jacoby-Sesmat-Blanché (JSB) hexagon [1,25,42] and the unconnectedness-4 (U4) hexagon [28], which do have multiple Boolean complexities.[6] In other words, since this paper deals with three-set diagrams, it will also need to take into account the issue of Boolean subtypes, as an additional layer of complexity for our analysis.

The paper is organized as follows. Section 2 briefly recapitulates the previous work on Euler and Aristotelian diagrams for two sets, and then describes a divide and conquer strategy to move to three sets, which looks promising but is ultimately found wanting. Section 3 then presents a better strategy, based on Venn diagrams, which does allow us to systematically describe the relationship between Euler and Aristotelian diagrams for three sets. Finally, Sect. 4 sketches how such three-set diagrams are related to the logical system of syllogistics.

2 The Divide and Conquer Strategy

Given a domain of discourse D, it is well-known that each pair (X, Y) of non-trivial[7] sets stands in exactly one of the following seven relations [44]:

1. contradiction (CD): $X \cap Y = \emptyset$ and $X \cup Y = D$, i.e., $X = \overline{Y}$,
2. contrariety (C): $X \cap Y = \emptyset$ and $X \cup Y \neq D$, i.e., $X \subset \overline{Y}$,
3. subcontrariety (SC): $X \cap Y \neq \emptyset$ and $X \cup Y = D$, i.e., $X \supset \overline{Y}$,
4. bi-implication (BI): $X \subseteq Y$ and $X \supseteq Y$, i.e., $X = Y$,
5. left-implication (LI): $X \subseteq Y$ and $X \not\supseteq Y$, i.e., $X \subset Y$,
6. right-implication (RI): $X \not\subseteq Y$ and $X \supseteq Y$, i.e., $X \supset Y$,
7. unconnectedness (UN): $X \cap Y \neq \emptyset$ and $X \cup Y \neq D$ and $X \not\subseteq Y$ and $X \not\supseteq Y$.

In [16] we systematically investigated how the two-set Euler diagrams for each of these seven binary relations give rise to well-defined Aristotelian diagrams. For example, the Euler diagram for $LI(A, B)$ in Fig. 1(b) gives rise to the classical square of opposition in Fig. 1(e). In general, see [16, Figs. 3–9].

Let us now move from two to three sets. The results above suggest an obvious 'divide and conquer' strategy for systematically transforming three-set Euler diagrams into well-defined Aristotelian diagrams. We start from (a) any Euler

---

[5] In particular, PCDs always have Boolean complexity 2, classical squares of opposition always have Boolean complexity 3, and degenerate squares always have Boolean complexity 4 [8,14].

[6] In particular, JSB hexagons can have Boolean complexities 3 and 4 [14,36], while U4 hexagons can have Boolean complexities 4 and 5 [12]. This issue is not restricted to hexagons of opposition; for example, Buridan octagons (which we encounter when studying four-set Euler diagrams) can have Boolean complexities 4, 5 and 6 [9].

[7] Given the domain of discourse D, a set X is said to be *non-trivial* iff $\emptyset \neq X \neq D$.

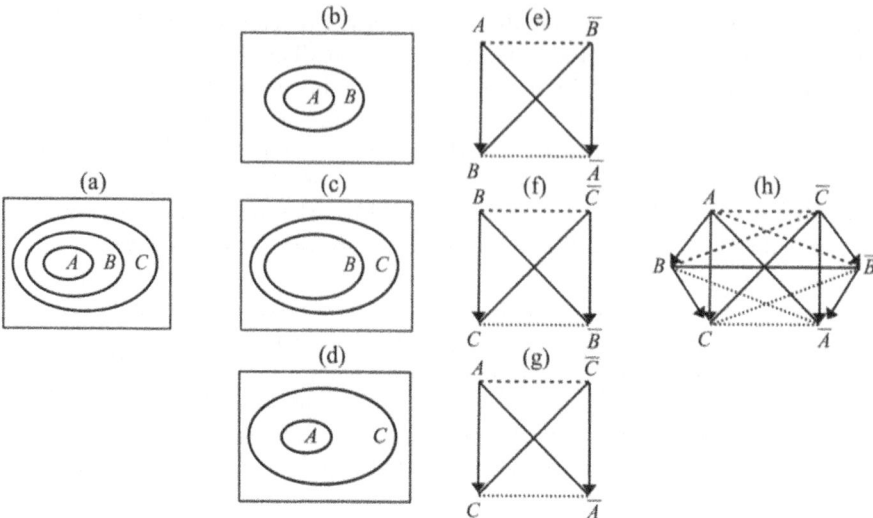

Fig. 1. Using the divide and conquer strategy to turn the Euler diagram for $A \subset B \subset C$ into a Sherwood-Czeżowski hexagon. As usual, contradiction, contrariety, subcontrariety and subalternation are visualized as resp. solid, dashed, dotted lines and arrows.

diagram for three non-trivial sets A, B and C, and decompose it into subdiagrams (b) for A and B, (c) for B and C, and (d) for A and C. Next, we use the results from [16] to transform these two-set diagrams into Aristotelian diagrams (e) for A and B, (f) for B and C, and (g) for A and C, respectively. Finally, we use these last three diagrams to compose (h) a single three-set Aristotelian diagram, which corresponds precisely to the original three-set Euler diagram. As an easy example, Fig. 1 shows how the Euler diagram for $A \subset B \subset C$ gives rise to a so-called Sherwood-Czeżowski hexagon of opposition [7,27,29]. The labels (a–h) in Fig. 1 correspond exactly to the steps laid out before.

To do this for *all* three-set diagrams, note that (A, B) stands in exactly one of 7 relations, and similarly for (B, C) and for (A, C), so in total, we have to consider $7 \times 7 \times 7 = 343$ combinations of pairwise relations. Out of these 343 combinatorial possibilities, many turn out to be set-theoretically inconsistent; for example, if we have $LI(A, B)$ and $LI(B, C)$, then we must have $LI(A, C)$, i.e., none of the six other relations is possible for (A, C). A tedious but straightforward calculation shows that out of the 343 combinations, 102 cases represent genuine set-theoretical possibilities.[8] These 102 three-set Euler diagrams give

---

[8] Historically speaking, this approach is analogous to what Gergonne already did in 1817 [20,22]. He only assumed the sets to be non-empty ($X \neq \emptyset$), rather than non-trivial ($\emptyset \neq X \neq D$), and therefore worked with 5 relations instead of 7 (using our notation, Gergonne's relations are BI, LI, RI, $CD \cup C$ and $SC \cup UN$). Gergonne thus considered $5 \times 5 \times 5 = 125$ combinations of pairwise relations, and showed that 54 of them represent genuine set-theoretical possibilities [20, pp. 211–213].

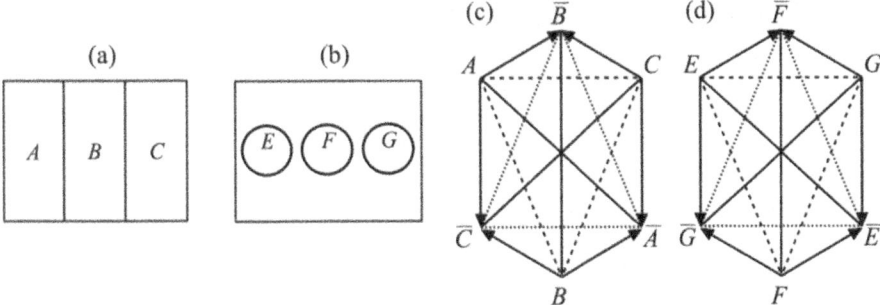

Fig. 2. Two Euler diagrams and their corresponding (strong and weak) JSB hexagons.

rise to the following Aristotelian diagrams (concrete examples will be provided in Sect. 3; formal definitions of all these diagram types can be found in [19]):

- 4 pairs of contradictories,
- 24 classical squares of opposition,
- 6 degenerate squares of opposition,
- 8 Jacoby-Sesmat-Blanché hexagons,
- 23 Sherwood-Czeżowski hexagons,
- 24 unconnectedness-4 hexagons,
- 12 unconnectedness-8 hexagons,
- 1 unconnectedness-12 hexagon.

The divide and conquer strategy seems to work quite well. However, it faces one major technical difficulty: it does not allow us to deal with the fact that hexagons of opposition can have multiple Boolean subtypes.[9] For example, the Euler diagrams in Fig. 2(a–b) have the same configuration of Aristotelian relations among their three sets: A, B and C are pairwise contrary, and so are E, F and G. Consequently, the divide and conquer strategy transforms them into the same type of Aristotelian diagram, viz., the Jacoby-Sesmat-Blanché (JSB) hexagons in Fig. 2(c–d), respectively. However, the JSB hexagon in Fig. 2(c) is called *strong* (and said to have Boolean complexity 3), since $A \cup B \cup C = D$, whereas the JSB hexagon in Fig. 2(d) is called *weak* (and said to have Boolean complexity 4), since $E \cup F \cup G \neq D$.[10] In a truly comprehensive analysis, we want to classify not only which Euler diagrams give rise to which Aristotelian diagrams, but also which Boolean subtypes these resulting Aristotelian diagrams belong to (e.g., a JSB hexagon of Boolean complexity 3 vs 4; a U4 hexagon of Boolean complexity 4 vs 5, etc.). The divide and conquer strategy does not have the expressive power that is required for this task.

---

[9] We did not encounter this issue in [16], since there, we only dealt with PCDs, classical squares and degenerate squares, which all have a unique Boolean complexity (recall Footnote 5). We did not encounter this problem in the example in Fig. 1 either, since Sherwood-Czeżowski hexagons have a unique Boolean complexity as well (viz., 4).

[10] Demey [11] already discussed a concrete historical example of this situation, without emphasizing its theoretical significance. In particular, Fig. 6(a–b) of [11, p. 196] is a three-set Euler diagram that gives rise to a *strong* JSB hexagon, while Fig. 10(a–b) of [11, p. 201] is a three-set Euler diagram that gives rise to a *weak* JSB hexagon.

3 Using Venn Diagrams to Obtain a Complete Account

To develop a new, more fine-grained account of three-set Euler and Aristotelian diagrams, we start by having another look at the theoretical foundations of our analysis of two-set diagrams. We started Sect. 2 with the observation that two non-trivial sets stand in exactly one of seven relations, each of which gives rise to a well-defined Aristotelian diagram. In [16,44], this is proved as follows:

1. Given two non-trivial sets A and B, we draw an Euler diagram that shows A and B as independently as possible, i.e., containing all possible intersections of $A/\overline{A}$ and $B/\overline{B}$ [44, p. 536, Fig. 4]. In the present paper, this is shown as Fig. 3(a). Note that this is a *Venn diagram*.
2. Since we start from 2 sets, the Venn diagram contains $2^2 = 4$ zones/minimal regions, viz., $A \cap B$, $A \cap \overline{B}$, $\overline{A} \cap B$ and $\overline{A} \cap \overline{B}$; cf. the labels 1–4 in Fig. 3(a).
3. Each of these 4 zones can either be empty or not, yielding $2^4 = 16$ possibilities. Out of these 16 possibilities, 9 entail that A and/or B is trivial after all, and thus have to be ruled out [44, p. 542–543, Theorem 2].
4. The remaining $16 - 9 = 7$ cases represent precisely the 7 possible relations that two non-trivial sets A and B can stand in. These can be visualized as Euler diagrams and also as Aristotelian diagrams. These 7 cases comprise 2 PCDs (for *CD* and *BI*), 4 classical squares of opposition (for *C*, *SC*, *LI* and *RI*), and 1 degenerate square of opposition (for *UN*) [16, Section 3].

This strategy via Venn diagrams can naturally be generalized from two to three sets. The major advantage of this approach is that the 'detour' via a Venn diagram allows us to take Boolean subtypes into account.[11] More concretely:

1. Given three non-trivial sets A, B and C, we draw an Euler diagram that shows A, B and C as independently as possible, i.e., containing all intersections of $A/\overline{A}$, $B/\overline{B}$ and $C/\overline{C}$. Again, this is a *Venn diagram*; cf. Figure 3(b).
2. Since we start from 3 sets, the Venn diagram contains $2^3 = 8$ zones, viz., $A \cap B \cap C$, $A \cap B \cap \overline{C}$, $A \cap \overline{B} \cap C$, $A \cap \overline{B} \cap \overline{C}$, $\overline{A} \cap B \cap C$, $\overline{A} \cap B \cap \overline{C}$, $\overline{A} \cap \overline{B} \cap C$ and $\overline{A} \cap \overline{B} \cap \overline{C}$; cf. the labels 1–8 in Fig. 3(b).
3. Each of these 8 zones can either be empty or not, yielding $2^8 = 256$ possibilities. Another tedious but straightforward calculation shows that out of these 256 possibilities, 63 entail that A and/or B and/or C is trivial after all, and thus have to be ruled out (cf. the Appendix of this paper).

---

[11] The tight connection that Boolean considerations have with Venn diagrams (more so than with Euler diagrams) has recently also been emphasized by Moktefi and Lemanski: "instead of orienting to *syllogistic* like *Euler* diagrams, *Venn* applied *Boolean algebra*" [35, p. 887, emphases added]. The connection between Euler diagrams, Aristotelian diagrams and syllogistics will be explored further in Sect. 4.

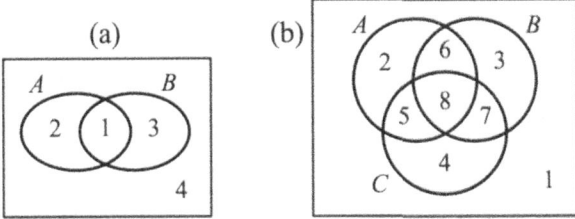

Fig. 3. Venn diagrams for two and three sets, incl. numeric labels for the zones.

4. The remaining 256 − 63 = 193 cases represent precisely the 193 possible configurations that three non-trivial sets A, B and C can stand in. These can be visualized as Euler diagrams and also as Aristotelian diagrams. These 193 cases comprise the following Aristotelian diagrams (cf. the Appendix):
    – 4 pairs of contradictories: all of them of Boolean complexity (BC) 2,
    – 24 classical squares of opposition: all of them of BC 3,
    – 6 degenerate squares of opposition: all of them of BC 4,
    – 16 Jacoby-Sesmat-Blanché hexagons: 8 of BC 3 and 8 of BC 4,
    – 24 Sherwood-Czeżowski hexagons: all of them of BC 4,
    – 48 unconnectedness-4 (U4) hexagons: 24 of BC 4 and 24 of BC 5,
    – 36 unconnectedness-8 (U8) hexagons: 24 of BC 5 and 12 of BC 6,
    – 35 unconnectedness-12 (U12) hexagons:
      2 of BC 4, 8 of BC 5, 16 of BC 6, 8 of BC 7, and 1 of BC 8.

All 256 cases are described in detail in the Appendix. In the remainder of this section, we will discuss examples of all (sub)types of Aristotelian diagrams and the Venn/Euler diagrams that give rise to them. For each example, we show how the Venn diagram described in the Appendix (as a configuration of 8 empty and non-empty zones) gives rise to an Euler diagram (which only shows the non-empty zones), and then determine the corresponding Aristotelian diagram. The latter's Boolean complexity corresponds to the number of non-empty zones in the Venn diagram, or equivalently, to the total number of zones in the Euler diagram.[12] In each Euler diagram, we include the zone numbers from Fig. 3(b), in order to facilitate the comparison with its description as a Venn diagram in the Appendix. Some Venn diagrams are much easier to turn into natural-looking Euler diagrams than others; in particular, in some cases we are forced to violate some of the so-called 'well-formedness properties' of Euler diagrams, such as *no multiple points* (i.e., no more than two curves should meet at a single point) and *no concurrency* (i.e., curve segments should not be concurrent) [18,41]. In general, when a type of Aristotelian diagrams has multiple Boolean subtypes (e.g., the JSB hexagons, which can be of Boolean complexities 3 and 4), the

---

[12] This is an interesting visual perspective on the notion of Boolean complexity, which has hitherto mainly been studied in logical geometry from a more abstract (logical/algebraic) perspective [8,15].

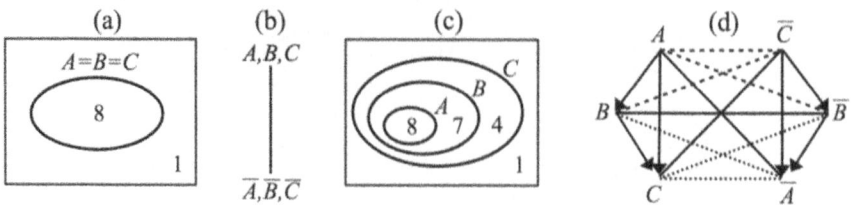

Fig. 4. Euler diagrams giving rise to a PCD and an SC hexagon.

cases with a higher Boolean complexity are easier[13] to visualize as an Euler diagram than those with a lower Boolean complexity. Therefore, in the sequence of examples below, we first discuss the case of higher Boolean complexity (e.g., a JSB hexagon of Boolean complexity 4) and only afterward that of lower Boolean complexity (e.g., a JSB hexagon of Boolean complexity 3).

PCD (of Boolean complexity 2). Case 130 from the Appendix involves zones 1 and 8 being non-empty, and all other zones being empty. It is shown as an Euler diagram in Fig. 4(a). Since zones 2 and 5 are empty, we have $A \subseteq B$, and since zones 3 and 7 are empty, we have $A \supseteq B$, and thus $A = B$, i.e., $BI(A, B)$. Completely analogously, we find $BI(B, C)$ and $BI(A, C)$. In total, this yields the PCD in Fig. 4(b).

Note that the Euler diagram in Fig. 4(a) violates *no concurrency*, since the curves for A, B and C entirely coincide with each other. This is of course a direct consequence of the fact that $A = B = C$, which is also reflected in the fact that the Aristotelian diagram in Fig. 4(b) is a PCD with 3 labels at each of its vertices (rather than a hexagon with only 1 label per vertex). Analogous remarks apply to the classical and degenerate squares below (cf. Fig. 5).

SC hexagon (of Boolean complexity 4). Case 148 from the Appendix involves zones 1, 4, 7 and 8 being non-empty, and all other zones being empty. The corresponding Euler diagram is shown in Fig. 4(c), and does not violate any well-formedness properties. Since zones 2 and 5 are empty, we have $A \subseteq B$, and since zone 7 is non-empty, we have $A \not\supseteq B$, and thus $LI(A, B)$. Completely analogously, we find $LI(B, C)$ and $LI(A, C)$. In total, this yields the SC hexagon in Fig. 4(d).[14]

Classical square (of Boolean complexity 3). Case 146 from the Appendix involves zones 1, 4 and 8 being non-empty, and all other zones being empty. It is shown as an Euler diagram in Fig. 5(a). Since zones 2, 3, 5 and 7 are empty, we have $BI(A, B)$. Since zones 3 and 6 are empty, we have $B \subseteq C$, and since zone 4 is non-empty, we have $B \not\supseteq C$, and thus $LI(B, C)$. Analogously, we find $LI(A, C)$. In total, this yields the classical square in Fig. 5(b).

---

[13] I.e., violating fewer well-formedness properties.
[14] Note that this is precisely the example that we already dealt with using the divide and conquer strategy in Sect. 2; in particular, cf. Fig. 1(a, h).

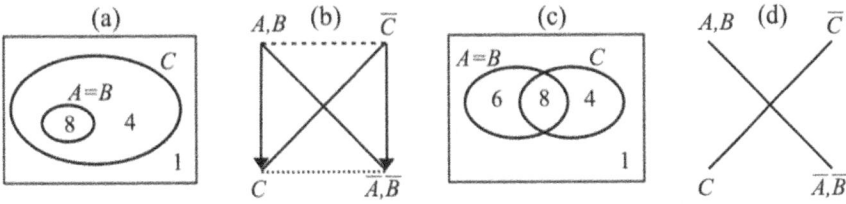

Fig. 5. Euler diagrams giving rise to a classical and a degenerate square

Degenerate square (of Boolean complexity 4). Case 150 from the Appendix involves zones 1, 4, 6 and 8 being non-empty, and all other zones being empty. It is shown as an Euler diagram in Fig. 5(c). Since zones 2, 3, 5 and 7 are empty, we have $BI(A, B)$. Since zones 1, 4, 6 and 8 are non-empty, we find that resp. $\overline{B} \cap \overline{C}$, $\overline{B} \cap C$, $B \cap \overline{C}$ and $B \cap C$ are non-empty, and thus $UN(B, C)$. Completely analogously, we find $UN(A, C)$. In total, this yields the degenerate square in Fig. 5(d).

JSB hexagon of Boolean complexity 4. Case 241 from the Appendix involves zones 1, 2, 3 and 4 being non-empty, and all other zones being empty. The corresponding Euler diagram is shown in Fig. 6(a), and does not violate any well-formedness properties. Since zones 6 and 8 are empty, we have $A \cap B = \emptyset$, and since zones 1 and 4 are non-empty, we have $A \cup B \neq D$, and thus $C(A, B)$. Completely analogously, we also find $C(B, C)$ and $C(A, C)$. In total, this yields the JSB hexagon in Fig. 6(b).

JSB hexagon of Boolean complexity 3. Case 113 from the Appendix involves zones 2, 3 and 4 being non-empty, and all other zones being empty. The corresponding Euler diagram is shown in Fig. 6(c). This case (113) is exactly like the one before (241), except that zone 1 is now empty. This means that the Euler diagram in Fig. 6(c) can be viewed as the result of removing zone 1 from the Euler diagram in Fig. 6(a). As a result, the new Euler diagram in Fig. 6(c) violates the wellformedness property of *no concurrency*, since the curves for A and B (and those for B and C) are partially concurrent. Just like in the previous case, we find $C(A, B)$, $C(B, C)$ and $C(A, C)$, and thus again obtain the JSB hexagon in Fig. 6(b).[15]

U4 hexagon of Boolean complexity 5. Case 249 from the Appendix involves zones 1, 2, 3, 4 and 5 being non-empty, and all other zones being empty. The corresponding Euler diagram is shown in Fig. 7(a), and does not violate any well-formedness properties. Since zones 6 and 8 are empty, we have $A \cap B = \emptyset$, and since zones 1 and 4 are non-empty, we have $A \cup B \neq D$, and thus $C(A, B)$. Completely analogously, we also find $C(B, C)$. Since zones 1, 2, 4 and 5 are non-empty, we find that resp. $\overline{A} \cap \overline{C}$, $A \cap \overline{C}$, $\overline{A} \cap C$ and $A \cap C$ are non-empty, and thus $UN(A, C)$. In total, this yields the U4 hexagon in Fig. 7(b).

---

[15] Note that the JSB hexagons of Boolean complexities 4 and 3 that we have just considered are precisely those that were already used in Sect. 2 to demonstrate the expressive inadequacy of the divide and conquer strategy; in particular; cf. Fig. 2.

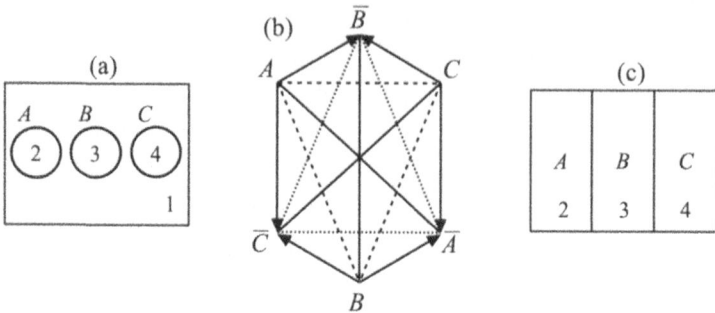

Fig. 6. Euler diagrams giving rise to JSB hexagons of Boolean complexities 4 and 3.

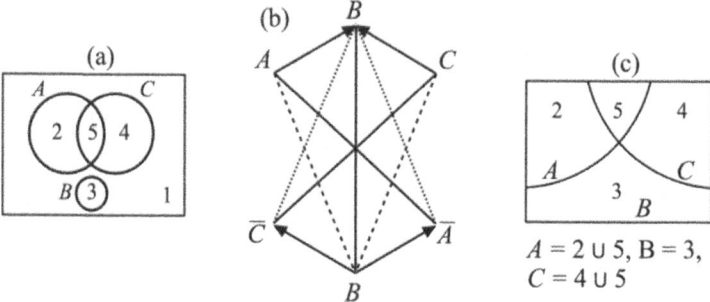

Fig. 7. Euler diagrams giving rise to U4 hexagons of Boolean complexities 5 and 4.

U4 hexagon of Boolean complexity 4. Case 121 from the Appendix involves zones 2, 3, 4 and 5 being non-empty, and all other zones being empty. The corresponding Euler diagram is shown in Fig. 7(c). This case (121) is exactly like the one before (249), except that zone 1 is now empty. This means that the Euler diagram in Fig. 7(c) can be viewed as the result of removing zone 1 from the Euler diagram in Fig. 7(a). As a result, the new Euler diagram in Fig. 7(c) violates the well-formedness property of *no concurrency*, since the curves for A and B (and those for B and C) are partially concurrent. Just like in the previous case, we find $C(A,B)$, $C(B,C)$ and $UN(A,C)$, and thus again obtain the U4 hexagon in Fig. 7(b).

U8 hexagon of Boolean complexity 6. Case 247 from the Appendix involves zones 1, 2, 3, 4, 6 and 7 being non-empty, and all other zones being empty. The corresponding Euler diagram is shown in Fig. 8(a), and does not violate any well-formedness properties. Since zones 2, 3, 4 and 6 are non-empty, we find that resp. $A \cap \overline{B}$, $\overline{A} \cap B$, $\overline{A} \cap \overline{B}$ and $A \cap B$ are non-empty, and thus $UN(A,B)$. Completely analogously, we also find $UN(B,C)$. Finally, since zones 5 and 8 are empty, we have $A \cap C = \emptyset$, and since the other zones are non-empty, we have $A \cup C \neq D$, and thus $C(A,C)$. In total, this yields the U8 hexagon in Fig. 8(b).

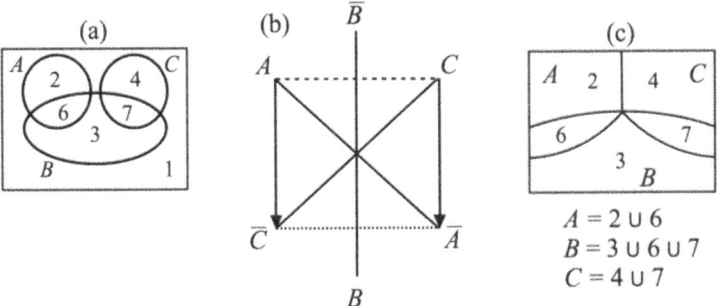

Fig. 8. Euler diagrams giving rise to U8 hexagons of Boolean complexities 6 and 5.

U8 hexagon of Boolean complexity 5. Case 119 from the Appendix involves zones 2, 3, 4, 6 and 7 being non-empty, and all other zones being empty. The corresponding Euler diagram is shown in Fig. 8(c). This case (119) is exactly like the one before (247), except that zone 1 is now empty. This means that the Euler diagram in Fig. 8(c) can be viewed as the result of removing zone 1 from the Euler diagram in Fig. 8(a). As a result, the new Euler diagram in Fig. 8(c) violates both the well-formedness property of *no multiple points* (since the curves for A, B and C all meet at a single point) and that of *no concurrency* (since the curves for A and C are partially concurrent). Just like in the previous case, we find $UN(A,B)$, $UN(B,C)$ and $C(A,C)$, and thus again obtain the U8 hexagon in Fig. 8(b).

U12 hexagon of Boolean complexity 8. Case 256 from the Appendix involves *all* zones being non-empty. Consequently, the corresponding Euler diagram, as shown in Fig. 9(a), simply *is* a Venn diagram, and does not violate any well-formedness properties. Since zones 2, 3, 4 and 6 are non-empty, we find that resp. $A \cap \overline{B}$, $\overline{A} \cap B$, $\overline{A} \cap \overline{B}$ and $A \cap B$ are non-empty, and thus $UN(A,B)$. Completely analogously, we also find $UN(B,C)$ and $UN(A,C)$. In total, this yields the U12 hexagon in Fig. 9(f).

U12 hexagon of Boolean complexity 7. Case 255 from the Appendix involves all zones being non-empty, except for zone 8. The corresponding Euler diagram is shown in Fig. 9(b), and can be viewed as the result of removing zone 8 from the Euler/Venn diagram in Fig. 9(a). As a result, the new Euler diagram in Fig. 9(b) violates the well-formedness property of *no multiple points*, since the curves for A, B and C all meet at a single point. Just like in the previous case, we find $UN(A,B)$, $UN(B,C)$ and $UN(A,C)$, and thus again obtain the U12 hexagon in Fig. 9(f).

U12 hexagon of Boolean complexity 6. Case 127 from the Appendix involves all zones being non-empty, except for zones 1 and 8. The corresponding Euler diagram is shown in Fig. 9(c), and can be viewed as the result of removing zone 1 from the Euler diagram in Fig. 9(b), or alternatively, as the result of removing zones 1 and 8 from the Euler/Venn diagram in Fig. 9(a). As a result, the new Euler diagram in Fig. 9(c) violates the well-formedness

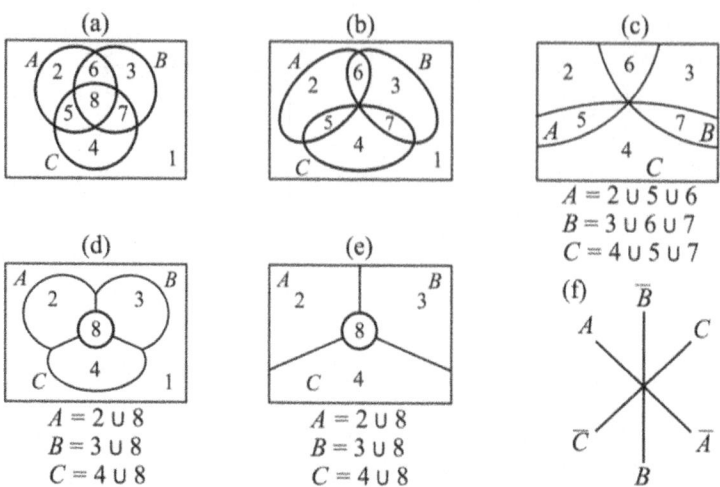

Fig. 9. Euler diagrams giving rise to U12 hexagons of BC 8, 7, 6, 5 and 4.

property of *no multiple points*, since the curves for A, B and C all meet at a single point. Just like in the previous case, we find $UN(A,B)$, $UN(B,C)$ and $UN(A,C)$, and thus again obtain the U12 hexagon in Fig. 9(f).

U12 hexagon of Boolean complexity 5. Case 242 from the Appendix involves all zones being non-empty, except for zones 5, 6 and 7. The corresponding Euler diagram is shown in Fig. 9(d), and can be viewed as the result of simultaneously removing zones 5, 6 and 7 from the Euler/Venn diagram in Fig. 9(a). As a result, the new Euler diagram in Fig. 9(d) violates the well-formedness property of *no concurrency*, since the curves for A and B (and those for B and C, and those for A and C) are partially concurrent. We still have $UN(A,B)$, $UN(B,C)$ and $UN(A,C)$, and thus again obtain the U12 hexagon in Fig. 9(f).

U12 hexagon of Boolean complexity 4. Case 114 from the Appendix involves zones 2, 3, 4 and 8 being non-empty, and all other zones being empty. The corresponding Euler diagram is shown in Fig. 9(e), and can be viewed as the result of removing zone 1 from the Euler diagram in Fig. 9(d), or alternatively, as the result of removing zones 1, 5, 6 and 7 from the Euler/Venn diagram in Fig. 9(a). As a result, the new Euler diagram in Fig. 9(e) violates the well-formedness property of *no concurrency*, since the curves for A and B (and those for B and C, and those for A and C) are partially concurrent. We still have $UN(A,B)$, $UN(B,C)$ and $UN(A,C)$, and thus again obtain the U12 hexagon in Fig. 9(f).

To conclude this systematic discussion, it is worth highlighting how intricately (the examples of) the various Boolean subtypes of U12 hexagons are related to each other. Starting from the Euler/Venn diagram in Fig. 9(a), which is of Boolean complexity 8, we can either (i) remove zone 8 (corresponding to the triple intersection $A \cap B \cap C$) to obtain the Euler diagram in Fig. 9(b), which is of BC 7, or (ii) remove zones 5, 6 and 7 (corresponding to the 'binary' intersections $A \cap B \cap \overline{C}$, $A \cap \overline{B} \cap C$ and $\overline{A} \cap B \cap C$) to obtain the Euler diagram in Fig. 9(d), which is of BC 5. In either scenario, we can subsequently remove zone 1: in scenario (i) this takes us from Fig. 9(b) to Fig. 9(c), which is of BC 6; in scenario (ii), this takes us from Fig. 9(d) to Fig. 9(e), which is of BC 4.

4 Outlook: Aristotelian Diagrams and Syllogistics

It is quite common to explain syllogistics, and in particular the validity of specific syllogisms, using Euler diagrams [26, p. 141ff.] or Venn diagrams; cf. [6, p. 197ff.], [26, p. 207ff.] and [39, p. 74ff.]. In light of the results from the previous section, we are now in a position to draw a connection with Aristotelian diagrams as well. For reason of space, we do not provide a detailed introduction to the system of syllogistics (using the traditional terminology of 'mood', 'figure', etc.), which can be found in many traditional as well as symbolic logic textbooks [6,26,39].

The semantic/model-theoretic validity of an argument is defined as *truth preservation*: every model that makes all premises true, also makes the conclusion true. In the case of a syllogism, which has precisely three terms (distributed over two premises and one conclusion), this universal quantification over models can be replaced with a universal quantification over the 256 Venn diagrams (and their corresponding Euler and Aristotelian diagrams) described in the Appendix. A syllogism is thus valid iff every diagram that makes its two premises true, also makes its conclusion true. As an example, let's consider the famous Barbara syllogism: 'all M are P, all S are M, so all S are P'. The first premise is true whenever $BI(M,P)$ or $LI(M,P)$, and similarly, the second premise is true whenever $BI(S,M)$ or $LI(S,M)$. We thus have to consider four cases in total:

BI(M,P) and BI(S,M). In this case, it follows that $BI(S,P)$, and thus the conclusion ('all S are P') is also true. This is case 130 from the Appendix, which gives rise to a PCD; cf. Fig. 4(a–b).[16]
LI(M,P) and BI(S,M). In this case, it follows that $LI(S,P)$, and thus the conclusion is also true. This is case 146 from the Appendix, which gives rise to a classical square of opposition; cf. Fig. 5(a–b).
BI(M,P) and LI(S,M). In this case, it follows that $LI(S,P)$, and thus the conclusion is also true. This is case 132 from the Appendix, which gives rise to a classical square of opposition.

---

[16] When referring to Figs. 4–5 in this context, all occurrences of the set labels 'A', 'B' and 'C' in those figures should be read as resp. 'S', 'M' and 'P'.

***LI(M,P)* and *LI(S,M)*.** In this case, it follows that $LI(S,P)$, and thus the conclusion is also true. This is case 148 from the Appendix, which gives rise to a Sherwood-Czeżowski hexagon; cf. Fig. 4(c–d).

Since all diagrams which make the two premises of Barbara true, also make its conclusion true, we find that Barbara is valid. Furthermore, we observe that this validity is exhibited by 1 PCD, 2 classical squares and 1 SC hexagon (for the four corresponding Euler diagrams, also see [26, p. 203]). Using the ranking numbers from the Appendix, we can write $[\![\text{Barbara}]\!] = \{130, 132, 146, 148\}$.

Continuing along this way, we obtain a diagrammatic semantics for syllogistics, which maps every syllogism σ onto a set $[\![\sigma]\!] \subseteq \{1, 2, \ldots, 255, 256\}$. In ongoing research, we are investigating the properties of this semantics. For example, one can show that if there is a U12 hexagon in $[\![\sigma]\!]$, then σ is invalid. This establishes an important connection between unconnectedness (as a relation between propositions [44]) and invalidity (as a property of arguments).

Appendix: Description of All 256 Cases

The table below describes all $2^8 = 256$ cases, as computed in Sect. 3. Each case is described as follows: a ranking number (from 1 to 256) that uniquely identifies the case, a description of which zones/minimal regions (cf. the numbering in Fig. 3(b)) are empty (indicated by 'o') and which are non-empty (indicated by '•'), and finally, a classification of the corresponding Aristotelian diagram. If a case entails that at least one of the sets A, B or C is trivial (i.e., equal to $\emptyset$ or to the domain D), this is indicated by '—'. The Boolean complexity of each case can simply be determined by counting the number of •'s in the case description.

	1	2	3	4	5	6	7	8	diagram
1	○	○	○	○	○	○	○	○	—
2	○	○	○	○	○	○	○	●	—
3	○	○	○	○	○	○	●	○	—
4	○	○	○	○	○	○	●	●	—
5	○	○	○	○	○	●	○	○	—
6	○	○	○	○	○	●	○	●	—
7	○	○	○	○	○	●	●	○	—
8	○	○	○	○	○	●	●	●	—
9	○	○	○	○	●	○	○	○	—
10	○	○	○	○	●	○	○	●	—
11	○	○	○	○	●	○	●	○	—
12	○	○	○	○	●	○	●	●	—
13	○	○	○	○	●	●	○	○	—
14	○	○	○	○	●	●	○	●	—
15	○	○	○	○	●	●	●	○	JSB hex.
16	○	○	○	○	●	●	●	●	JSB hex.
17	○	○	○	●	○	○	○	○	—
18	○	○	○	●	○	○	○	●	—
19	○	○	○	●	○	○	●	○	—
20	○	○	○	●	○	○	●	●	—
21	○	○	○	●	○	●	○	○	PCD
22	○	○	○	●	○	●	○	●	class. sq.
23	○	○	○	●	○	●	●	○	class. sq.
24	○	○	○	●	○	●	●	●	SC hex.
25	○	○	○	●	●	○	○	○	—
26	○	○	○	●	●	○	○	●	—
27	○	○	○	●	●	○	●	○	—
28	○	○	○	●	●	○	●	●	—
29	○	○	○	●	●	●	○	○	class. sq.
30	○	○	○	●	●	●	○	●	SC hex.
31	○	○	○	●	●	●	●	○	U4 hex.
32	○	○	○	●	●	●	●	●	U4 hex.
33	○	○	●	○	○	○	○	○	—
34	○	○	●	○	○	○	○	●	—
35	○	○	●	○	○	○	●	○	—
36	○	○	●	○	○	○	●	●	—
37	○	○	●	○	○	●	○	○	—
38	○	○	●	○	○	●	○	●	—
39	○	○	●	○	○	●	●	○	—
40	○	○	●	○	○	●	●	●	—
41	○	○	●	○	●	○	○	○	PCD
42	○	○	●	○	●	○	○	●	class. sq.
43	○	○	●	○	●	○	●	○	class. sq.
44	○	○	●	○	●	○	●	●	SC hex.
45	○	○	●	○	●	●	○	○	class. sq.
46	○	○	●	○	●	●	○	●	SC hex.
47	○	○	●	○	●	●	●	○	U4 hex.
48	○	○	●	○	●	●	●	●	U4 hex.
49	○	○	●	●	○	○	○	○	—
50	○	○	●	●	○	○	○	●	JSB hex.
51	○	○	●	●	○	○	●	○	—
52	○	○	●	●	○	○	●	●	JSB hex.
53	○	○	●	●	○	●	○	○	class. sq.
54	○	○	●	●	○	●	○	●	U4 hex.
55	○	○	●	●	○	●	●	○	SC hex.
56	○	○	●	●	○	●	●	●	U4 hex.
57	○	○	●	●	●	○	○	○	class. sq.
58	○	○	●	●	●	○	○	●	U4 hex.
59	○	○	●	●	●	○	●	○	SC hex.
60	○	○	●	●	●	○	●	●	U4 hex.
61	○	○	●	●	●	●	○	○	degen. sq.
62	○	○	●	●	●	●	○	●	U8 hex.
63	○	○	●	●	●	●	●	○	U8 hex.
64	○	○	●	●	●	●	●	●	U8 hex.
65	○	●	○	○	○	○	○	○	—
66	○	●	○	○	○	○	○	●	—
67	○	●	○	○	○	○	●	○	PCD
68	○	●	○	○	○	○	●	●	class. sq.
69	○	●	○	○	○	●	○	○	—
70	○	●	○	○	○	●	○	●	—
71	○	●	○	○	○	●	●	○	class. sq.
72	○	●	○	○	○	●	●	●	SC hex.
73	○	●	○	○	●	○	○	○	—
74	○	●	○	○	●	○	○	●	—
75	○	●	○	○	●	○	●	○	class. sq.
76	○	●	○	○	●	○	●	●	SC hex.
77	○	●	○	○	●	●	○	○	—
78	○	●	○	○	●	●	○	●	—
79	○	●	○	○	●	●	●	○	U4 hex.
80	○	●	○	○	●	●	●	●	U4 hex.
81	○	●	○	●	○	○	○	○	—
82	○	●	○	●	○	○	○	●	JSB hex.
83	○	●	○	●	○	○	●	○	class. sq.
84	○	●	○	●	○	○	●	●	U4 hex.
85	○	●	○	●	○	●	○	○	class. sq.
86	○	●	○	●	○	●	○	●	U4 hex.
87	○	●	○	●	○	●	●	○	degen. sq.
88	○	●	○	●	○	●	●	●	U8 hex.
89	○	●	○	●	●	○	○	○	—
90	○	●	○	●	●	○	○	●	JSB hex.
91	○	●	○	●	●	○	●	○	SC hex.
92	○	●	○	●	●	○	●	●	U4 hex.
93	○	●	○	●	●	●	○	○	SC hex.
94	○	●	○	●	●	●	○	●	U4 hex.
95	○	●	○	●	●	●	●	○	U8 hex.
96	○	●	○	●	●	●	●	●	U8 hex.
97	○	●	●	○	○	○	○	○	—
98	○	●	●	○	○	○	○	●	JSB hex.
99	○	●	●	○	○	○	●	○	class. sq.
100	○	●	●	○	○	○	●	●	U4 hex.
101	○	●	●	○	○	●	○	○	—
102	○	●	●	○	○	●	○	●	JSB hex.
103	○	●	●	○	○	●	●	○	SC hex.
104	○	●	●	○	○	●	●	●	U4 hex.
105	○	●	●	○	●	○	○	○	class. sq.
106	○	●	●	○	●	○	○	●	U4 hex.
107	○	●	●	○	●	○	●	○	degen. sq.
108	○	●	●	○	●	○	●	●	U8 hex.
109	○	●	●	○	●	●	○	○	SC hex.
110	○	●	●	○	●	●	○	●	U4 hex.
111	○	●	●	○	●	●	●	○	U8 hex.
112	○	●	●	○	●	●	●	●	U8 hex.
113	○	●	●	●	○	○	○	○	JSB hex.
114	○	●	●	●	○	○	○	●	U12 hex.
115	○	●	●	●	○	○	●	○	U4 hex.
116	○	●	●	●	○	○	●	●	U12 hex.
117	○	●	●	●	○	●	○	○	U4 hex.
118	○	●	●	●	○	●	○	●	U12 hex.
119	○	●	●	●	○	●	●	○	U8 hex.
120	○	●	●	●	○	●	●	●	U12 hex.
121	○	●	●	●	●	○	○	○	U4 hex.
122	○	●	●	●	●	○	○	●	U12 hex.
123	○	●	●	●	●	○	●	○	U8 hex.
124	○	●	●	●	●	○	●	●	U12 hex.
125	○	●	●	●	●	●	○	○	U8 hex.
126	○	●	●	●	●	●	○	●	U12 hex.
127	○	●	●	●	●	●	●	○	U12 hex.
128	○	●	●	●	●	●	●	●	U12 hex.
129	●	○	○	○	○	○	○	○	—
130	●	○	○	○	○	○	○	●	PCD
131	●	○	○	○	○	○	●	○	—
132	●	○	○	○	○	○	●	●	class. sq.
133	●	○	○	○	○	●	○	○	—
134	●	○	○	○	○	●	○	●	class. sq.
135	●	○	○	○	○	●	●	○	JSB hex.
136	●	○	○	○	○	●	●	●	U4 hex.
137	●	○	○	○	●	○	○	○	—
138	●	○	○	○	●	○	○	●	class. sq.
139	●	○	○	○	●	○	●	○	JSB hex.
140	●	○	○	○	●	○	●	●	U4 hex.
141	●	○	○	○	●	●	○	○	JSB hex.
142	●	○	○	○	●	●	○	●	U4 hex.
143	●	○	○	○	●	●	●	○	U12 hex.
144	●	○	○	○	●	●	●	●	U12 hex.
145	●	○	○	●	○	○	○	○	—
146	●	○	○	●	○	○	○	●	class. sq.
147	●	○	○	●	○	○	●	○	—
148	●	○	○	●	○	○	●	●	SC hex.
149	●	○	○	●	○	●	○	○	class. sq.
150	●	○	○	●	○	●	○	●	degen.sq.
151	●	○	○	●	○	●	●	○	U4 hex.
152	●	○	○	●	○	●	●	●	U8 hex.
153	●	○	○	●	●	○	○	○	—
154	●	○	○	●	●	○	○	●	SC hex.
155	●	○	○	●	●	○	●	○	JSB hex.
156	●	○	○	●	●	○	●	●	U4 hex.
157	●	○	○	●	●	●	○	○	U4 hex.
158	●	○	○	●	●	●	○	●	U8 hex.
159	●	○	○	●	●	●	●	○	U12 hex.
160	●	○	○	●	●	●	●	●	U12 hex.
161	●	○	●	○	○	○	○	○	—
162	●	○	●	○	○	○	○	●	class. sq.
163	●	○	●	○	○	○	●	○	—
164	●	○	●	○	○	○	●	●	SC hex.
165	●	○	●	○	○	●	○	○	—
166	●	○	●	○	○	●	○	●	SC hex.
167	●	○	●	○	○	●	●	○	JSB hex.
168	●	○	●	○	○	●	●	●	U4 hex.
169	●	○	●	○	●	○	○	○	class. sq.
170	●	○	●	○	●	○	○	●	degen. sq.
171	●	○	●	○	●	○	●	○	U4 hex.
172	●	○	●	○	●	○	●	●	U8 hex.
173	●	○	●	○	●	●	○	○	U4 hex.
174	●	○	●	○	●	●	○	●	U8 hex.
175	●	○	●	○	●	●	●	○	U12 hex.
176	●	○	●	○	●	●	●	●	U12 hex.
177	●	○	●	●	○	○	○	○	—
178	●	○	●	●	○	○	○	●	U4 hex.
179	●	○	●	●	○	○	●	○	—
180	●	○	●	●	○	○	●	●	U4 hex.
181	●	○	●	●	○	●	○	○	SC hex.
182	●	○	●	●	○	●	○	●	U8 hex.
183	●	○	●	●	○	●	●	○	U4 hex.
184	●	○	●	●	○	●	●	●	U8 hex.
185	●	○	●	●	●	○	○	○	SC hex.
186	●	○	●	●	●	○	○	●	U8 hex.
187	●	○	●	●	●	○	●	○	U4 hex.
188	●	○	●	●	●	○	●	●	U8 hex.
189	●	○	●	●	●	●	○	○	U8 hex.
190	●	○	●	●	●	●	○	●	U12 hex.
191	●	○	●	●	●	●	●	○	U12 hex.
192	●	○	●	●	●	●	●	●	U12 hex.
193	●	●	○	○	○	○	○	○	—
194	●	●	○	○	○	○	○	●	class. sq.
195	●	●	○	○	○	○	●	○	class. sq.
196	●	●	○	○	○	○	●	●	degen. sq.
197	●	●	○	○	○	●	○	○	—
198	●	●	○	○	○	●	○	●	SC hex.
199	●	●	○	○	○	●	●	○	U4 hex.
200	●	●	○	○	○	●	●	●	U8 hex.
201	●	●	○	○	●	○	○	○	—
202	●	●	○	○	●	○	○	●	SC hex.
203	●	●	○	○	●	○	●	○	U4 hex.
204	●	●	○	○	●	○	●	●	U8 hex.
205	●	●	○	○	●	●	○	○	JSB hex.
206	●	●	○	○	●	●	○	●	U4 hex.
207	●	●	○	○	●	●	●	○	U12 hex.
208	●	●	○	○	●	●	●	●	U12 hex.
209	●	●	○	●	○	○	○	○	—
210	●	●	○	●	○	○	○	●	U4 hex.
211	●	●	○	●	○	○	●	○	SC hex.
212	●	●	○	●	○	○	●	●	U8 hex.
213	●	●	○	●	○	●	○	○	SC hex.
214	●	●	○	●	○	●	○	●	U8 hex.
215	●	●	○	●	○	●	●	○	U8 hex.
216	●	●	○	●	○	●	●	●	U12 hex.
217	●	●	○	●	●	○	○	○	—
218	●	●	○	●	●	○	○	●	U4 hex.
219	●	●	○	●	●	○	●	○	U4 hex.
220	●	●	○	●	●	○	●	●	U8 hex.
221	●	●	○	●	●	●	○	○	U4 hex.
222	●	●	○	●	●	●	○	●	U8 hex.
223	●	●	○	●	●	●	●	○	U12 hex.
224	●	●	○	●	●	●	●	●	U12 hex.
225	●	●	●	○	○	○	○	○	—
226	●	●	●	○	○	○	○	●	U4 hex.
227	●	●	●	○	○	○	●	○	SC hex.
228	●	●	●	○	○	○	●	●	U8 hex.
229	●	●	●	○	○	●	○	○	—
230	●	●	●	○	○	●	○	●	U4 hex.
231	●	●	●	○	○	●	●	○	U4 hex.
232	●	●	●	○	○	●	●	●	U8 hex.
233	●	●	●	○	●	○	○	○	SC hex.
234	●	●	●	○	●	○	○	●	U8 hex.
235	●	●	●	○	●	○	●	○	U8 hex.
236	●	●	●	○	●	○	●	●	U12 hex.
237	●	●	●	○	●	●	○	○	U4 hex.
238	●	●	●	○	●	●	○	●	U8 hex.
239	●	●	●	○	●	●	●	○	U12 hex.
240	●	●	●	○	●	●	●	●	U12 hex.
241	●	●	●	●	○	○	○	○	JSB hex.
242	●	●	●	●	○	○	○	●	U12 hex.
243	●	●	●	●	○	○	●	○	U4 hex.
244	●	●	●	●	○	○	●	●	U12 hex.
245	●	●	●	●	○	●	○	○	U4 hex.
246	●	●	●	●	○	●	○	●	U12 hex.
247	●	●	●	●	○	●	●	○	U8 hex.
248	●	●	●	●	○	●	●	●	U12 hex.
249	●	●	●	●	●	○	○	○	U4 hex.
250	●	●	●	●	●	○	○	●	U12 hex.
251	●	●	●	●	●	○	●	○	U8 hex.
252	●	●	●	●	●	○	●	●	U12 hex.
253	●	●	●	●	●	●	○	○	U8 hex.
254	●	●	●	●	●	●	○	●	U12 hex.
255	●	●	●	●	●	●	●	○	U12 hex.
256	●	●	●	●	●	●	●	●	U12 hex.

References

1. Blanché, R.: Sur l'opposition des concepts. Theoria **19**, 89–130 (1953)
2. Bolz, R.: Logical diagrams, visualization criteria, and Boolean algebras. In: Beziau, J.Y., Vandoulakis, I. (eds.) The Exoteric Square of Opposition, pp. 195–224. Studies in Universal Logic. Birkhäuser, Cham (2022). https://doi.org/10.1007/978-3-030-90823-2_9
3. Bourou, D., Schorlemmer, M., Plaza, E.: Euler vs Hasse diagrams for reasoning about sets: a cognitive approach. In: Giardino, V., Linker, S., Burns, R., Bellucci, F., Boucheix, J.M., Viana, P. (eds.) Diagrammatic Representation and Inference. Diagrams 2022. LNCS, vol. 13462, pp. 151–167. Springer, Cham (2022). https://doi.org/10.1007/978-3-031-15146-0_13
4. Bourou, D., Schorlemmer, M., Plaza, E.: An image-schematic analysis of Hasse and Euler diagrams. In: Hedblom, M.M., Kutz, O. (eds.) ISD7 – Proceedings of the 7th Image Schema Day 2023, pp. 1–8. CEUR-WS 3511, CEUR-WS (2023)
5. Christensen, R.: The first square of opposition. Phronesis **68**, 371–383 (2023)
6. Copi, I.M., Cohen, C.: Introduction to Logic, Eighth Edition. Prentice Hall, Hoboken (1990)
7. Czeżowski, T.: On certain peculiarities of singular propositions. Mind **64**, 392–395 (1955)
8. Demey, L.: Computing the maximal Boolean complexity of families of Aristotelian diagrams. J. Log. Comput. **28**, 1323–1339 (2018)
9. Demey, L.: Boolean considerations on John Buridan's octagons of opposition. Hist. Philos. Log. **40**, 116–134 (2019)
10. Demey, L.: Metalogic, metalanguage and logical geometry. Logique et Anal. (N.S.) **248**, 453–478 (2019)
11. Demey, L.: From Euler diagrams in Schopenhauer to Aristotelian diagrams in logical geometry. In: Lemanski, J. (ed.) Language, Logic, and Mathematics in Schopenhauer. SUL, pp. 181–205. Springer, Cham (2020). https://doi.org/10.1007/978-3-030-33090-3_12
12. Demey, L., Erbas, A.: Boolean subtypes of the U4 hexagon of opposition. Axioms **13**, 1–20 (2024)
13. Demey, L., Smessaert, H.: The relationship between Aristotelian and Hasse diagrams. In: Dwyer, T., Purchase, H., Delaney, A. (eds.) Diagrams 2014. LNCS (LNAI), vol. 8578, pp. 213–227. Springer, Heidelberg (2014). https://doi.org/10.1007/978-3-662-44043-8_23
14. Demey, L., Smessaert, H.: Combinatorial bitstring semantics for arbitrary logical fragments. J. Philos. Log. **47**, 325–363 (2018)
15. Demey, L., Smessaert, H.: Geometric and cognitive differences between Aristotelian diagrams for the Boolean algebra $\mathbb{B}_4$. Ann. Math. Artif. Intell. **83**, 185–208 (2018)
16. Demey, L., Smessaert, H.: From Euler diagrams to Aristotelian diagrams. In: Giardino, V., Linker, S., Burns, R., Bellucci, F., Boucheix, J.M., Viana, P. (eds.) Diagrammatic Representation and Inference. Diagrams 2022. LNCS, vol. 13462, pp. 279–295. Springer, Cham (2022). https://doi.org/10.1007/978-3-031-15146-0_24
17. Edwards, A.W.F.: An eleventh-century Venn diagram. BSHM Bull. **21**, 119–121 (2006)
18. Fish, A., Khazaei, B., Roast, C.: User-comprehension of Euler diagrams. J. Vis. Lang. Comput. **22**, 340–354 (2011)
19. Frijters, S., Demey, L.: The modal logic of Aristotelian diagrams. Axioms **12**, 1–26 (2023)

20. Gergonne, J.D.: Essai de dialectique rationelle. Annales des Mathématiques Pures et Appliquées **7**, 189–228 (1817)
21. Geudens, C., Demey, L.: On the Aristotelian roots of the modal square of opposition. Logique et Anal. (N.S.) **255**, 313–348 (2021)
22. Giard, L.: La Dialectique rationnelle de Gergonne. Revue d'Histoire des Sciences **25**, 97–124 (1972)
23. Hammer, E., Shin, S.J.: Euler's visual logic. Hist. Philos. Log. **19**, 1–29 (1998)
24. Howse, J., Stapleton, G., Flower, J., Taylor, J.: Corresponding regions in Euler diagrams. In: Hegarty, M., Meyer, B., Narayanan, N.H. (eds.) Diagrams 2002. LNCS (LNAI), vol. 2317, pp. 76–90. Springer, Heidelberg (2002). https://doi.org/10.1007/3-540-46037-3_7
25. Jacoby, P.: A triangle of opposites for types of propositions in Aristotelian logic. New Scholasticism **24**, 32–56 (1950)
26. Keynes, J.N.: Studies and Exercises in Formal Logic. MacMillan, New York (1884)
27. Khomskii, Y.: William of Sherwood, singular propositions and the hexagon of opposition. In: Béziau, J.Y., Payette, G. (eds.) New Perspectives on the Square of Opposition. Peter Lang, Bern (2011)
28. Kraszewski, Z.: Logika stosunków zakresowych. Stud. Log. **4**, 63–116 (1956)
29. Kretzmann, N.: William of Sherwood's Introduction to Logic. Minnesota Archive Editions (1966)
30. Lemanski, J.: Periods in the use of Euler-type diagrams. Acta Baltica Historiae et Philosophiae Scientiarum **5**, 50–69 (2017)
31. Lemanski, J.: Logic diagrams in the Weigel and Weise circles. Hist. Philos. Log. **39**, 3–28 (2018)
32. Lemanski, J.: Euler-type diagrams and the quantification of the predicate. J. Philos. Log. **49**, 401–416 (2020)
33. Lemanski, J., Demey, L.: Schopenhauer's partition diagrams and logical geometry. In: Basu, A., Stapleton, G., Linker, S., Legg, C., Manalo, E., Viana, P. (eds.) Diagrams 2021. LNCS (LNAI), vol. 12909, pp. 149–165. Springer, Cham (2021). https://doi.org/10.1007/978-3-030-86062-2_13
34. Londey, D., Johanson, C.: Apuleius and the square of opposition. Phronesis **29**, 165–173 (1984)
35. Moktefi, A., Lemanski, J.: On the origin of Venn diagrams. Axiomathes **32**(Suppl 3), S887–S900 (2022)
36. Pellissier, R.: Setting n-opposition. Log. Univers. **2**(2), 235–263 (2008)
37. Priss, U.: A semiotic-conceptual analysis of Euler and Hasse diagrams. In: Pietarinen, A.-V., Chapman, P., Bosveld-de Smet, L., Giardino, V., Corter, J., Linker, S. (eds.) Diagrams 2020. LNCS (LNAI), vol. 12169, pp. 515–519. Springer, Cham (2020). https://doi.org/10.1007/978-3-030-54249-8_47
38. Priss, U.: Set visualisations with Euler and Hasse diagrams. In: Cochez, M., Croitoru, M., Marquis, P., Rudolph, S. (eds.) GKR 2020. LNCS (LNAI), vol. 12640, pp. 72–83. Springer, Cham (2021). https://doi.org/10.1007/978-3-030-72308-8_5
39. Quine, W.V.O.: Methods of Logic (Revised Edition). Holt, Rinehart and Winston, New York (1966)
40. Rival, I.: The diagram. In: Rival, I. (ed.) Graphs and Order: The Role of Graphs in the Theory of Ordered Sets and Its Applications, pp. 103–133. Springer, Dordrecht (1985)
41. Rodgers, P.: A survey of Euler diagrams. J. Vis. Lang. Comput. **25**, 134–155 (2014)
42. Sesmat, A.: Logique II. Hermann (1951)
43. Smessaert, H.: On the 3D visualisation of logical relations. Log. Univers. **3**, 303–332 (2009)

44. Smessaert, H., Demey, L.: Logical geometries and information in the square of opposition. J. Logic Lang. Inform. **23**, 527–565 (2014)
45. Stalnaker, R.C.: Inquiry. MIT Press, Cambridge (1984)

What Does It Mean that Diagrams Represent Constructions?

Piotr Kozak

University of Bialystok, Swierkowa 20 B, 15-328 Bialystok, Poland
piotr.kozak1@gmail.com

Abstract. Although it is commonly held that geometrical diagrams can represent constructions, the nature of the construction representations has rarely been closely investigated. There is no clarity on the subject of construction nature nor the nature of the representation relation. In this paper, I address the question of how geometrical diagrams can represent constructions. I describe constructions as the procedures for arriving at a target. Diagrams exemplify these procedures.

Keywords: diagrams · semantics · constructions · exemplification · geometry

1 Introduction

Let us take an example of a triangle construction. Take three line segments a, b, and c, where any two of them taken together are longer than the third, and a straight line l. Next, take the line segment a and construct a corresponding in-length line segment DE lying on l. Do the same with the line segments b, c and corresponding line segments EF, FG. Now, construct two circles. The first one has a centre marked by E and a radius ED. The second one has a centre marked by F and a radius FG. Both circles intersect at H. If you connect points EFH, you arrive at the triangle EFH. The construction is complete. It is represented by the diagram (Fig. 1).

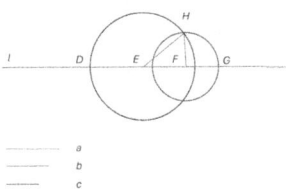

Fig. 1. The diagram represents the construction of the triangle EFH. © The author

This geometrical example marks a vital feature of diagrams. One of their key functions is to show how to connect elements to reach the goal. In other words, diagrams can represent constructions.

What it means that diagrams represent constructions is, however, far from clear. It is not a trivial question, for it appears that representing constructions causes problems for the standard depiction models. For one thing, representing constructions seems to differ from depicting objects, such as apples and trees, for there is no object that can be depicted. For another, it seems that representing constructions cannot be simply taken under resemblance-based accounts of pictorial content [1]. These are, however, only temporary and negative descriptions of the subject matter.

Let us start with some clarifications. Firstly, by a diagram, I understand a spatial representation that spatial features can be exploited in a cognitively non-trivial way [2–6]. Obviously, this functional definition is far from being exhaustive and would not be a good starting point if this paper were on the topic of what diagrams are. However, this definition seems to be sufficiently inclusive, metaphysically neutral, and close to the everyday use of the word in actual scientific practice [4].

Secondly, for the sake of brevity and the clarity of the argument, I focus on geometric diagrams only (considered both in continuous and discrete spaces) [7]. In principle, the theoretical framework I present here should apply to other diagrammatic representations, such as charts and graphs [8]. However, in this paper, I do not offer any independent argument for the extension of my analysis to other forms of diagrams.

Thirdly, I hold that diagrams can represent constructions. Yet, I am not saying that diagrams do not represent mathematical objects. Particularly, I am not holding that diagrams do not refer to any platonic-like objects. In this paper, I do not offer any argument in favour or against Platonism in mathematics.

Fourthly, I do not offer any argument regarding the relationship between the concept of construction and proof, particularly the construction proof. This would require another kind of paper.

Finally, to answer how diagrams represent constructions, one needs to address two issues. On the one hand, one must have something to say about the object of representation. In other words, one needs to say what construction is. On the other, one has to specify the properties of the representation relation.

The plan of the paper is as follows. In the next section, I address the problem of the nature of the construction. I argue that constructions can be understood as admissible procedures for reaching a target. Consequently, diagrams represent these procedures. Next, I specify the nature of the representation relation. I hold that diagrams represent constructions by exemplifying them.

2 What Are Constructions?

Let us start with the concept of construction. In mathematics, it is commonly understood as a set of procedures (or outcomes of such procedures) of using idealized instruments, such as idealized rulers and compasses, to represent geometric objects, e.g., creating a line segment by connecting two points.

However, originally, in the works of Euclid and Proclus, construction is understood more broadly as procedures of adding something that is "lacking in the given for finding what is sought" [9]. To explain it, let us consider two examples.

Firstly, consider the case of constructive reasoning [10]. Here, we start from a set of initial objects and a set of permissible operations (e.g., algorithms) over these objects. Construction is based on conducting permissible operations over the set of initial objects, determining the relation between them to arrive at a new object. This new object is a product of applying permissible operations over the set of initial objects.

Secondly, let us go back to the Fig. 1. We began with the set of initial objects, i.e., the line segments a, b, and c, given beforehand, and the concept of a triangle, which was the construction target. The goal of the diagram is to show how to arrive at the triangle. We need to know the permissible procedures (e.g., not bending straight lines) leading from initial objects to a triangle. These permissible procedures determine the relation between initial objects. A diagram represents these procedures.

Drawing on these examples, one can notice the following: Primarily, constructions can be reduced neither to the elements nor to the definition of the constructed object. On the one hand, knowing that we seek a polygon with three edges and three vertices is not the same as knowing how to arrive at such a figure. On the other hand, the same elements can be arranged differently, leading to mathematical objects with different properties. For instance, we can arrive at a hexagon and a tetrahedron by performing different operations over six equal line segments.

Next, knowledge of constructions is general. Knowing how to construct a specific triangle implies that you know how to arrive at the same type of figure with different elements, e.g., with different lengths of line segments. By the same token, knowing how to apply an operation to a specific mathematical object implies that one knows how to apply it to any object that is given beforehand.

Moreover, the concept of construction is irreducible to actually producing an object. Using actual tools, such as a physical ruler, is a matter of conducting certain procedures with physical means. Certain physical objects, such as a physical line segment conducted with a ruler, represent idealized mathematical objects, such as the Euclidean line segment, defined as an interval between two points. However, understanding the Euclidean line segment's concept is knowing how to find this interval. For instance, in the case of the diagram represented by Fig. 1, one does not have to create a triangle. One has to know how to arrive at it. Knowing how to construct an object is dissociated from the actual ability to create it. One can understand that the diagram represents a triangle construction even if one has no drawing skills or knowledge of how to use physical tools. However, one has to know how to apply some rules to given objects. Thus, knowing the properties of a construction is a kind of procedural knowledge of employing certain admissible procedures over the set of initial objects, leading to arriving at some new mathematical objects.

Drawing on these analyses, let us informally define the concept of construction. In terms of operation, it denotes admissible procedures of localizing the target in some geometric space by determining the parameters of this space [8, 11]. In terms of the result of operation, construction is an outcome of applying these procedures. Let me unpack this definition.

Let us take the information space (S) to be a set of all information. In the case of geometry, the information space would be a space made of points. The parameters of the space determined by a construction (C) are the basic units of construction (b) taken in the most neutral metaphysical terms as the states of some space together with the order between b. A partially ordered set (to determine the subsequence of operations) of permissible operations (Φ) over the ordered set of b in S determines the relation (R) between b. The goal of C is to reach the sought-for object. Let us call this object the target (T). In the case of a construction problem, the target is the object possessing the properties stated in the problem's assumptions. Yet, the target can also be discovered in the course of a successful construction.

Importantly, not *any* partially ordered set of operations Φ is admissible for constructing a mathematical object, and not every operation φ counts as a construction of T. For example, in a straightedge-and-compass construction, only operations conducted with a straightedge and a pair of compasses are allowed. However, it does not imply that there are only straightedge-and-compass constructions. There are non-Euclidean constructions, e.g., neusis constructions, too [12]. It follows that for every construction C, there is a set of operations Φ that belongs to some system Σ that determines admissible operations (and means of operations), i.e., these operations (and means) that count as admissible constructions of T in Σ and guarantee that the outcome of these operations is also admissible in Σ. For example, the system Σ_1 allows using only a straightedge and a compass, while the system Σ_2 allows using conic sections, etc.

To wrap it up, in terms of operation, a partially ordered set of operations $\Phi = \langle \varphi 1, \varphi 2, \ldots, \varphi n \rangle$ in a system Σ is a construction C of T iff $\varphi: \varphi \in \Phi$ takes the ordered set of the basic units of construction $\langle b1, b2, \ldots, bn \rangle$ in the domain S and determines the relation R between $(b1, b2, \ldots, bn)$ in order to get to T. In terms of the result of operation, construction C can be interpreted as the ordered set $\langle b1, b2, \ldots, bn \rangle$ in S with relations R defined over them.

Solving a construction problem requires determining the procedural steps that take us from a given set of initial objects to the target. It involves defining a partially ordered set of admissible operations Φ in Σ carried out over $\langle b1, b2, \ldots, bn \rangle$ in the domain S in order to get to T. In the case of the triangle construction in Fig. 1, we start from the basic construction units, i.e., the line segments a, b, and c given beforehand. The properties of T are given by the analysis of the properties of the sought-for object, i.e., a polygon with three edges and three vertices. The construction represented by the diagram specifies admissible operations over the basic construction units and determines the order and the relation between these units. For instance, the construction of the triangle in Fig. 1 determines that the line segments linking two vertices of a triangle cannot be curved or broken since construction procedures allow only these operations that result in straight line segments. We can prove that the constructed object is the sought-for object by showing that the resulting object possesses the properties given in the problem's formulation [13–15].

For our purposes, it is important to note that the provided description of construction reduces the concept of construction neither to the concepts of construction elements nor the definition of a constructed object. According to this description, constructions are not some objects represented by diagrams. Instead, constructions are certain procedures required for arriving at some objects.

As a consequence, we must sharply distinguish the properties of construction and the properties of the represented object. In other words, the properties of construction do not describe constructed objects. Construction properties allow us to arrive at constructed objects and localize them in some space by determining the properties of this space. The properties of a triangle construction in Fig. 1 do not describe the sought-for object as if there were some object, i.e., a triangle, that has to be depicted in the diagram. In particular, the triangle construction does not predicate anything of a triangle. It shows how to arrive at it by identifying admissible geometric transformations and determining the parameters of the space, i.e., by determining basic units of construction, their order, and the relation between them. The properties of the construction, e.g., the curvature of lines, are the properties of the searching procedures by means of which you arrive at the target.

To summarize, constructions are admissible procedures required for arriving at the target. The properties of construction are not the properties of constructed objects. Instead, they are the properties of the procedures for arriving at these objects.

3 How Are Constructions Represented?

What does it mean that diagrams represent constructions? The concept of representation refers to the idea that something stands for something another. However, granted that constructions are not objects that a diagram can represent, they cannot be represented like chairs and tables. Thus, we have to be more specific when we want to describe how diagrams represent constructions.

Drawing on the works of Goodman and Elgin [16–19], we can distinguish between two modes of representation. Either something denotes the object of representation or exemplifies it. Denotation is a dyadic relation between a representation and something it stands for. Exemplification runs in the opposite direction. It refers back from the selected properties of the object to its representation. Moreover, exemplification requires instantiation. Consider a case of a lake water sample. On the one hand, it represents the quality of the water. On the other, it instantiates this quality. By investigating the properties of the sample, one investigates the properties of the water.

The relation between constructions and diagrammatic representations cannot be a denotation. The denotation separates the representation and something that is represented. Thus, the properties of representation do not have to fit the properties of the represented object. For instance, the term 'dog' denotes dogs. Yet, this term does not possess the properties of dogs. In contrast, the properties of diagrams must indicate the properties of constructions. Otherwise, diagrams would be useless.

Even though diagrams cannot denote construction, they can exemplify them. On the one hand, a diagram instantiates the properties of construction. For instance, the diagram represented in Fig. 1 possesses the construction properties of a triangle. On the other hand, by investigating the properties of the diagram, one learns something about the properties of construction, e.g., we can learn how to arrive at a triangle by investigating the diagram.

Moreover, by holding that diagrams exemplify constructions, we can point to the rationale behind the intuitive belief that diagrams are necessary for representing constructions. Particularly, it helps us to argue that diagrams are irreducible to descriptions of constructions.

Obviously, constructions can be described. For example, I can describe the procedures leading from the elements to the triangle figure. However, there is a striking asymmetry between the descriptions of these procedures and the diagrammatic exemplifications of these procedures. In the case of descriptions, we can say that we understand a construction description only if we are able to identify such a construction on a diagram, for diagrams exemplify constructions. If we hold that we understand the procedure description but cannot recognize these procedures on a diagram, then we do not understand this description. In the case of diagrams, we do not have to understand any kind of construction description. It is enough to identify the construction exemplified by the diagram. If the diagram properties exemplify the properties of the construction, then the diagram properties are the objects we want to understand.

Consider the diagram represented in Fig. 1. I can describe its construction, specifying successive steps in arriving at the triangle. In fact, that is the thing we have started our investigations with. However, the diagram represented in Fig. 1 does not illustrate the construction description. Instead, it is an exemplification of such a construction. To be able to say that I understand the construction description, I have to be able to understand the diagram, for the construction properties exemplified by the diagram are the object of my understanding.

What does it mean that diagrams exemplify constructions? If constructions are not objects, then diagrams do not exemplify the properties of constructions in the same way a water sample represents water quality. Instead, let us think about exemplifying constructions in terms of exemplifying procedures leading to the goal.

To illustrate this crude description, let us consider the case of a tangram. Tangram is a game of putting together flat figures to arrive at some pattern. Figure 2 represents one of the possible solutions to this game.

Fig. 2. The diagram represents the rules of tangram. © The author

Now, if someone were to ask how to play tangram, the answer could be a description of the procedures leading to the desired pattern. However, one can also show how to play tangram by showing Fig. 2. In this case, the outcome of applying these procedures exemplifies these procedures. Thus, if one wants to point out these procedures, one can point at the particular tangram diagram. By the same token, if one wants to show how to construct a triangle, one can point at the triangle diagram in Fig. 1. The final outcome of the construction procedures, i.e., the triangle diagram, is indistinguishable from the procedures leading to creating it.

There can be a few possible objections to this picture. Firstly, the tangram diagram is a single figure and a final outcome of applying a set of different procedures. Thus, one may argue that it does not represent a construction, for by representing constructions, we understand representing admissible procedures leading to a target. Only a sequence of diagrams can represent a set of these procedures. This argument can even be strengthened by pointing to historical examples. For example, straightedge-and-compass constructions can be represented by the application of basic constructions using the points, lines, circles, and intersections between them.

However, if one holds that a diagram can represent a single admissible operation conducted over a set of initial objects, then it does not follow that it cannot represent two or three of them, etc. We would need another kind of argument to hold that. Granted, such a composed diagram can be more complicated and difficult to understand than a single construction diagram. That is the reason for decomposing it into a sequence of single diagrams. However, that teaches us more about the nature of human cognitive skills than the nature of diagrams.

Secondly, one can argue that to understand constructions, we must introduce time as an important parameter. For example, to understand the properties of construction in Fig. 1, we need to understand which step follows another. Moreover, that is the core information given in the description of the geometric construction problem. By the same token, the tangram diagram can represent constructions only if we can visualize subsequent procedural steps.

This possible objection indicates an important feature of understanding procedures. It is not enough to know how to perform a procedure. It is essential to know what is the correct sequence of these procedures. However, the subsequence of operations does not need to involve time. Any partially ordered set determines the sequence of its elements in a 'timeless' manner. Introducing the temporal factor is often required to *understand* the sequence of construction procedures. Yet, it does not affect the content of these procedures. The content of the diagram represented in Fig. 1 and the representation consisting of animated moves resulting in Fig. 1 would be the same, although the second representation would be easier to grasp.

To sum up, I hold that diagrams represent constructions by exemplifying them. If that is the case, then diagrams are necessary for understanding constructions. If diagrams exemplify constructions, then (selected) properties of the diagram exemplify the construction properties. Therefore, to understand the properties of construction, one has to be able to understand the properties of a diagram.

Acknowledgments. I want to thank anonymous reviewers for their substantial comments, valuable feedback, and, most importantly, for their priceless time.

References

1. Kulvicki, J.: Images. Routledge, New York (2013)
2. De Toffoli, S.: 'Chasing' the diagram—the use of visualizations in algebraic reasoning. Rev. Symb. Log. **10**, 158–186 (2017). https://doi.org/10.1017/S1755020316000277
3. De Toffoli, S.: What are mathematical diagrams? Synthese **200**, 86 (2022). https://doi.org/10.1007/s11229-022-03553-w
4. Johansen, M.W., Misfeldt, M., Pallavicini, J.L.: A typology of mathematical diagrams. In: Chapman, P., Stapleton, G., Moktefi, A., Perez-Kriz, S., Bellucci, F. (eds.) Diagrams 2018. LNCS, vol. 10871, pp. 105–119. Springer, Cham (2018). https://doi.org/10.1007/978-3-319-91376-6_13

5. Larkin, J.H., Simon, H.A.: Why a diagram is (sometimes) worth ten thousand words. Cogn. Sci. **11**, 65–100 (1987). https://doi.org/10.1111/j.1551-6708.1987.tb00863.x
6. Stenning, K.: Seeing Reason. Image and Language in Learning to Think. Oxford University Press, Oxford (2002)
7. Jamnik, M.: Mathematical Reasoning with Diagrams: from Intuition to Automation. Center for the Study of Language and Information, Stanford (2001)
8. Kozak, P.: Thinking in Images: Imagistic Cognition and Non-propositional Content. Bloomsbury Academic, Bloomsbury Publishing Plc, London/New York (2023)
9. Morrow, G.R.: Proclus: A Commentary on the First Book of Euclid's Elements. Princeton University Press, Princeton (1970)
10. Vandoulakis, I.M., Stefaneas, P.: Proof-events in history of mathematics. Ganita Bharati **35**, 119–157 (2013)
11. Tichy, P.: Constructions. Philos. Sci. **53**, 514–534 (1986). https://doi.org/10.1086/289338
12. Brzeziński, J.: Geometric constructions. In: Galois Theory Through Exercises. Springer Undergraduate Mathematics Series, pp. 81–83. Springer, Cham (2018). https://doi.org/10.1007/978-3-319-72326-6_14
13. Djoric, M.: Constructions, instructions, interactions. Teach. Math. Its Appl. **23**, 69–88 (2004). https://doi.org/10.1093/teamat/23.2.69
14. Hintikka, J., Remes, U.: The Method of Analysis: Its Geometrical Origin and Its General Significance. D. Reidel Pub. Co., Dordrecht (1974)
15. Vandoulakis, I.M.: The readings of Apollonius' on the cutting off of a ratio. Arab. Sci. Philos. **22**, 137–149 (2012). https://doi.org/10.1017/S0957423911000130
16. Elgin, C.Z.: True Enough. MIT Press, Cambridge (2017)
17. Goodman, N.: Languages of Art: An Approach to a Theory of Symbols. Hackett, Indianapolis (20)
18. Goodman, N.: Of Mind and Other Matters. Harvard University Press, Cambridge (1984)
19. Goodman, N., Elgin, C.Z.: Reconceptions in Philosophy and Other Arts and Sciences. Hackett Pub. Co., Indianapolis (1988)

The Topology of Assertion: A Diagrammatic Rationale for Our Enduring Love of Truth

Dave Beisecker[✉]

University of Nevada, Las Vegas, USA
beiseckd@unlv.nevada.edu

Abstract. Our practices prioritize truth. We take free-standing utterances of a statement to be normed on the speaker's *belief* along with the *truth* of that statement's content. But why is this so natural and universal? Why do we think it would be so absurd to have a communicative practice in which free-standing utterances are instead understood to be *denials*, and so normed to falsity or warranted *disbelief*? In this paper, I draw upon Peirce's discussion of the diagrammatic naturalness of the existential graphs over the entitative graphs to provide a possible answer to this much more fundamental question.

Keywords: Truth · Assertion · Rejection · Denial · Peirce · Graphs

Our conversational practices prioritize truth. We take free-standing utterances of a statement to be normed on the speaker's *belief* along with the *truth* of that statement's content. When somebody utters a declarative sentence, and then they utter another, then unless there are indications otherwise, it is commonplace – if not universal – in natural language to interpret them as having *asserted* both of those sentences, and to have done so more or less independently. Free-standing utterances of statements are typically understood as *assertions* – and so understood to be normed to truth, knowledge, or warranted belief. This default in favor of assertion – which is so natural as to be almost invisible – would seem to go far to explain the tendency amongst philosophers of language to accord that particular speech act some sort of conceptual primacy over others in our language games. As Brandom puts it in a way so as to evoke (or perhaps provoke!) Wittgenstein's geist, assertion belongs at the heart – or in the "downtown" of – our understanding of discursive practice [2: 43].

This aspect of our language use was not entirely invisible to Frege. That is why he included the judgement stroke (the *urteilstrich*) in his logical notation. However, aside from a similarly functioning stroke of definition, he makes no use of other forces in his notation, leading other logicians mostly to dispense with the *urteilstrich* as a quaint embellishment, even though they too were sensitive to the distinction between force and content. Instead, they've followed Peirce, and taken free-standing inscriptions on a page to be asserted. But why is this so natural and universal? Why do we think it would be so absurd or perverse to have an inverted or topsy-turvy *sprachspiel* in which free-standing utterances are instead understood to be *denials*, and so normed to falsity or warranted *dis*belief?

The conceivability of a topsy-turvy *sprachspiel* has been countenanced before. Gerald Massey raised the possibility in his attempts to clarify Quine's indeterminacy thesis [5], and, more recently, Kensuke Ito has raised it to challenge both a particular form of truth-deflationism, and a naïve conception of Davidson's Principle of Charity [4]. Though both these authors endorse the possibility of a topsy-turvy sprachspiel, neither of them raise the question that I'm posing here: *just why do we so naturally and overwhelmingly favor assertion and norming our talk to truth?*

1 A Little Motivation: The Logic of Falsity

By way of motivation, let's take a look at some alternate Tableau-style proofs of the same sequent: constructive dilemma:

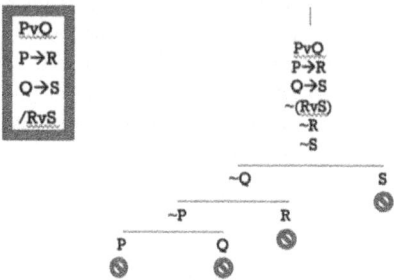

Fig. 1. Truth Tree (Jeffreys, Smullyan)

In Fig. 1, we have a familiar "truth-tree" demonstration. In such Smullyan Trees, the consequences of asserting premises are systematically worked out, along with the consequences of asserting the conclusion's negation, with the aim of exposing inconsistencies marked by closed branches. A fully-closed tree exposes the inconsistency of affirming all of the formulas that start out at the tree's root. By a familiar rule of entailment for negation (Premises $\models$ conclusion j.i.c. Premises, ~conclusion $\models$), that is sufficient to prove the validity of the sequent in question.

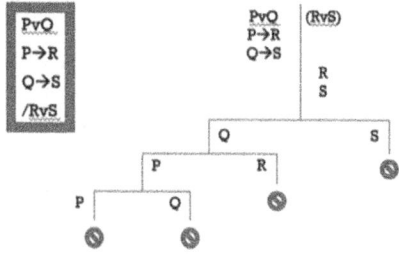

Fig. 2. Two-Sided Tree (Perloff, Fitts and Beisecker) [3]

In Fig. 2, we have the same demonstration, but with what I call a two-sided or bilateral proof tree. Instead of negating conclusions, this type of proof tree places premises on the left-hand side of the root, and conclusion(s) on the right. It then proceeds by systematically developing the consequences of asserting (holding true) compounds on the left along with the consequences of denying (holding false) compounds on the right. Incoherencies, and so branch closures, arise when one and the same formula appears on both the left- and right-hand sides of a branch, signifying its simultaneous affirmation and denial. This system essentially turns the sequent calculus upside down. For our purposes, however, the thing to notice is that it has basically the same structure as the one before.

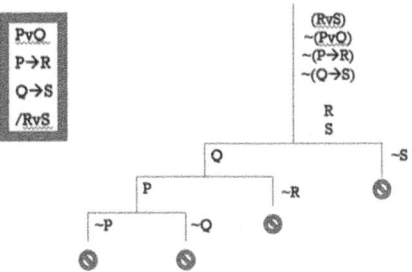

Fig. 3. Falsity Tree (Topsy-Turvians)

But why stop there, with just the conclusion moved to the right? Figure 3 is the same demonstration, only with all of the premises, now negated, shifted to the right-hand side: that of denial. This tree, which one could well think of as a *falsity-tree,* proceeds by systematically developing the logical consequences of *denying* statements, with an aim now of showing that the formulas making up the root is a *sub-contrary* set; they cannot collectively be false (or it is impermissible to take them all to be). Notice once again that even though the specific rules of tree development for such a system seem awkward and involuted to us, the structure of the proof itself looks very much like the other two. The system can be generalized to do exactly the same work as the other two.

The point of all this is that at least from the perspective of formal logic it is perfectly coherent to develop a system that operates completely on the side of falsity. Valid sequents in a system that prioritizes truth (and truth-preserving rules of inference) can also be shown to be such in an equivalent system that prioritizes falsity (and falsity-preserving rules of inference). While such a system of logic might seem hopelessly strange to us, topsy-turvy practices might not be as logically alien as one might suppose. We might be able to detect, against all expectation, that a particular *sprachspiel* (or system of coded messages) is actually topsy-turvy, and prioritizes denial, by looking at the involuted forms of their inference rules (as embodied in their practices).

2 The Question Again: Why Truth, Why Assertion?

So we will have to look to something other than formal logic to find the reason or reasons why we have such a predilection for speaking truly, rather than falsely. And surely we would expect there to be such reasons. Norming our statements to truth appears to be common to all human languages. However, you won't find the answer to this question in, for instance, the entry on assertion in the *Stanford Encyclopedia of Philosophy* or the recent *Oxford Handbook on Assertion* (2020). While there is plenty of discussion about how best to characterize the norm(s) of assertion – whether they be directly normed to truth, or rather indirectly normed to it through a speaker's epistemic situation – there is virtually no discussion about why it is *assertion* that accords such pride of place, or why it should be *truth* to which our bare utterances should be normed.

While it is entertaining both to imagine the practical limitations of topsy-turvian discursive practice and also to consider whether such practice could be learnable *ab initio*, a fully satisfactory explanation for why we norm to truth rather than falsity, and so why we speak in assertions rather than denials, might prove a little more elusive than one at first supposes, for the symmetries between assertion and denial run deep. For instance, one might argue that denial is hopelessly ambiguous, because denials come in both strong and weak flavors. However, assertion turns out to have both strong and weak forms as well [7]. Or one might think that our preference for truth is vindicated simply by looking to the observational dimension of language use. After all, we're primarily and immediately interested in what's actually "out there", not with what's not "out there" (thus prioritizing a "metaphysics of presence" over that of absence). But I beg of us to use some imagination on behalf of our imagined topsy-turvians. A topsy-turvian practice might understand some form of negation to be bound up within their regular sign of observation, such as "Lo!" or "Look!" Thus, under conditions where we might assert "Lo, there's a rabbit!" the topsy-turvians would just as easily use an observative exclamation to *deny* something like "Not - a rabbit over there!" Indeed, just as Frege sought to dispense with the consideration of denial altogether by unpacking it as the assertion of a negation, a topsy-turvian Frege might just introduce negation to go the other way around, showing in effect, that anything one can do with an asserted statement could be done just as well with the (strong) denial of that statement's negation. *Dis*believing that logic is *not* the study of falsity, our topsy-turvian Frege would *deny* Frege's insistence that logic is the study of truth.[1] Topsy-turvians would be flummoxed by the paradox of the Truth-teller! And topsy-turvian psychologists would sincerely deny belief-psychology, by pointing out in effect that their actions and expressions arise out of the contents of their *dis*belief boxes.

3 Turning to Peirce

For help in addressing our question, I propose to pursue a line of reasoning initiated by that great Sage of Arisbe, Charles Sanders Peirce. As mentioned earlier, though he was fully aware of the force-content distinction, Peirce dispensed with an explicit sign of assertion in his graphical notation for logic. Instead, any free-standing statement written down on his diagrammatic expanse, the "phemic sheet," was thereby to be understood as asserted. Hence, he explicitly understood this expanse to be the "sheet of assertion." So if "There's a raccoon in the tree." (or a propositional symbol thereof) is scribed on the sheet, we are to be understand that sentence as being asserted. And if that sheet also

---
[1] And he would be partly correct; one should rather disbelieve that logic is not the study of entailment.

bears the inscription "The dog is barking" (or symbol thereof), then that statement is also to be understood as asserted.

But what are we to do if we wish to scribe a statement in a *non*-assertive context, such as when it is within the antecedent or consequent of some conditional? Here Peirce uses a system of nested ovals, which effectively "cuts" such statements out of, or removes them from, the sheet of assertion. For instance, Peirce cuts off the antecedent and consequent of a conditional from the sheet of assertion by drawing an oval around both. He then separates the items in the antecedent position from the items in the consequent by drawing an inner loop around the latter (forming what he calls a "scroll"). In the mature forms of his simplest system(s) – the alpha and beta graphs – these loops or ovals act as a sort of prohibition: one may not collectively assert the items within. Thus an oval around a single propositional icon amounts to a prohibition from its assertion: it is a form of denial, while an oval enclosing several items amounts to a prohibition of their joint assertion: it amounts to a joint incompatibility claim. Ovals may nest within one another, and so express prohibitions on patterns of assertion and denial. Peirce's sign for the conditional thus amounts to the prohibition of asserting (the joint contents of) the antecedent while also denying (the joint contents of) the consequent. So with mere juxtaposition acting as a sign of conjunction and the ovals signifying a prohibition, which doubles both as a sign of negation and a scope indicator or collection device, Peirce's alpha system of graphs can be seen to be an expressively adequate notation for sentential logic.

However, before he developed this graphical system of logical notation, Peirce briefly flirted with an alternate notation for expressing the "illative relation" between antecedents and consequents, or premises and conclusions [1]. Earlier notations had expressed this relation with a complex sign, roughly ⌐/ or ⌐., which Peirce decomposed into two distinct parts. The first is the top line, or *vinculum*. Peirce took this part to apply to, and extend over, the antecedent. It is a forerunner of his "cut." The second, roughly in the shape of a v or a+, Peirce took to separate the antecedent from the consequent. Since a material conditional is equivalent to a disjunction of the negated antecedent and the consequent, Peirce thereby took the vinculum as a sign of both negation and scope (as it still operates in some notations to this day), and the other as a sign of disjunction (echoes of this form of symbolization also persist). Simplifying this further, Peirce modifies the vinculum into the cut, and dispenses with the second sign altogether, treating juxtaposition outside of cuts to be disjunctive. Note, however, that at this point juxtaposition within a single cut is conjunctive, while justification outside is disjunctive. Finally, Peirce notices that he can provide a unified treatment of juxtaposition, were he instead to break apart the vinculum so that multiple lines (or cuts) are drawn over (or around) each respective premise individually. This works, since by DeMorgan's Rule: A B C D[where juxtaposition is treated conjunctively and the box indicates the scope of the vinculum] is equivalent to A̅ B̅ C̅ D̅[where the juxtaposition is treated disjunctively]. Thus we have arrived at Peirce's notation of "entitative graphs," which he later observes are basically his subsequent system of existential graphs "turned inside out."

Though the entitative graphs, in which juxtaposition is to be understood disjunctively, is expressively equivalent to the existential graphs, in which juxtaposition is understood conjunctively, Peirce quickly abandons them as hauntingly unnatural and "aniconic." Here's what he had to say about the matter:

It is a subsidiary recommendation of a mode of diagrammatization, but one which ought to be accorded some weight, that it is one that nature and habits of our minds will cause us at once to understand, without our being put to the trouble of remembering a rule that has no relation to our natural and habitual ways of expression. Certainly, no convention of representation could possess this merit in a higher degree than the plan of writing both of two assertions in order to express the truth of both. It is so very natural, that all who have ever used letters or almost any method of graphic communication have resorted to it. It seems almost unavoidable, although in my first invented system of graphs, which I call *entitative graphs*, propositions written on the sheet together were not understood to be independently asserted but to be *alternatively* asserted [...] One system seems to be about as good as the other, except that the unnaturalness and aniconicity haunt every part of the system of entitative graphs, which is a curious example of how late a development simplicity is [6: 138-9].

There are (at least!) two things to observe about this passage. First is a characteristically Peircean focus upon written (or scribed) systems of communication, rather than spoken ones. The second is that Peirce doesn't really explain the naturalness of defaulting to assertion, so much as he reaffirms it all over again. Both the entitative and the existential graphs are scribed on a sheet with assertive import. The naturalness of defaulting to assertion in the case of diagrammatic expression seems to derive from a similar defaulting to assertion that is already present as a "habitual" means of expression in the spoken medium.

Still, I think we can push Peirce's line a little further into something of a justification for why it is so natural to prioritize assertion over denial more generally. It was important for Peirce, that the independent scribing of two propositions on a logical diagram be logically equivalent to writing them down alongside one another. What is required, then, is a pairing of a force and a logical operation in which the latter distributes into the former. As Peirce points out, assertion and conjunction perform this trick admirably. The assertion of a conjunction of two statements is equivalent to asserting them both separately. Assertion distributes into conjunction, but not disjunction. That's why his entitative graphs are so aniconic. They have at their very heart a mismatch between the diagram's fundamental force and logical operation.

Since assertive force distributes in and out of conjunction, it is natural to think of a logical diagram in which juxtaposition or co-location of propositional signs amounts to their conjunction to be inscribed on a sheet of assertion. Still, that doesn't yet amount to a full-fledged defense of the priority of assertion, because denial distributes into disjunction. So why couldn't we just turn our logical diagrams inside out more thoroughly, and consider entitative graphs scribed on a sheet of denial? After all, isn't this what happens on the right-hand sides of our bilateral tree-proofs, as well as on our falsity trees [as well as the obverse of Peirce's Phemic Sheet!]?

4 The Topology of Assertion

At this point, I recommend that we imagine actually playing a "topsy-turvy" *sprachspiel*, which norms to *falsity* and *disbelief*, and so has its "downtown" occupied by denial rather than assertion. Some of you might have played (or tried to play) such a game on "opposite day," in which the challenge was to say exactly the opposite of what you mean. Such childish games, of course, break down pretty quickly (and hilariously), if only because it runs against all force of habit and convention. However, in order to capture at least one reason why such a practice would break down naturally, let's first imagine this game being played on special pieces of paper, posted perhaps on a bulletin board called the "board of denial." That is, anytime we encounter a piece of paper with some declarative sentence written on the board, we conventionally understand that paper to express the *denial* of that statement's content. Thus when I encounter messages on which my partner has written "The dishes have been done." and "The clothes have been folded.", I take them to express the falsity of each.

Denial, however, distributes disjunctively, not conjunctively. Together, then, the *two* notes mentioned above would amount to the denial of "Either the dog has been fed OR the clothes have been folded." Suppose now that I wanted to deny something weaker – that both the dog has been fed and the clothes have been folded. The natural and convenient thing to do would be to put *both* of those sentences on *the same note* – amounting to a denial of their conjunction.

Once again, we have found good reason for thinking of juxtaposition or collocation on the same note connoting or signifying conjunction. But now we have done so without appealing to the fact that assertion distributes into conjunction. Quite the opposite; it is motivated by the fact that denial does *not* distribute into conjunction. As in Peirce's cuts, the use of distinct pieces of paper would be an easy and natural device for expressing distinct incompatibility sets – sets of contrary sentences not all of which can be true at once. We could do so simply by listing all of the members of such a set on a single piece of paper, and then posting it on the board.

Unlike Peirce's cuts, however, the boundaries of the pieces of paper are not also operating as a negation sign or to indicate a change in force. The default force is that of denial, meaning that any note on the sheet of denial is thereby understood as denied, and that distinct pieces of paper deny their contents separately and independently. The juxtaposition of distinct notes thus operates conjunctively as well, though it does so in a manner that requires that two distinct notes will have a different significance from a single note that lists the contents of them both. That seems to be the heart of the problem Peirce would have with starting out with something like denial as the dominant force behind his diagrams. In order to make messages on a board of denial work, one would also have to avail themselves to something like a multitude of distinct pieces of paper. It would be simpler to post or inscribe things directly to the board without having to put them to paper first. But in order to avoid that, the inscription of A and B on the board separately will have to have the same logical significance of writing them down together. That is the great advantage of assertion over denial. So Peirce's decision to conceive of his diagrams inscribed, specifically, on a sheet of *assertion* comes down to *a topological choice* when he conceives of the expanse upon which his logical diagrams are inscribed to be a singular undifferentiated expanse – one that is not cut up into independent discrete regions, as signified by the distinct pieces of paper.

This justification for prioritizing assertion in Peirce's logic diagrams would seem to apply even more naturally to the spoken medium. For the expanse into which our spoken utterances are projected (that is, the air around is) is similarly undifferentiated. Suppose someone utters one statement, and then they go on to utter another. In a topsy-turvy *sprachspiel* – one that norms to falsity – there lurks a systematic ambiguity. The denying of two statements jointly is not equivalent to their joint denial. Should this speaker be interpreted as denying both statements independently, or rather denying their conjunction? The trouble is that without some explicit manner of punctuation to indicate where individual speech acts begin and end, or an explicit sign of force marking the boundaries between speech acts, one cannot say. That of course was the function of the *separate pieces of paper* in our earlier game (as well as Frege's *urteilstrich*). The spoken medium, however, lacks such resources.

In short, we have found how a rather mundane, topological feature of the spoken medium might contribute to the explanation of our nearly irresistible predilection to speak truly. Ours is not a comic book world; our spoken utterances don't neatly sort themselves out into discrete talk or thought boxes, which could then be used to frame distinct speech acts. And so it behooves us to make our statements against a backdrop in which they can be made independently, without the need for such framing. Assertion allows us to do this; denial does not. This humble fact, I would suggest, accounts for why speaking falsely seems so abominably unnatural to us. If this is correct, the great Zoroastrian contest between the forces of Asha (truth) and the forces of Druj (falsity) is not a struggle over the Cosmic World Order. Nor does it turn on any alleged superiority of a metaphysics of presence over that of absence. Nor even does it turn on logic. Rather, our enduring love of truth – enshrined in the name of academia's second oldest profession (behind geometry) – is a product of a most undramatic pragmatic dimension of our discursive practice.

References

1. Bellucci, F., Pietarinen, A.-V.: From Mitchell to Carus: fourteen years of logical graphs in the making. Trans. Charles S. Peirce Soc. **52**, 539–375 (2016)
2. Brandom, R.: Between Saying and Doing. Oxford University Press, Oxford (2007)
3. Fitts, J., Beisecker, D.: Teach. Philos. **42**, 41–56 (2019)
4. Ito, K.: Truth and falsity in communication: assertion, denial, and interpretation. Erkenntnis **88**, 657–674 (2023)
5. Massey, G.: The indeterminacy of translation: a study in philosophical exegesis. Philos. Top. **20**, 317–345 (1992)
6. Peirce, C.S.: Logical tracts. No. 2 (R492): on existential graphs, Euler's diagrams. And logical algebra. In: Pietarinen, A.-H. (ed.) Logic of the Future, vol. 2/1. De Gruyter (2021)
7. Incurvati, L., Schlöder, J.: Weak assertion. Philos. Q. **69**, 741–770 (2019)

Schopenhauer's Sorites Diagram

Christina Kittsteiner

FernUniversität in Hagen, Hagen, Germany
`christina.kittsteiner@studium.fernuni-hagen.de`

Abstract. Arthur Schopenhauer was one of the first logicians of the 19th century to develop a visual representation of the sorites with which it is possible to depict multiple terms in one diagram. With the help of an example from Seneca, he explains how the sorites diagram works and provides two reading interpretations for the diagram. Since Schopenhauer's sorites diagram in particular or sorites diagrams in general have never been dealt with in research before, this paper will introduce and analyse Schopenhauer's sorites diagram and its different reading possibilities. Thereby, it will not only be shown how the sorites diagram can help to dissolve the sorites back into individual syllogisms, but also other topics such as the reading directions, the extension and intension of the sorites diagram will be discussed in more detail. It turns out that Schopenhauer's sorites was just the foundation for further works in his philosophical system.

Keywords: Arthur Schopenhauer · Logic diagrams · Sorites

1 Introduction

In recent times, the research on Schopenhauer turns to his logic works and figures out many unexpected considerations that show on the one hand that Schopenhauer engaged very comprehensive with logic (see e.g. [3, pp. 1–2], [4] and [5]) and on the other hand that he also tried to teach logic by showing intuitively understandable examples as in his *Berlin Lectures* from 1820 [8], where he created various diagrammatic representations in the chapters about logic and eristic.

This paper will take a look on Schopenhauer's sorites diagram, which is located in the chapters on logic in the *Berlin Lectures*. To the best of our knowledge, sorites diagrams in general were never treated in research before and there is no author who ever discussed Schopenhauer's sorites diagram.

The sorites diagram was significant for Schopenhauer, because he wanted to connect more than three terms together, a circumstance that is especially noteworthy, since diagrammatic reasoning involving n-terms was long considered a modern innovation [6, p. 122] and not common in the early 19th century. Because of this, the depiction of the sorites can be seen as a diagrammatic novelty in Schopenhauer's period.[1]

---

[1] Other diagrammatic representations from Schopenhauer can even depict up to 30 terms, but they concern eristic considerations and are thus not treated in this paper.

It is important for the further course of this work, to distinguish the sorites syllogism, which is depicted with the sorites diagram, from the widely known sorites paradox, which is also known as the heap paradox [7]. Where the paradox denotes problems regarding the formulation of vagueness, the structure of the sorites syllogism in the diagram has the purpose of connecting multiple terms together to form complex inferences.

In the following, this paper will introduce Schopenhauer's sorites diagram in Sect. 2 and then explore their diverse interpretation possibilities in greater detail. Three key aspects of the sorites diagram are its potential to help with the dissolution of a sorites in Sect. 3, the reading direction in Sect. 3.1 and the extension and intension of the sorites diagram in Sect. 3.2. Section 4 will conclude with a short summary and an outlook for further research possibilities.

2 The Sorites Diagram in Schopenhauer's Work

Prior to the introduction of the sorites diagram, Schopenhauer had already used various circle-shaped Euler diagrams for the depiction of logical relationships like terms, judgements and inferences (see [8, p. 211–305] and [5]). To eventually evolve the sorites diagram out of this exploratory work, Schopenhauer specifically introduces the *categorical* sorites as a syllogism that chains together more than three terms (which is why a sorites can also be called a *chain-syllogism* or a *polysyllogism* [8, p. 314–315]). Within a categorical sorites, the predicate of each sentence is the subject of the following one and can look like the example in Table 1, which was taken by Schopenhauer from Seneca's epistle no. 85.[2]

Table 1. Epistle No. 85 from Seneca's *Epistuale Morales Ad Lucilium* [8, p. 314] and its translation.

Qui prudens est, et temperans est:	Those who are prudent, are also moderate:
qui temperans est, et constans:	those who are moderate, are also consistent:
qui constans est, et impertubatus est:	those who are consistent, are also unshaken:
qui impertubatus est, sine tristitia est:	those who are unshaken, are also free from sadness:
qui sine tristitia est, beatus est:	those who are free from sadness, are happy:
ergo prudens beatus est:	that is why those who are prudent are happy:
(et prudentia ad vitam beatam satis est)	(and prudence suffices for a happy life)

Schopenhauer then illustrates this sorites as a circle-shaped diagram that builds up on the diagrams he already entrenched in his earlier works. The emerging sorites diagram consists of multiple circles, which are nested inside one another as it is shown in Fig. 1.

---

[2] All translations in this work were made by the author.

(a) Schopenhauer's sorites diagram taken from *Berlin Lectures*, StB PK, Na 50, NL Schopenhauer, 1428, Bl. 150 (urn:nbn:de:hebis:30:2-417557) Also at [8, S. 314].

(b) The sorites diagram and its translated description in clear text.

Fig. 1. Schopenhauer's sorites diagram. Instead of variables, he presents the circles filled with terms.

3 Dissolving the Sorites with Help of the Sorites Diagram

Dissolving a sorites into its different sub-syllogisms can be especially important, if the sorites is intricately arranged and therefore difficult to overlook. The sorites diagram is a useful diagrammatic tool for supporting this process, because it visualizes the matter and yields a more intuitive understanding, as will be shown in the following: Because of its circle-like structure, a sorites diagram can be read from the center to its periphery or vice versa. These two reading variants are called the *Aristotelian* and the *Goclenian* [8, pp. 314–315]. Both are visualized in Fig. 2. To dissolve the sorites, Schopenhauer uses these two reading variants to describe directions:

> The common Aristotelian, or regressive sorites goes upwards from the narrower to the broader term, can be dissolved into individual inferences, which are insofar called prosyllogisms, the whole chain of the same polysyllogism. The progressive or Goclenian sorites goes downwards from the broadest term to the narrowest; is dissolved in episyllogisms. [8, pp. 314-315]

The Aristotelian reading starts by taking into account that the term *prudent* is not only the center of the sorites diagram in Fig. 2, but also the subject of the first line of Seneca's sorites example in Table 1 (*"Qui prudens est [. . .]"*). Because of this, *prudent* (P) must only occur in the second (or minor) premise of all the dissolving prosyllogisms. Starting at the center of the sorites diagram

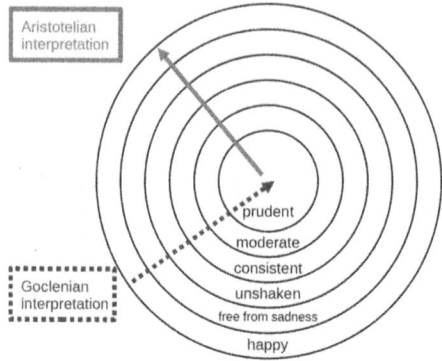

Fig. 2. Depiction of the different reading variants of a sorites diagram.

in Fig. 2, the dissolution can be made according to the instructions of Table 2 where every middle-circle is covered until the periphery *happy* (H) is reached. The terms *moderate* (M), *consistent* (C), *unshaken* (U) and *free from sadness* (F) are the middle terms, which is why four inferences must be build.

Table 2. Dissolution process with prosyllogisms.

Inference 1:	Inference 2:	Inference 3:	Inference 4:
M is C	C is U	U is F	F is H
P is M	P is C	P is U	P is F
Thus P is C	Thus P is U	Thus P is F	Thus P is H

To dissolve the sorites according to the Goclenian variant, it must be taken into consideration that this variant starts with the outer circle, *happy*, which is the predicate of the sorites ("[...] *beatus est*"). Because of this, the term *happy* must exclusively occur in the first (or major) premise of the episyllogisms like it is shown in Table 3.

Table 3. Dissolution process with episyllogisms.

Inference 1:	Inference 2:	Inference 3:	Inference 4:
F is H	U is H	C is H	M is H
U is F	C is U	M is C	P is M
Thus U is H	Thus C is H	Thus M is H	Thus P is H

Table 4 shows the summary reading after the dissolution process was applied.

Table 4. Results of the dissolution according to the two reading variants.

Aristotelian reading:	Goclenian reading:
the one who is prudent is moderate,	the one who is free fr. sadn. is happy,
the one who is moderate is consistent,	the one who is unshaken is free fr. sadn.,
the one who is consistent is unshaken,	the one who is consistent is unshaken,
the one who is unshaken is free fr. sadn.,	the one who is moderate is consistent,
the one who is free fr. sadn. is happy,	the one who is prudent is moderate,
the one who is prudent is happy	the one who is prudent is happy

At this point, it can be observed very well that the dissolution process that was constructed with help of the Tables 2 and 3 can be made much faster and much more intuitive, if the sorites diagram from Fig. 2 is used.

3.1 The Interpretation of the Sorites Diagrams' Reading Direction

Schopenhauer's use of prosyllogisms and episyllogisms fits very well into his line of thinking, since he already discusses the methods to gain knowledge on the basis of the dissection of terms as *analysis* (or *definition*) and the addition of terms as *synthesis* in an earlier chapter of the *Berlin Lectures* [8, p. 67–68]. This is important, because, as a passionate Kant reader, Schopenhauer must have been considering Kant's *Jäsche Logic*: The prosyllogism, or regressive variant, can be understood as an analytic method and the episyllogism, or progressive variant, can be understood as a synthetic method [2, p. 148–149, p. 162–163; §87, §88, §117].

The reading variants and the assertions concerning the pro- and episyllogisms can therefore be interpreted reasonably, but Schopenhauer's designation of the reading direction as "upwards" and "downwards" from the above given quote seems problematic. According to Schopenhauer, the Aristotelian reading should proceed upwards (from the center to the periphery), which is depicted correctly in Fig. 2. But already the downwards directing Goclenian variant does not follow Schopenhauer's presets. The arrow in Fig. 2 is directing upwards, although it is clearly following the direction from periphery to the center.

However, if the sorites diagram would be thought of as a 3-dimensional funnel, then the direction towards the center would always go downwards. Likewise, the reading direction towards the periphery would always go upwards. But this way is not how a sorites diagram must be understood, because it shows 2-dimensional circles and not a 3-dimensional figure. Either the sorites diagram is not suitable to make a conclusive visual assertion about the respective direction of the two different reading variants as it was done by Schopenhauer, or the designation "upwards" and "downwards" is problematic, since it implies a spatial direction. If Schopenhauer meant to describe the direction of the scope, another designation would be less ambiguous. For this, one could use, e.g. the direction specification "centered" and "peripheral" in the case of the sorites diagram.

3.2 Extension and Intension of the Sorites Diagram

Schopenhauer wrote the following comment alongside his drawing of the sorites diagram:

> Each of these terms has initially the one that encloses it as its predicate, and thus indirectly all the others including the outermost; so they are actually all already thought in prudent, but implicitly; the sorites states it explicitly. [8, p. 314]

With this, Schopenhauer claims that ultimately all terms in the sorites diagram are taken together in the centered term *prudent*. But this claim is problematic, since the sorites diagram does not show this relation. Schopenhauer himself called the center of the sorites (*prudent*), the "narrowest" term of the diagram, what makes it difficult for it to embrace other terms. The term *happy* (labeled the "broadest" term) seems the most likely to contain the other terms. To analyze this issue further, both the extension and the intension of the sorites diagram with a focus on the term *prudent* shall be analysed.

The extension of the term *prudent* is the amount of all terms that fall under *prudent*, that is, all terms to which *prudent* refers to [10, p. 365]. Regarding Schopenhauer's sorites diagram, the extension of *prudent* would be the entirety of adjectives that describe being prudent: *moderate, consistent, unshaken, free from sadness* and *happy*. Looking at Fig. 3a, where the extensional interpretation is depicted, it can be observed that this formulation corresponds exactly to the sorites diagram of Fig. 1b. But following this thought, all individuals, who are prudent would only be a subset of those, who are *moderate, consistent*, etc., which in the end implies that there are more individuals, who are moderate or consistent (etc.) than there are individuals, who are prudent. Therefore it can be stated that Schopenhauer's comment was indeed correct when it came to the extensional consideration of the term *prudent* as a diagrammatic representation in the sorites diagram, but with regard to its content, the assertion is not correct.

The intension is the content of a term, that is, the attributes which are shared by its extensional terms (*moderate, consistent*, etc.) [10, p. 365]. So, the intension of the term *prudent* includes the properties of being moderate, being consistent, being unshaken, being free from sadness and being happy. This way, the intensional interpretation of the term *prudent* is becoming inverse to the extensional one: Because the term *prudent* includes the terms *moderate, consistent* (etc.), the circles of the sorites diagram are inverted as depicted in Fig. 3b. Here, Schopenhauer's comment would be correct regarding the terms (*moderate, consistent* etc.), which are "included *in*" or "thought *in*" prudent, but the depiction of the sorites diagram does not fit. It seems that the intensional meaning of the term *prudent* is mixed up with its extensional diagrammatic representation. But what seems to be a mistake, can be explained with help of the *Jäsche Logic* from Kant again. Since Kant did have a big influence on Schopenhauer's thinking, Schopenhauer's understanding of the concepts of extension and intension may originate there.

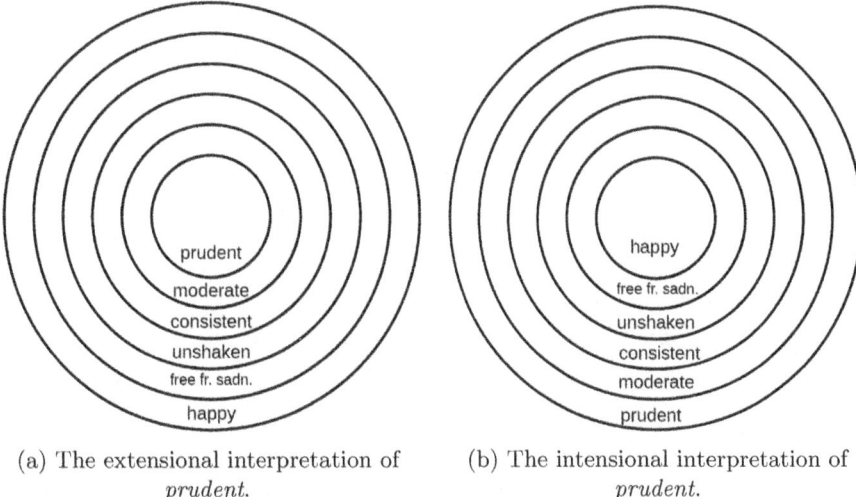

(a) The extensional interpretation of *prudent*.

(b) The intensional interpretation of *prudent*.

Fig. 3. Different interpretations of the term *prudent* (see also [1, p. 58]). These circles show also the different hierarchical reading from Sect. 3.1.

In §7 of the *Jäsche Logic* can be read that the extension and the intension of a term are in an inverse relationship to each other [2, p. 104; §7]. Figure 3 supports this assertion. However, the next paragraph, §8, is of crucial importance, because Kant writes: "The extent [...] of a term is the greater, the more things are under it and can be thought through it." [2, p. 104; §8]. This is almost exactly the same formulation that Schopenhauer uses. It can thus be concluded that Schopenhauer probably had in mind to attribute an extensional reading to the term *prudent*, but that he made a mistake in his formulation so that the two different interpretations were mixed up.

4   Summary and Outlook

Schopenhauer's own interpretative attempts on the sorites diagram are not always clear and leave room for questions and different interpretative approaches. Nevertheless, diagrammatic reasoning on multiple terms does not only become easier to execute with the help of the sorites diagram, but also yields the advantage of intuitive usage that comes with visual representation. This could be observed particularly well regarding the dissolution of a sorites with help of the sorites diagram. Thereby, the sorites diagram is a logical advancement from Schopenhauer's initially used circle-shaped diagrams. Where this early developed diagrams displayed concepts, judgements and inferences, the combination of multiple circles enabled Schopenhauer to go further in his analysis by showing more than three terms.

All in all, the structure of the sorites syllogism allows multiple interruption-free chains of terms to form complex assertions that can be used as successive parts of an argumentation or presentations of proofs. Schopenhauer also surmises that each scientific and also philosophical consideration emanates from a sorites-like chain of thought [8, p. 355]. He uses the sorites also to lay the foundation for a new direction, where, according to the intention of the individual, a sorites can be used to lead a chain-like argumentation (or proof). This can lead to correctly guided results to deliberately confusing or mislead argumentation and especially the latter is often used to persuade someone from an opinion. At this point, Schopenhauer attached the first further investigations into a topic, which is now known as the Eristic Dialectic (see [8, pp. 351-355, especially p. 354], also [9]).

It is beyond the scope of this work to ask whether Schopenhauer used the sorites diagram in his remarks on the Eristic Dialectic or not, and, if so, which form these remarks have. Therefore, it remains the task of further research, to connect Schopenhauer's diagrammatic reasoning and the sorites diagram with the topic of the Eristic Dialectic.

Acknowledgement. I thank Jens Lemanski for his help on the development of this paper.

References

1. Bernhard, P.: Euler-Diagramme. Zur Morphologie einer Repräsentationsform in der Logik. Paderborn: mentis (2001)
2. Kant, I; Jäsche G.B.: Logik. Ein Handbuch zu Vorlesungen, herausgegeben von G. B. Jäsche. Erläutert von J.H.v. Kirchmann. Berlin: L. Heimann (1869)
3. Lemanski, J.: An introduction to language, logic and mathematics in Schopenhauer. In: Lemanski, J. (ed.) Language, Logic, and Mathematics in Schopenhauer. SUL, pp. 1–11. Springer, Cham (2020). https://doi.org/10.1007/978-3-030-33090-3_1
4. Lemanski, J.: World and Logic. Rickmannsworth: College Publications, Norcross (2021)
5. Lemanski, J.: Schopenhauer. Logic and Dialectic. In: Internet Encyclopedia of Philosophy, https://iep.utm.edu/schopenhauer-logic-and-dialectic/. Accessed 10 Feb 2024
6. Moktefi, A.: Schopenhauer's Eulerian diagrams. In: Lemanski, J. (ed.) Language, Logic, and Mathematics in Schopenhauer. SUL, pp. 111–127. Springer, Cham (2020). https://doi.org/10.1007/978-3-030-33090-3_8
7. Stanford Encyclopedia of Philosophy. https://plato.stanford.edu/entries/sorites-paradox/. Accessed 10 Feb 2024
8. Schopenhauer, A.: Vorlesung über Die gesamte Philosophie. 1. Teil: Theorie des gesamten Vorstellens, Denkens und Erkennens. Vol. 4 Ed. Schubbe, D. et al., Hamburg: Meiner (2022)
9. Schopenhauer, A.: Die Kunst, recht zu behalten. Stuttgart: Reclam (2023)
10. Stapleton, G., Moktefi, A., Howse, J., Burton, J.: Euler diagrams through the looking glass: from extent to intent. In: Chapman, P., Stapleton, G., Moktefi, A., Perez-Kriz, S., Bellucci, F. (eds.) Diagrams 2018. LNCS (LNAI), vol. 10871, pp. 365–381. Springer, Cham (2018). https://doi.org/10.1007/978-3-319-91376-6_34

Category Theory for Aristotelian Diagrams: The Debate on Singular Propositions

Alexander De Klerck, Leander Vignero, and Lorenz Demey

Center for Logic and Philosophy of Science, KU Leuven, Leuven, Belgium
{alex.deklerck,leander.vignero,lorenz.demey}@kuleuven.be

Abstract. The theoretical study of Aristotelian diagrams is at an all-time high since the conception of logical geometry. This framework studies Aristotelian diagrams in a systematic way, revealing many links with contemporary mathematics (esp. algebra). Most recently, this has led to the introduction of several notions of morphism between Aristotelian diagrams, which we are studying in the context of category theory. This is not merely a mathematical enterprise, but also carries major philosophical importance. As a proof of concept of this claim, we investigate the historically rich discussion on the status of singular propositions. It has been debated for centuries whether these should be viewed as a special kind of universal propositions or particular propositions, or as a third, completely separate kind. Interpreting each of these views as a morphism in one of our categories, we obtain a clean picture of the entire discussion in a single image. Additionally, we apply the machinery from category theory (in casu, the notion of equalizer) to make some interesting comparative observations regarding the three views on singular propositions.

Keywords: Aristotelian diagrams · Logical diagrams · Logical geometry · Category theory · Categorification · Singular propositions

1 Introduction

Aristotelian diagrams, such as the square of opposition, have a rich history in philosophy and logic [1,2]. Their theoretical investigation has flourished in the past two decades [3], esp. with the arrival of logical geometry [1,4]. This framework studies Aristotelian diagrams in a systematic fashion, and currently often involves the mathematical area of category theory [5,6]. More precisely, it can be shown that Aristotelian diagrams, together with a suitable notion of morphism between diagrams, constitute a category. This categorification of logical geometry entails a change of perspective: rather than focusing on the Aristotelian diagrams by themselves, we now view them as objects within a category, shifting our attention to the various morphisms that exist between them.[1]

The program of categorification is not only driven by internal mathematical considerations, but also carries a great potential for logical geometry itself. The present paper can be viewed as a proof of concept of this logical fruitfulness. More concretely, we will consider the historical debate on the logical status of singular propositions (e.g., 'this S is P') vis-à-vis universal propositions (e.g., 'all S are P') and particular propositions (e.g., 'some S are P'), and demonstrate how category theory can help us visualize and investigate this discussion.

In Sect. 2 we provide some background on Aristotelian diagrams and the program of categorifying logical geometry. Then, in Sect. 3, we present a brief overview of the historical debate on singular propositions, and place this debate in the framework of category theory. Finally, Sect. 4 wraps things up.

2 Aristotelian Diagrams and Category Theory

In contemporary logical geometry, we often use a general definition of Aristotelian diagrams in terms of Boolean algebras, because of various mathematical and conceptual considerations [5,6]. We will follow this trend in this paper too.

Definition 1. *(Aristotelian diagram) An Aristotelian diagram D is a pair $(\mathcal{F}, B)$, where B is a Boolean algebra $(B, \wedge_B, \vee_B, \neg_B, 1_B, 0_B)$ and $\mathcal{F}$ is a fragment of B, i.e., $\mathcal{F} \subseteq B$. When the Boolean algebra B is clear from context, it is usually omitted as a subscript to $\wedge$, $\vee$, etc.*

When such a diagram is visualized, the elements of $\mathcal{F}$ are connected using the four Aristotelian relations, as in Fig. 1. These naturally belong to two other sets of relations, namely the opposition and implication relations [1].

Example 1. Two very different, elementary examples of Aristotelian diagrams are shown in Fig. 1. The one on the left is an informal version of an instance of the traditional square of opposition that was mentioned before, and it has as its Boolean algebra B the Lindenbaum-Tarski algebra of syllogistics. The one on the right is situated in the bitstring Boolean algebra $\{0,1\}^3$.

The categorification program of logical geometry was first initiated by Vignero in [5], and significantly expanded in [6]. In the remainder of this paper, we make heavy use of both the terminology of category theory and the notions we introduced in [6], the most important one for this paper being the notion of increasing information $\mathcal{OR} \times \mathcal{IR}$ morphism (which we will call 'increasing infomorphism' in the present paper, for terminological simplicity).

Categories in which the objects are Aristotelian diagrams are denoted by the letter $\mathbb{D}$, together with some sub-/superscripts that indicate which kind of morphisms are present. In [6], we defined and studied ten such categories. A category

---

[1] While some promising and important first steps have been taken [6], this research line is still far from finished. We are currently preparing a book-length study that investigates and compares the most important categories of Aristotelian diagrams.

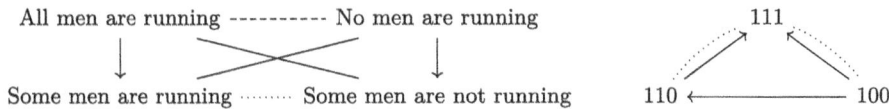

Fig. 1. Two examples of Aristotelian diagrams. Contradiction, contrariety, subcontrariety and subalternation are visualized as resp. solid, dashed, dotted lines and arrows.

always gives rise to some notion of 'isomorphism', which captures exactly what it means for two objects to have 'the same structure'. For example, the category $\mathbb{D}^{Inc}_{\mathcal{OR} \times \mathcal{IR}}$ is a category in which the objects are Aristotelian diagrams, the morphisms are increasing information $\mathcal{OR} \times \mathcal{IR}$ morphisms, and two Aristotelian diagrams are isomorphic iff they have the same Aristotelian structure.

In [6], we concluded that, of the ten categories we defined, $\mathbb{D}^{Inc}_{\mathcal{OR} \times \mathcal{IR}}$ and $\mathbb{D}^{Dec}_{\mathcal{OR} \times \mathcal{IR}}$ are the most worthy of further study. We are currently investigating these two categories, together with a new one that focuses on the Boolean structure, on a deeper level of category theory (limits and colimits, adjunctions, etc.) (cf. Footnote 1). This work has serious implications for previous as well as future research in logical geometry. As for previous research, this framework allows us to look back and reinterpret it in the language of category theory. For example, we have been able to summarize entire philosophical discussions in a simple way in these categories. As for future research, this framework also allows us to make new concrete observations that arise from theoretical insights.

This paper is meant as a proof of concept of the categorification program. More concretely, we will show that we can summarize the historical discussion on the status of singular propositions in one of our categories. In this way, we acquire an illuminating overview of this debate, which is desirable from a cognitive point of view. Additionally, by lifting the debate to this framework, we enable ourselves to use tools from category theory to make further observations.

3 The Debate on Singular Propositions

The traditional square of opposition in Aristotelian logic was already briefly touched upon in Example 1 and Fig. 1. It comprises four categorical propositions that contain the same subject and predicate (S and P). These propositions can be classified according to their quantity (universal/particular) and quality (affirmative/negative), as shown in the following table:

	Affirmative	Negative
Universal	(A) All S are P	(E) No S are P
Particular	(I) Some S are P	(O) Some S are not P

Next to the universal and particular propositions, Aristotelian logic also considers indefinite propositions (which lack an explicit quantifier altogether, like

'S is P', e.g., 'Men are running') and singular propositions (in which the subject term refers to a single entity, like 'This S is P' or 's is P', e.g., 'This man is running' or 'Socrates is running').[2] While indefinite propositions are standardly taken to be equivalent to particular propositions [2], the quantificational status of singular propositions has been under debate for centuries. We start by summarizing this debate in Sect. 3.1. For a more extensive historical overview, including quotations from primary sources, see [7]. In Sect. 3.2, we then show that, in the category $\mathbb{D}_{\mathcal{OR} \times \mathcal{IR}}^{Inc}$ from Sect. 2, this entire debate can be summarized in a single illuminating image.

3.1 Different Views on Singular Propositions

Historically speaking, by far the most popular position has been that singular propositions are simply a special case of either universal or particular propositions; we will call these the u-view and the p-view, respectively.[3] Such a view is needed to maintain the position that syllogisms consist exclusively of universal and particular propositions (as specified in their so-called 'mood'), while also accepting the famous 'Socrates' argument as an actual syllogism.[4] However, some authors argue that singular propositions are neither universal nor particular, but are genuinely *sui generis*; we will call this the s-view.

The u-view: Singular Propositions are a Kind of Universal Propositions. This view has been most popular throughout history. It was held by authors such as Lambert of Auxerre, John Wallis, Leibniz, Euler, Whately, Carroll and Keynes [7], and even a contemporary champion of symbolic logic like Quine called this view "artificial but not incorrect" [9, p. 78]. Proponents of the u-view argue that in singular propositions, the entire subject is predicated. This is obviously the case in universal propositions as well, but not in particular ones. Consequently, singular propositions must be a special kind of universal propositions. Diagrammatically speaking, this means that the singular proposition 'This man is running' should coincide with the upper left corner (A) of the square from Fig. 1.

Thus far we have only focused on the u-view's treatment of *affirmative* singular propositions. When we turn to *negative* singular propositions, however, we are faced with some interpretative difficulties. On the one hand, most proponents of the u-view claim (with various degrees of explicitness) that their view applies to all singular propositions, affirmative and negative alike. They are thus committed to viewing 'this man is not running' as an E-proposition, just like they view 'this man is running' as an A-proposition. On the other hand, the resulting

---

[2] The last example is to be considered under the assumption that Socrates is a man, or in general, that s is an S.

[3] Khomskii [7] refers to the u-view as 'view 2', and to the p-view as 'view 1'.

[4] The 'Socrates' argument runs as follows: 'All men are mortal; Socrates is a man; therefore Socrates is mortal'. The u-view and the p-view interpret this argument as a Barbara and a Darii syllogism, respectively [8, p. 108].

position is logically inconsistent, since affirmative and negative universal propositions (A and E) are *contrary* to each other, whereas affirmative and negative singular propositions are standardly taken to be *contradictory* to each other. Most proponents of the u-view do not seem to have been aware of this inconsistency, most likely because they simply did not pay much attention to negative singular propositions; this is illustrated by the fact that nearly all examples that they gave of singular propositions were affirmative. The only exception is Leibniz, who gave both affirmative and negative examples of singular propositions, and explicitly acknowledges that, in order to maintain the contradictoriness between affirmative and negative singular propositions, the identification of affirmative singular propositions with universal (A) propositions commits us to identifying negative singular propositions with particular (O) propositions:

> "How is it that [contradictory] opposition is valid in the case of singular propositions—e.g. 'The Apostle Peter is a soldier' and 'The Apostle Peter is not a soldier'—since elsewhere a universal affirmative and a particular negative are [contradictorily] opposed? Should we say that a singular proposition is equivalent to a *particular* [viz., in case it is negative] and to a *universal* proposition [viz., in case it is universal]? Yes, we should." [10, p. 115, emphases added]

In this way, we obtain a consistent version of the u-view, which will henceforth be called the u*-view. This u*-view can be ascribed to Leibniz directly, and can be seen as a rational reconstruction of other historical proponents of the u-view.

The p-view: Singular Propositions are a Kind of Particular Propositions. This view was held by R. F. Clarke [7], but not many other explicit proponents have been identified. However, it must have had some historical popularity, because an author like Euler, for example, explicitly wrote that certain people defended the p-view (without specifying any names), and then presented the u-view in opposition to this [7]. Proponents of the p-view focus on the fact that affirmative singular propositions are concerned with an individual $s \in S$. Now, particular propositions are concerned with *at least one* of the S, whereas universal propositions are concerned with *all* of the S. Since an individual is just one, singular propositions must be a special kind of particular propositions. Diagrammatically speaking, this means that the singular proposition 'This man is running' should coincide with the lower left corner (I) of the square from Fig. 1.

When we turn to negative singular propositions, entirely analogous remarks can be made regarding the p-view as were already made above regarding the u-view. More specifically, the p-view turns out to be inconsistent (because it views affirmative and negative singular propositions as subcontraries, rather than contradictories), but it can be turned into a consistent p*-view, which holds that affirmative singular propositions are particular (I) propositions, while negative singular propositions are universal (E) propositions.

The s-view: Singular Propositions are Sui Generis. The reasoning behind this view can be traced back to William of Sherwood and Tadeusz Czeżowski [7, 11, 12].

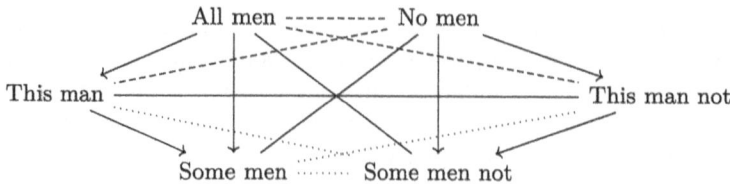

Fig. 2. An instance of a Sherwood-Czeżowski (SC) hexagon

Proponents of the s-view focus on the fact that 'This S is P' and 'This S is not P' are contradictory to each other. Consequently, they cannot be both universal in nature (as the naïve u-view would hold), nor both particular in nature (as the naïve p-view would hold). Rather than viewing one of the singular propositions as universal and the other as particular (as the u*- and p*-views hold), the s-view maintains that (both) singular propositions are *sui generis*, i.e., they are neither universal nor particular. Diagrammatically speaking, this means that the square needs to be extended to a *hexagon* of opposition, as shown in Fig. 2.[5] Aristotelian diagrams of this form are appropriately called 'Sherwood-Czeżowski (SC) hexagons' [1,4]. Since the s-view is already consistent, it does not need to be be 'repaired'; however, to maintain notational consistency, we will also refer to it as the s*-view (which is identical to the s-view).

3.2 Using Category Theory to Visualize The Debate

In this subsection, we describe how each of the five aforementioned views can be represented by a function between Aristotelian diagrams. Interestingly, each of the three *consistent* views (the *-views) corresponds to an increasing infomorphism in the category $\mathbb{D}_{\mathcal{OR} \times \mathcal{IR}}^{Inc}$, whereas the two inconsistent views do not.

Let us start with the (inconsistent) u- and p-views. Both are represented by a function from the SC hexagon of Fig. 2 to the square of Fig. 1. The u-view maps the singular propositions 'This man is running' and 'This man is not running' to the universal A- and E-propositions at the top of the square. This turns the contradiction between both singular propositions into a contrariety, which is less informative. Analogously, the p-view maps these singular propositions to the particular I- and O-propositions at the bottom of the square. This turns the contradiction between both singular propositions into a subcontrariety, which is again less informative. In Fig. 3, the u- and p-views are shown as the functions u and p. The dashed/dotted way in which these arrows are drawn is a reminder of what happens with the contradiction between the two singular propositions. It is immediately clear that neither u nor p is an increasing infomorphism.

The u*-view and the p*-view are also represented as functions between the same two diagrams. The u*-view maps the proposition 'This man is running' to the A-corner of the square, and the contradictory proposition 'This man is not running' to the O-corner. This can be visualized as the function u^* in Fig. 3,

---

[5] We leave out 'is/are running' throughout the hexagon, for reasons of visual simplicity.

where we use green ovals to indicate where this function maps the singular propositions. Note that this function is in fact an increasing infomorphism. For example, the subcontrariety between 'This man is running' and 'Some men are not running' gets mapped by u^* onto the contradiction between 'All men are running' and 'Some men are not running', and contradiction is more informative than subcontrariety. As another example, the left-implication (i.e., subalternation) from 'This man is not running' to 'Some men are not running' gets mapped to the bi-implication (i.e., equivalence) between 'Some men are not running' and itself, and bi-implication is more informative than left-implication. In a completely analogous way, we can visualize the p*-view as an increasing infomorphism. This is done by the function p^* in Fig. 3, where we use yellow rectangles to indicate where this function maps the singular propositions.

Finally, the SC hexagon already represents in its entirety the perspective of the s*-view (which is the same as the s-view). This is visualized by the function s^* in Fig. 3, which is simply the identity function on the SC hexagon (and thus trivially an increasing infomorphism). For the sake of completeness, we use blue underlining to indicate where this function takes the singular propositions.

The visualization in Fig. 3 offers certain cognitive advantages, allowing us to summarize an entire philosophical discussion in a single picture. Another major point is that we can now use the tools of category theory to compare the increasing infomorphisms of the story, which are exactly the consistent views. For example, since u^* and p^* have the same domain (viz., the SC hexagon from Fig. 2) and the same codomain (viz., the classical square from Fig. 1), we can try to look for an equalizer of these two morphisms. One of our unpublished results (cf. Footnote 1) is that all equalizers (and even all finite limits) exist in the category $\mathbb{D}^{Inc}_{\mathcal{OR} \times \mathcal{IR}}$, and they are given by the subdiagram of their domain on which both morphisms agree. More concretely, in the example of Fig. 3, the equalizer is the original square of opposition that we started with, and it can be viewed as the part on which the u*- and p*-views completely agree.

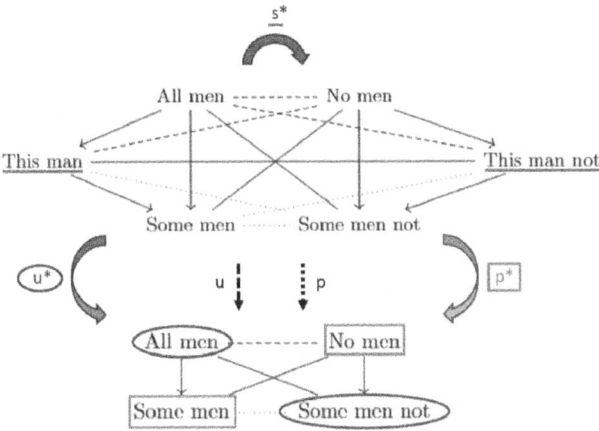

Fig. 3. Different views on singular propositions, visualized as morphisms in $\mathbb{D}^{Inc}_{\mathcal{OR} \times \mathcal{IR}}$

We can also embed the codomain of u^* and p^* inside the SC hexagon, so that we can take the equalizer of all three consistent views at the same time. Unsurprisingly, it turns out that they all agree on precisely the classical square.

4 Conclusion

The categorification of logical geometry has a major impact on previous and future research on Aristotelian diagrams. This paper served as a proof of concept of this point of view, by summarizing different historically influential accounts on singular propositions as different increasing infomorphisms between Aristotelian diagrams. Thereby, we were able to visualize a centuries-long discussion in a single picture. Additionally, we were able to use equalizers, an important concept in category theory, to quickly draw conclusions from this picture.

Acknowledgements. Funded by the European Union (ERC, STARTDIALOG, 101040049). Views and opinions expressed are however those of the authors only and do not necessarily reflect those of the European Union or the European Research Council Executive Agency. Neither the European Union nor the granting authority can be held responsible for them. Demey's work was also funded through project 3H220024/G063622N of the Research Foundation-Flanders (FWO). Thanks to Atahan Erbas, Stef Frijters, Hans Smessaert and three reviewers for their valuable feedback.

References

1. Smessaert, H., Demey, L.: Logical geometries and information in the square of oppositions. J. Logic Lang. Inform. **23**, 527–565 (2014)
2. Keynes, J.N.: Studies and Exercises in Formal Logic, 4th edn. MacMillan and Co, New York (1906)
3. Béziau, J.-Y.: New light on the square of oppositions and its nameless corner. Logical Invest. **10**, 218–232 (2003)
4. Demey, L., Smessaert, H.: Combinatorial bitstring semantics for arbitrary logical fragments. J. Philos. Log. **47**, 325–363 (2018)
5. Vignero, L.: Combining and relating Aristotelian diagrams. In: Basu, A., Stapleton, G., Linker, S., Legg, C., Manalo, E., Viana, P. (eds.) Diagrams 2021. LNCS (LNAI), vol. 12909, pp. 221–228. Springer, Cham (2021). https://doi.org/10.1007/978-3-030-86062-2_20
6. De Klerck, A., Vignero, L., Demey, L.: Morphisms Between Aristotelian Diagrams. Logica Universalis (forthcoming)
7. Khomskii, Y.: William of Sherwood, singular propositions and the hexagon of opposition. In: Béziau, J.-Y. & Payette, G. (eds.) The Square of Opposition. A General Framework for Cognition, pp. 43–60. Peter Lang, Bern (2012)
8. Englebretsen, G.: Something to Reckon With. The Logic of Terms. University of Ottawa Press, Ottawa (1966)

9. Quine, W.V.O.: Methods of Logic. Holt, Rinehart and Winston, New York (1966)
10. Parkinson, G.H.R.: Leibniz. Logical Papers. Oxford Univ. Press, Oxford (1966)
11. Kretzmann, N. (ed.).: William of Sherwood's Introduction to Logic. University of Minnesota Press, Minneapolis (1966)
12. Czeżowski, T.: On certain peculiarities of singular propositions. Mind **64**, 392–395 (1955)

Open Access This chapter is licensed under the terms of the Creative Commons Attribution 4.0 International License (http://creativecommons.org/licenses/by/4.0/), which permits use, sharing, adaptation, distribution and reproduction in any medium or format, as long as you give appropriate credit to the original author(s) and the source, provide a link to the Creative Commons license and indicate if changes were made.

The images or other third party material in this chapter are included in the chapter's Creative Commons license, unless indicated otherwise in a credit line to the material. If material is not included in the chapter's Creative Commons license and your intended use is not permitted by statutory regulation or exceeds the permitted use, you will need to obtain permission directly from the copyright holder.

Euler and Venn Diagrams

Rectangular Euler Diagrams and Order Theory

Uta Priss[1] and Dominik Dürrschnabel[2]

[1] Faculty of Computer Science, Ostfalia University, Wolfenbüttel, Germany
u.priss@ostfalia.de
[2] Knowledge and Data Engineering Group, University of Kassel and Interdisciplinary Research Center for Information System Design, Kassel, Germany
duerrschnabel@cs.uni-kassel.de

Abstract. This paper discusses the relevance of order-theoretical properties, such as order dimension, for determining properties of Euler diagrams, such as whether a given poset can be represented with or without shading. The focus is on linear, tabular and rectangular Euler diagrams with shading and without split attributes and constructions with subdiagrams and embeddings. Euler diagrams are distinguished from geometric containment orders. Basic layout strategies are suggested.

1 Introduction

Euler diagrams are a well-known means for visualising sets and their partial orders (posets). Automated drawing of Euler diagrams, however, is still a difficult challenge (Alsallakh et al. 2016). Not all posets can be represented with all types of Euler diagrams. Thus it is desirable to determine properties of posets that affect which strategies of diagram construction and layout are suitable. This paper considers order-theoretical properties for such purposes.

Users tend to perceive Euler diagrams as more intuitive to read than Hasse[1] diagrams (Priss 2020), but Hasse diagrams are much easier to automatically draw with a computer due to a number of well-known layout algorithms (Dürrschnabel & Stumme 2021). Algorithms for Euler diagrams exist (Alsallakh et al. 2016), but there are still many open questions about construction and layout. Maybe Hasse diagrams are in some sense closer to order theory than Euler diagrams because they appear to depend less on the 2-dimensional geometrical space in which they are drawn than Euler diagrams.

Two negative results are presented in this paper: first, even though the order dimension has relevance for Euler diagrams, it is not as helpful in determining whether a diagram can be drawn or not as one might wish. Second, rectangular Euler diagrams are not geometric containment orders. The reason why these two results are of interest is because order theory is the mathematical foundation of both Hasse and Euler diagrams. Thus one might assume that questions about Euler diagrams can be easily reduced to order-theoretical questions which may

---

[1] The diagrams should not be named after Hasse because he did not invent them, but the name is widely established in the literature.

have already been solved. In some respects that is true. In particular, Formal Concept Analysis (FCA) which models binary relations as mathematical lattices using order theory (Ganter & Wille 1999) has developed a large repertoire of models and algorithms that are relevant for Euler diagrams. Nevertheless this paper argues that applying FCA, and order theory in general, to Euler diagrams is beneficial but not always sufficient.

All types of visualisations tend to have particular purposes and particular limits. Euler diagrams are effective even for large amounts of data, if it is possible to extract a small set of features (e.g. 5–10 features) which is then utilised to aggregate the data. Euler diagrams are particularly suitable for displaying implications and dependencies amongst features that relate to a small number of concepts, for example, for teaching mathematics (Priss 2023). In other applications, Euler diagrams might not be an appropriate visualisation.

More details about and an overview of Euler diagrams are provided by Rodgers (2014) and Alsallakh et al. (2016). This paper focuses on linear, tabular and rectangular Euler diagrams with shading and without split attributes. Linear Euler diagrams are considered to have superior usability by Chapman et al. (2014). Petersen (2010) investigates conditions for determining whether a linear Euler diagram exists, however, she uses a different terminology and does not refer to her diagrams as Euler diagrams. The advantages of rectangular Euler diagrams were independently discussed by Paetzold et al. (2023) and Priss (2023). Paetzold et al. (2023) present algorithms for automated drawing but they ignore whether or not zones should be shaded.

The following section defines Euler diagrams and related terminology. Section 3 and 4 discuss the relevance of order dimension, lattice theory and geometric containment orders. Section 5 elaborates on layout strategies. Section 6 concludes the paper.

2 Euler Diagrams

Venn, Euler and Hasse diagrams are different means for representing partially ordered sets (cf. Priss, 2020). In the following definitions, two sets A and O are used. The elements of A are called *attributes*, the elements of O are called *objects*. The powerset of a set S is denoted by $P(S)$ in this paper.

Definition 1. *A* zone set $Z(A)$ *over a set* A *is a subset of the powerset* $P(A)$. *Elements of* $Z(A)$ *are called* zones *and elements of* $P(A) \setminus Z(A)$ *are called* gaps.

The running example for this section is a zone set $Z^*(A) = \{\{\}, \{a\}, \{b\}, \{c\}, \{a,b\}, \{b,c\}, \{a,b,c,d\}\}$ over $A = \{a,b,c,d\}$ with gaps $\{d\}, \{a,d\}, \{a,c\}$ and so on. A challenge for a discussion about Euler diagrams is that an attribute contains zones or, dually, is contained in zones if a zone is defined as a set of attributes. The same holds for objects: they belong to a zone with its attributes or, dually, contain a set of attributes. It is important to always be aware of which of the two dual perspectives is currently employed. The notions of "attribute" and "object" are meant to differentiate the perspectives and to avoid confusion with any meta-level use of the word "set" in a discussion about Euler diagrams. In the running example, the attributes are considered sets of objects as follows: $a = \{2, 3, 6\}$, $b = \{3, 5, 6, 7\}$, $c = \{4, 5, 6\}$ and $d = \{6\}$ with $O = \{1, 2, 3, 4, 5, 6\}$.

As usual, a *partially ordered set (or poset)* $(A, \leq)$ is a binary relation that is reflexive, symmetric and transitive (i.e., $\forall (a, b, c \in A)\, a \leq a \wedge (a \leq b \wedge b \leq a \Rightarrow a = b) \wedge (a \leq b \wedge b \leq c \Rightarrow a \leq c)$). A *linear order* is a poset in which any two elements are comparable (i.e., $\forall (a, b \in A)\, a \leq b \vee b \leq a$). It is well-known from set theory that $(Z(A), \subseteq)$ is a poset. In the context of Euler diagrams, it is actually more natural to consider the dual of $(Z(A), \subseteq)$, in this paper denoted by $(Z(A), \leq)$ with $z_1 \leq z_2 :\iff z_1 \supseteq z_2$.

Figure 1 shows different visualisations of the running example $(Z^*(A), \leq)$. The Hasse diagram in 1C is drawn with a node for each element of $Z^*(A)$ and objects attached to the lowest zone which they belong to. The other two types of diagrams in Fig. 1 which are formally defined below are Euler diagrams (1B and 1D) and a formal context (1A) which has objects as rows, attributes as columns and a cross if an object is associated with an attribute. The relationship between formal contexts and Hasse diagrams is explained by FCA and not further discussed in this paper. The equivalence between the Euler and Hasse diagrams in Fig. 1 should be visually deducible from the zones and their objects. The shaded region in the Euler diagrams is explained below.

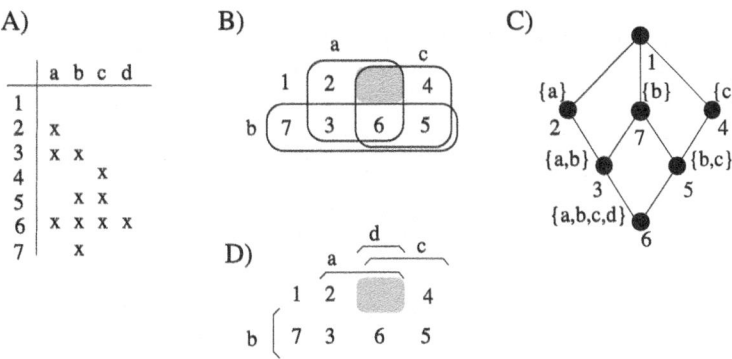

Fig. 1. $Z^*(A)$ as A) formal context, B) and D) tabular Euler diagrams, C) Hasse diagram

Definition 2. *A formal context (O, A, I) is a binary relation $I \subseteq O \times A$ with two functions $\phi_O : O \to P(A)$ and $\phi_A : A \to P(O)$ with $\phi_O(o) = \{a \mid a \in A, (o, a) \in I\}$ and $\phi_A(a) = \{o \mid o \in O, (o, a) \in I\}$.*

The two functions determine which attributes are associated with which object and which objects are associated with which attribute. That means they read columns and rows from a formal context represented as a table as in Fig. 1, such as $\phi_A(\{a\}) = \{2, 3, 6\}$. The next definition connects zone sets with formal contexts.

Definition 3. *A zone set $Z(A)$ is represented as a formal context (O, A, I) if $\forall (o \in O)\, \phi_O(o) \in Z(A)$ and $\forall (z \in Z(A))\, \exists (o \in O)\, \phi_O(o) = z$. It is said that zone z accommodates all objects o for which $\phi_O(o) = z$ holds.*

In other words in a formal context, zones correspond to distinct rows. Each object is accommodated by exactly one zone and each zone accommodates at least one object. For example for $Z^\star(A)$, zone $\{a\}$ accommodates object 2. Because several rows of a formal context can have the same crosses, a zone can also accommodate more than one object. A gap $\{a, b, c\}$ is not a row of a formal context. Zones accommodate all objects that belong to an intersection of the zone's attributes without objects that are accommodated by lower zones according to $(Z(A), \leq)$. We distinguish between *contain* and *accommodate* in this paper: the intersection $a \cap b = \{3, 6\}$ contains 3 and 6. But zone $\{a, b\}$ accommodates only object 3 because 6 is accommodated by zone $\{a, b, c, d\}$. Even though the attributes of a formal context do not need to be sets in all applications, the function ϕ_A provides a natural means for interpreting them as sets. In general, FCA assumes no structural difference between objects and attributes and discusses many consequences arising from a dual relationship between objects and attributes. FCA usually completes partial orders to lattices. But that is not required for this paper. The following definitions establish Euler diagrams as used in Fig. 1 after first defining the types of graphical elements that are used.

Definition 4. *A* curve *is a Jordan curve (i.e. divides a plane into an interior and an exterior) that does not intersect itself. A* minimal region *is an area of a diagram that is enclosed by curve segments without containing any curve segments itself.* Shading *is a graphical darkening applied to minimal regions so that a minimal region can have exactly two states: either shaded or not shaded.*

Definition 5. *An* Euler diagram $E(Z(A))$ *is a graphical representation of $Z(A)$ where each $a \in A$ is represented as a curve and each zone z is represented by a single non-shaded minimal region that is contained in all curves a with $a \in z$. Shaded minimal regions represent gaps. Each gap corresponds to at most one shaded minimal region. $\mathcal{E}$ and $\mathcal{E}^-$ are the sets of all zone sets that can be represented as Euler diagrams with and without shading, respectively.*

Because of the dual perspectives, the condition $a \in z$ is correct because zones are sets of attributes even though graphically a zone is contained in all of the curves of its attributes. A similar definition can be constructed for *Euler diagrams with split attributes* by allowing an attribute to correspond to several curves. Unless stated differently below, in this paper the notion *Euler diagram* always refers to a diagram *without* split attributes.

Venn diagrams are special kinds of Euler diagrams that contain exactly as many minimal regions as the elements of a powerset $P(A)$. While Euler diagrams often have fewer shaded minimal regions than gaps, Venn diagrams have as many shaded minimal regions as gaps. Other authors use other definitions for Euler diagrams and discuss several conditions for when to consider Euler diagrams *well-formed* (Flower, Fish, & Howse 2008). In this paper, Definition 5 prohibits zones or gaps to be split into several minimal regions but otherwise does not require any further conditions.

Figure 1B shows an Euler diagram with 3 curves, 7 zones (including the outer zone which accommodates object 1) and a shaded minimal region for the gap $\{a,c\}$. Curves are allowed to be concurrent, but then they are drawn slightly dislocated as in Fig. 1B so that they can be visually distinguished. Priss (2023) argues that users will naturally regard the tiny extra regions that sometimes occur between dislocated concurrent curves as drawing errors and ignore them. Some concurrency can be avoided if curves that coincide exactly with a zone set are textually added, such as adding "$d := a \cap b \cap c$" in Fig. 1B instead of drawing a curve for d. Figure 1D displays the same information as 1B but uses a different notation. While Fig. 1B and 1D are structurally equivalent, it is often possible to represent a single zone set with structurally differing Euler diagrams. Therefore equivalence of diagrams is discussed further below. At a minimum, diagrams that are obtained from each other by changing sizes without changing relative locations should always be considered equivalent or identical. The next definition and lemma explain that shading allows to use the same layout of an Euler diagram for different zone sets.

Definition 6. A *subdiagram* of an Euler diagram is obtained by shading 0 or more zones. A *superdiagram* of an Euler diagram is obtained by turning 0 or more shaded minimal regions into (non-shaded) zones. An *utmost diagram* of an Euler diagram is a superdiagram that does not contain any shaded minimal regions.

Lemma 1. *I) Every Euler diagram has a unique utmost diagram.*
II) If $E_1(Z_1(A))$ is a subdiagram of $E(Z(A))$ then $Z_1(A) \subseteq Z(A)$.
III) If $Z_1(A) \subseteq Z(A)$, then $E_1(Z_1(A))$ need not be a subdiagram of $E(Z(A))$.

Proof. I) Because the utmost diagram is obtained by removing any shading which is a deterministic process. II) Shading turns zones into gaps and therefore reduces a set $Z(A)$ to a subset. III) Fig. 2 displays $Z^* \setminus \{\{b\}\}$, but the Euler diagram in Fig. 2B is not a subdiagram of Fig. 1D because in both diagrams the arrangement of the zones is different (2 rows vs 1 row). The difference is not just a matter of shading. □

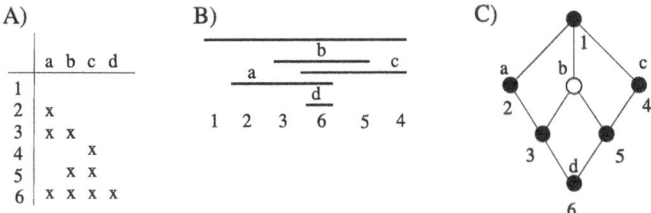

Fig. 2. For $Z^* \setminus \{\{b\}\}$: A) Formal context, B) Linear Euler diagram, C) Hasse diagram

Figure 2 shows an example where the corresponding Hasse diagram requires an element corresponding to gap ($\{b\}$), but the Euler diagram does not require a shaded minimal region. Priss (2020) calls elements of posets that occur but do not accommodate any objects *supplemental*. In Hasse diagrams, supplemental elements are drawn as non-filled circles and fulfil a similar role as shading in Euler diagrams. Priss (2020) provides a more detailed discussion of the drawability of Hasse diagrams with or without supplemental elements compared to Euler diagrams. It should be noted that in the Hasse diagrams in the remainder of this paper, only the attributes (and not the zones) are written into the diagrams. Euler diagrams tend to be easier to read if shading is avoided but that is not always possible:

Lemma 2. *I) All zone sets can be represented as Euler diagrams with shading. II) Not all zone sets can be represented as Euler diagrams without shading, i.e. $\mathcal{E}^- \subset \mathcal{E}$.*

Proof. I) Venn diagrams are Euler diagrams according to Definition 5. Because Venn already showed that powersets of any size can be represented as Venn diagrams (Baron 1969) all subsets of powersets can be represented as subdiagrams of Venn diagrams. II) Depending on what shapes of curves are allowed, there are always zone sets that can only be represented with shading. The example in Fig. 1B cannot be represented without shading if the curves are rectangles. □

Hasse diagrams are equivalent to each other if they represent the same zone set and contain the same supplemental nodes or, visually, if they can be obtained from each other by moving nodes freely around as long as the upwards direction of the edges is not changed. For Euler diagrams it is not as easy to define equivalence because a zone set can be represented as a subdiagram of different utmost diagrams and it is not as clear what it means to transform one representation into another one. Thus, different kinds of equivalence can be defined. The least and most restrictive ones are:

Definition 7. *Two Euler diagrams $E_1(Z_1(A))$ and $E_2(Z_2(A))$ are zone-equivalent if $Z_1(A) = Z_2(A)$ and shading-equivalent if they are zone-equivalent and have the same utmost diagram.*

The following types of Euler diagrams are discussed in this paper:

Definition 8. *A rectangular (Euler) diagram is an Euler diagram where all curves are rectangles (ignoring the rounded corners) and all edges are either horizontal or vertical. A tabular (Euler) diagram is a rectangular diagram where all curves are placed into a single rectangle t so that for each curve (other than t) either the top and bottom edges are concurrent with t or the left and right edges are concurrent with t. A linear (Euler) diagram is a rectangular diagram where all curves are placed into a single rectangle t so that for all curves the top and bottom edges are concurrent with t. $\mathcal{E}_1$, $\mathcal{E}_{1x1}$ and $\mathcal{E}_2$ are the sets of all zone sets that can be represented as subdiagrams of linear, tabular and rectangular diagrams, respectively.*

An important question is to determine $\mathcal{E}_1$, $\mathcal{E}_{1x1}$ and $\mathcal{E}_2$. As mentioned in the introduction, Chapman et al. (2014) conclude that linear diagrams have superior usability compared to other Euler diagrams, but they do not investigate rectangular Euler diagrams. Priss (2023) argues that rectangular diagrams have similar usability as linear diagrams because users only need to trace a horizontal and a vertical direction. Priss (2021) suggests that tabular diagrams in the styles of Fig. 1B or 1D are optimal because they combine the advantages of linear diagrams with a more compact design. In particular Fig. 1D is just a table similar to many other kinds of tables that users are familiar with. Rectangular diagrams are furthermore easy to automatically process with a computer because computing curve intersections (as needed for shading) is straightforward for rectangles. Linear diagrams are uniquely identified by their linear sequence of zones. Tabular diagrams are uniquely identified by their sequence of zones for their rows and columns. Definition 8 requires that a single attribute (other than one containing everything) does not occur both as a row and as a column label for tabular diagrams because otherwise its shape would not be a rectangle. The relationship between linear and tabular diagrams can be described as a direct product.

Definition 9. *For zone sets $Z_1(A_1)$ and $Z_2(A_2)$ with $A_1 \cap A_2 = \emptyset$, the direct product is defined as a zone set $Z_1(A_1) \otimes Z_2(A_2) = \{z_1 \cup z_2 \mid z_1 \in Z_1(A_1), z_2 \in Z_2(A_2)\}$. A direct product of two Euler diagrams $E_1(Z_1(A_1))$ and $E_2(Z_2(A_2))$ with $A_1 \cap A_2 = \emptyset$ is an Euler diagram that is zone-equivalent to $Z_1(A_1) \otimes Z_2(A_2)$.*

Normally, direct products are Cartesian products, resulting in sets of tuples instead of sets of sets. But because the zone sets $Z_1(A)$ and $Z_2(A)$ do not have any attributes in common, the difference between sets and tuples is just a matter of notation. For Hasse diagrams FCA drawing algorithms exist that construct a direct product from its factors. For Euler diagrams such a graphical construction may not always be possible. But the direct product of two linear diagrams always yields a tabular diagram.

Lemma 3. *I) The utmost diagram of a tabular diagram is a direct product of two linear diagrams. Thus $\mathcal{E}_{1x1}$ is determined by direct products of elements of $\mathcal{E}_1$.*
II) A linear diagram with n attributes contains at most $2n$ zones.
III) A tabular diagram with n row attributes and m column attributes contains at most $4mn$ zones.

Proof. I) follows directly from Definition 8 because every cell of a table belongs to exactly one row and one column. II) because each curve splits at most one zone at each of its ends. III) because the linear diagram of the rows contains at most $2n$ and the linear diagram of the columns at most $2m$ zones, thus $2n2m = 4mn$. □

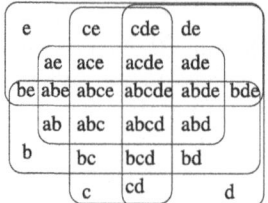

Fig. 3. Rectangular diagrams with 5 attributes with 26 zones (left) and 27 zones (right) including the outer zone. Brackets and commas are omitted for the zones, i.e. 'ce' instead of $\{c,e\}$.

It follows that a powerset with 5 attributes (i.e. $2^5 = 32$ elements) cannot be represented with a tabular diagram which would have at most $4 \times 2 \times 3 = 24$ zones. Figure 3 displays two rectangular diagrams with 5 attributes and slightly more than 24 zones. Rectangular diagrams are thus more powerful than tabular diagrams. But Fig. 3 also demonstrates that even small rectangular diagrams can already be difficult to read. Venn diagrams without shading correspond to zone sets that are powersets without any gaps. Venn already showed that at most 3 attributes can be represented with circular curves (Baron 1969). A Venn diagram with at most 4 attributes can be represented with rectangular curves (as shown in Fig. 3 if the curve for a is deleted). For larger numbers of attributes the curves would need to be ellipses or concave shapes.

Lemma 4. *I) All zone sets with at most 4 attributes are in $\mathcal{E}_{1x1}$.*
II) If 3 attributes of a zone set each contain an object that the others do not contain and a zone containing all 3 attributes exists, then the zone set is not in $\mathcal{E}_1$.
III) If 5 attributes of a zone set each contain an object that the others do not contain and a zone containing all 5 attributes exists, then the zone set is not in $\mathcal{E}_{1x1}$.
IV) $\mathcal{E}_1 \subset \mathcal{E}_{1x1} \subset \mathcal{E}_2 \subset \mathcal{E}$.
V) For any number of attributes, there still exist some zone sets that are in $\mathcal{E}_2$.

Proof. I) Can be represented as a subdiagram of a Venn diagram for a powerset of 4 attributes. II) In a linear diagram, a zone $\{a, b\}$ must be between zones $\{a\}$ and $\{b\}$. For 3 attributes it is not possible to arrange them so that $\{a, b, c\}$ is between all three attributes without one of the attributes being contained in the union of the other attributes (as in Fig. 2). III) The utmost diagram would be a direct product of a linear diagram with 2 and one with 3 attributes. The latter is not possible according to II). IV) The inclusions $\subseteq$ are obvious. They are proper subsets because: $\mathcal{E}_1 \subset \mathcal{E}_{1x1}$ follows from II) and III). Figure 3 shows examples that are in $\mathcal{E}_2$ but not in $\mathcal{E}_{1x1}$. A zone set that is a powerset of a set of 5 attributes is in $\mathcal{E}$ but not in $\mathcal{E}_2$. V) Fig. 8B below shows a rectangular diagram for a zone set with 6 attributes. The principle underlying that figure can be extended to any number of attributes. □

Euler diagrams can represent hundreds or thousands of objects if these are aggregated in some form. But Lemma 4 indicates that 5 attributes might already pose a problem. In order to increase the number of possible representations, Euler diagrams with split attributes can be defined. For example, the utmost diagram of Fig. 8C is a powerset of 6 attributes where 2 attributes (c and f) are split. It should be noted that in that case it is still required that every zone occurs at most once in a diagram. Thus a split attribute must be combined with other attributes in each instance in which it occurs.

Lemma 5. *If Euler diagrams are defined with split attributes (modifying Def. 5), then all zone sets can be drawn as tabular or even linear diagrams.*

Proof. Because the attributes can be split, even a linear diagram can be drawn where the zones occur in any random sequence and each column is labelled by all attributes that belong to a zone. □

Obviously, a random arrangement of zones as indicated in the proof is not useful because it will not render diagrams that are easy for humans to interpret. Furthermore, splitting attributes too many times certainly reduces readability. Therefore questions arise as to how to minimise shading in cases where a representation is possible and how to minimise splitting and shading otherwise. Criteria for determining whether representations are possible at all are essential. The next two sections discuss in how far order-theoretical notions might contribute to answering such questions. While splitting attributes may be necessary for applications, it is of interest to first investigate what is possible without splitting attributes as discussed in the next two sections.

3 Order Dimension

Very "complex" zone sets cannot be represented as rectangular Euler diagrams. But it is not easy to characterise exactly how this complexity can be measured. One possible measure is the *order dimension* of a poset (e.g. Ganter & Wille 1999). First the notions of *order embedding* and *direct product* are required:

Definition 10. *For two posets $(P, \leq_p)$ and $(Q, \leq_q)$, a map $\psi : P \to Q$ is called an* order embedding *if $\forall (p, q \in P)\ p \leq_p q \iff \psi(p) \leq_q \psi(q)$.*
The direct product *of two posets $(P_1, \leq)$ and $(P_2, \leq)$ is a poset $(P_1 \times P_2, \leq)$ with $(x_1, x_2) \leq (y_1, y_2) \iff x_1 \leq y_1$ and $x_2 \leq y_2$.*

According to Ganter & Wille (1999) a finite poset $(P, \leq_p)$ can be order embedded into a powerset of a set and can be represented as a zone set using a formal context according to Defnition 3. Thus at least for finite sets, every zone set is a poset and every poset can be modelled as a zone set. The direct product in Defnition 9 then coincides with what is stated in Defnition 10.

Definition 11. A poset $(P, \leq)$ has an order dimension $dim(P, \leq) = n$ if $(P, \leq)$ can be embedded into a direct product of n linear orders and n is the smallest such number.

A poset $(P, \leq)$ has a k-dimension $dim_k(P, \leq) = n$ if $(P, \leq)$ can be embedded into a direct product of n linear orders of length k and n is the smallest such number.

For example, Fig. 4A shows a direct product of 2 linear orders. Any poset that can be embedded into 4A has order dimension ≤ 2. The equivalent Euler diagram in Fig. 4B is constructed by rotating the Hasse diagram 45° to the left and then drawing rectangles instead of edges. Embedding a poset (as in 4C) into another poset (such as 4A) entails arranging the nodes in similar positions and leaving sufficient edges so that the $\leq$-relation still holds between all nodes. As shown in 4D this may require some shading. Furthermore, contrary to a subdiagram which has the same attributes as its utmost diagram, *embedding may require different attributes*. In Fig. 4 the embedded poset has two attributes that are not in the original poset ($g := b \cap d$ and $h := f \cap a$). The Euler diagram in 4D is rectangular, not tabular. In this case, it can be represented as a tabular diagram (4E). The question arises as to which zone sets can be represented using embedding.

Definition 12. Let $\mathcal{E}^+_{1x1}$ denote the set of zone sets that can be represented by embedding into tabular diagrams.

Lemma 6. $\mathcal{E}^+_{1x1} = \mathcal{E}_2$

Proof. The vertical and horizontal curve segments of rectangular diagrams can be projected onto a vertical and a horizontal axis, respectively. The projections are linear diagrams. Every zone of a rectangular diagram is thus representable as embedded into a direct product of two linear diagrams. □

Thus embedding into $\mathcal{E}_{1x1}$ allows to represent as much as $\mathcal{E}_2$ whereas creating subdiagrams is less powerful because of $\mathcal{E}_{1x1} \subset \mathcal{E}_2$. But as mentioned above, in applications embedding is only suitable if it is acceptable to change attributes. An algorithm for drawing rectangular diagrams might consist of determining two linear diagrams in which the diagram can be embedded or which it is a subdiagram of.

The next lemma discusses the relevance of order dimension for Euler diagrams. For example, zone sets in $\mathcal{E}_1$ can have an order dimension larger than 1 as demonstrated by Fig. 2.

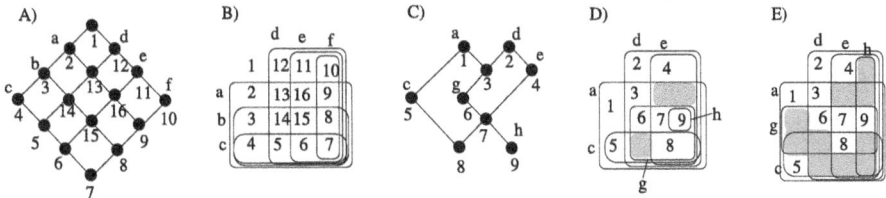

Fig. 4. Direct product of 2 linear orders and example of embedding

Lemma 7. *I)* $dim(P, \leq) = 1 \Rightarrow$ *zone set in* $\mathcal{E}_1 \Rightarrow dim(P, \leq) \leq 2$.
II) $dim(P, \leq) = 2 \Rightarrow$ *zone set in* $\mathcal{E}_2 \Rightarrow dim(P, \leq) \leq 4$.
III) Not every poset of $dim(P, \leq) = 2$ *is in* $\mathcal{E}_1$.
IV) Not every poset of $dim(P, \leq) = 2$ *is in* $\mathcal{E}_{1x1}$.
V) Not every poset of $dim(P, \leq) = 3$ *is in* $\mathcal{E}_2$.

Proof. I) If $dim = 1$ then the poset can be embedded into a linear order and is thus a linear diagram. If the poset is in $\mathcal{E}_1$ then it has a planar Hasse diagram and thus $dim \leq 2$. II) If $dim \leq 2$ then the construction mechanism described for Fig. 4 can be used. If the poset is in $\mathcal{E}_2$ then it can be represented by a rectangle containment order (as explained in the next section) for which it is known that $dim \leq 4$. III) For example, even if the linear orders in Fig. 4A were one element shorter, i.e. a 3×3 grid, then it cannot be represented as a linear diagram for the same reason as in Fig. 2. The node in the middle would have to be supplemental. IV) For example, if the linear orders in Fig. 4A were one element longer, i.e. a 5×5 grid, then it cannot be represented as a tabular diagram unless the node in the centre is supplemental. V) For example, a $3 \times 3 \times 3$ grid cannot be represented as a rectangular diagram. Again, 26 zones can be represented, but the 27th node in the centre of the cube would have to be supplemental. □

The 2-dimension corresponds to the number of atoms of the smallest powerset which a poset can be embedded into.

Lemma 8. *I)* $dim_2(P, \leq) \leq 4 \Longrightarrow$ *zone set in* $\mathcal{E}_2$.
II) $dim_k(P, \leq) \leq 2 \Longrightarrow$ *zone set in* $\mathcal{E}_2$.

Proof. I) according to Lemma 4 powersets of 4 attributes are in $\mathcal{E}_{1x1}$ and $dim_2 = 4$ means an embedding into a powerset of 4 attributes. II) According to Lemma 7II). □

In summary, the order dimension of a poset provides some hints as to whether it might be representable as a linear, tabular or rectangular Euler diagram. Posets arising from applications, however, often have at least order dimension 3 or 4. And in those cases, it is not clear whether or not the poset can be represented by a rectangular or tabular diagram but Dürrschnabel & Priss (2024) provide some further criteria.

4 Lattices and Geometric Containment Orders

A finite poset is called a *lattice* if each set of elements has a supremum and an infimum. Priss (2020) explains that some zone sets of Euler diagrams only form lattices if supplemental nodes are added (e.g. Fig. 2C and Fig. 5). Supplemental nodes correspond to intersections of curves whose smallest larger container is a union. For example, $\{c,d\}$ in 5B is a gap, but $c \cap d$ is visible in the linear diagram even though it can only contain objects that are also in a or b, i.e. $c \cap d \subseteq a \cup b$. Furthermore, the empty set cannot be satisfactorily represented in Euler diagrams because it corresponds to every gap and could simultaneously be inserted everywhere in a diagram. In a Hasse diagram (as in 5A), the empty set can always be added as a supplemental node at the bottom if it is not already there. In applications the bottom element is often supplemental and can be omitted because it represents attributes which exclude each other. The Hasse diagrams in Fig. 5 demonstrate that with supplemental nodes both posets are lattices. Without supplemental nodes, some elements would not have suprema and infima. In fact, Dürrschnabel & Priss (2024) show that a zone set together with its supplemental nodes always forms a lattice. FCA provides algorithms for calculating such lattices.

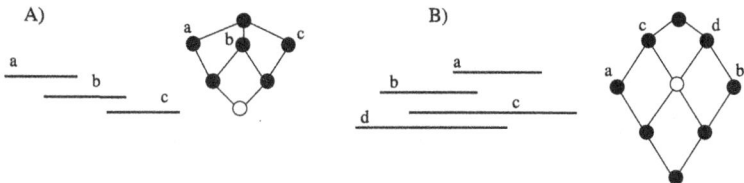

Fig. 5. Zone sets with supplemental nodes are always lattices

From the viewpoint of order theory, an important negative result is that geometric containment orders are not Euler diagrams. On first sight, it might seem as if interval containment orders are the same as linear diagrams and rectangle containment orders are the same as tabular diagrams. If that was the case, then obtaining rectangular Euler diagrams would be mostly trivial because it is a well-known fact that any poset of order dimension 2 (or less) can be represented by an interval containment order and any poset of order dimension 4 (or less) can be represented by a rectangle containment order and data from sets in applications usually have order dimension 4 (or less). But the bounds presented in Lemma 7 cannot be reduced which is an indication that Euler diagrams are not geometric containment orders.

Definition 13. *A* geometric containment order *is an inclusion order in an Euclidean space where $x < y$ if and only if the shape corresponding to x is properly included in the shape of y. An* interval containment order *is a geometric containment order of intervals on the real line. A* rectangle containment order *is* *a geometric containment order of rectangles on a real valued coordinate system with two axes.*

In an interval containment order, $b < a$ means that the left endpoint of a is smaller than the left endpoint of b and the right endpoint of a is larger than the right endpoint of b. The same principle is applied to the four cornerpoints of a rectangle in a rectangle containment order. Even though the interval containment order in Fig. 6A is the same picture as in Fig. 2B (apart from the added label u), the corresponding Hasse diagrams are different in the case of a linear Euler diagram and an interval containment order. The difference is that geometric containment orders only consider containment and incomparability, but not intersections and unions. Thus a, b and c in Fig. 6A are mutually incomparable. They are all contained in u and all contain d. But the fact that b is contained in the union of a and c is ignored. In other words, geometric containment orders only represent an ordering on attributes, but ignore objects. For example in Fig. 6A, it is not possible for $b \setminus (a \cup c)$ to accommodate any objects but that is irrelevant if the diagram is read as a geometric containment order.

Lemma 9. *Euler diagrams and geometric containment orders are not the same.*

Proof. As explained in the previous paragraph and because of Lemma 7. □

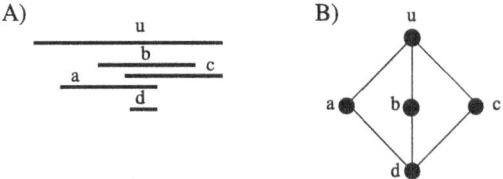

Fig. 6. A) Interval containment order and B) its Hasse diagram

Instead of embeddings into direct products of linear orders a different, more commonly used (and equivalent) definition of order dimension involves intersections of linear orders. In that case, order dimension 2 means that a poset can be represented as an intersection of two linear orders. For example, the interval containment order of Fig. 6 can be represented as an intersection of $d < c < b < a < u$ and $d < a < b < c < u$. In Fig. 7 these two linear orders are utilised to produce different graphical representations. Each diagram has two axes which are labelled with the attributes in a sequence according to the two linear orders. In 7A and 7B the axes form right angles, in 7C a 180° angle. A line then connects the two locations of each attribute resulting in rectangular curves in 7A1, triangular curves in 7A2 and B and linear curves in 7C. Essentially, Fig. 7A1 follows the same principle as Fig. 4. Figure 7 demonstrates that different curve shapes produce different numbers of zones. Interestingly, if the curve shape is modified from rectangle to triangle (as in 7A2 and 7B), the line connecting b can be moved to either produce the Hasse diagram in 7A or in 7B whereas using rectangles only the Hasse diagram in 7A is possible. The figure also shows that posets of order dimension ≤ 2 (as in 7A1) can always be represented as interval containment orders (7C) requiring only to change the angle between the axes. But 7C contains fewer zones than 7A and cannot accommodate any objects in zone $\{b\}$.

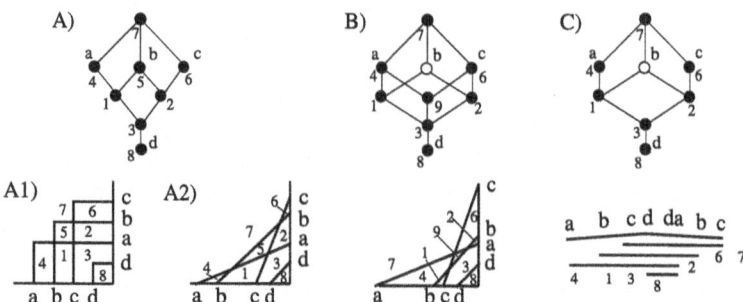

Fig. 7. Different Euler diagrams corresponding to the same interval containment order

5 Layout Strategies for Rectangular Euler Diagrams

A motivation for this research is to develop software for generating diagrams from databases using existing FCA algorithms but with Euler diagrams instead of Hasse diagrams. The aim is to automatically produce diagrams for very small data sets which are easy to read for users who have not been trained in FCA. This section does not intend to provide an overview of the literature on diagram design and layout but to sketch the design strategies that we have identified for our software. The usability of diagrammatic representations has been studied by many authors for many purposes but small changes can impact the results. For example, a study by Blake et al. (2014) derives 9 guidelines for good Euler diagrams one of which states that circles are preferable to rectangles. But contrary to their rectangles ours are dislocated and with rounded corners (adhering to their guideline of using smooth curves). Therefore Blake's negative result about rectangular diagrams is not applicable to our research.

An advantage of tabular diagrams is that users know how to read tables and only rows and columns matter. But if one compares the rectangular diagram in Fig. 4D with the tabular diagram in Fig. 4E, then the rectangular diagram is probably easier to read because it contains fewer shaded minimal regions. In this case the diagram in 4D is mostly tabular. Its two non-tabular rectangles (for g and h) do not extend across a large number of rows and columns and are thus not difficult to visually discern. Furthermore, the relationships $h < a$ and $g < d$ are clear in 3D and obscured in 3E. In summary, we have identified the following design strategies for obtaining diagrams with respect to our software:

S1 produce tabular diagrams if possible
S2 minimise shading
S3 try to visually preserve structures such as containment relationships and symmetries
S4 group shaded areas into large blocks instead of scattering them across the diagram
S5 avoid splitting attributes
S6 use added textual explanations if they simplify a diagram
S7 if the diagram is too large: reduce or split the data set

Unfortunately, some of the strategies are mutually contradicting. Figure 8 presents six different Euler diagrams for the Hasse diagram in 8A. The universal set corresponding to the top node of the Hasse diagram is only drawn in 8D and 8E and omitted in the other diagrams. The embedding into a product of two linear orders in 8B follows the constructions in Fig. 4. A subdiagram of a powerset poset with 6 atoms is shown in 8C as a tabular diagram with two split attributes. Omitting rows and columns in 8C that are fully shaded results in 8F. An embedding into a powerset poset with 4 atoms is displayed in 8G. Figure 8G adheres to S2 and S5 by reducing shading and avoiding splitting but contradicts S3 because of scattered shaded minimal regions and difficult to see rectangles. The linear diagrams in 8D and 8E contradict S5 (split attributes) and S3 (containment and symmetries) but optimise S2. They contradict S3 in different ways: in 8D all attributes are structurally equal whereas 8E preserves one containment relationship. Usability experiments could determine which diagrams are most suitable for certain tasks. In this case, a tabular diagram without split attributes is not possible. Amongst the tabular diagrams, Fig. 8D appears to be least cluttered. The meaning of "a...f" in this diagram can be easily explained to users with an added short textual statement.

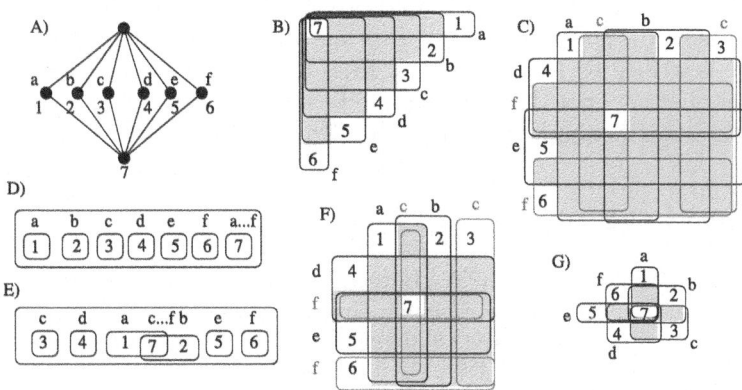

Fig. 8. Seven different representations of the same zone set

The examples in Fig. 8 are all subdiagrams of or embeddings into standard diagrams, such as products of linear orders or powersets which are easy to generate. The number of shaded minimal regions can be counted and compared. An FCA method called "conceptual partitioning" supports decisions about which attributes might be most suitable for splitting (Priss, 2023). The grouping of shaded areas can be optimised by permuting the sequence of the rows and columns. Last but not least, algorithms need not be fully automated. Drawing high quality Euler diagrams completely by hand is very time consuming. For the production of high quality diagrams, a software that provides a selection of possible layouts for additional manual modification with the support of a suitable drawing software is already very desirable.

6 Conclusion

This paper was written in the context of developing software for generating Euler diagrams with comfortable drawing capabilities so that the diagrams can be manually edited to some degree, but also automatically generated from binary relations. It builds on previous work by Priss (2020, 2021 and 2023) and continues with the goal of building software that visualises mathematical concepts for educational purposes. The currently existing educaJS software[2] only provides Hasse and Venn diagrams and is intended to also produce Euler diagrams.

Even for small data sets, automatically generating "good" diagrams is surprisingly difficult. A strategy from Sect. 5 is to mostly generate tabular Euler diagrams because rectangular shapes are geometrically easy to construct and users know how to read tables. Section 4 elaborates on order-theoretical properties that are more of mathematical relevance and may not directly support construction algorithms whereas the order-theoretical properties presented in Sect. 3 directly support decisions about whether or not tabular diagrams without split attributes are possible at all. An important step is to determine whether an attribute set can be partitioned into two sets so that a tabular diagram can be produced as a direct product of two linear diagrams. Algorithms for such tasks are described by Dürrschnabel & Priss (2024). If a diagram without split attributes does not exist then the data can be reduced or split, for example using "conceptual partitioning" as discussed by (Priss, 2023).

References

Alsallakh, B., Micallef, L., Aigner, W., Hauser, H., Miksch, S., Rodgers, P.: The State-of-the-Art of Set Visualization. Comput. Graph. Forum **35**(1), 234–260 (2016)

Baron, M.E.: A note on the historical development of logic diagrams: Leibniz, Euler and Venn. Math. Gazette **53**(384), 113–125 (1969)

Blake, A., Stapleton, G., Rodgers, P., Cheek, L., Howse, J.: The impact of shape on the perception of Euler diagrams. In: Dwyer, T., Purchase, H., Delaney, A. (eds.) Diagrams 2014. LNCS (LNAI), vol. 8578, pp. 123–137. Springer, Heidelberg (2014). https://doi.org/10.1007/978-3-662-44043-8_16

---

[2] https://upriss.github.io/educaJS/.

Chapman, P., Stapleton, G., Rodgers, P., Micallef, L., Blake, A.: Visualizing sets: an empirical comparison of diagram types. In: Dwyer, T., Purchase, H., Delaney, A. (eds.) Diagrams 2014. LNCS (LNAI), vol. 8578, pp. 146–160. Springer, Heidelberg (2014). https://doi.org/10.1007/978-3-662-44043-8_18

Dürrschnabel, D., Stumme, G.: Force-directed layout of order diagrams using dimensional reduction. In: Braud, A., Buzmakov, A., Hanika, T., Le Ber, F. (eds.) ICFCA 2021. LNCS (LNAI), vol. 12733, pp. 224–240. Springer, Cham (2021). https://doi.org/10.1007/978-3-030-77867-5_14

Dürrschnabel, D., Priss, U.: Realizability of Rectangular Euler Diagrams. arXiv:2403.03801 (2024)

Flower, J., Fish, A., Howse, J.: Euler diagram generation. J. Vis. Lang. Comput. **19**(6), 675–694 (2008)

Ganter, B., Wille, R.: Formal Concept Analysis. Mathematical Foundations. Springer, Heidelberg (1999). https://doi.org/10.1007/978-3-642-59830-2

Paetzold, P., Kehlbeck, R., Strobelt, H., Xue, Y., Storandt, S., Deussen, O.: RectEuler: visualizing intersecting sets using rectangles. Comput. Graph. Forum **42** (2023)

Petersen, W.: Linear coding of non-linear hierarchies: revitalization of an ancient classification method. In: Fink, A., et al. (eds.) Advances in Data Analysis, Data Handling and Business Intelligence. Studies in Classification, Data Analysis, and Knowledge Organization, pp. 307–316. Springer, Heidelberg (2010). https://doi.org/10.1007/978-3-642-01044-6_28

Priss, U.: Set visualisations with Euler and Hasse diagrams. In: Cochez, M., Croitoru, M., Marquis, P., Rudolph, S. (eds.) GKR 2020. LNCS (LNAI), vol. 12640, pp. 72–83. Springer, Cham (2021). https://doi.org/10.1007/978-3-030-72308-8_5

Priss, U.: Visualising lattices with tabular diagrams. In: Basu, A., Stapleton, G., Linker, S., Legg, C., Manalo, E., Viana, P. (eds.) Diagrams 2021. LNCS (LNAI), vol. 12909, pp. 378–386. Springer, Cham (2021). https://doi.org/10.1007/978-3-030-86062-2_38

Priss, U.: Representing concept lattices with Euler diagrams. In: Dürrschnabel, D., López Rodríguez, D. (eds.) ICFCA 2023. LNCS, vol. 13934, pp. 183–197. Springer, Cham (2023). https://doi.org/10.1007/978-3-031-35949-1_13

Rodgers, P.: A survey of Euler diagrams. J. Vis. Lang. Comput. **25**(3), 134–155 (2014)

Reference by Occurrence

Francesco Bellucci

University of Bologna, Bologna, Italy
francesco.bellucci4@unibo.it

Abstract. In the literature on Euler diagrams, both classical and contemporary, it has sometimes been claimed that free rides emerge because of certain "spatial properties" of this system of representation. In this paper, I argue that this claim is insufficient. I show that a variety of Euler diagrams is possible in which the spatial conditions for the emergence of free rides are met but in which free rides do not emerge. This implies that another condition has to be met in order to have free rides, which I call "reference by occurrence" or "occurrence-referentiality".

Keywords: Euler Diagrams · Peirce · Reference · Spatial Constraints

1 Introduction

In some systems of representation, expressing a certain piece of information may result in the expression of other pieces of information that are consequences of the chosen information. This is the idea that Charles S. Peirce connected to the notion of "icon". He says: "a great distinguishing property of the icon is that by the direct observation of it other truths concerning its object can be discovered than those which suffice to determine its construction" [11, p. 63]. As applied to logic, the idea is that the representation of logical contents allows the observation of other contents that are necessary consequences of the former. Here is how Peirce defines logical diagrams: "A diagram composed of dots, lines, etc., in which logical relations are signified by such *spatial relations* that the necessary consequences of these logical relations are at the same time signified, or can, at least, be made evident by transforming the diagram in certain ways which conventional 'rules' permit" [12, 3.347; my emphasis]. A logical diagram is a representation of certain logical relations that is *ipso facto* the representation of the necessary consequences of those relations. In Peirce scholarship, this idea has become known as "operational iconicity" [22].

Peirce invented a system for the representation of first-order quantificational logic that he called "existential graphs" [2, 3, 15]. Yet he also conducted research on Euler diagrams and proposed modifications of them [10, 14]. One point that has not been addressed in the literature is Peirce's argument in support of operational iconicity. Why is it that certain systems of representation, and in particular Euler diagrams, have this property? In the definition given above he appears to motivate it by reference to "spatial relations". However, in a paper on Euler diagrams written in 1903 [12, 4.350–371] he offers a different argument: it is because the logical relations of inclusion and exclusion have the same mathematical form as the spatial relations of inclusion and exclusion that Euler diagrams are operationally iconic.

Outside Peirce scholarship, the idea of "operational iconicity" is connected to the notion of "free rides" [15, 16], which has recently been generalized to that of "observational advantage" [19, 20]. A free ride is "where a reasoner attains a semantically significant fact s in a diagram site, while the instructions of operations that the reasoner has followed do not entail the realization of s" [16, p. 32]. With reference to Euler diagrams, free rides have not infrequently been motivated by reference to the spatial nature of the representation. In the entry "Diagram" in the *Stanford Encyclopedia of Philosophy* [17], free rides are considered "consequences of spatial properties of diagrams" (Sect. 3). Likewise, Smessaert, Shimojima and Demey say: "Suppose we take the target types $B \subseteq C$ (θ_1) and $C \cap A = \emptyset$ (θ_2) as the premises of a syllogism [Celarent]. In order for the Euler diagram in Fig. 1 to express these two pieces of information, the semantic conventions require us to realise two source types [...], namely the circle labeled 'B' is inside the circle labeled 'C' (σ_1) and the circle labeled 'C' is outside the circle labeled 'A' (σ_2). By virtue of the *natural spatial (geometrical and topological) constraints on the arrangements of symbols* in Euler diagrams, the realisation of σ_1 and σ_2 automatically realises a third source type, namely that the circle labeled 'B' is outside the circle labeled 'A' (σ_3). Although this is a side effect of the original operation, σ_3 has an independent semantic value, namely that $B \cap A = \emptyset$. This target type θ_3 is a piece of information that we get 'for free'. Hence, to check the validity of the syllogism, we do not have to infer conclusion θ_3 from the premises (θ_1, θ_2). The constraint governing Euler diagrams takes over the work of making the necessary inference, a mechanism called Free Ride." [18, p. 421, my emphasis; circles lettering changed].

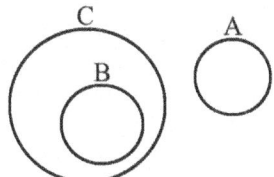

Fig. 1. Free rides in Euler diagrams

Just as in Peirce's definition of logical diagram quoted above, here the reason for the emergence of a free ride is said to reside in the "natural spatial (geometrical and topological) constraints on the arrangements of symbols".

In this paper, I argue that this claim is insufficient. I show that a system of representation is possible in which the "spatial conditions" for the emergence of free rides are met but in which free rides do not emerge. This implies that free rides do not emerge because of those spatial conditions, or at least not because of those spatial conditions *alone*. Another condition has to be met in order to have free rides, which I call "reference by occurrence" or "occurrence-referentiality".

The paper is organized as follows. In the second section I introduce the notion of non-unitary Euler diagram. In the third section I argue that non-unitary Euler diagrams undermine the claim that certain spatial conditions explain free rides in both its Peircean

2 Non-unitary Euler Diagrams

Beside being capable of free-rides, Euler diagrams are notoriously "over-specific" [15]. If I have the syllogistic premises "All B is C" and "No B is A", there is no way to unambiguously represent both in the same Euler diagram. For once the first premise is represented by drawing circle B inside circle C, I have three options as to how to represent the second premise: I may draw circle A outside of circle C—but this means that "No A is C", which does not follow from the premises; or I may draw circle A so as to have an intersection with circle C but not with that portion of circle C which is circle B—but this means that "Some A is C", which also does not follow from the premises; or, finally, I may draw circle A inside circle C but outside circle B—but this, again, means that "All A is C", which does not follow from the premises. The system of representation is over-specific, because it forces us to express information that does not follow from the premises.

This problem was clearly perceived by Johann H. Lambert in the 18th century. He worked with a system of linear diagrams that roughly follow the same principles as Euler diagrams [4, 7, 9, pp. 210–216, 310–318]. Lambert says: "I begin by drawing the middle term, and then I draw either of the other two terms. If the third is capable of being drawn, then the representation gives me anything that follows immediately from the premises. If the third term cannot be drawn, then nothing follows therefrom" [7, p. 152]. For Lambert, if the diagram of two premises cannot be constructed, this means that from those premises nothing follows syllogistically.

However, nothing prevents us from modify the system of representation so as to make it capable of representing non-syllogistic premises (i.e. premises from which nothing follows syllogistically). In [6] the notion of "compound diagram" is introduced as follows: "Given two unitary diagrams D1 and D2, we can connect D1 and D2 with a straight line to produce a diagram D = D1–D2. […] The connection operation is commutative, D1–D2 = D2–D1. Hence, if a diagram has n unitary components, then these components can be placed in any order. The semantics predicate of a compound diagram D is the disjunction of the semantics predicates of its component unitary diagrams" [6, p. 29]. As applied to Euler diagrams, the idea is the following. Since I have three possible options as to where circle A has to be drawn with respect to circle C, I can represent each option in a single unitary Euler diagram, i.e. a standard Euler diagram, and then connect the three unitary Euler diagrams by a line which represents their disjunction (Fig. 2). While each of the unitary diagrams of Fig. 2 represents information that does not follow by necessity from the premises, their disjunction does: for of course if "All B is C" and "No B is A", then it follows that "Either no A is C, or some A is C", or all "A is C".

That neither Euler nor his followers ever employed non-unitary diagrams of this sort is no argument against their theoretical *possibility*. As it will become evident in the following sections, the mere possibility of conceiving non-unitary Euler diagrams is sufficient to undermine the claim that the source of free rides is the obtaining of certain spatial conditions.

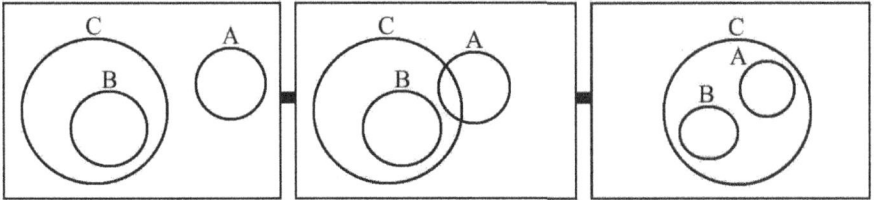

Fig. 2. Non-unitary Euler diagrams

3 Reference by Occurrence

Let us consider the premises of the Celarent syllogism of Sect. 1 ("All B is C", "No C is A") and let us draw each of the two premises in two separate unitary Euler diagrams, which we then compound into a non-unitary, conjunctive Euler diagram, as in Fig. 3 (where the line means conjunction rather than disjunction as in Fig. 2).

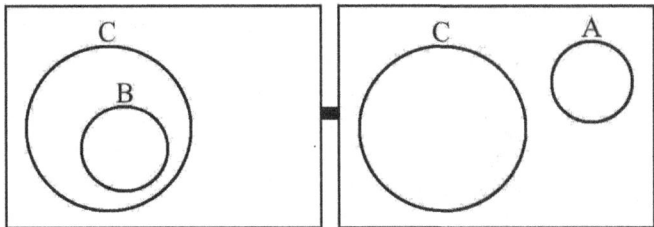

Fig. 3. Premises of *Celarent* in non-unitary Euler diagrams

Let us review the matter from the point of view of the theoretical framework used in [18]. They distinguish source types and target types. Given a representation s and an object or fact represented t, a source type σ is a property of s and a target type θ is a property of t. Now, the target types are "All B is C" (θ_1) and "No C is A" (θ_2), which are the premises of Celarent; these are represented by the source types consisting in the circle B being inside the circle C (σ_1) and in the circle C being outside the circle A (σ_2). These representation relations (σ_1 represents θ_1 and σ_2 represents θ_2) are fulfilled by the non-unitary Euler diagram in Fig. 3 just as they were fulfilled by the unitary Euler diagram in Fig. 1. However, unlike what happens in Fig. 1, in Fig. 3 the realization of σ_1 and σ_2 does not automatically realize a third source type, namely that the circle B is outside the circle A (σ_3). Why does σ_3 fail to be realized? No one would deny that "natural spatial (geometrical and topological) constraints on the arrangements of symbols" are active in Fig. 3, if with this it is meant that logical inclusion (θ_1) and exclusion (θ_2) are represented spatially, i.e. by means of spatial relations (σ_1 and σ_2). What really distinguishes Fig. 1 from Fig. 3 is a further condition, over and above the spatial constraints on the arrangement of the symbols. What distinguishes Fig. 1 from Fig. 3 is that in Fig. 1 there is a one-to-one correspondence between objects represented (classes) and symbol occurrences (occurrences of circles), while in Fig. 3 this is not the case. To put it simply, in Fig. 3 the symbol for class C occurs *twice* (once in the unitary diagram on the left and once in the unitary diagram on the right), while in Fig. 1 it occurs *once*.

Extending a suggestion by John Etchemendy to Keith Stenning [21, p. 134], I call a system of representation in which the sameness of object represented (e.g. a class) is represented by the sameness of the symbol type "type-referential": in a type-referential system, each occurrence of a symbol type refers to one and the same object. I call a system in which the sameness of object represented is represented by the sameness of the symbol occurrences "occurrence-referential": in an occurrence-referential system, each occurrence of a symbol type refers to a distinct object.

Unitary Euler diagrams (Fig. 1) are occurrence-referential: there is one symbol type, the circle; each occurrence of the symbol, i.e. each distinct circle, represents a distinct class; two distinct occurrences of the symbol represent distinct classes. By contrast, non-unitary Euler diagrams (Fig. 3) are type-referential: here, again, there is one single type, the circle; but what counts in the determination of the sameness or distinctness of the classes represented is not the symbol occurrence; indeed, in Fig. 3 there are four circle-occurrences but three classes represented. What counts in the determination of the sameness or distinctness of the classes represented is the sameness of labeled circles, each labeled circle being a distinct sub-type of the circle type. In unitary Euler diagrams (Fig. 1), labeling is useful, but not necessary: there is only one way in which Fig. 1 may represent the premises of a syllogism. By contrast, in non-unitary Euler diagrams (Fig. 3), labeling is necessary, for otherwise we could not know how to determine the identity of the circles. Thus, in the system of Fig. 3 each labeled circle is a sub-type of the circle type, and each represents a distinct class; there are three labeled circles in Fig. 3, and the classes represented are three. Non-unitary Euler diagrams are type-referential.

Since the only thing that distinguishes unitary Euler diagrams (Fig. 1) from non-unitary ones (Fig. 3) is their "method" of reference (reference by occurrence or reference by type), this difference must be what explains why in Fig. 1 the realization of σ_1 and σ_2 automatically realizes σ_3 while in Fig. 3 this does not happen. In both these systems there are "spatial" constraints that determine how circles represent certain information about classes. But only in the system of Fig. 1 there is also the further constraint that imposes reference by occurrence. Thus the emergence in Fig. 1 of free rides must be imputed to this further constraint.

Is occurrence-referentiality in Euler diagrams a "spatial constraint"? In a certain sense, it is: when I decide that there must be a one-to-one correspondence between spatial and logical objects, I am imposing a constraint on the spatial conventions of the system. But the same is true of non-unitary Euler diagrams: here, too, I impose a constraint (albeit a different one) on the spatial conventions of the system.

Perhaps the matter should be put as follows: occurrence-referentiality *must* be spatially realized, i.e. realized by spatial conventions, objects, etc. Consider the existential graph in Fig. 4 and the equivalent FOL sentences 4. In Fig. 4, the fact that two possibly distinct individuals are referred to is expressed by the two distinct occurrences of the variable, namely the line of identity, while in sentence 4 the same fact is expressed by two distinct variable types [12, 4.388, 3].

$$(x)(\exists y)(Lxy) \tag{1}$$

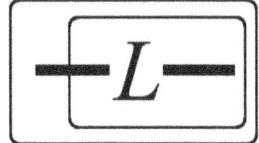

Fig. 4. A Beta existential graph

As far as quantification is concerned, while existential graphs are occurrence-referential, FOL is type-referential.[1] Now, occurrence-referentiality (at the quantificational level) in existential graphs is indeed realized by spatial relations; the line of identity is a topological object, while the variables "x", "y" etc. of FOL are spatial objects in no obvious sense. Thus, existential graphs are another example in which occurrence-referentiality is spatially realized.

It may well be that this is a general rule about systems of representation, i.e. that if they are occurrence-referential at some level, occurrence-referentiality at that level has to be spatially realized. It is a nice hypothesis that requires further investigation. My argument supports a weaker conclusion: both unitary and non-unitary Euler diagrams are "spatial" in the sense that certain spatial properties (σ) represent certain logical properties (θ); but only unitary Euler diagrams allow free rides, so that free rides cannot be imputed to the spatial nature of system (in the sense specified). By contrast, the emergence of free rides in unitary Euler diagrams and their non-emergence in non-unitary Euler diagrams must be imputed to the method of reference (occurrence- vs type-referential), which is the only thing that distinguishes the two systems.

4 Peirce's Argument

In the second section of *Logical Tract No. 2* (1903) Peirce asks the following question: "What is it, then, that these diagrams are supposed to accomplish? Is it to prove the validity of the syllogistic formula? That sounds rather ridiculous—as if anything could be more evident than a syllogism—yet that is not far from the opinion of Friedrich Albert Lange, a thinker of no ordinary force. Suppose we ask ourselves why it is that, if a circle P wholly encloses a circle M which itself wholly encloses a circle S, the circle P necessarily wholly encloses the circle S" [12, 4.354]. The reference to Lange is significant, because Lange had claimed in his *Logische Studien* that the validity of the syllogism depends on the intuition of space relations [8, pp. 9–20; cf. 1]. Peirce cannot agree. Neither does logical validity depends on the holding of certain space relations nor vice versa. By a long and technical argument Peirce reaches the conclusion that "as far as logical dependence goes, the validity of the syllogism and the property of the Eulerian diagram depend upon a common principle. They are analogous phenomena neither of which is, properly speaking, the cause or principle of the other" [12, 4.355]. The common principle in question is transitivity, which Peirce expresses as "At once r of and r of whatever is r'd

---

[1] Notice, however, that existential graphs are occurrence-.referential in the representation of the variables of quantification, but are type-referential in the representation of predicates (in Beta) and sentential variables (in Alpha) [5].

by". Both the validity of syllogism and the spatial relations that hold of Euler diagrams depend on transitivity: the transitivity of universal predication ("at once predicates of and predicates of whatever is predicated by") and the transitivity of spatial inclusion ("at once includes and includes whatever is included by") are just two manifestations of transitivity "in itself", which grounds both logical and spatial inclusion. The same applies to exclusion: when a class excludes another, it excludes whatever the latter includes; thus exclusion "transitates" to whatever the excluded element includes. (The two principles—transitivity of inclusion, and transitivity of exclusion over inclusion—are what medieval logicians called the "dictum de omni et nullo").

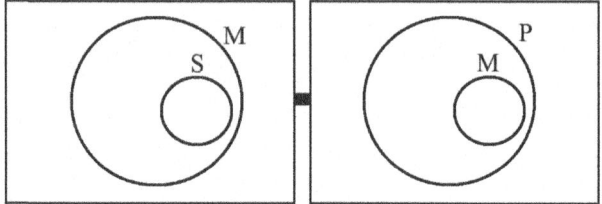

Fig. 5. Non-unitary Euler diagram

The upshot of Peirce's argument may be summed up as follows. It is because the logical relations of inclusion and exclusion have the same mathematical form (i.e. transitivity) as the spatial relations of inclusion and exclusion that Euler diagrams allow us to observe the necessary consequences of the relations represented. This argument, however, seems to be susceptible of the same reply as I provided above. Take Peirce's formulation: "why it is that, if a circle P wholly encloses a circle M which itself wholly encloses a circle S, the circle P necessarily wholly encloses the circle S" [12, 4.354]. Now, if circle M is represented in this system by two separate occurrences of the circle, as it happens in Fig. 5, it remains true that P wholly encloses a circle M and that M wholly encloses a circle S; but that circle P wholly encloses circle S is not observable anymore. Spatial inclusion is transitive, but in order to observe the consequences of transitivity the condition has also to be satisfied that *the geometrical object that mediates the transition be identically the same*, i.e. that it is the *same occurrence* of the circle M that is wholly enclosed by circle P and wholly encloses circle S. If this condition is not met, as in Fig. 5, while inclusion continues to be transitive, the consequences of transitivity are not observable. The answer to Peirce's question is thus that the phenomenon he seeks to explain happens not only because spatial inclusion is, like logical inclusion or universal predication, transitive, but also and crucially because in this system of representation (Euler diagrams) circles refer by their occurrences, not by their types. In other words, Euler diagrams "show" the transitivity of predication because they are occurrence-referential.

References

1. Bellucci, F.: Diagrammatic reasoning. Some notes on C.S. Peirce and F.A. Lange. History Philos. Logic **34**, 293–305 (2013)

2. Bellucci, F., Pietarinen, A.-V.: Existential graphs as an instrument for logical analysis. Part 1: alpha. Rev. Symbolic Logic **9**, 209–237 (2016)
3. Bellucci, F., Pietarinen, A.V.: An analysis of existential graphs–part 2: beta. Synthese **199**(3), 7705–7726 (2021)
4. Bellucci, F., Moktefi, A., Pietarinen, A.-V.: Diagrammatic autarchy. Linear diagrams in the 17th and 18th centuries. In: Burton, J. and Choudhury, L. eds. Proceedings of the International Workshop on Diagram Logic and Cognition, pp. 23–30. CEUR Proceedings (2014)
5. Bellucci, F., Burton, J.: Observational advantages and occurrence referentiality. In: Pietarinen, A.V., Chapman, P., Bosveld-de Smet, L., Giardino, V., Corter, J., Linker, S. (eds.) Diagrams 2020. LNCS, vol. 12169, pp. 202–215. Springer, Cham (2020). https://doi.org/10.1007/978-3-030-54249-8_16
6. Howse, J., Molina, F., Taylor, J.: On the Completeness and Expressiveness of Spider Diagram Systems. In: Anderson, M., Cheng, P., Haarslev, V. (eds.) Diagrams 2000. LNCS, vol. 1889, pp. 26–41. Springer, Dordrecht (2000). https://doi.org/10.1007/3-540-44590-0_8
7. Lambert, J.H.: Neues Organon. Leipzig, Wendler (1764)
8. Lange, F.A. Logische Studien. Iserlohn (1877)
9. Lemanski, J.: World and Logic. College, London (2021)
10. Moktefi, A., Pietarinen, A.-V.: Negative terms in Euler diagrams: Peirce's solution. In: Jamnik, M., Uesaka, Y., Elzer Schwartz, S. (eds.) Diagram 2016. LNCS, vol. 9781, pp. 286–288. Springer, Cham (2016). https://doi.org/10.1007/978-3-319-42333-3_25
11. Peirce, C.S.: Selected Writings on Semiotics, ed. F. Bellucci. Berlin, De Gruyter (2020)
12. Peirce, C.S.: Collected Papers of Charles Sanders Peirce. Hartshorne, C., Weiss, P., Burks, A.W., eds. Harvard University Press, Cambridge, MA (1933–1958)
13. Pietarinen, A.-V.: Exploring the beta quadrant. Synthese **192**, 941–970 (2015)
14. Pietarinen, A.-V.: Extensions of Euler diagrams in Peirce's four manuscripts on logical graphs. In: Jamnik, M., Uesaka, Y., Elzer Schwartz, S. (eds.) Diagram 2016. LNCS, vol. 9781, pp. 39–154. Springer, Cham (2016). https://doi.org/10.1007/978-3-319-42333-3_11
15. Shimojima, A.: On the efficacy of representation. Ph.D. thesis, Indiana University (1996a)
16. Shimojima, A.: Operational constraint in diagrammatic reasoning. In: Allwein, G., Barwise, J. (eds.). Logical Reasoning with Diagrams, 27–48. Oxford University Press, Oxford (1996b)
17. Shin, S.-J., Lemon, O., Mumma, J.: Diagrams. In: The Stanford Encyclopedia of Philosophy (Winter 2018 Edition), ed. by Edward N. Zalta. URL = <https://plato.stanford.edu/archives/win2018/entries/diagrams/>
18. Smessaert, H., Shimojima, A., Demey, L.: Free rides in logical space diagrams versus Aristotelian diagrams. In: Pietarinen, A.V., Chapman, P., Bosveld-de Smet, L., Giardino, V., Corter, J., Linker, S. (eds.) Diagram 2020. LNCS, vol. 16169, pp. 419–435. Springer, Cham (2020). https://doi.org/10.1007/978-3-030-54249-8_33
19. Stapleton, G., Jamnik, M., Shimojima, A.: What makes an effective representation of information: a formal account of observational advantages. J. Logic Lang. Inform. **26**, 143–177 (2017)
20. Stapleton, G., Shimojima, A., Jamnik, M.: The observational advantages of Euler diagrams with existential import. In: Chapman, P., Stapleton, G., Moktefi, A., Perez-Kriz, S., Bellucci, F. (eds.) Diagram 2018. LNCS, vol. 10871, pp. 313–329. Springer, Cham (2018). https://doi.org/10.1007/978-3-319-91376-6_29
21. Stenning, K.: Distinctions with differences: comparing criteria for distinguishing diagrammatic from sentential systems. In: Anderson, M., Cheng, P., Haarslev, V. (eds.) Diagram 2000. LNCS, vol. 1889, pp. 132–148. Springer, Heidelberg (2000). https://doi.org/10.1007/3-540-44590-0_15
22. Stjernfelt, F.: On operational and optimal iconicity in Peirce's diagrammatology. Semiotica **186**, 395–419 (2011)

EulerMerge: Simplifying Euler Diagrams Through Set Merges

Xinyuan Yan[1](✉)[iD], Peter Rodgers[2](✉)[iD], Peter Rottmann[3][iD],
Daniel Archambault[4][iD], Jan-Henrik Haunert[3][iD], and Bei Wang[1](✉)[iD]

[1] University of Utah, Salt Lake City, USA
{xinyuan.yan,beiwang}@sci.utah.edu
[2] University of Kent, Canterbury, UK
p.j.rodgers@kent.ac.uk
[3] University of Bonn, Bonn, Germany
{rottmann,haunert}@igg.uni-bonn.de
[4] Newcastle University, Newcastle upon Tyne, UK
daniel.archambault@newcastle.ac.uk

Abstract. Euler diagrams are an intuitive and popular method to visualize set-based data. In an Euler diagram, each set is represented as a closed curve, and set intersections are shown by curve overlaps. However, Euler diagrams are not visually scalable and automatic layout techniques struggle to display real-world data sets in a comprehensible way. Prior state-of-the-art approaches can embed Euler diagrams by splitting a closed curve into multiple curves so that a set is represented by multiple disconnected enclosed areas. In addition, these methods typically result in multiple curve segments being drawn concurrently. Both of these features significantly impede understanding. In this paper, we present a new and scalable method for embedding Euler diagrams using set merges. Our approach simplifies the underlying data to ensure that each set is represented by a single, connected enclosed area and that the diagram is drawn without curve concurrency, leading to wellformed and understandable Euler diagrams.

Keywords: Euler diagrams · Set visualization · Hypergraph visualization · Scalability

1 Introduction

Set-based data are found in many real-world examples. In personalized recommendation systems, sets capture multivariate relationships among users, query topics, item features [11], and reasoning [4,20]. Set-based data are also prevalent in biological systems to encode multiway relationships among entities in protein complexes, transcription factor and microRNA regulation networks, protein function annotations, and metabolic processes [36].

An intuitive way to visualize set-based data is through an Euler diagram, which captures sets and their relationships. In an Euler diagram, sets are represented by closed curves that enclose regions. The way the regions overlap reveals the intersections between the sets. Representing sets using an Euler diagram is visually intuitive; however, these approaches can suffer from comprehensibility issues even with a small number of sets, and scaling is considered to be limited to 10 sets [2]. Recall that a planar graph is a graph that can be drawn on a plane without edge crossings, finding a visualization of a set system as an Euler diagram is equivalent to finding a planar embedding of its dual graph [9]. However, for many set systems, no planar embedding exists. In these cases, previous algorithms to embed Euler diagrams represent a set by splitting it into two or more closed curves [25,32], resulting in the set not being represented by a single *connected enclosed area*. This has the advantage that all instances of set systems are embeddable, but has the disadvantage that the diagram is much harder to understand, as the same set can be represented in different parts of the diagram.

In addition, these prior layout methods typically introduce *concurrency*, where multiple curves share a line segment. Both concurrency and disconnected enclosed areas are violations of important *wellformedness conditions*, which are known to impede understanding [26].

We present a new method that simplifies an Euler diagram via set merges. A set merge takes the union of two or more sets in a set system and represents the resulting set as a single closed curve. This increases the scalability of data that the method can successfully visualize compared to previous methods. Our contributions are as follows:

- We produce Euler diagrams that satisfy a number of wellformedness conditions. In particular, our simplification process ensures that each set (or merged group of sets) is represented by a single connected enclosed area and that there is no concurrency.
- We demonstrate via experiments that, typically, a small number of set merges leads to Euler diagrams that are wellformed and understandable.

2 Related Work

The visualization of set-based data has been an active area of research. We use the terms "set system" and "hypergraph" interchangeably with the understanding that these terms arise from different research communities. Hypergraphs generalize graphs by allowing *hyperedges*, that is, edges that contain more than two nodes. Hence, a hyperedge is a set and a hypergraph is a set system.

Many set visualization techniques use geometric elements such as lines, circles, ovals and closed curves to connect or enclose set elements. Other concepts use tables or matrices (see [2,8] for surveys). However, in this section, we concentrate on the closely related background in Euler diagrams and simplification.

Euler and Venn Diagrams. Various methods for embedding a set system using closed curves have been developed. These methods vary by the type of shape applied, for instance, constraining the Euler diagram curves to circles [14,33–35] or hexagonal/square grid cells [28]. In these cases, the diagram is represented using only these shapes. However, only a subset of set systems are embeddable with such shape restrictions, a limitation not present in our work.

Other work takes elements that have been previously embedded and superimpose polygons on top of them [6]. Similarly, GMap [10] creates one or more polygons around areas of a graph that are in the same set. GMap relies on a prior layout of data to be grouped, which our work does not require.

Mäkinen [21] introduced an edge-based and a subset-based approach to draw hypergraphs, where hyperedges are drawn as smooth curves *connecting* or *enclosing* their nodes, respectively. However, an Euler diagram is not embeddable for all set systems [25,32] with a single closed curve representing each set, as dual graphs derived from these set systems cannot be drawn in a planar way. Other methods, such as SPEULER [13], instead arrange set elements using a circular layout. However, in order to have an embeddable diagram, SPEULER produces overlaps of two curves when there are no elements within them.

Techniques also exist to refine an Euler diagram drawing once it has been generated, improving its readability. For instance, eulerForce [22] uses a force-directed algorithm to refine the set, whereas EulerSmooth [31] uses curve shortening flows to achieve the same objective.

Graph and Hypergraph Simplification. A number of approaches have been proposed that simplify and summarize graphs (e.g., [29]). Visualization of simplified graphs has been explored [7,16,18], which may be applicable to the node-link diagrams of set systems. Our approach is related to graph coarsening driven by visual criteria [1,3], which aggregates subgraphs into single nodes based on topological properties. Instead of set merges, Euler diagrams may also be simplified via element removal and optimization [27]. Zhou et al. [37] simplified hypergraphs by combining nodes if they belong to almost the same set of hyperedges, and merging hyperedges if they share almost the same set of nodes. However, it is different from our work in the simplification criteria, algorithm, and visualization perspectives. Oliver et al. [24] recently proposed a framework for visualizing scalable hypergraphs, with a convex polygon-based layout. Their approach incorporates an iterative, reversible simplification process and layout optimization. They have several merging operations, including hyperedge merging. As with [37], the simplification criteria, underlying algorithms and resultant visualization differ markedly. Whereas their method attempts to reduce the unwanted overlap of the convex polygons representing hyperedges, our approach guarantees no unwanted overlaps for the curves representing sets.

3 Technical Background

Euler Diagrams and Zones. A closed curve γ in the plane $\mathbb{R}^2$ is a continuous function $\gamma\colon [0,1] \to \mathbb{R}^2$, where $\gamma(0) = \gamma(1)$. An *Euler diagram* is a pair $\mathcal{E} = (\Gamma, \pi)$, where Γ is a finite set of closed curves in $\mathbb{R}^2$, L is a set of labels, and $\pi\colon \Gamma \to L$ is a mapping that assigns to each curve $\gamma \in \Gamma$ a label in L. A *minimal region* of an Euler diagram is a connected component of $\mathbb{R}^2 - \bigcup_{\gamma \in \Gamma} \text{image}(\gamma)$. A *zone* of an Euler diagram is a set of minimal regions that represent the intersection of sets. The set of zones is called the *abstract description* of the diagram.

Wellformedness Conditions. An Euler diagram $\mathcal{E}$ may have a number of desirable properties, referred to as *wellformedness conditions* [34]. These include:

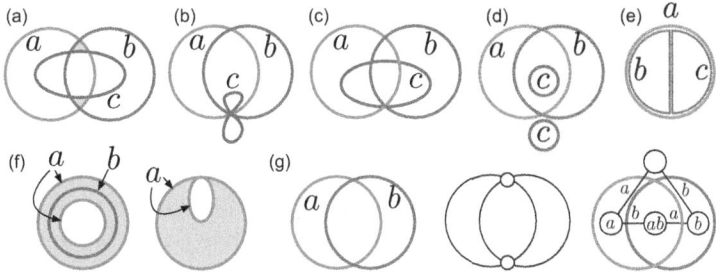

Fig. 1. Key wellformedness conditions: (a) a disconnected zone in pink; (b) a non-simple curve c; (c) a triple point; (d) duplicated curve labels c with an unconnected closed area; (e) concurrency: a and b, a and c, b and c; (f) two examples of duplicated curve label (left and right): label a represents a set with a connected enclosing area that contains a hole. (g) An Euler diagram (left), its Euler graph (middle), and its dual graph (right). (Color figure online)

- **Simplicity**: if all curves in Γ are simple curves (i.e., no self-crossings);
- **No triple points**: if there are no triple points of intersection among the curves in Γ;
- **Transversality**: if two curves in Γ intersect, they intersect transversally;
- **Connected zones**: if each zone of $\mathcal{E}$ is connected.

Our construction method guarantees that the diagram has simple curves and connected zones. There is evidence to show that triple points and transversality have limited impact on user understanding [26]. See Fig. 1 for an illustration. Of particular importance to this paper are the two wellformedness conditions:

- **No concurrency**: if no pairs of curves in Γ run concurrently.
- **Unique curve labels**: if π is an injective function.

Unique to this paper, we further refine the condition of unique curve labels into two conditions:

- **Genus free**: if the area enclosed by curves of the same label in Γ does not contain any genus (that is, a hole).
- **Connected enclosed area**: if the area enclosed by curves of the same label in Γ is connected.

For example, in Fig. 1(f), the connected enclosed area condition is not violated because the curve "a" is duplicated, but the region represented by "a" is not split into two components. However, the diagram is not genus free as "a" has a hole. However, in Fig. 1(d) curve "c" is duplicated and the region represented is split into two components, one inside "a" and "b", with the other region outside them. As a result, "c" does not have a connected enclosed area.

Euler Graph and Dual Graph. An *Euler graph* $G_\mathcal{E}$ constructed from an Euler diagram $\mathcal{E}$ has vertices defined at all curve intersection points, and edges defined as the curve segments that connect the vertices. By construction, each face of $G_\mathcal{E}$ is a minimal region of $\mathcal{E}$. The *dual graph* of an Euler diagram $\mathcal{E}$

is the standard dual graph of the Euler graph $G_{\mathcal{E}}$. The vertices of the dual graph represent the zones in the Euler diagram and the edges of the dual graph connect adjacent zones. Vertices are labeled by the curves that enclose their corresponding zones, and edges are labeled by the symmetric difference of their endpoints. See Fig. 1(g).

We consider the main cause of poor interpretation with duplicate curve labels to be caused by violations of connected enclosed area rather than genus free. Having genus present in a diagram means that curves with the same label are inside other curves with the same label, whereas the presence of a disconnected enclosed area means that the curves with the same label can be anywhere in the diagram, and therefore users may not spot all such curves when interpreting the diagram. Our approach is designed to ensure that the resulting Euler diagrams have simple curves and connected zones. We then further simplify with set merges to ensure connected enclosed areas and avoid concurrency.

4 Algorithm

In this section, we describe our novel algorithm for simplifying Euler diagrams with set merging. The code is available under a GPL open source license from https://github.com/tdavislab/EulerMerge. We use JGraphT [23], which provides planarity testing.

4.1 Algorithm Preliminaries

An Euler diagram is a visual representation of a set system. A *set system* is a set of sets $\mathcal{S} = \{S_1, S_2, ..., S_m\}$, where each set $S_i \in \mathcal{S}$ ($1 \leq i \leq m$) is a nonempty subset of a universe $U = \bigcup_{i=1}^{m} S_i$. With an abuse of notation, for an element $u \in U$, $\mathcal{S}(u)$ contains sets from $\mathcal{S}$ that contain u, i.e., $\mathcal{S}(u) = \{S_i \in \mathcal{S} \mid u \in S_i\}$. We assume a set system is always given with a label-assigning map l and $l(S_i)$ is the label associated with the set S_i.

Our algorithm merges sets in $\mathcal{S}$ through an iterative process. In each iteration, two sets are selected from $\mathcal{S}$ and replaced in $\mathcal{S}$ by their union. During this process, the algorithm maintains a set of zones Z, which at any time is uniquely defined as follows: Z contains the empty set $z_0 := \emptyset$ and multiple nonempty sets $z_1, z_2, \ldots, z_n$ that partition U, such that any two elements $u, v \in U$ are contained in the same zone if and only if $\mathcal{S}(u) = \mathcal{S}(v)$. In other words, elements of U are in the same zone if they are contained in the same subset of sets from $\mathcal{S}$.

We work with a graphical representation G of the set of zones. We design our algorithm such that, preferably after few iterations, G will be the dual graph of a wellformed Euler diagram of the simplified set system. Although, initially, G may lack the property of a dual graph of a wellformed Euler diagram, for simplicity, we refer to it as *dual graph* throughout the whole merging process.

Let $G = (Z, E, l_Z, l_E)$. With an abuse of notation, the vertices Z represent the zones and the edges E model pairwise relationships among the zones, where l_Z and l_E are (mappings of) zone labels and edge labels, respectively. The zone label $l_Z(z)$ of each $z \in Z$ is formed by a subset of $\mathcal{S}$ that constitutes the zone.

The edge label $l_E(e)$ of each edge $e = \{z_i, z_j\} \in E$ is the symmetric difference of the two zone labels, i.e., $l_E(e) := l_Z(z_i) \triangle l_Z(z_j)$. These labels are updated along with the set system as the algorithm progresses. As noted in Sect. 3, the set of zone labels is the *abstract description* of the set system.

4.2 A Running Example

We illustrate the set merging process with a running example. This is a set system $\mathcal{S}$ of a director from a movie database, see Sect. 6. Each set is a movie, and set elements are the actors that appear in the movie. Therefore in an Euler diagram, curves represent movies, and the curves overlap if the movies share at least one actor.

For the running example, the director is *Bonowicz, Brett Ryan*. He has seven movies, forming a set system of seven sets:

(a) Garriage; actors: Caps, Bonowicz, Fox, Kessler, Kostenbaudor, Kozlow.
(b) Last Days of Ki, The; actors: Herbst, Stilwell, Trad-DeStefano, Ashkin, Bonowicz, Chai, Chernyak, Dixon, Harpole, Lindo, Peters, Sawyer, Suppa.
(c) Interview for a Night Job; actors: Dastoli, James, Vergara, Edwin.
(d) Pressing the Public Opinion; actors: DeVries, Yeager, Bonowicz, Chernyak, Coolman, Lindo, Moore (Michael), Nelson.
(e) Baseball and Glory; actors: Caffrey, Dienstag, Seabright, Shults, Chernyak, Coolman, Dastoli, Denniberg, Garcia, Grant, Leery, Myers, Reiber, Sawyer, Shields, Tompkins, Weinstein.
(f) Signs and Voices; actors: Hecht, Moore (Julianne), Shepherd, Bonowicz, Dean.
(g) Banana Shell, The; actors: Baksh, Ashkin, Coolman, Fernandez, Grant, Gunn, Sawyer, Zawacki, Niki.

Its corresponding abstract description (set of zone labels) can be produced by finding the nonempty intersections in the set system. That is, if an actor $u \in U$ is in a collection of movies $\mathcal{S}(u)$ (and no other movies), $\mathcal{S}(u)$ is added to the abstract description, giving: $\{\emptyset, \{a\}, \{b\}, \{c\}, \{d\}, \{e\}, \{f\}, \{g\}, \{b,d\}, \{b,g\}, \{c,e\}, \{e,g\}, \{b,d,e\}, \{b,e,g\}, \{d,e,g\}, \{a,b,d,f\}\}$.

For example, the actor *Dastoli* is in movies "c" and "e": "Interview for a Night Job" and "Baseball and Glory" and no other sets, which gives rise to an element $\{c, e\}$ in the abstract description, and a zone with a label of $\{c, e\}$ (for simplicity, also referred to as "ce").

4.3 An Overview of EulerMerge Algorithm

Our set merging algorithm (Algorithm 1) takes an input set system and generates an initial dual graph. It then selectively merges pairs of sets to produce a planar dual graph and finally applies additional set merges to remove concurrency.

The input to our EulerMerge algorithm is a set system $\mathcal{S}$ equipped with (a mapping of) set labels l, and the output is a dual graph of an Euler diagram G

Algorithm 1: EulerMerge

Input   : Set system $\mathcal{S} = \{S_1, S_2, \ldots, S_m\}$
Output: dual graph $G = (Z, E)$
$G \leftarrow \text{InitialDualGraph}(\mathcal{S})$;
$G \leftarrow \text{NonPlanarToPlanar}(G)$;
$G \leftarrow \text{ConcurrencyRemoval}(G)$;
return G
---

with zone labels l_Z and edge labels l_E; for simplicity, these labels are sometimes omitted in the pseudocode.

The InitialDualGraph algorithm creates an initial dual graph G for an input set system $\mathcal{S}$. First, the algorithm derives the abstract description by computing $\mathcal{S}(u)$ for each $u \in U$. Second, it creates edges and edge labels of G. Two zones z_i and z_j in G are connected with an edge if their zone labels differ by one. Third, the algorithm computes an induced graph for each set $S_i \in \mathcal{S}$. If this is connected then there are no duplicate curve labels for that set in the corresponding Euler diagram. If the induced graph is not connected, the algorithm adds edges to connect it. However, this operation will introduce concurrency wherever the labels of the two connected zones differ by more than one element.

The initial dual graph G may not correspond to a wellformed Euler diagram. First, there may not be a planar dual graph for the initial set system. Second, the constructive process for initializing a dual graph is heuristic and may not produce a planar dual even if one exists. However, using a heuristic process is justifiable as deciding whether a given set system can be drawn as an Euler diagram is NP-complete [12]. Additionally, G may not correspond to a wellformed Euler diagram because the diagram may have concurrent edges.

For the running example, the initial dual graph can be seen in Fig. 2(a). The InitialDualGraph procedure ensures that zones that have labels with single symmetric difference are connected by edges. For example, "b" and "bd" are connected by an edge. However, this leaves the subgraph induced from the set "a" disconnected. Hence "a" and "abdf" are linked with an edge. This dual graph is nonplanar, so set merges via NonPlanarToPlanar (Sect. 4.4) are applied.

4.4 Set Merging for Planarity

Once we have an initial dual graph, we then find a planar dual by merging sets. We prioritize the planarity objective because we cannot embed an Euler diagram without a planar dual. Furthermore, we can move toward planarity and reduce concurrency simultaneously.

Every nonplanar graph contains a Kuratowski subdivision [17] (i.e., a subdivision of K_5 or $K_{3,3}$) as a subgraph, denoted G^K. Moreover, such a subgraph can be found in linear time [5]. As shown in Algorithm 2, we merge two sets present in such a subgraph until G becomes planar.

Our process for merging a pair of sets is shown in Algorithm 2. We first replace any zone label and edge label containing $l(S_2)$ with $l(S_1)$. We then merge vertices in the dual graph with identical labels.

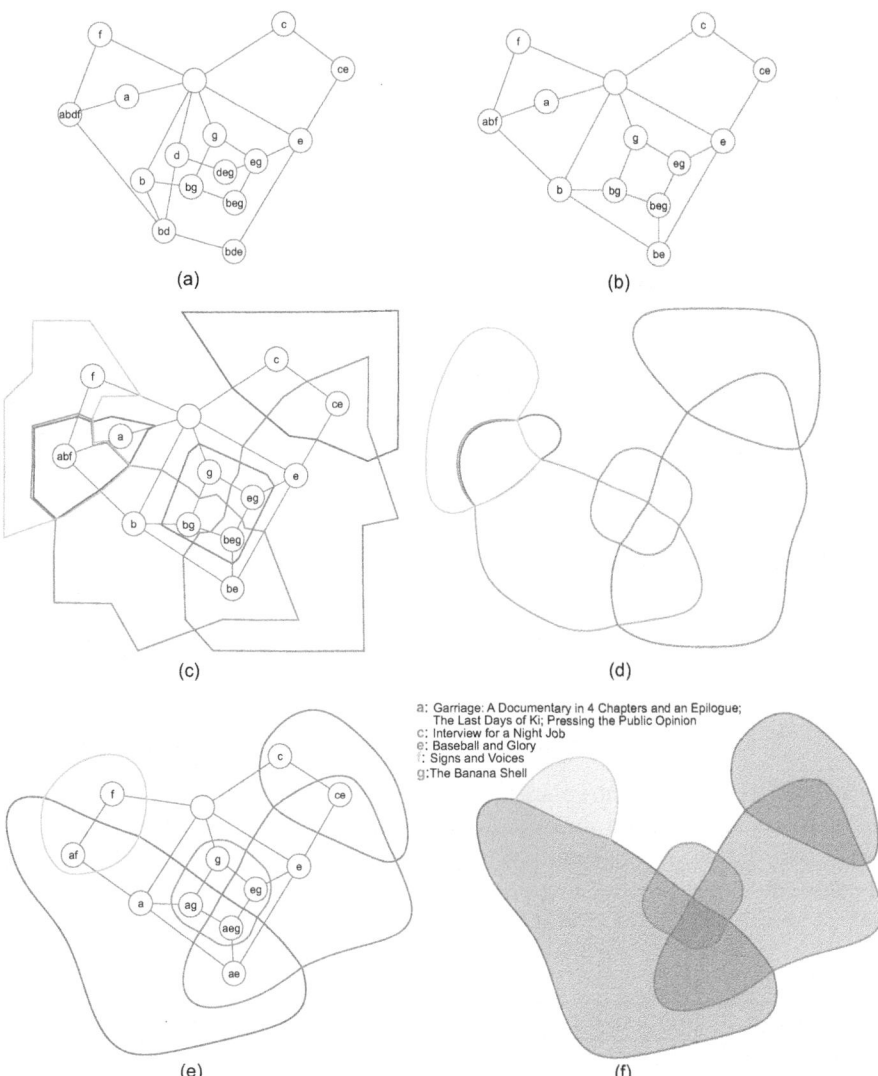

Fig. 2. Steps in the merging process: (a) The initial dual graph. (b) The first planar dual graph after merging sets "b" and "d". (c) The first planar dual graph and Euler diagram without smoothing. (d) The first Euler diagram with smoothing. (e) The dual graph and concurrency free final Euler diagram after merging sets "a" and "b". (f) The final Euler diagram with set labels.

Although achieving planarity is our top priority, we decide on the order of pairwise set merges based on the reduction of concurrency because the two sets are in a Kuratowski subdivision and merging them very likely also removes the subdivision. As a result we can remove concurrency whilst finding a planar dual.

Algorithm 2: PairwiseSetMerge

Input : dual graph $G = (Z, E)$, Sets S_1 and S_2
Output: dual graph $G' = (Z', E')$
$G' \leftarrow G$;
for $z \in Z'$ **do**
    if $l(S_2) \in l'_Z(z)$ **then**
        $l'_Z(z) \leftarrow l'_Z(z) \cup l(S_1) \backslash l(S_2)$
for $e \in E'$ **do**
    if $l(S_2) \in l'_E(e)$ **then**
        $l'_E(e) \leftarrow l'_E(e) \cup l(S_1) \backslash l(S_2)$
for $z, z' \in Z'$ **do**
    if $l'_Z(z) = l'_Z(z')$ **then**
        $G' \leftarrow$ ZoneMerge(G', z, z')
return G';

In a limited number of cases (for instance, when the sets to be merged are both in exactly the same zones), the subdivision remains, whereas concurrency is greatly reduced by the set merges and so the second aim of the merging process is satisfied. In this case, planarity will be achieved through subsequent merges.

We introduce a measure that quantifies the concurrency in a dual graph:

$$\text{Concurrency}(G) = \sum_{e \in E} |l_E(e)| - |E|.$$

Recall that concurrent curve segments appear in the Euler diagram because there are multiple sets on an edge label. Concurrency(G) measures the overall size of edge labels. Concurrency$(G) = 0$ means that G has no concurrency as all edges are labeled with a single set. In Algorithm 3, we merge two sets in G^K that cause the most reduction in concurrency.

Algorithm 3: NonplanarToPlanar

Input : dual graph $G = (Z, E)$
Output: dual graph $G' = (Z', E')$
$G' \leftarrow G$
while !IsPlanar(G') **do**
    $G^K = (Z^K, E^K) \leftarrow$ KuratowskiSubdivision(G')
    $R \leftarrow \bigcup_{z \in Z} l_Z^K(z)$;
    $G^M \leftarrow G'$;
    for $R_i \in R$ **do**
        for $R_j \in R$ **do**
            $G^S \leftarrow$ PairwiseSetMerge(G', R_i, R_j);
            if Concurrency$(G^S) <$ Concurrency(G^M) **then**
                $G^M \leftarrow G^S$;
    $G' \leftarrow G^M$;
return G'

In our running example, the initial dual graph is nonplanar and so we must apply Algorithm 3. In this case, we only need a single iteration as merging sets "b" and "d" results in a planar dual, shown in Fig. 2(b). We retain the set label with the lowest lexicographical order during set merges, in this case, "b".

This set merge also leads to reduced concurrency. The dual graph in Fig. 2(a) has a Concurrency of 6, whereas the dual graph in Fig. 2(b) has a Concurrency of 2. Once we have a planar dual graph, it can be used to embed an Euler diagram as shown in Fig. 2(c) and with improved layout in Fig. 2(d).

4.5 Set Merging to Remove Concurrency

We can apply additional set merges to our planar dual to remove the remaining concurrency, see Algorithm 4. We apply a greedy approach, that is, by merging pairs of sets that reduce the most amount of concurrency at each step. We note that alternative strategies for ordering pairwise set merges are also possible, as discussed in Sect. 7. The final if statement, which ensures planarity by calling Algorithm 3, is for rare cases where nonplanar duals have been produced by set merging. We have not encountered any example that contains such a rare case.

Algorithm 4: ConcurrencyRemoval

Input : dual graph $G = (V, E)$
Output: dual graph $G' = (V', E')$
$G' \leftarrow G$
while Concurrency$(G') > 0$ **do**
$\quad R \leftarrow \bigcup_{z \in Z'} l_{Z'}(z)$;
$\quad G^M \leftarrow G'$;
$\quad$ **for** $R_i \in R$ **do**
$\quad\quad$ **for** $R_j \in R$ **do**
$\quad\quad\quad G^S \leftarrow$ PairwiseSetMerge(G', R_i, R_j);
$\quad\quad\quad$ **if** Concurrency$(G^S) <$ Concurrency(G^M) **then**
$\quad\quad\quad\quad G^M \leftarrow G^S$;
$\quad G' \leftarrow G^M$;
if !IsPlanar(G') **then**
$\quad G' \leftarrow$ NonPlanarToPlanar(G')
return G'

In the running example, the dual graph in Fig. 2(c) still contains concurrency, e.g., between the connected zones "a" and "abf", as seen in the Euler diagram curve where segments "b" (orange) and "f" (yellow) run concurrently. To remove all concurrency in this case, we need only a single iteration by merging sets "a" and "b", shown in Fig. 2(e). The merge renames "abf" as "af" removing the concurrency as they have a single symmetric difference with "a". With no concurrency, the merging process is complete. The final Euler diagram is given in Fig. 2(f). Here, one curve represents the merging of three movies. We have now simplified the abstract representation and the Euler diagram, embedding an Euler diagram without concurrency, at the cost of losing some detail.

Once we have formed an embeddable dual, we apply existing algorithms for embedding the dual graph [25], followed by smoothing using EulerSmooth [31] to refine the diagram boundaries.

Finally, we compare against the prior state-of-the-art general Euler diagram embedding [25], which we refer to as EulerGeneral. As shown in Fig. 3, EulerGeneral produces an Euler diagram where sets "a" and "f" are represented by disconnected enclosed areas. It has Concurrency of 5.

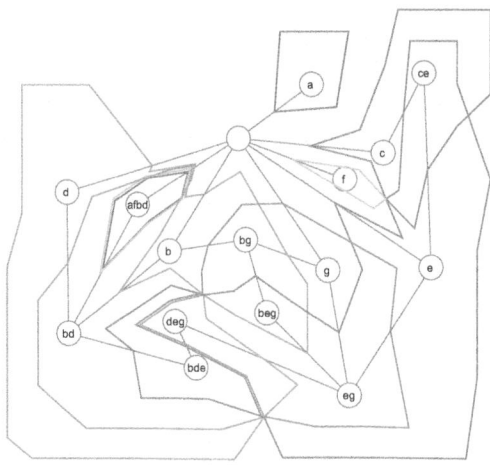

Fig. 3. The result of the running example with prior state-of-the-art EulerGeneral [25].

5 Evaluation

We compare our EulerMerge algorithm with the previous general Euler diagram embedder EulerGeneral [25]. Our algorithm produces Euler diagrams that have connected enclosed areas and no concurrency (see Sect. 3). It reduces the number of sets to be visualized. The EulerGeneral algorithm [25] visualizes all sets at the cost of containing disconnected enclosed areas and concurrency.

For EulerMerge, we need to consider the number of set merges required to produce a planar dual graph as well as those required to remove concurrency. We therefore count the number of set merging operations in Algorithm 3 and Algorithm 4. EulerGeneral might not obtain a wellformed diagram as duplicated curve labels and concurrency appear in many cases. Hence we quantify the number of duplicated curve labels as well as the amount of concurrency in the diagram. We note that in some complex cases, EulerGeneral does not produce an embedding at all. These cases have been removed from the data.

We use two collections of real-world set systems for evaluation (Table 1). First, the MovieDB data from the 2007 InfoVis contest [15]. A set system is formed from movies directed by a director with the movies as sets and the actors that appear in a movie as elements in the set. Second, a set system from the Twitter Circles [19] collection contains sets that are interests groups formed by Twitter users. We include only diagrams that contain duplicated curve labels or concurrency.

For EulerMerge, Table 2 shows the average number of set merges required to produce a planar dual graph as well as those required to achieve concurrency. Concurrency reduction is over ten times more common than planarity reduction for both collections.

Table 1. Real-world data summary, reporting the number of set systems per collection, the mean number of sets (Mean Sets) and mean number of zones (Mean Zones).

Collection	#Set Systems	Mean Sets	Mean Zones
MovieDB	225	4.4	7.78
Twitter Circles	59	5.92	8.24

Table 2. EulerMerge: the average number of merges for achieving planarity (Planarity) and removing concurrency (Concurrency), with the average of both (Both).

Collection	Planarity	Concurrency	Both
MovieDB	0.11	1.23	1.33
Twitter Circles	0.07	1.97	2.04

Table 3. With EulerGeneral, the average number of duplicated curve labels (#Duplicated Curve Labels) and the average Concurrency count (Concurrency).

Collection	#Duplicated Curve Labels	Concurrency
MovieDB	0.35	4.63
Twitter Circles	0.41	6.19

For EulerGeneral, Table 3 shows the average number of duplicated curves and the average Concurrency count. Duplicated curve labels occur in less than half of the diagrams generated, which may be artificially reduced by the failure of the algorithm to visualize some set systems, particularly complex ones.

In Table 2 and Table 3, it is shown that a smaller number of set merges are required to remove concurrency with EulerMerge, compared to the amount of concurrency in EulerGeneral.

6 Examples of Use Cases

Twitter Data Example. Figure 4 illustrates the process of applying our method to an example group of users from the Twitter data, where each user

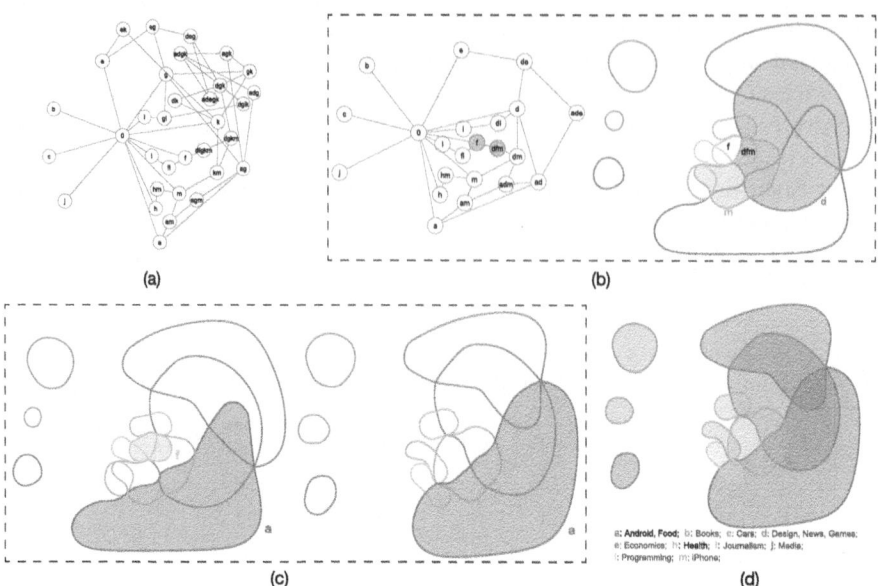

Fig. 4. The set merging process of a Twitter data set. (a) The original dual graph. (b) The first planar dual graph (left) and its corresponding Euler diagram (right) with concurrency. (c) From left to right: sets "a" and "f" merge into "a". (d) The final Euler diagram without concurrency.

may belong to multiple interest groups. The set system consists of 13 sets and 32 intersections. The names of sets are shortened to single letters.

We first construct an initial dual graph, as shown in Fig. 4(a) where vertices represent zones, namely, the 32 nonempty intersections of sets. However, the initial dual graph is nonplanar and, thus, it is not possible to generate an Euler diagram from it. We therefore employ Algorithm 3 to merge sets until the graph can be embedded without edge crossings. The first iteration merges sets "d" and "k". A second merge is required before reaching planarity, so "d" and "g" are merged. The two merges produce a planar dual graph with 11 sets and 21 intersections, see Fig. 4(b). However, the result exhibits concurrency. For example, the vertices "f" and "dfm" are connected, but differ by two sets. To eliminate concurrency, we continue to merge sets using Algorithm 4, which merges "a" and "f", as shown in Fig. 4(c). The resulting Euler diagram has no concurrency.

This example demonstrates that EulerMerge has achieved a desirable Euler diagram: after three set merges, two for planarity and one for concurrency, we can visualize this as an Euler diagram with connected enclosed areas and without concurrency, see Fig. 4(d).

Movie Data Example (Director: Hooker, Keith). This data set consists of five sets in total. Each set represents a movie. We show set intersections where one or more actor appears in those films and no other films. As shown in Fig. 5(a),

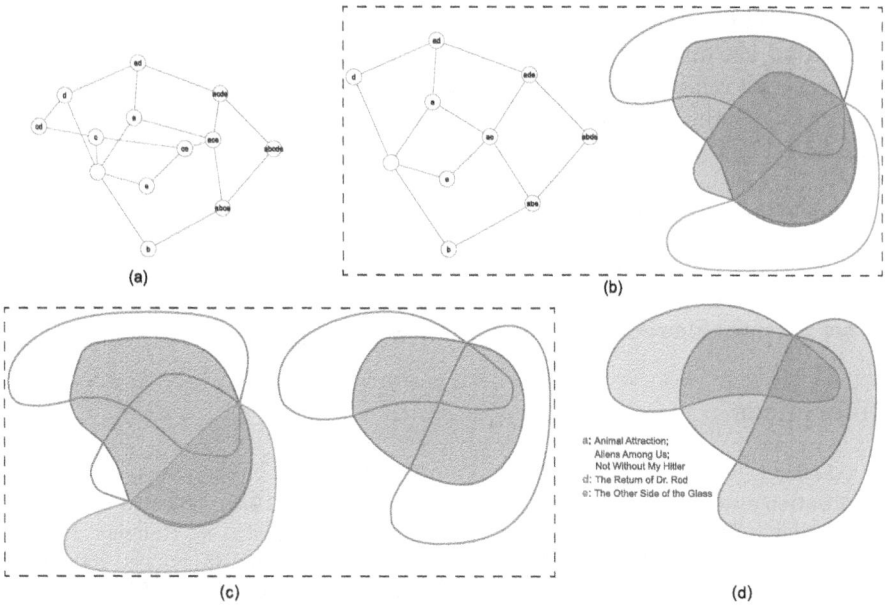

Fig. 5. The set merging process of a movie data set involving the director Hooker, Keith. (a) The initial dual graph. (b) Planarity merges: merging sets "a" and "c" into "a" and then merging "a" and "b" into "a". (c) Concurrency merges: merging "a" and "b" in to "a". (d) The final diagram with the original set names.

the initial dual graph is not planar. The simplification process merges once for planarity (sets "a" and "c" merge into "a") and merges once for concurrency removal (sets "a" and "b" merge into "a"). Figure 5(b) left shows the first planar dual graph. Figure 5(b) right shows the corresponding Euler diagram, which exhibits concurrency. Figure 5(c) shows the concurrency removal step, where sets "a" and "b" in the left Euler diagram merge into "a" in the right Euler diagram. Figure 5(d) gives the final Euler diagram with the original set names.

7 Conclusion

This paper provides, for the first time, an algorithm that uses set merges to simplify the layout of Euler diagrams to meet multiple wellformedness conditions. Merging just a few sets (often three or less) using EulerMerge results in a diagram without split sets or concurrency, thus producing a simplified diagram that is easier to comprehend than the non-wellformed alternative. Merging two sets into one reduces the amount of detail available to the user, which may be seen as a disadvantage. However, we see the simplification of a complex Euler diagram as a potential benefit. If needed, with the integration of a suitable user interface, the accepted visualization approach of "overview first, zoom and filter, then details-on-demand" [30] can be employed to reveal the missing details.

We further perceive a simplified, wellformed diagram as a basis for hierarchical exploration via interactively expanding/collapsing merged sets, thus reflecting the precise set relations required by well-matchedness [2]. Finally, a future direction is to study the balance between visual readability and information loss, built upon additional set simplification approaches via element removal [27].

Acknowledgements. We thank Dagstuhl Seminar 22462 "Set Visualization and Uncertainty" (November 13–18, 2022) for initializing this research. This research was partially supported by grants DOE DE-SC0021015 and NSF IIS-2145499. For the purpose of open access, the author has applied a CC BY public copyright license to any Author Accepted Manuscript version arising from this submission.

References

1. Abello, J., van Ham, F., Krishnan, N.: ASK-GraphView: a large scale graph visualization system. IEEE Trans. Visual Comput. Graph. **12**(5), 669–676 (2006)
2. Alsallakh, B., Micallef, L., Aigner, W., Hauser, H., Miksch, S., Rodgers, P.: The state-of-the-art of set visualization. Comput. Graph. Forum **35**(1), 234–260 (2016)
3. Archambault, D., Munzner, T., Auber, D.: Tugging graphs faster: efficiently modifying path-preserving hierarchies for browsing paths. IEEE Trans. Visual Comput. Graphics **17**(3), 276–289 (2010)
4. Bourou, D., Schorlemmer, M., Plaza, E.: Euler vs Hasse diagrams for reasoning about sets: a cognitive approach. In: Giardino, V., Linker, S., Burns, R., Bellucci, F., Boucheix, J.M., Viana, P. (eds.) Diagrams 2022. LNCS, vol. 13462, pp. 151–167. Springer, Cham (2022). https://doi.org/10.1007/978-3-031-15146-0_13
5. Boyer, J.M., Myrvold, W.J.: On the cutting edge: simplified $O(n)$ planarity by edge addition. J. Graph Algorithms Appl. **8**, 241–273 (2004)
6. Collins, C., Penn, G., Carpendale, S.: Bubble sets: revealing set relations with isocontours over existing visualizations. IEEE Trans. Visual Comput. Graphics **15**(6), 1009–1016 (2009)
7. Dunne, C., Shneiderman, B.: Motif simplification: improving network visualization readability with fan, connector, and clique glyphs. In: Proceedings of the SIGCHI Conference on Human Factors in Computing Systems, pp. 3247–3256 (2013)
8. Fischer, M.T., Frings, A., Keim, D.A., Seebacher, D.: Towards a survey on static and dynamic hypergraph visualizations. In: 2021 IEEE Visualization Conference (VIS), pp. 81–85 (2021)
9. Flower, J., Howse, J.: Generating Euler diagrams. In: International Conference on Diagrammatic Representation and Inference (DIAGRAMS), pp. 61–75 (2002)
10. Gansner, E.R., Hu, Y., Kobourov, S.: GMap: visualizing graphs and clusters as maps. In: 2010 IEEE Pacific Visualization Symposium, pp. 201–208 (2010)
11. Huang, X., Liu, X.: Incorporating a topic model into a hypergraph neural network for searching-scenario oriented recommendations. Appl. Sci. **12**(15), 7387 (2022)
12. Johnson, D.S., Pollak, H.O.: Hypergraph planarity and the complexity of drawing Venn diagrams. J. Graph Theory **11**(3), 309–325 (1987)
13. Kehlbeck, R., Gortler, J., Wang, Y., Deussen, O.: SPEULER: semantics-preserving Euler diagrams. IEEE Trans. Visual Comput. Graph. **28**(1), 433–442 (2022)
14. Kestler, H.A., et al.: VennMaster: Area-proportional Euler diagrams for functional go analysis of microarrays. BMC Bioinform. **9** (2008)

15. Kosara, R., Jankun-Kelly, T.J., Chlan, E.: IEEE InfoVis 2007 contest: InfoVis goes to the movies. https://eagereyes.org/blog/2007/infovis-contest-2007-data
16. Koutra, D., Kang, U., Vreeken, J., Faloutsos, C.: VoG: summarizing and understanding large graphs. In: Proceedings of the 2014 SIAM International Conference on Data Mining pp. 91–99 (2014)
17. Kuratowski, C.: Sur le probleme des courbes gauches en topologie. Fundam. Math. **15**(1), 271–283 (1930)
18. Lee, K., Jo, H., Ko, J., Lim, S., Shin, K.: SSumM: sparse summarization of massive graphs. In: Proceedings of the 26th ACM SIGKDD International Conference on Knowledge Discovery and Data Mining, pp. 144–154 (2020)
19. Leskovec, J., Krevl, A.: SNAP Datasets: Stanford large network dataset collection (2014). http://snap.stanford.edu/data
20. Linker, S.: Intuitionistic Euler-Venn diagrams. In: Pietarinen, A.-V., Chapman, P., Bosveld-de Smet, L., Giardino, V., Corter, J., Linker, S. (eds.) Diagrams 2020. LNCS (LNAI), vol. 12169, pp. 264–280. Springer, Cham (2020). https://doi.org/10.1007/978-3-030-54249-8_21
21. Mäkinen, E.: How to draw a hypergraph. Int. J. Comput. Math. **34**(3-4), 177–185 (1990)
22. Micallef, L., Rodgers, P.: eulerForce: force-directed layout for Euler diagrams. J. Vis. Lang. Comput. **25**(6), 924–934 (2014)
23. Michail, D., Kinable, J., Naveh, B., Sichi, J.V.: JGraphT—a java library for graph data structures and algorithms. ACM Trans. Math. Softw. **46**(2) (2020)
24. Oliver, P., Zhang, E., Zhang, Y.: Scalable hypergraph visualization. IEEE Trans. Visual Comput. Graph. **30**(1), 595–605 (2024)
25. Rodgers, P., Zhang, L., Fish, A.: General Euler diagram generation. In: Stapleton, G., Howse, J., Lee, J. (eds.) Diagrams 2008. LNCS (LNAI), vol. 5223, pp. 13–27. Springer, Heidelberg (2008). https://doi.org/10.1007/978-3-540-87730-1_6
26. Rodgers, P., Zhang, L., Purchase, H.: Wellformedness properties in Euler diagrams: which should be used? IEEE Trans. Visual Comput. Graph. **18**(7), 1089–1100 (2011)
27. Rottmann, P., Rodgers, P., Yan, X., Archambault, D., Wang, B., Haunert, J.H.: Generating Euler diagrams through combinatorial optimization. Comput. Graph. Forum **43**(3) (2024)
28. Rottmann, P., Wallinger, M., Bonerath, A., Gedicke, S., Nöllenburg, M., Haunert, J.H.: MosaicSets: embedding set systems into grid graphs. IEEE Trans. Visual Comput. Graph. **29**(1), 875–885 (2023)
29. Shin, K., Ghoting, A., Kim, M., Raghavan, H.: SWeG: lossless and lossy summarization of web-scale graphs. In: WWW Conference, pp. 1679–1690 (2019)
30. Shneiderman, B.: The eyes have it: a task by data type taxonomy for information visualizations. In: Proceedings of the IEEE Symposium on Visual Languages, pp. 336–343 (1996)
31. Simonetto, P., Archambault, D., Scheidegger, C.: A simple approach for boundary improvement of Euler diagrams. IEEE Trans. Visual Comput. Graph. **22**(1), 678–687 (2016)
32. Simonetto, P., Auber, D.: Visualise undrawable Euler diagrams. In: 12th International Conference Information Visualisation, pp. 594–599 (2008)
33. Stapleton, G., Flower, J., Rodgers, P., Howse, J.: Automatically drawing Euler diagrams with circles. J. Vis. Lang. Comput. **23**(3), 163–193 (2012)
34. Stapleton, G., Rodgers, P., Howse, J., Zhang, L.: Inductively generating Euler diagrams. IEEE Trans. Visual Comput. Graphics **17**(1), 88–100 (2011)

35. Wilkinson, L.: Exact and approximate area-proportional circular Venn and Euler diagrams. IEEE Trans. Visual Comput. Graphics **18**(2), 321–331 (2012)
36. Zhou, W., Nakhleh, L.: Properties of metabolic graphs: biological organization or representation artifacts? BMC Bioinform. **12**(132) (2011)
37. Zhou, Y., Rathore, A., Purvine, E., Wang, B.: Topological simplifications of hypergraphs. IEEE Trans. Vis. Comput. Graph. (TVCG) **29**(7), 3209–3225 (2023)

Representing Uncertainty with Expanded Ueberweg Diagrams

Amirouche Moktefi[1(✉)], Reetu Bhattacharjee[2], and Jens Lemanski[2,3]

[1] Ragnar Nurkse Department of Innovation and Governance, Tallinn University of Technology, Tallinn, Estonia
amirouche.moktefi@taltech.ee
[2] Philosophisches Seminar, University of Münster, Münster, Germany
{reetu.bhattacharjee,jenslemanski}@uni-muenster.de
[3] Faculty of Humanities and Social Sciences, FernUniversität in Hagen, Hagen, Germany

Abstract. Euler diagrams often require several figures to adequately represent propositions and syllogisms. Euler's followers, notably Friedrich Ueberweg, endeavored to overcome this difficulty with the use of dotted lines to express uncertainty about the relation between the terms of a proposition. Subsequently, Venn regarded such attempts as ineffectual and went to construct his own celebrated scheme. In this paper, we argue that Ueberweg's method could be expanded to meet Venn's expectations, and hence, produce alternative Venn-like diagrams.

Keywords: Euler diagram · Venn diagram · dotted lines · uncertainty · Ueberweg

1 Introduction

John Venn invented his celebrated diagrams due to his dissatisfaction with older (Eulerian) schemes [14]. His main criticism pertained to their inability to express uncertainty as to the precise relation between the terms of a proposition [9]. Prior to Venn, some logicians used dotted lines to tackle this difficulty. This convention was known to Johann Heinrich Lambert (1764) [6], Johann Maass (1793) [7], Arthur Schopenhauer (around 1820) [11], William Thomson (1849) [12], and others. Friedrich Ueberweg expanded the use of such dotted lines in his *System of Logic*, first published in German in 1857, revised and translated to English in 1871 [13]. Subsequent logicians commonly referred to him in this respect [15, p. 50; 5, p. 132; 17, p. 219].

Venn knew this dotting technique but regarded it as "wholly mis-aimed and ineffectual" [15, p. 50]. Against Venn, we argue here that this technique makes an important advance towards the kind of diagrams that Venn was searching for, that is a scheme "which shall be competent to indicate imperfect knowledge on our part; for this will at once enable us to appeal to it step by step in the process of working out our conclusion" [14, p. 4]. For the purpose, we amend and expand Ueberweg's ideas to construct a new scheme that meets Venn's expectations. The paper is organized as follows. In Sect. 2, we present Ueberweg's original scheme. Then, in Sect. 3, we introduce some amendments to ease the representation of propositions. In Sect. 4, we use this amended scheme to solve syllogisms. Finally, we briefly discuss, in the conclusion section, some characteristics of this scheme.

2 Ueberweg's Diagrams

Ueberweg acknowledged that the representation of traditional categorical propositions (whose terms are S and P) with Eulerian diagrams often requires the "combination" of several figures among the five diagrams that depict the strict relation between circles S and P, shown in (Fig. 1). A proposition 'All S is P' requires the combination of two figures: (d) and (e). 'Some S is P' requires four figures: (a), (b), (d) and (e). 'Some S is not P' requires three figures: (a), (b) and (c). Only propositions of the form 'No S is P' require a single figure: (c) [13, p. 217].

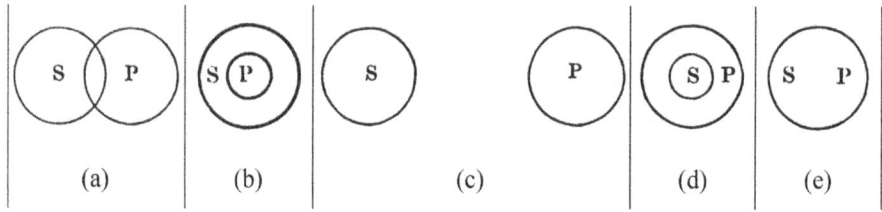

Fig. 1. Relations between two terms S and P [13, p. 217]

Consequently, Ueberweg introduced a "presupposition" to reduce the representation of each proposition to a single figure, where "the definite [is] denoted by a continuous and the indefinite by a dotted line" [13, p. 217]. The introduction of this convention produces the diagrams shown in (Fig. 2)[1]. Figure (f) combines relations (d) and (e) to represent the proposition 'All S is P'. Indeed, if the dotted line is made continuous, one obtains the relation (d), and if it is removed, it produces (e). Ueberweg's figure (g) for propositions 'Some S is P' has two dotted lines, one on the left, the other on the right. The four combinatorial possibilities regarding the existence of these lines produce the four required relations (a), (b), (d) and (e). Ueberweg's figure (h) for propositions 'Some S is not P' is inelegant and problematic, as it makes it possible for circle S to become a mere

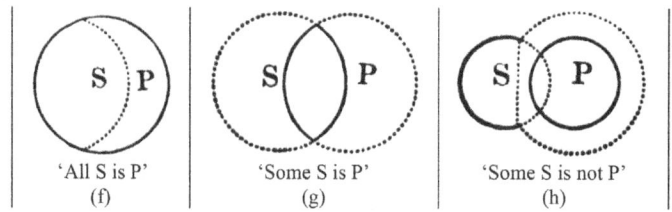

Fig. 2. Ueberweg's diagrams for categorical propositions [13, p. 217–218]

---

[1] The proposition 'No S is P' is not included as it can already be represented with a single figure with the Eulerian scheme, and hence does not require the introduction of dotted lines.

arc when all dotted lines are removed. Amendments to this figure will be subsequently discussed.

Ueberweg did not systematically use this convention for syllogisms but his examples inform us about his method. Consider the case of a pair of premises: 'No M is B' and 'Some A is M'. Ueberweg presented three figures which cover various combinations of A, B and M, permitted by the premises (Fig. 3). All figures show disjoint circles M and B in accordance with the premise 'No M is B' and intersecting circles A and M in accordance with the other premise 'Some A is M'. Various relations between A and B are permitted: they might be disjoint (i), they might intersect (j), or A includes B (k).

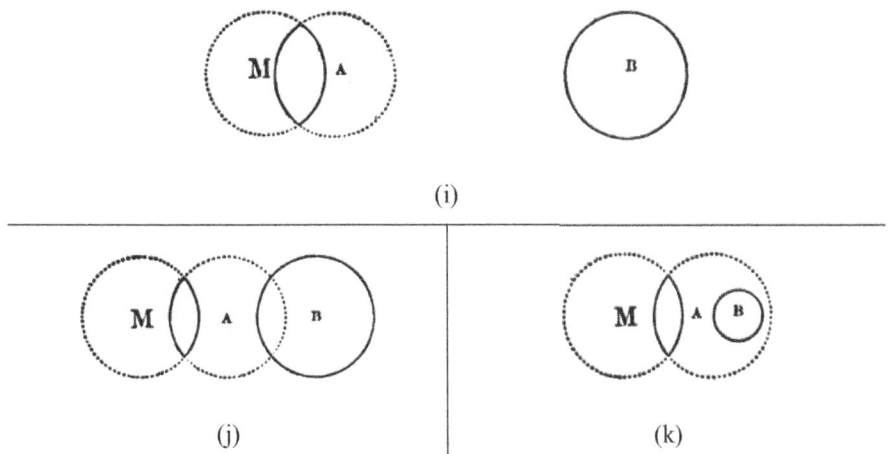

Fig. 3. Ueberweg's diagrams for 'No M is B' and 'Some A is M' [13, p. 388]

Then, Ueberweg identified the conclusion of the syllogism to be 'Some A is not B', on the ground that B "is quite separated from all M, and must therefore, also be separated from those A which coincide with M" [13, p. 388]. Unlike his treatment of propositions, Ueberweg did not 'combine' the diagrams that stand for a syllogism within a single figure. In the following, we achieve this combination by expanding Ueberweg's method to obtain dotted Venn-like diagrams.

3 The Representation of Propositions

Before addressing syllogisms, we suggest amendments that would ease the construction and manipulation of Ueberweg's scheme. As indicated earlier, a key weakness is its poor representation of propositions 'Some S is not P'. James Welton rightly commented that Ueberweg's figure "can scarcely be regarded as sufficiently simple and obvious to be satisfactory" [17, p. 219]. An alternative representation has been suggested by William S. Jevons. He introduced a "broken line", shown in (Fig. 4), to indicate that "some violets are known to be outside the odorous things, but that it is doubtful whether some violets are inside or not" [4, p. 47]. If this plan would be adopted for the propositions 'Some

S is not P', it would still be unsatisfactory because it does not make it possible for P to be strictly contained in S. A better solution, shown in (Fig. 5), was offered by John N. Keynes who believed it to be "simpler [than Ueberweg's], but equally effective". It suffices to "fill in" and "strike out" the dotted lines to produce the required relations of the proposition 'Some S is not P' [5, p. 133]. In the following, we adopt Keynes' solution.

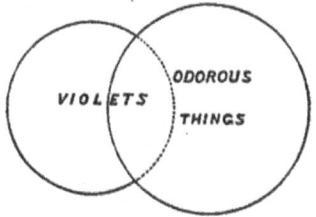
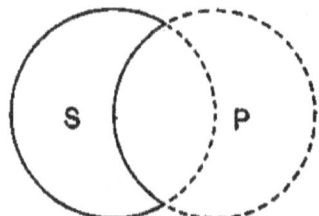

Fig. 4. Jevons' solution [4, p. 47]   **Fig. 5.** Keynes' solution [5, p. 133]

To improve the manipulation of our diagrams, we also need a strict labelling practice. Indeed, when dotted lines are removed, labels are crucial to indicate what region(s) remain(s). In the following, we use letters to label regions and convene that a label covers all the regions which it denotes. For instance, in (Fig. 6), labels indicate that: in (l), the term A is represented by the whole circle; in (m), A is depicted by the whole circle while B covers only the right half area; in (n), A is represented by the whole circle, B covers the two right areas and C refers only to the right-bottom area.

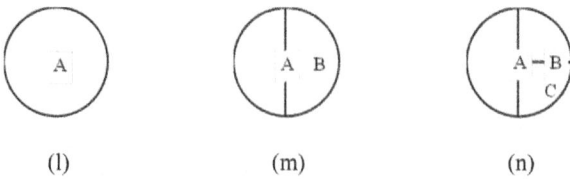

Fig. 6. Examples of labels on diagrams

For our purpose, we also incorporate the possibility that the terms of our propositions are empty. This contrasts with the view generally "assumed in the use of Euler's diagrams that [the two terms of a proposition] both exist in the universe of discourse, while neither of them exhausts that universe" [5, p. 131]. However, our perspective aligns with Venn diagrams which, similarly, permit the representation of empty terms. A consequence of this assumption is that, unless their existence is asserted by a proposition, terms are represented with dotted circles. For instance, 'No S is P' is represented with two disjoint dotted (rather than continuous, as in Euler's and Ueberweg's original schemes) circles S and P.

In the following, we introduce our system of dotted diagrams that we name 'Ueberweg diagrams' (which should not be confused with the original scheme to which we referred to as 'Ueberweg's diagrams' so far). To construct an Ueberweg diagram, we first sketch a primary figure which does not represent any specific proposition, but rather serves as a framework for the subsequent introduction of information. For a proposition with two terms, this primary diagram consists in two intersecting dotted circles and depicts the possible borders separating the regions produced by our terms (Fig. 7). This framework evidently resembles Venn's primary diagram (Fig. 8). The main difference is that Ueberweg diagrams exhibit the uncertainty of the borders while Venn diagrams exhibit the uncertainty of the areas.

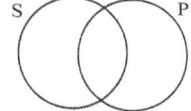

Fig. 7. Ueberweg framework   **Fig. 8.** Venn framework

To represent propositions with Ueberweg diagrams, one simply needs to change the borders of the primary diagram: a border becomes continuous if it is known to exist and is erased if it is known not to exist. Borders which are not affected by a proposition remain dotted. This procedure makes it easy to effectively represent the categorical propositions of syllogistic with our diagrams, as shown in (Fig. 9). For reference, we added the representation of those propositions with Venn diagrams[2].

	Ueberweg diagrams	Venn diagrams
All S is P		
No S is P		
Some S is P		
Some S is not P		

Fig. 9. Ueberweg diagrams for categorical propositions

It can be readily seen that the resulting diagrams resemble those of Ueberweg's original figures, except for 'Some S is not P' which produces Keynes' solution.

---

[2] It is easy to observe the similarity of procedure in Ueberweg diagrams with Venn diagrams where syntactic devices are introduced to mark the emptiness or existence of the regions to which propositions refer, while unrelated regions are left unmarked. We used shading to indicate emptiness and a cross to mark existence.

4 The Representation of Syllogisms

To treat syllogisms with Ueberweg diagrams, we introduce a primary diagram for three terms (similar to Venn's), then we represent propositions progressively by filling in, striking out or leaving unchanged our dotted lines. In (Fig. 10), two examples of syllogisms are tackled with this procedure. For reference, Venn diagrams for those syllogisms are also provided. The resulting Ueberweg diagrams for syllogisms readily exhibit their conclusions. Also, each syllogism is depicted with a single diagram as one obtains with Venn diagrams, unlike Euler's and Ueberweg's original schemes.

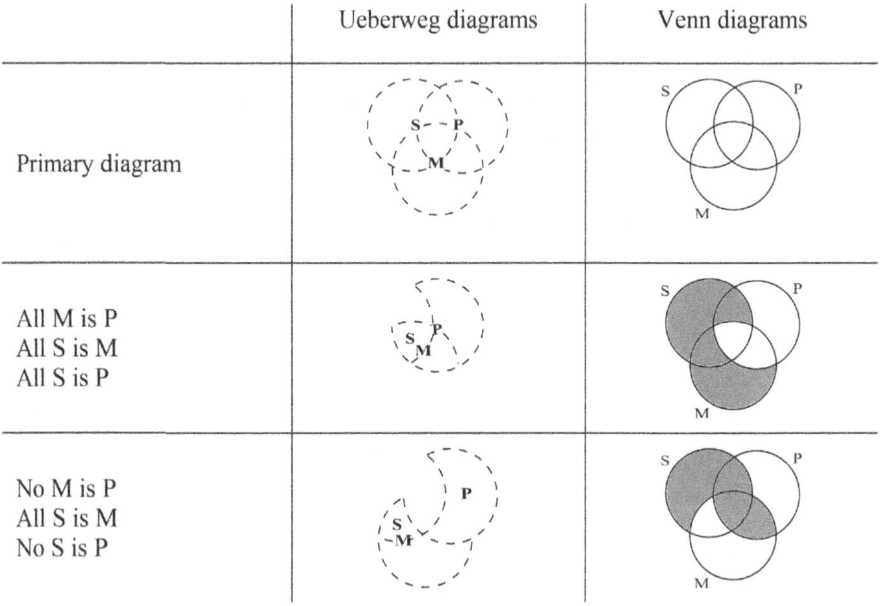

	Ueberweg diagrams	Venn diagrams
Primary diagram		
All M is P All S is M All S is P		
No M is P All S is M No S is P		

Fig. 10. Examples of syllogisms with Ueberweg diagrams

Previous examples depicted syllogisms with universal premises. Let us now tackle a syllogism with an existential premise. For the purpose, we consider the pair of premises, discussed earlier, for which Ueberweg provided three distinct figures (Fig. 3). The premises are: 'No M is B' and 'Some A is M'. In (Fig. 11), we show the steps required for the construction of our Ueberweg diagram. First, we use the 3-term primary diagram (r). Then, we introduce the premises, in accordance with the conventions explained earlier. The first premise 'No M is B' is represented by striking out the border of M which is inside B, as shown in (s). Then, the second premise 'Some A is M' is added by filling in the borders of the region that is shared by A and M. We obtain figure (t), which readily exhibits the conclusion 'Some A is not B'.

We observe that Ueberweg's three diagrams for this syllogism are reduced here to one figure. For reference, we also represent the syllogism with Venn diagrams (Fig. 12).

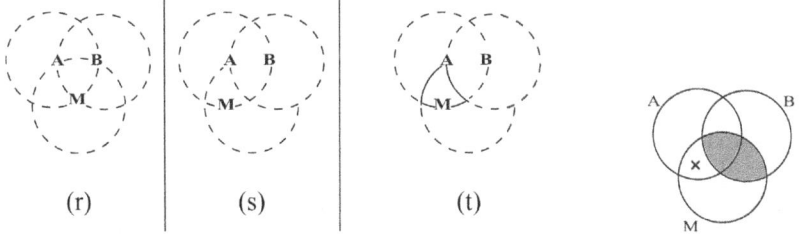

Fig. 11. Ueberweg diagram for 'No M is B' and 'Some A is M'. **Fig. 12.** Solution with Venn diagrams

5 Conclusion

We amended and expanded Ueberweg's diagrams which used dotted lines to express imperfect knowledge. We produced a new scheme (Ueberweg diagrams) which adequately represents categorical propositions and traditional syllogisms. This scheme fulfills Venn's requirements: it is "competent to indicate imperfect knowledge" and we do "appeal to it step by step in the process of working out our conclusion" [14, p. 4]. In this respect, it may be considered as an alternative to Venn diagrams.

It remains to explore the specific merits and weaknesses of this new system of diagrams and how it relates to other schemes. Both Venn and Ueberweg diagrams exhibit imperfect knowledge, but they differ in the object whose uncertainty is conveyed. While Venn exhibits the uncertainty of regions, our scheme focuses on the borders between those regions. Our diagrams can also, indirectly, be interpreted with reference to terms: a term exists if its corresponding region has continuous borders, is empty is it has no corresponding region and is uncertain if its corresponding region has (at least) some dotted border(s)[3]. Interestingly, Ueberweg diagrams share some of the shortcomings of Venn's scheme. For instance, the open outer region makes it difficult to represent some propositions with negative terms [2]. Other areas of improvement include the definition of formal rules of construction and manipulation. Venn himself did not present such rules for his scheme, and it was his immediate followers (notably Lewis Carroll [18, p. 105] and Charles S. Peirce [1]) who undertook this task. Finally, one needs to investigate the potential of Ueberweg diagrams for complex problems involving a higher number of terms. This issue played a decisive role in the development of logic diagrams in Venn's time [8].

One might rightly point out the difficulty of inventing such a system of diagrams without prior knowledge of Venn's scheme[4]. Indeed, our familiarity with the latter facilitated our endeavor and inspired our procedures. But that does not undermine our claim

---

[3] Venn's approach here is more direct since it is rooted in Boole's logic of terms, while ours calls for a logic of borders that would address relations of opposite terms (separated by a border). [3] might be seen as step in this direction.

[4] Interestingly, Venn made a similar objection against Keynes' methods for solving complex problems without the usage of symbolic notations [10]. Venn recognized the ingenuity of Keynes' procedures but argued that "Keynes himself would not have worked out his scheme unless he had been a thorough adept in the more symbolic methods" [16, p. 304].

that a Venn-like scheme, using dotted lines, was possible and presently exists. It is not claimed that Ueberweg diagrams are to be preferred over Venn's. Our aim was merely to demonstrate that, against Venn's scepticism, it was possible to design dotted diagrams that meet Venn's expectations. As such, for our purpose, the primary merit of Ueberweg diagrams is their very existence.

Acknowledgments. The third author of this paper is supported by the Fritz Thyssen Foundation (project: History of Logic Diagrams in Kantianism). The second and third authors are supported by the ViCom-project Gestures and Diagrams in Visual-Spatial Communication funded by the Deutsche Forschungsgemeinschaft (DFG, German Research Foundation) - (RE 2929/3-1).

References

1. Bhattacharjee, R., Moktefi, A.: Revisiting Peirce's rules of transformation for Euler-Venn diagrams. In: Basu, A., Stapleton, G., Linker, S., Legg, C., Manalo, E., Viana, P. (eds.) Diagrams 2021. LNCS, vol. 12909, pp. 166–182. Springer, Cham (2021). https://doi.org/10.1007/978-3-030-86062-2_14
2. Bhattacharjee, R., Moktefi, A., Pietarinen, A.-V.: The representation of negative terms with Euler diagrams. In: Béziau, J.Y., Desclés, J.P., Moktefi, A., Pascu, A.C. (eds.) Logic in Question, pp. 43–58. Birkhaüser, Cham (2023). https://doi.org/10.1007/978-3-030-94452-0_3
3. Demey, L., Smessaert, H.: From Euler diagrams to Aristotelian diagrams. In: Giardino, V., Linker, S., Burns, R., Bellucci, F., Boucheix, J.M., Viana, P. (eds.) Diagrams 2022. LNCS, vol. 13462, pp. 279–295. Springer, Cham (2022). https://doi.org/10.1007/978-3-031-15146-0_24
4. Jevons, W.S.: Logic. D. Appleton and Company, New York (1876)
5. Keynes, J.N.: Formal Logic. Macmillan, London (1894)
6. Lambert, J. H.: Neues Organon. Johann Wendler, Leipzig (1764)
7. Maass, J. G. E.: Grundriss der Logik. Michaelis und Compagnie, Halle (1793)
8. Moktefi, A., Edwards, A.W.F.: One more class: Martin Gardner and logic diagrams. In: Burstein, M. (ed.) A Bouquet for the Gardener, pp. 160–174. The Lewis Carroll Society of North America, New York (2011)
9. Moktefi, A., Lemanski, J.: On the origin of Venn diagrams. Axiomathes **32**, 887–900 (2022)
10. Moktefi, A., Schang, F.: Another side of categorical propositions: the Keynes-Johnson octagon of oppositions. Hist. Philos. Logic **44**(4), 459–475 (2023)
11. Schopenhauer, A.: Philosophische Vorlesungen (Sämtliche Werke IX). Piper & Co, München (1913)
12. Thomson, W.: An Outline of the Necessary Laws of Thought. William Pickering, London (1849)
13. Ueberweg, F.: System of Logic and History of Logical Doctrines. Longmans, Green, and Co., London (1871)
14. Venn, J.: On the diagrammatic and mechanical representation of propositions and reasonings. Philos. Mag. **10**, 1–18 (1880)
15. Venn, J.: On the employment of geometrical diagrams for the sensible representation of logical propositions. Proc. Camb. Philos. Soc. **4**, 47–59 (1880)
16. Venn, J.: Review of John N. Keynes' studies and exercises in formal logic. Mind **9**(34), 301–304 (1884)
17. Welton, J.: A Manual of Logic. vol. 1, 2nd ed, W. B. Clive, London (1896)
18. Wilson, R., Moktefi, A.: The Mathematical World of Charles L. Dodgson (Lewis Carroll). Oxford University Press, Oxford (2019)

Indeterminate Set Space Diagrams

Björn Gottfried$^{(\boxtimes)}$ [iD]

Hamburg University of Applied Sciences, Hamburg, Germany
`bjoern.gottfried@haw-hamburg.de`

Abstract. A diagrammatic system is presented, which combines benefits of Euler and Venn diagrams. As a tool intended for teaching, easy-to-use diagrams are desired. Their underlying geometry is motivated by insights from cognitive psychology, to make the diagrams easily accessible to humans. The price of simple diagrams is a restriction in expressiveness. But, enriching the diagrams with a syntactical refinement increases their expressiveness, enabling the distinction between determinate and indeterminate information. This facilitates their employment as *minimal abstract*, and simultaneously, as *limited abstract systems*.

As a consequence, this new representation offers both, a focus on the relevant relationships, like in Euler diagrams, and the possibility to take into account all set intersections, like in Venn diagrams. The former enables a simple visual mode, that makes it easy to comprehend relations of determined information – at the minimal abstraction level. The latter allows the exhaustive analysis of all potential relationships of indeterminate information – at the limited abstraction level.

Keywords: linear diagrams · categorical logic · cognitive adequacy · minimal abstractions · limited abstractions · indeterminacies · inferences

1 Introduction

Research in diagrammatic reasoning focuses on several issues. E.g., formal and cognitive aspects are dealt with in the engineering and cognitive science communities, respectively. Researchers in the former field are interested in accurately defined diagrammatic systems, which make them even applicable to formal proofs (e.g. [25]). Researchers in the latter area are interested in finding diagrammatic systems which aid human reasoners in problem solving (e.g. [2]).

This paper presents a diagrammatic system which is defined on a formal basis, and simultaneously, considers evidence from cognitive psychology to provide a useful reasoning tool. Hence, we are going to relate our work to [24], who provide a categorisation scheme which concerns both the structure of diagrammatic systems and the cognitive level of users. This categorisation affects the degree of abstraction, which in turn has an influence on the efficacy of diagrams.

Applications include the support of engineers and training in linear reasoning. Well-known examples are syllogisms, which have been analysed in a myriad of contexts [1,7,8,17,20,24]. One example is as follows:

All limited representations are easy-to-use systems. (premise 1)
Some limited representations are linear diagrams. (premise 2)
Thus, some linear diagrams are easy-to-use systems. (conclusion)

Such syllogisms show two premises and a conclusion. In linear reasoning tasks, the latter is either to be drawn from the premises or to be verified as to be true or false. Yet simpler inferences can be drawn from single statements, such that from the first premise it immediately follows that: *No limited representations are non-easy-to-use systems.* This paper describes a limited diagrammatic system, which represents such inference tasks.

1.1 Structure

Initially, a number of constraints are established which help in the definition of visually simple diagrams that aid human reasoners. Next, set space diagrams motivated by those constraints are recapitulated. Their limitations are shown and used to motivate their extension towards a representation of indeterminate information. The extended system enables the representation of all valid syllogisms, simpler to Venn diagrams in the sense that not all set intersections for the given classes are to be represented. Classifying the new diagrams with respect to their mode of abstraction, ranging from minimal to unlimited, eventually facilitates their comparison with Euler circles and Venn diagrams.

2 Constraints for Diagrammatic System Design

This section collects a number of constraints which are intended to support the design of a diagrammatic system. Its expressiveness should at least allow for a subclass of linear reasoning tasks which encompass syllogisms. The required expressiveness of a system can be easily reached by including as many syntactical features as necessary. The more features deployed, the more complex the proofs for the soundness and completeness of the resulting system become. More interesting for us, and not less demanding, is the search for a suitable set of features and layouts, when we are interested in a system that supports human perception and reasoning. Suggestions for appropriate criteria are taken from the cognitive sciences as follows.

In [6] it has been pointed out that researchers in cognitive psychology provide different models that explain the reasoning strategies of human beings who are confronted with linear reasoning problems: While [14] assumes that the same reasoning strategies hold for all individuals and problems, [1] found individual differences in the representations and processes that people use in reasoning. Inconsistent results indicate that it is hardly possible to derive criteria for diagrammatic system design that would closely align with findings from cognitive psychology about human reasoning strategies. Nevertheless, cognitive psychology provides at least some evidence for particular criteria.

Our focus lies on efficient reasoning systems, which permit an effortless understanding of the state of affairs of a given problem. Thus, the simplicity of diagrams, in the sense of having less visual content, is a first important characteristic: Problems with high visual content have been identified to be less efficiently solved by humans; this has been referred to as the visual impedance effect [15]. The visual content in the context of this effect concerns the mental level during problem solving and the explanation for this effect concerns the cognitive effort for managing that visual content. We assume that a corresponding effect can apply to complex diagrams, too, as such diagrams also require the mental processing of the rich visual content while being distracting from the actual task.

Thus, instead of creating arbitrarily complex diagrams, one should look from the very beginning for diagrams which provide just as little visual content as necessary, in order to satisfy particular purposes; in our case, to be able to represent linear reasoning tasks. For this purpose, a number of geometrical constraints can be found, which aid in maintaining simple diagrams (contrary to [24] we think that besides spatial properties, visual properties are also supportive). Without being able to motivate each of the following constraints in detail, perceptual psychology provides good evidence about well organised visual fields, avoiding visual ambiguities and complexity [18]. Accordingly, the following syntactical constraints for representations are formulated:

(1) Employ different geometrical entities parsimoniously.
(2) Draw entities, which are to be related for a conclusion, close to each other.
(3) Avoid intersections of any geometrical entities, in order to avoid ambiguities.
(4) Orient the entities clearly regarding both axes left-right and top-bottom.

Besides the organisation of the visual field, the question arises which information should be represented by a diagram that supports linear reasoning problems. It is beyond the scope of this paper to discuss different models in human reasoning. Hence, we refer to one of the predominant models, namely the mental model theory [13], since it has been confirmed that the mental model theory is the only account for specific performance differences in syllogistic reasoning [8]. It suggests that mental models represent what is true, but not what is false. Although [3] argues, that mental models can also represent what is false, but only by temporarily assuming that the false information is true.

Having this model of human reasoning in mind, we argue for the following semantical constraints about what to represent in diagrams:

(1) Something that is depicted in a diagram holds.
(2) If more items are depicted, they all hold simultaneously.
(3) Something that holds and its opposite should not be depicted together, since one of them does not hold.
(4) Set emptiness should be depicted by the absence of those kinds of features, which usually represent existent elements.

(1) The first constraint refers to the existential import of elements, asserting that everything exists which is represented in the diagram. With regard to a

represented class this means that this class contains at least one element. (2) The second constraint says that only conjunctive information is depicted. Or to put it differently: Alternative models cannot be shown in a single diagram. (3) This is closely related to the third constraint, which states that inconsistent scenarios should be avoided. Being more specific than the second constraint, the third one is about the avoidance of contradictory statements. (4) In opposition to the first constraint, the last one emphasises that an empty class should be represented by the absence of geometrical features, howsoever this could be achieved.

Note that we do not claim that it would make sense to always assert that anything exists, solely on the basis of universal quantifiers. It is just for the purpose of our simplified view on diagrams that we, within a first step, assume the existential import to everything that is depicted. Later we shall drop this assertion when the representation of indeterminate information is introduced to our diagrams. Otherwise indeterminate information could not be represented.

There is a trade-off between expressiveness (going hand in hand with a high level of abstraction) and visual power (going hand in hand with a low level of abstraction): the more expressive a diagrammatic system, the more it loses its visual power, and vice versa. A system which sticks to the mentioned constraints is clearly restricted regarding its expressiveness. But we are aiming at designing diagrams, that exhibit a high visual power. We present such a system in the following and show its effectiveness in representing linear reasoning tasks.

3 Set Space Diagrams

The motivation of set space diagrams has been to provide a diagrammatic reasoning aid with diagrams that can be easily drawn using paper and pencil. Their simplicity restricts their expressiveness in such a way that they can be conceived as an example of what [24] refer to as a *minimal abstraction representational system*: under any intended interpretation, there exists exactly one model. The advantage of this restriction is the ease of processing such diagrams, since distractive alternatives play no role.

A second motivation for set space diagrams has been the possibility to integrate as much information into a single diagram as required, in particular information about more than three sets, frequently a challenge for other systems which often lack a clear appearance when integrating four or more sets into a single diagram [9]. This is a requirement which is already necessary for immediate inference tasks that are even simpler than syllogisms, as shown below.

Eventually, set space diagrams are linear diagrams which stick to the constraints of Sect. 2. Previous approaches to linear diagrams in the context of syllogistic reasoning do not satisfy those constraints [7], others conceive them as an effective tool for set-based data [5]. An overview of them can be found in [7]. It has already been shown that linear diagrams are superior to region-based diagrams, in the sense that students make fewer errors when employing linear diagrams and process them faster [12]. In [20] it has even been shown that a simple variant of them are as efficient as Euler circles in syllogistic reasoning.

3.1 Set Space Diagrams

This section summarises the syntax and semantics of set space diagrams [9].

Syntax. There are two kinds of objects, a diagram stripe and line segments. The former consists of two vertical and parallel lines which enclose a zone (Fig. 1).

Fig. 1. The diagram stripe.

All line segments are horizontal, straight, finite, and, in general, disconnected. The diagram stripe and segments define set space diagrams as follows:

Definition 1 (Set space diagram)
A set space diagram $\mathcal{D}$ is a zone in the Euclidean plane, which is bounded by two vertical lines. This zone is called the diagram stripe. It contains a finite number of horizontal, straight, finite, and disconnected segments $\mathbb{S}$.

In order to refer to the segments of $\mathbb{S}$ they are denoted by $l_1, l_2, \ldots, l_k, k \in \mathbb{N}$. For any diagram $\mathcal{D}$ to be a well-formed diagram it is required that:

- all segments are drawn between both vertical lines of the diagram stripe,
- all segments are arranged horizontally, and thus in parallel, and
- no two segments are concurrent, i.e. have no point in common.

Each segment $l_i \in \mathbb{S}(\mathcal{D})$ can be labelled by any symbolic string, written left of the diagram stripe. The names of the segments $l_1, \ldots, l_k$ are to refer to the geometrical entities and are to be distinguished from the labels, though names of segments can also be used as labels if no confusions arise.

There are two segments which limit the range of possible segment lengths: The longest lower bound is the improper segment, l_{inf}, of length 0. The shortest upper bound, l_{sup}, spans over the entire diagram stripe.

Relations. Any segment l_i can be divided into as many sections as desired which are referred to as $l_i^k \in l_i, k \in \mathbb{N}$. Then, the coincidence between two segments, $l_i, l_j \in l_{sup}$ contained in a diagram $\mathcal{D}$, can be defined as follows:

Definition 2 (Coincidence relation)
If l_i and l_j share a common section, denoted by l_i^k or l_j^m, when orthogonally projected onto each other, it holds $l_i^k \parallel l_j \wedge l_j^m \parallel l_i \wedge l_i^k \parallel l_j^m$. In this case, l_i and l_j are said to be coincident.

Note how $\parallel$ denotes the parallelism of finite segments, not of infinite lines. Hence, we define more strictly, that $l_i \parallel l_k$ is true only when l_i and l_j share at least one point, when orthogonally projecting both segments onto each other.

A number of different relations between two segments are distinguished based on the degree of their coincidence. In order to define them the mereological concept of a *proper part* is adopted:

Definition 3 (Proper part)
That l_i is a proper part of l_j is expressed by $l_i \sqsubset l_j$.

(R1) $\qquad l_i \sqsubset l_j \equiv \forall_{l_i^k \in l_i} \, l_i^k \parallel l_j \,\wedge\, \exists_{l_j^m \in l_j} \, l_j^m \nparallel l_i$

Each section of l_i is shared with l_j. Simultaneously, there is a section on l_j which is not part of the common section shared with l_i. Based on this primitive, the following binary relations between two segments can be defined:

Definition 4 (Equivalence)
l_i and l_j are equivalent iff they completely coincide.[1]

(R2) $\qquad l_i = l_j \equiv \forall_{l_k \in l_{sup}} \, l_k \sqsubset l_i \leftrightarrow l_k \sqsubset l_j$

Definition 5 (Improper part)
l_i is an improper part of l_j iff l_i is a proper part of l_j or equal to l_j.

(R3) $\qquad l_i \sqsubseteq l_j \equiv l_i \sqsubset l_j \vee l_i = l_j$

Definition 6 (Proper overlap)
l_i and l_j properly overlap iff there is a section l_k both segments share. Simultaneously, both l_i and l_j include a section l_m and l_n, respectively, which is not shared with the other segment.

(R4) $\qquad \begin{aligned} l_i \circ l_j \equiv\; &\exists_{l_k \in l_{sup}} \, l_k \sqsubset l_i \wedge l_k \sqsubset l_j \,\wedge \\ &\exists_{l_m \in l_{sup}} \, l_m \sqsubset l_i \wedge \neg(l_m \sqsubseteq l_j) \,\wedge \\ &\exists_{l_n \in l_{sup}} \, l_n \sqsubset l_j \wedge \neg(l_n \sqsubseteq l_i) \end{aligned}$

Definition 7 (Disjointness)
l_i and l_j are disjoint iff they share no section.

(R5) $\qquad l_i \bowtie l_j \equiv \neg(l_i \sqsubseteq l_j) \wedge \neg(l_i \circ l_j) \wedge \neg(l_j \sqsubseteq l_i)$

Each segment is uniquely defined by its vertical position in the diagram stripe and by the horizontal positions and lengths of its connected components (i.e. a segment l_i is either a single connected component or it contains a number of disconnected components at the same vertical level). Two segments l_i and l_j at different vertical positions in $\mathcal{D}$ which are equal with respect to the other two parameters are syntactically equivalent, $l_i = l_j$, as given by Definition 4.

Figure 3 shows examples for all four relations, which are employed to represent certain information: proper part, equivalence, proper overlap, and disjointness.

---

[1] The symbol '=' is employed to represent the equivalence of segments, i.e. that they entirely coincide when projected onto each other, as the two segments of '=' themselves do. This symbol is not used to show the identity of two geometrical objects.

Semantics. While each segment represents a set, the relations are interpreted in the set-theoretic sense as defined in [9]. Therefore, the model of a diagram is defined as follows:

Definition 8 (Model of a set space diagram)
Let $\mathcal{D}$ be a set space diagram and let the model of $\mathcal{D}$, $\mathcal{M}(\mathcal{D})$, be a pair (Ω, Ψ), where Ω is a domain and Ψ is a function

(M1) $$\Psi : \$(\mathcal{D}) \to 2^\Omega$$

mapping each segment of diagram $\mathcal{D}$ to a subset of the domain Ω, so that

(M2) $$\Psi(l_{inf}) = \emptyset$$

(M3) $$\Psi(l_{sup}) = \Omega$$

$\mathcal{M}$ is a model of $\mathcal{D}$, referred to as $\mathcal{M} \models \mathcal{D}$, if

(M4) $$\forall_{l_i, l_j \in \mathcal{D}}\ l_i = l_j \leftrightarrow \Psi(l_i) = \Psi(l_j)$$

(M5) $$\forall_{l_i, l_j \in \mathcal{D}}\ l_i \sqsubset l_j \leftrightarrow \Psi(l_i) \subset \Psi(l_j)$$

(M6) $$\forall_{l_i, l_j \in \mathcal{D}}\ l_i \circ l_j \leftrightarrow \Psi(l_i) \cap \Psi(l_j) \neq \emptyset \wedge \Psi(l_i) \nsubseteq \Psi(l_j) \wedge \Psi(l_i) \nsupseteq \Psi(l_j)$$

(M7) $$\forall_{l_i, l_j \in \mathcal{D}}\ l_i \bowtie l_j \leftrightarrow \Psi(l_i) \cap \Psi(l_j) = \emptyset$$

One could in principle conceive each horizontal position in the diagram stripe as to belong to exactly one element e_i of the universe Ω. Then, if desired, a subset or even all elements of Ω could be depicted in one of the rows, as shown in Fig. 2. This also illustrates that the elements can be distributed arbitrarily within the diagram[2]. Depending on their distribution, the segments, representing the sets, will be disconnected or shifted.

Instead of interpreting segments as sets, they could also represent monadic predicates, which apply to subsets of the underlying universe. For example, the predicates $M(x)$ and $P(x)$ represent *x are limited representations* and *x are easy-to-use systems*, respectively. Then, the relation *All limited representations are easy-to-use systems* is given on the left hand side of Fig. 3 where $l_M \sqsubset l_P$ holds, with the segment l_M representing $M(x)$ and l_P representing $P(x)$, as given by the labels. Taking M and P as predicates, however, the limitations of set space diagrams become quickly apparent.

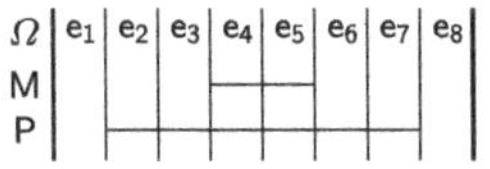

$$M = \{e_4, e_5\}$$
$$P = \{e_2, e_3, e_4, e_5, e_6, e_7\}$$
$$M \cap P = \{e_4, e_5\}$$

Fig. 2. *All* M *are* P.

---
[2] Sets are unordered collections.

```
M|  ─────  |    M|  ═════  |    M|  ─────  |    M|  ──  ──  |
P|          |    P|          |    P|  ─────  |    P|          |
    Def. 3           Def. 4           Def. 6           Def. 7
```

Fig. 3. The proper part, equivalence, proper overlap, and disjointness relations.

3.2 Limitations of Set Space Diagrams

Being confined to only represent certain information, the expressiveness of set space diagrams is limited. For two sets, M and P, Fig. 3 shows the proper part, equivalence, proper overlap, and disjointness relations (cf. Sect. 3.1).

Analysing these relations their distinctions to logical relations become apparent. E.g., the proper overlap represents two overlapping sets, M and P, and according to Definition 6, all three subsets are non-empty: $M \cap P$, $M \backslash P$, and $P \backslash M$. It is then not possible to represent statements with the particular affirmative in the sense of Aristotelian logic: *Some* M *are* P means that M contains at least one element that is simultaneously in P. However, this can be the case under several circumstances: when M is equivalent to P, when M is a proper subset of P (or vice versa), or when M and P properly overlap. Indeed, the particular affirmative *some* has been discussed controversially in the context of human reasoning [17]. In logic *Some* M *are* P can mean *Some (possibly all)* M *are* P, while in natural language and regarding our proper overlap relation, we almost invariably mean that it stands for *Some are and some are not* [16].

Seeing the limitations already with regard to single statements, how would the diagrams represent syllogisms, e.g. Datisi, which includes the particular affirmative just discussed? It would be diagrammed as shown in Fig. 4: On the left hand side there is the first premise, *All M are P*, followed by the second premise, *Some M are S*, in the middle. The conclusion, *Some S are P*, is given on the right hand side. This figure emphasises how both premises are diagrammed before the conclusion can be drawn. Usually, one would employ only a single diagram as is given in Fig. 5, inasmuch as the conclusion is then given as a free ride, which can be read off the diagram, as soon as both premises have been drawn [21].

For Datisi it holds $S \cap M \cap P \neq \emptyset$, as is correctly given in the set space diagram (Fig. 5). But non-emptiness does not necessarily hold for the other represented sets, namely S, P, and $M \cap P$, as incorrectly demanded by that set space diagram. Although set space diagrams allow the derivation of the correct conclusion, their drawback is a lack of representing indeterminate information. However, this becomes necessary, if we would like to diagram Datisi in the sense of Aristotelian logic. That is, all of the sets S, P, and $M \cap P$ are indeterminate.

```
M|  ─────  |        M|  ─────  |           P|  ─────  |
P|          |   ∧   S|  ─────  |    ⇒     S|  ─────  |
 premise 1: MaP      premise 2: MiS          conclusion: SiP
```

Fig. 4. Modus Datisi shown as a set space diagram. For easier comparison, segments representing the same sets are drawn at the same height in each diagram.

Fig. 5. Left: Modus Datisi drawn in one step. Right: The represented set intersections.

While the conclusion SiP can be correctly drawn from this diagram, the semantics of both premises differ with regard to Aristotelian logic, in which *All M are P* either means *M is a subset of P* or *M is equal to P*, whereas in Fig. 5 only the former is represented. The right hand side of Fig. 5 shows all the set intersections represented by this diagram. It is obvious that it is necessary to extend the expressiveness of set space diagrams in order to represent syllogisms.

One such extension is discussed in [10] where compound diagrams are introduced, that represent a disjunctive normal form for set space diagrams. They pertain to the category of *limited abstraction representational systems* [24]: each set space diagram of a disjunction represents one possible model. As there are 2 (MaP) · 4 (MiS) = 8 disjunctive diagrams to be taken into account for diagramming Aristotelian Datisi by means of set space relations, such compound diagrams become rather unhandy (Fig. 6); although their advantage is that they do not require any counterpart relation, in contrast to compound Venn diagrams [22] (p. 53). While this advantage maintains their visual power, the many possibilities represented in a compound diagram cancel out this benefit.

It is therefore reasonable to look for an alternative, that integrates the many possibilities concisely in a single diagram. Thus, we shall augment the diagrams with a new type of line, to be able to diagram the indeterminacy as to whether subsets are empty or not – sticking to the geometrical constraints of Sect. 2.

4 Indeterminate Set Space Diagrams

In set theory, the membership of an element in a set is assessed in binary terms according to a bivalent condition: an element either belongs or does not belong to the set. In fuzzy set theory [26], bivalent sets are referred to as *crisp* sets. Accordingly, we shall refer to set space diagrams defined so far as *crisp* (set space/linear) diagrams. In contrast to fuzzy sets, which consider infinite many membership values in the interval $[0, 1]$, we shall employ only one intermediate category which represents our uncertainty about the bivalent membership of an element and which replaces the certain membership representation of a solid segment by a curly line ∿, increasing the degree of abstraction [24]. Consequently,

Fig. 6. Modus Datisi represented as a compound diagram: MaP ∧ MiS ⇒ SiP

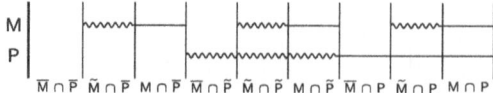

Fig. 7. A diagram with all possible intersections between two segments M and P.

the subset of M which contains those elements with an indeterminate membership status is denoted by M̃. As opposed to crisp diagrams we call the extended diagrams *indeterminate* (set space/linear) diagrams. Their formal semantics for sets and their relationships have been introduced in [11]. We consider their extension towards their application to categorical logic in the following sections.

The representation of non-emptiness by solid line segments is maintained, as is the representation of emptiness by the absence of line segments. Additionally, curly lines are introduced to represent indeterminate information, in other words, that there is either at least one element in that set which is represented by the curly line or that it is empty. Curly lines give the impression (similar as in a sketch, inasmuch as everything else is straight and very tidy) that the depicted information is given less straightforward than the information represented by straight segments (which suggest certainty: The person who drew the diagram was certain about the pieces of information depicted by straight segments). As noted above, [3] argues that mental models can also represent what is false, but only by temporarily assuming that the false information is true. In our representation this temporal assumption of true information is represented by curly lines as well: they can be introduced everywhere without changing the validity of the represented information, as they only increase its uncertainty. After their introduction they can be interpreted towards both directions.

As soon as more information is available, curly lines might disappear or be replaced by straight segments. Overall, the constraints introduced in the second section can be maintained, when diagramming all 9 possible subsets for two, partially indeterminate, sets M and P, as shown in Fig. 7. The Aristotelian categorical statements are represented in Fig. 8 (cf. Fig. 3). Note that certain information is still represented by means of solid segments, as for the particular statements MiP and MoP which both assert the existence of at least one element.

The next subsections demonstrate how immediate inferences can be drawn with these diagrams. Afterwards, the representation of syllogisms is discussed.

Fig. 8. Representing Aristotelian statements by indeterminate diagrams.

4.1 Conversion

Interchanging subject and predicate terms, the converse of a categorical statement is obtained. Conversion is only valid for MeP and MiP. This becomes apparent when considering the symmetries of both diagrams in the middle of Fig. 8. By contrast, the other two diagrams for MaP and MoP are asymmetric, and indeed, the converse relations of these categorical statements are not valid.

In the Aristotelian tradition there is also the inference called *conversion by limitation*: subject and predicate terms of a statement are exchanged and the quantity is changed from universal to particular. This allows the inference of PiM from MaP. In the diagram (left-most in Fig. 8) this inference relates to the overlapping section. However, one must concede that conversion by limitation claims the existence of at least one element, from nowhere.

All limited representations are easy-to-use systems. (MaP)
$\xrightarrow{converse}$ *Some easy-to-use systems are limited representations.* (PiM)

4.2 Obversion

The obverse of a statement is obtained by changing the quality of that statement (from affirmative to negative and vice versa). Additionally, the predicate is to be replaced by its term-complement, i.e. the term identifying its class complement.

In [11], a visual reasoning rule is defined for constructing $\overline{P}$: (i) curly lines remain curly; (ii) straight and empty segments are to be exchanged. (i) relates to keeping indeterminacy indeterminate, while (ii) concerns the actual complement of certain information. Applying this rule, the term-complement of the predicate $\overline{P}$ is explicitly shown (Fig. 9). Comparing M and $\overline{P}$, the obverse can be directly obtained, instead of comparing the curly line of M with the empty section of P.

All limited representations are easy-to-use systems. (MaP)
$\xrightarrow{obverse}$ *No limited representations are non-easy-to-use systems.* (Me$\overline{P}$)

Note that the diagram with $\overline{P}$ is disconnected from the top diagram with M and P: Whenever indeterminate information gets determinate, a curly line changes either to an empty segment or a solid segment. This change may be either equal or unequal for two sets, putting those sets either into the same diagram or in different ones, respectively. Suppose information becomes available that a specific element is definitely an element of P. Then, as far as at the same vertical position there is also a curly line for M, both representations, for P and M, get a solid segment, whereas for $\overline{P}$, at that vertical position, an empty segment is

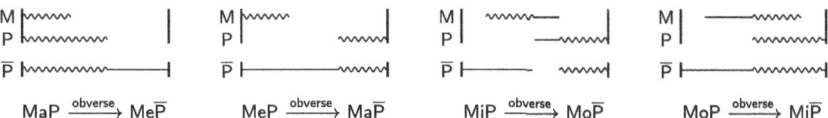

Fig. 9. The obverse including the full consideration of indeterminacy.

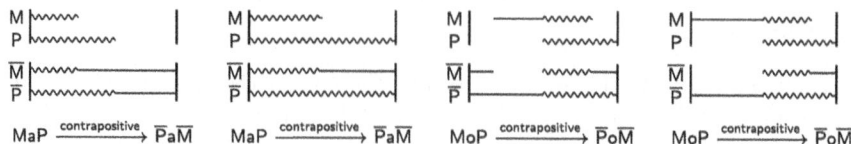

Fig. 10. The contrapositions of MaP and MoP are valid. Both inferences are shown for the general case and a specific case, in which the background set could be empty.

to be introduced. Such changes of curly lines are valid transformations, since all sets and diagrams are vertically aligned to the same universe (cf. Fig. 2).

We observe: as we assume elements in the universe which are neither in M nor in P, there exists at least one element in $\overline{P}$. This means that the obverse of a universal statement in the context of a non-empty background set implies an existential quantification. This justifies the solid segment derived at the syntactical level. Moreover, obversion is valid for all four categorical statements.

4.3 Contraposition

Replacing the subject with the term-complement of the predicate and the predicate with the term-complement of the subject, the contrapositive of a statement is obtained. It is valid only for MaP and MoP (cf. Fig. 10). But for MeP it is valid if the background set is empty ($\overline{M \cup P} = \emptyset$) and for MiP it is valid if the background set is non-empty ($\overline{M \cup P} \neq \emptyset$) (cf. Fig. 11).

All limited representations are easy-to-use systems. (MaP)
$\xrightarrow{contrapositive}$ *All non-easy-to-use systems are non-limited representations.* ($\overline{P}a\overline{M}$)

Similar to conversion, there exists an inference rule called *contraposition by limitation*: replacing subject with the term-complement of predicate and predicate with the term-complement of subject; moreover, the quantity is changed from universal to particular. This allows the inference of $\overline{P}o\overline{M}$ from MeP.

The left-most diagram in Fig. 11 contains this inference: it appears at the overlapping part between $\overline{P}$ and $\overline{M}$ where $\overline{P}$ is straight and $\overline{M}$ is curly. This shows that the *contraposition by limitation* holds only as long as M is not empty.

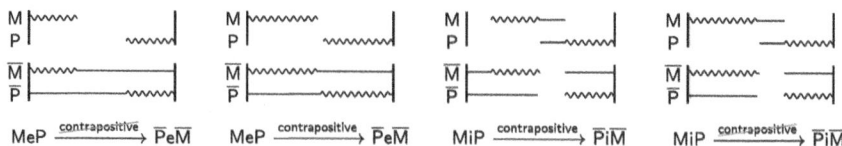

Fig. 11. The contrapositions of MeP and MiP are generally not valid. Specific valid cases are shown in the two middle diagrams.

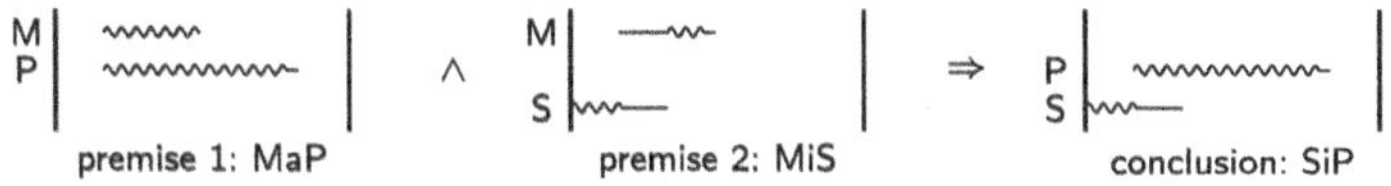

Fig. 12. Modus Datisi. For easier comparison segments representing the same sets are drawn at the same height in each diagram.

4.4 Syllogisms

Referring back to Fig. 5 we discussed the limitations of set space diagrams through the example of Modus Datisi. We pick up this example again in Fig. 12. This time employing curly lines to represent indeterminate information.

Once a solid segment appears, curly lines (of set intersections) that are coincident with it, can be transformed into solid segments as well (excluded are curly lines in vertically disconnected diagrams, like set complements). E.g., in Fig. 12 a part of set P overlaps with a solid segment of S in the conclusion. This is a valid transformation because each column in the diagram stripe stands for one particular element. If its existence becomes certain, this also holds for all other sets represented by segments at the very same vertical position. For reasons of space, we have not carried out this transformation in any of our diagrams.

From the 256 logically distinct syllogisms only 15 types are valid. 9 other syllogisms become valid, as soon as the existence of at least one element for one specific class is explicitly stated. This makes for a total of 24 syllogisms which are presented in Fig. 13. Note that it is assumed that there are further elements apart from those in classes M, P, and S. Therefore, it holds $\overline{M} \cap \overline{P} \cap \overline{S} \neq \emptyset$. The diagrams can be easily changed when dropping this assumption. For this purpose, the parts within the diagram stripe that do not overlap with any segment (as e_1 and e_8 in Fig. 2) can be omitted.

When only interested in the conclusion of a particular syllogism, it is sufficient to draw the set intersections within the premises that are relevant for that conclusion. This is the case for all syllogisms represented in Fig. 13. For Datisi this amounts to restricting set S to a solid segment that is parallel to M (see Datisi in Fig. 13). However, as soon as one intends to consider all set intersections, one needs to consider what [22] (p. 57) has referred to as the *partial-overlapping rule*. Adapted to our purposes: a new (probably disconnected) segment representing another set in a new row should overlap a proper (probably curly) part of every existent segment in the other rows. This has been done for a selection of syllogisms shown in Fig. 14. They represent the same information as Venn diagrams, which always encompass all set intersections.

Different syllogisms integrate different figures, which contain the same quantifiers. The resulting diagrams appear equal and demonstrate why the according syllogisms draw the same conclusions. This holds, for example, for Cesare and Celarent as well as for Calemes and Camestres. Both pairs even show their semantic similarity by means of similar diagrams in which just the roles of P and S in relation to M are exchanged.

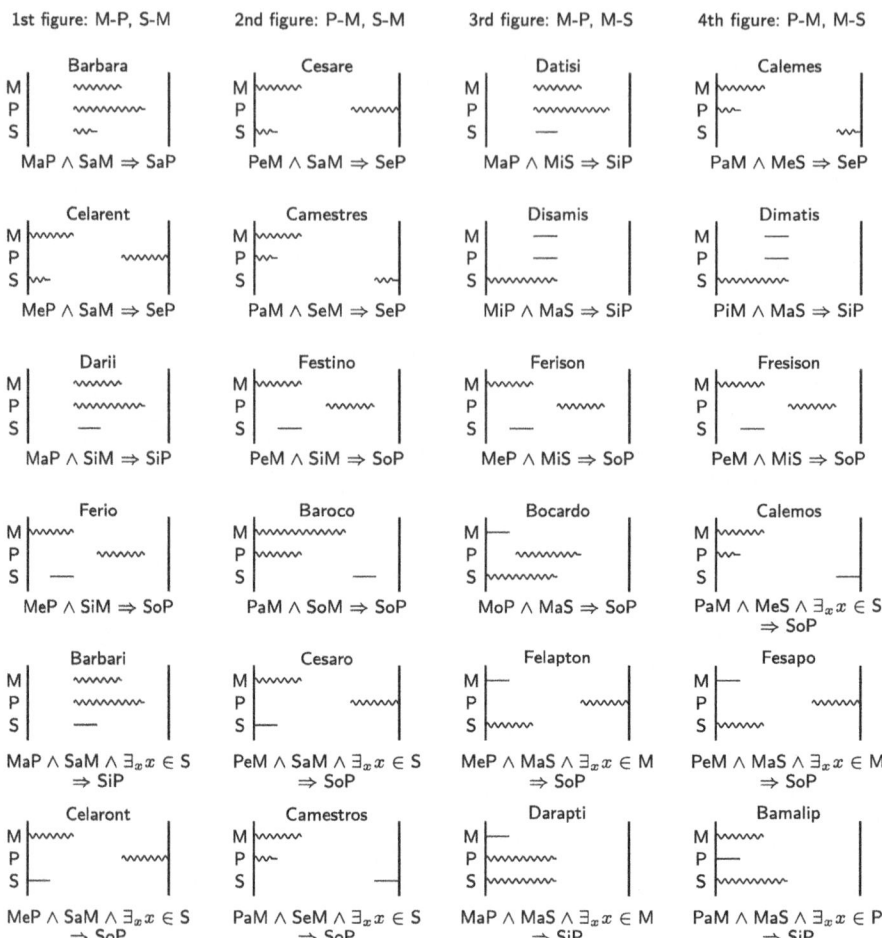

Fig. 13. The 24 valid syllogisms restricted to represent those set intersections that are relevant for establishing the conclusion.

5 Discussion and Future Work

5.1 Euler and Venn Diagrams

It has been the goal to combine benefits of Euler circles and Venn diagrams. An advantage of Euler circles is the existential import to the diagrams, providing an immediate relationship between topological relations at the syntactical level and set-theoretic relations at the semantical level, making their comprehension intuitive; elsewhere discussed as well-matchedness [23], a notion that relates to the homomorphism between syntax and semantics. However, our focus lies on the degree of abstraction: Euler circles are confined to represent one model at a time. They are representations of minimal abstraction [24]. By contrast,

Fig. 14. Some syllogisms including all possibly nonempty set intersections.

Venn diagrams do not assert that anything exists, solely on the basis of existent regions. Rather empty regions represent indeterminate information, which makes them a representation of limited abstraction. Their higher expressiveness is due to their higher degree of abstraction, making them less accessible [24].

5.2 Set Space Diagrams

While Euler and Venn diagrams are based on two different levels of abstraction, set space diagrams work with a mix of both: They emphasise their minimal abstraction proportion by solid segments. These are clearly separated from curly lines, representing their limited abstraction proportion.

That curly lines are built up quite similar to solid lines, having the same horizontal orientation and using the same relations alike solid lines, shows how set space diagrams increase their degree of abstraction carefully: curly lines are added in such a way that the appearance of relational information, such as inclusion and exclusion, is maintained [20], requiring no added conventions to work with indeterminate information. Yet other approaches exist to represent indeterminacies by means of linear diagrams [4].

5.3 Future Work

Figure 15 contrasts the different systems with respect to Datisi. Preferences for Euler circles and linear diagrams as opposed to Venn diagrams exist [20]. These findings should be compared with further evaluations including set space diagrams. This concerns, in particular, the distinction between figure 1 and figure 4 syllogisms (cf. Fig. 13), for which participants, given in textual form, show different performances [8]. The use of the diagrams could level out those differences,

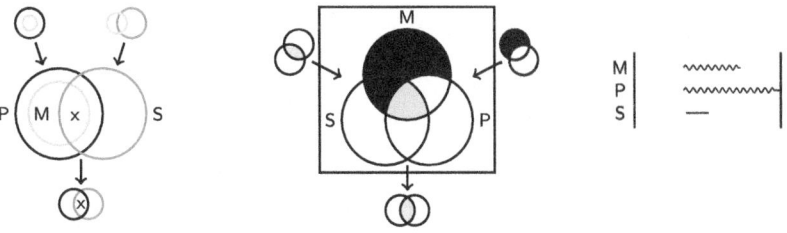

Fig. 15. Datisi: MaP ∧ MiS ⇒ SiP: Euler circles, Venn diagram, set space diagram.

since the positions of the middle term does not play a significant role anymore in set space diagrams (cf. Fig. 13). An additional improvement could be achieved by guidelines, as in [19] and in Figs. 2, 5, and 7. Eventually, it is of interest whether set space diagrams are more performant when comprising only solid segments. This investigates their efficacy depending on their degree of abstraction.

References

1. Bacon, A., Handley, S., Newstead, S.: Individual differences in strategies for syllogistic reasoning. Thinking Reason. **9**, 133–168 (2003)
2. Bourou, D., Schorlemmer, M., Plaza, E.: Euler vs Hasse diagrams for reasoning about sets: a cognitive approach. In: Giardino, V., Linker, S., Burns, R., Bellucci, F., Boucheix, J.M., Viana, P. (eds.) Diagrams 2022. LNCS, vol. 13462, pp. 151–167. Springer, Cham (2022). https://doi.org/10.1007/978-3-031-15146-0_13
3. Byrne, R.M.J.: The Rational Imagination: How People Create Counterfactual Alternatives to Reality. MIT Press, Cambridge (2005)
4. Chapman, P., Stapleton, G., Rodgers, P.: PaL diagrams: a linear diagram-based visual language. J. Visual. Lang. Comput. **25**, 945–954 (2014)
5. Chapman, P.: Interactivity in linear diagrams. In: Basu, A., Stapleton, G., Linker, S., Legg, C., Manalo, E., Viana, P. (eds.) Diagrams 2021. LNCS (LNAI), vol. 12909, pp. 449–465. Springer, Cham (2021). https://doi.org/10.1007/978-3-030-86062-2_47
6. DeLeeuw, K.E., Hegarty, M.: What diagrams reveal about representations in linear reasoning, and how they help. In: Stapleton, G., Howse, J., Lee, J. (eds.) Diagrams 2008. LNCS (LNAI), vol. 5223, pp. 89–102. Springer, Heidelberg (2008). https://doi.org/10.1007/978-3-540-87730-1_11
7. Englebretsen, G.: Figuring It Out: Logic Diagrams. De Gruyter (2020)
8. Espino, O., Santamaria, C., Meseguer, E., Carreiras, M.: Early and late processes in syllogistic reasoning: evidence from eye-movements. Cognition **98**, B1–B9 (2005)
9. Gottfried, B.: Set space diagrams. J. Visual. Lang. Comput. **25**, 518–532 (2014)
10. Gottfried, B.: T. diamond of contraries. J. Visual. Lang. Comput. **26**, 29–41 (2015)
11. Gottfried, B.: The systematic design of visual languages applied to logical reasoning. J. Visual. Lang. Comput. **28**, 212–225 (2015)
12. Gottfried, B.: A comparative study of linear and region based diagrams. J. Spat. Inf. Sci. **10**, 3–20 (2015)
13. Johnson-Laird, P.N.: The history of mental models. In: Psychology of Reasoning, pp. 179–212. Psychology Press (2004)
14. Knauff, M., Johnson-Laird, P.N.: Visual imagery can impede reasoning. Memory Cogn. **30**, 363–371 (2002)
15. Knauff, M., May, E.: Mental imagery, reasoning, and blindness. Q. J. of Exp. Psychol. **59**, 161–177 (2006)
16. Larvor, B.: Three is a magic number. Philosophers' Mag. **44**, 83–88 (2009)
17. Newstead, S.E.: Can natural language semantics explain syllogistic reasoning? Cognition **90**, 193–199 (2003)
18. Palmer, S.E.: Vision Science: Photons to Phenomenology. MIT Press, Cambridge (1999)
19. Rodgers, P., Stapleton, G., Chapman, P.: Visualizing sets with linear diagrams. ACM Trans. Comput. Hum. Interact. **22**, 1–39 (2015)

20. Sato, Y., Mineshima, K.: The efficacy of diagrams in syllogistic reasoning: a case of linear diagrams. In: Cox, P., Plimmer, B., Rodgers, P. (eds.) Diagrams 2012. LNCS (LNAI), vol. 7352, pp. 352–355. Springer, Heidelberg (2012). https://doi.org/10.1007/978-3-642-31223-6_49
21. Shimojima, A.: On the Efficacy of Representation. Indiana University, Bloomington (1996)
22. Shin, S.-J.: The Logical Status of Diagrams. Cambridge University Press, Cambridge (1994)
23. Stapleton, G., Rodgers, P., Touloumis, A., Blake, A.: Well-matchedness in euler and linear diagrams. In: Pietarinen, A.-V., Chapman, P., Bosveld-de Smet, L., Giardino, V., Corter, J., Linker, S. (eds.) Diagrams 2020. LNCS (LNAI), vol. 12169, pp. 247–263. Springer, Cham (2020). https://doi.org/10.1007/978-3-030-54249-8_20
24. Stenning, K., Oberlander, J.: A cognitive theory of graphical and linguistic reasoning: logic and implementation. Cogn. Sci. **19**, 97–140 (1995)
25. Weisgerber, S.: Visual proofs as counterexamples to the standard view of informal mathematical proofs? In: Giardino, V., Linker, S., Burns, R., Bellucci, F., Boucheix, J.M., Viana, P. (eds.) Diagrams 2022. LNCS, vol. 13462, pp. 37–53. Springer, Cham (2022). https://doi.org/10.1007/978-3-031-15146-0_3
26. Zadeh, L.A.: Fuzzy sets. Inf. Control **8**, 338–353 (1965)

Can Euler Diagrams Improve Syllogistic Reasoning in Large Language Models?

Risako Ando[1]([✉]), Kentaro Ozeki[1,2], Takanobu Morishita[1], Hirohiko Abe[1], Koji Mineshima[1], and Mitsuhiro Okada[1]

[1] Keio University, 2-15-45 Mita, Minato-ku, Tokyo 108-8345, Japan
{risakochaan,morishita,hirohiko-abe}@keio.jp,
{minesima,mitsu}@abelard.flet.keio.ac.jp
[2] The University of Tokyo, 7-3-1 Hongo, Bunkyo-ku, Tokyo 113-8654, Japan

Abstract. In recent years, research on large language models (LLMs) has been advancing rapidly, making the evaluation of their reasoning abilities a crucial issue. Within cognitive science, there has been extensive research on human reasoning biases. It is widely observed that humans often use graphical representations as auxiliary tools during inference processes to avoid reasoning biases. However, currently, the evaluation of LLMs' reasoning abilities has largely focused on linguistic inferences, with insufficient attention given to inferences using diagrams. In this study, we concentrate on syllogisms, a basic form of logical reasoning, and evaluate the reasoning abilities of LLMs supplemented by Euler diagrams. We systematically investigate how accurately LLMs can perform logical reasoning when using diagrams as auxiliary input and whether they exhibit similar reasoning biases to those of humans. Our findings indicate that, overall, providing diagrams as auxiliary input tends to improve models' performance, including in problems that show reasoning biases, but the effect varies depending on the conditions, and the improvement in accuracy is not as high as that seen in humans. We present results from experiments conducted under multiple conditions, including a Chain-of-Thought setting, to highlight where there is room to improve logical diagrammatic reasoning abilities of LLMs.

Keywords: Euler diagrams · Syllogisms · Large language models

1 Introduction

In recent years, large language models (LLMs) have made significant progress in AI and Natural Language Processing (NLP) research, and their reasoning abilities have been evaluated in various ways. In particular, LLMs such as BERT [7] and GPT [3] are often claimed to perform natural language inferences at a level comparable to human reasoning. However, it remains unclear whether LLMs can perform logical reasoning accurately. Previous studies [1,6,8,19] have shown that LLMs often exhibit reasoning biases similar to those identified in humans by cognitive science research [10,15,24]. This suggests that there is room for improvement in the logical reasoning abilities of current LLMs.

In cognitive science, extensive research has been conducted to elucidate the cognitive role of diagrammatic representations, as opposed to sentential representations [2,12,23]. These studies explore the cognitive effects of diagrammatic representations and the reasons behind these effects, with a specific focus on the role of the Euler diagrams in syllogistic reasoning. Sato et al. [22] empirically tested the efficacy of Euler diagrams in human syllogistic reasoning. Their findings suggest that humans use diagrammatic representations as aids in logical reasoning, thereby mitigating the effects of reasoning biases that can impede accurate logical reasoning.

To date, the evaluation of the reasoning ability of LLMs has concentrated on linguistic reasoning, and sufficient research has not been done on reasoning with diagrams. Based on the findings in cognitive science research on diagrams, we may expect that LLMs perform logical reasoning tasks better when aided by diagrams, potentially avoiding reasoning biases just like humans. In this paper, therefore, we systematically examine how accurately LLMs can perform logical reasoning by using diagrams as auxiliary input, and whether they exhibit reasoning biases similar to those of humans. In particular, we focus on syllogistic reasoning, which is known to include various reasoning biases, and examine whether Euler diagrams are effective for error-prone syllogistic reasoning. In doing so, we construct a syllogism dataset where each premise sentence is paired with corresponding Euler diagrams, and each syllogism is annotated with labels indicating reasoning biases (Sect. 3). We will report experimental results and analyses from evaluating the current state-of-the-art LLMs using this dataset under various conditions (Sect. 4).

2 Background

In this section, we first introduce syllogisms and their representation in Euler diagrams (Sect. 2.1). Then, we explain the background of the cognitive science study of human reasoning using Euler diagrams (Sect. 2.2) and the recent research on LLMs' reasoning ability with sentences and diagrams (Sect. 2.3).

2.1 Syllogistic Reasoning and Euler Diagrams

A syllogism is an inference that consists of two premises and one conclusion, where the premises and the conclusion are composed of four basic types of quantified sentences as shown in Table 1.

Each syllogism is classified by the order of terms in the premises, traditionally called *figure*. Given three terms, S, P, and M, we assume that the terms in the conclusion always follow the order S–P. Based on this, there are four possible figures, as shown in Table 2. For instance, in figure 1, the terms appear in the premises in the order M–P and S–M. This implies that each form of syllogism is identified by the figure and the types of its two premises. For example, (1) below

Table 1. Four types of categorical sentences and Euler diagrams.

Type	Pattern	Set Theory	Euler diagram	Mood
A	All S are P	$S \subseteq P$		Universal Affirmative
E	No S are P	$S \cap P = \emptyset$		Universal Negative
I	Some S are P	$S \cap P \neq \emptyset$		Particular Affirmative
O	Some S are not P	$S \setminus P \neq \emptyset$		Particular Negative

Table 2. Four figures of syllogisms.

	Figure			
	1	2	3	4
Premise 1 (Major Premise)	M–P	P–M	M–P	P–M
Premise 2 (Minor Premise)	S–M	S–M	M–S	M–S
Conclusion	S–P	S–P	S–P	S–P

is the syllogism where Premise 1 (**P1**) is of A-type, Premise 2 (**P2**) is of A-type, and the conclusion (**C**) is also of A-type, with its figure being 1 (the M–P, S–M order). This is abbreviated as AA1A. (2) is a syllogism where Premise 1 (**P1**) is of A-type, Premise 2 (**P2**) is of E-type, and the conclusion (**C**) is also of E-type, with its figure being 2 (the P–M, S–M order). This is abbreviated as AE2E. Both are valid syllogisms.

(1) AA1A: **P1**: All M are P
            P2: All S are M
            C: All S are P

(2) AE2E: **P1**: All P are M
            P2: No S are M
            C: No S are P

Each sentence is assigned a set-theoretic meaning as shown in Table 1 and represented in its corresponding Euler diagram. In this paper, we adopt the representation system of Euler diagrams presented in Sato et al. [21,22]. The sentence "All S are P" is represented by the inclusion relation between circles, while "No S are P" is represented by the circles that are disjoint. In addition, this system uses the symbol × to represent the existence of elements in a region. Consequently, the sentences "Some S are P" and "Some S are not P" can be uniquely represented by the diagrams shown in Table 1.

2.2 Human Reasoning with Euler Diagrams

Syllogistic reasoning is often challenging for humans, and it is widely studied in cognitive science which types of syllogisms are more likely to cause human reasoning errors, i.e., which involve reasoning biases [10,15,24]. Some typical biases of syllogism will be introduced in Sect. 3.

Based on prior studies on the cognitive role of diagrams in human deductive reasoning [2,12], Sato et al. [22] empirically investigated the cognitive differences between reasoning with Euler diagrams and reasoning with Venn diagrams, and analyzed the role that each diagram plays in the reasoning process. Sato et al. [22] conducted an experiment with human subjects to test whether the auxiliary use of diagrams suppresses reasoning biases. The results showed that among the two types of diagrams, Euler diagrams not only contribute to subjects' correct interpretation of categorical sentences, but also play an important role in the reasoning process itself. In particular, the results suggested that Euler diagrams work effectively in problems that cause reasoning biases.

2.3 Logical and Diagrammatic Reasoning Abilities of LLMs

Advanced deep learning models including LLMs have made significant progress in recent AI research. Given their considerable natural language reasoning abilities, extensive research has been devoted to solving complex logical inferences including syllogisms using deep learning algorithms in the field of natural language processing [20,29]. However, the extent to which these models can accurately perform logical reasoning remains uncertain.

Dasgupta et al. [6] showed that in reasoning tasks including syllogism and Wason's selection task, LLMs reason more accurately about believable or realistic situations. According to their study, LLMs tend to judge inferences with believable content as valid and those with content inconsistent with our ordinary beliefs as invalid regardless of forms of inferences, thus failing to separate *forms* from *contents* (the content effects). Eisape et al. [8] demonstrated that LLMs exhibit error tendencies similar to those of humans, including the figural effect, in syllogistic reasoning. Ando et al. [1] and Ozeki et al. [19] worked on a syllogism dataset called NeuBAROCO, a bilingual parallel dataset of syllogisms in English and Japanese with annotations related to three types of reasoning biases. Evaluating LLMs on this dataset revealed that they exhibit reasoning biases similar to humans, indicating a need for improvement in their logical reasoning capabilities.

As one of the few pioneering studies in diagrammatic reasoning combined with deep learning research, Wang et al. [26] and Wang [25] introduced a method for utilizing deep learning models to learn continuous vector representations of diagrams, and presented Euler-Net, a visual reasoning system for solving syllogisms with Euler diagrams. Euler-Net takes diagrammatic premises as input and generates either categorical or diagrammatic conclusions with high accuracy (99.5%) without human intervention.

While Wang [25] focuses on purely visual (spatial) reasoning with Euler diagrams, we focus on hybrid reasoning combining sentences and diagrams, with focus on the effectiveness of diagrams as auxiliary means of linguistic inference. We also focus on the reasoning abilities of current state-of-the-art LLMs such as GPT-4 [17] and GPT-4-Vision [16] in the zero-shot learning setting with and without Chain-of-Thought prompting [27], which is currently being actively studied. We systematically examine whether Euler diagrams are effective in reasoning in which humans and LLMs are likely to make mistakes.

3 Dataset for Reasoning with Euler Diagrams

We present our constructed syllogism dataset, where the premises of each syllogism are associated with Euler diagrams.[1] We synthetically generated syllogism problems, as explained in Sect. 3.1, assigning correct answer labels (gold labels), inference types, and labels related to reasoning biases to each problem.

3.1 Two Types of Inference Problems

The dataset we constructed consists of two types of inference problems: *Multiple Choice* (MC) problems and *Validity Checking* (VC) problems. Figure 1 shows some examples. Multiple Choice problems are a format frequently used in psychological experiments on syllogistic reasoning (see [4,11,14] for an overview). This format is also employed in studies on diagrammatic reasoning, such as those conducted in [21,22]. It involves selecting the correct answer from five possible conclusions (hypotheses), given two premises. The hypotheses are sentences that combine two terms in forms based on the four quantifiers, *All* (A), *No* (E), *Some* (I), *Some ... not* (O). Additionally, it is possible to answer with 'None of them', which means that there is no valid conclusion drawn from the two premises.

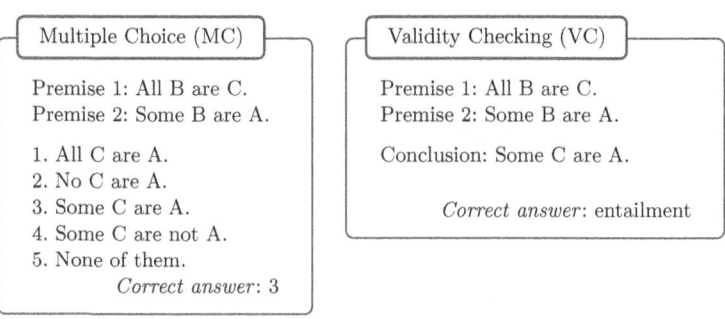

Fig. 1. Two types of syllogism problems: Multiple Choice and Validity Checking

Validity Checking problems are known as the Recognizing Textual Entailment (RTE) [5] or Natural Language Inference (NLI) [28] task in NLP research.

---
[1] The dataset is available at https://github.com/kmineshima/euler-diagrams-llm.

Table 3. Examples of syllogisms labeled as *Symbolic*, *Congruent*, and *Incongruent*. All examples are an instance of *entailment*.

Type	Example
Symbolic	**P1:** All A are B
	P2: All B are C
	C: All A are C
Congruent	**P1**: One friend of Taro is a friend of Paul
	P2: All of Paul's friends are German
	C: One of Taro's friends is German
Incongruent	**P1:** Some animals are human beings
	P2: All animals are tomatoes
	C: Some humans are tomatoes

In the case of syllogism, this format requires checking whether the set of two premises (P1, P2) entails the conclusion (C). There are three types of answers: *entailment*, *contradiction*, and *neutral*. If the premises entail the conclusion (i.e., if the premises are true, then the conclusion is also true), the inference is labeled *entailment*. If the premises contradict the conclusion (i.e., if the premises are true, then the conclusion is false), the inference is labeled *contradiction*. If the relationship between the premises and the conclusion is neither entailment nor contradiction, the inference is labeled *neutral*.

As shown in Table 2, with three terms S, M, and P forming a syllogism and the conclusion where the order of the terms is fixed to S–P, there are a total of 64 inference patterns. Out of these, 15 patterns are considered valid according to modern predicate logic, while the rest are considered invalid.[2] For the MC problems, we adopt a total of 32 patterns to ensure a balance between valid and invalid patterns, following Sato et al. [22]. This includes the 15 valid patterns and the 17 invalid patterns that share the same figure as any of those valid patterns (See Table 11 for all 32 patterns). For the VC problems, the 15 valid inference patterns are labeled as *entailment*. From these patterns, those that are obtained by exchanging the forms of conclusions between *All* (A-type) and *Some...not* (O-type), or between *Some* (I-type) and *No* (E-type), result in contradictions and are labeled as *contradiction*. The rest of the patterns are labeled as *neutral*. We describe the detailed construction of the dataset in Sect. 3.3.

3.2 Inference Types and Reasoning Biases

We distinguish three types of inferences, *symbolic*, *congruent*, and *incongruent*, in terms of what kind of content the sentences appearing in the inference have as follows. Table 3 shows some examples.

---

[2] In this study, we do not take into account the so-called existential presuppositions (existential import) posited by traditional Aristotelian logic. Thus, we do not assume that "All S are P" implies "Some S are P".

Table 4. Examples of conversion errors: Both are labeled as *neutral*, but when **P1** is interpreted as shown in parentheses, the label changes to *entailment*.

AO1O-type	OA4O-type
P1: All B are A (⇝ All A are B)	**P1:** Some A are not B (⇝ Some B are not A)
P2: Some C are not B	**P2:** All B are C
C: Some C are not A	**C:** Some C are not A

Symbolic. When all terms are abstract symbols such as A, B, and C, the inference is labeled as *symbol*.

Congruent. If there is no inconsistency with common-sense beliefs in all premises and conclusions, the inference is labeled *Congruent*.

Incongruent. If at least one of the premises or the conclusion does not align with common-sense beliefs, the inference is labeled *Incongruent*.

The class of *Incongruent* problems may cause *belief-bias* effects in human reasoning, one of the well-known biases in cognitive psychology [9,10]. This is the tendency of humans to endorse inferences whose conclusions they believe and to reject inferences whose conclusions they do not believe, regardless of their logical validity. In addition to *Incongruent* problems, we consider two other types of reasoning biases in syllogistic reasoning.

Conversion Errors. Conversion errors are the errors that occur by mistakenly reversing the order of the two terms appearing in a quantified sentence [10,11]. For example, *All A are B* and *Some A are not B* are misinterpreted as *All B are A* and *Some B are not A*, respectively. Each of these two sentences may appear similar, but their logical meanings are different. We labeled as *Conversion* those *neutral* inferences where A-type (*All*) or O-type (*Some...not*) sentences appear in the premises and their correct answer changes from *neutral* to *entailment* if the order of the terms appearing in at least one premise is converted.[3] Typical examples are shown in Table 4.

In Euler diagrams, A-type and O-type sentences are represented as shown in Fig. 2. These diagrams directly illustrate that the relationships between the two circles are asymmetric in both the A-type and the O-type pairs, indicating that the two terms cannot be substituted interchangeably. Consistent with this observation, Sato et al. [22] reported the results of human experiments showing that providing Euler diagrams with syllogisms can mitigate conversion errors.

Figural Effects. Human reasoning is sensitive to the order in which terms appear, and it has been observed that deriving the conclusion of a syllogism in the order of C–A is easier with figure 1 (the B–A; C–B order) than with figure 4 (the A–B; B–C order), an effect known as the *figural effect* [13]. Following [22],

---
[3] The class of syllogisms that have the *Conversion* label consists of AA2A, AA3A, AA4A, AI2I, AI4I, IA1I, IA2I, AE1E, AE3E, EA3E, EA4E, AO1O, OA1O, AO4O, OA4O, AO3O, OA2O types.

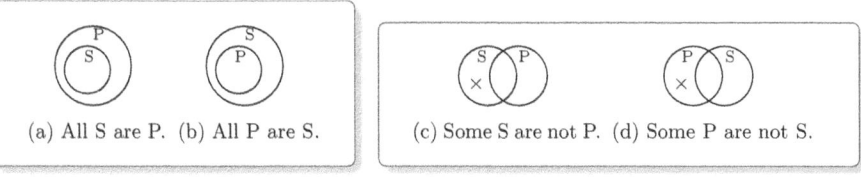

Fig. 2. Euler diagrams for A-type sentences and O-type sentences: (a) and (b) for A-type (*All*) and (c) and (d) for O-type (*Some...not*).

Table 5. Examples of figural effects.

EI1O-type	EI4O-type
P1: No B are A .	**P1**: No A are B.
P2: Some C are B.	**P2**: Some B are C .
C: Some C are not A .	C: Some C are not A .

we focus on the difference between the EI1O-type and the EI4O-type. Examples of the two types of syllogisms are shown in Table 5. In EI1O, the order of terms when combining the two premises by identifying the middle term B matches the order of terms in the conclusion (both in C–A order). However, in EI4O, the order differs: the premises are arranged in A–C order when combining them by identifying the middle term B, while the conclusion is in C–A order.

As noted in [22], one reason for the difference for humans between EI1O and EI4O is the difficulty in understanding the logical equivalence between the E-type sentences *No A are B* and *No B are A*, as well as between the I-type sentences *Some A are B* and *Some B are A*. Sato et al. [22] reported that figural effects can be mitigated by using the Euler diagrams, as shown in Fig. 3. This is because it can be immediately seen that the two circles in the E-type pair and the O-type pair are symmetric, thereby conveying the logically equivalent information.

While the belief bias is a bias related to the *content* of syllogisms, conversion errors and figural effects are biases related to their *form*. We examine whether these biases can be mitigated in the context of reasoning by LLMs, as observed in human reasoning.

3.3 Overview of the Dataset

For MC problems, we first created a list of term triples, consisting of one triple of abstract symbols (A, B, C) and 21 common noun triples, such as (*fruits, foods, stones*). These triples were used to fill prepared sentence templates to formulate two premises. The templates are based on all 64 syllogistic patterns. For each problem, four hypotheses determined by the figure of the syllogism, plus an option for 'none of them', were numbered from 1 to 5. The number of the correct answer was also included as a gold label. The order of the options was

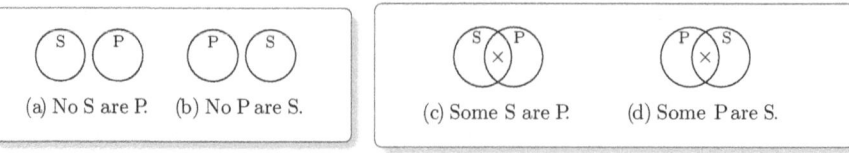

Fig. 3. Euler diagrams for E-type sentences and I-type sentences: (a) and (b) for E-type (*No*) and (c) and (d) for I-type (*Some*).

Table 6. The classification of sentences according to noun-pairs and sentence types in terms of belief congruence. The sentences in the gray cells are labeled *Incongruent* and the others are labeled *Congruent*.

Pattern	Subset	Superset	Overlap	Disjoint
Relation	$N_1 \subseteq N_2$	$N_1 \supseteq N_2$	$N_1 \cap N_2 \neq \emptyset$	$N_1 \cap N_2 = \emptyset$
Example: (N_1, N_2)	(fruits, foods)	(fruits, foods)	(students, males)	(dogs, robots)
A-type	All fruits are foods.	All foods are fruits.	All students are males.	All dogs are robots.
E-type	No fruits are foods.	No foods are fruits.	No students are males.	No dogs are robots.
I-type	Some fruits are foods.	Some foods are fruits.	Some students are males.	Some dogs are robots.
O-type	Some fruits are not foods.	Some foods are not fruits.	Some students are not males.	Some dogs are not robots.

randomized so that the gold labels were evenly distributed. In order to classify the non-symbolic problems into *Congruent* and *Incongruent* ones, we annotated the relations between the common nouns of every triple in the initial list into four categories: *subset*, *superset*, *overlap*, and *disjoint*. We sorted out whether the sentences of the types *A*, *E*, *I*, and *O* are consistent with common belief or not, as shown in Table 6. Finally, we adopted the 32 syllogistic patterns of problems discussed in Sect. 3.1 from the resulting problems. Given that *Incongruent* problems outnumbered *Congruent* problems in the dataset, we balanced it by randomly sampling from the former.

For VC problems, we created four distinct problems from each MC problem described above. These problems were derived by combining the premises with a hypothesis from the MC problems. We then annotated the problems with gold labels (*entailment*, *contradiction*, and *neutral*) along with other labels. Moreover, within each class of an inference type, we ensured an even distribution of gold labels by randomly sampling the problems, with the constraint that all possible combinations of sentence and figure types were covered in the final dataset.

Our dataset consists of 194 MC and 285 VC problems. Table 7 shows the counts of the MC problems for each inference type, as well as the numbers of valid syllogisms and those with no valid conclusion in the options. Table 8 shows the numbers of the VC problems for each label. Each VC problem is annotated with (i) entailment labels (*entailment, contradiction,* or *neutral*) and (ii) inference type labels (*Symbolic, Congruent, Incongruent*).

Table 7. Number of MC problems by type and label. **Valid** means that the problem has a valid conclusion in the options. **NoValid** means that the problem has no valid conclusion in the options (i.e., the correct answer is 'none of them'.)

Type	#Total	#Valid	#NoValid	#Conversion
Symbolic	32	15	17	17
Congruent	81	40	41	41
Incongruent	81	44	37	37
All	194	99	95	95

Table 8. Number of VC problems by type and label.

Type	#Total	#Entailment	#Contradiction	#Neutral	#Conversion
Symbolic	45	15	15	15	5
Congruent	120	40	40	40	9
Incongruent	120	40	40	40	5
All	285	95	95	95	19

4 Experiments

4.1 Evaluated Models and Experimental Settings

In our experiment, we used GPT-4-Vision (GPT-4V) [16] as the main subject of the evaluation. GPT-4V is a multimodal language model capable of accepting both text and images as input. For comparison, in experiments without diagrams, we also evaluated using GPT-3.5 [18] and GPT-4 [17], which only accept text as input. These models were accessed through OpenAI's API.[4] Regarding the hyperparameters of the LLMs, to prevent excessively long responses, the maximum token length was set to 350. Default values were used for all other hyperparameters.

The experiments were conducted in settings with and without Euler diagrams. In the setting without diagrams, in addition to ch1GPTsps4V, GPT-3.5 and GPT-4 were also evaluated. In the setting with diagrams, experiments were conducted with three different types of prompts:

1. The basic prompt is the same as the one used in the setting without diagrams.

---

[4] The versions of the models used were gpt-4-1106-vision-preview, gpt-3.5-turbo-0125, and gpt-4-0125-preview, respectively.

2. The *Extended* prompt is a prompt that adds a minimal description and instructions about the diagrams to the basic prompt. For both the basic and Extended prompts, instructions are given to output only the answer (Figs. 4 and 5 show examples).
3. The *CoT* prompt is based on the Chain-of-Thought prompting technique [27], instructing the language model to first explain the diagrams and then provide an answer (Fig. 6 shows an example output). The explanation of the diagrams is instructed to be 200 words or less.

To indicate the (non-basic) prompts used, we append subscripts to the names of the models: "Ext" for the Extended prompt and "CoT" for the CoT prompt.

In the setting without diagrams, a single input (prompt) includes instructions regarding the problem and answer, along with a single problem from the dataset. In the setting with diagrams, the diagrams corresponding to the problems from the dataset are further added as image inputs. The output of the language model is non-deterministic, so the results may vary with re-experimentation, but in our preliminary experiments, the models showed almost similar tendencies.

4.2 Tasks

We conducted experiments on the following two tasks.

Multiple Choice (MC) Task. In this task, in addition to the instructions, two premises and four hypotheses that are candidates for the conclusion (each of which is one of the codes A, E, I, or O), plus 'none of them' are presented, numbered from 1 to 5, and the language models are asked to answer with one of the numbers. Figure 4 shows an example Extended prompt for this task.

Validity Checking (VC) Task. In this task, two premises and one hypothesis that is a candidate for a conclusion are presented along with instructions about the problem. The models are asked to answer either *entailment, contradiction,* or *neither*. Figure 5 shows an example Extended prompt for this task.

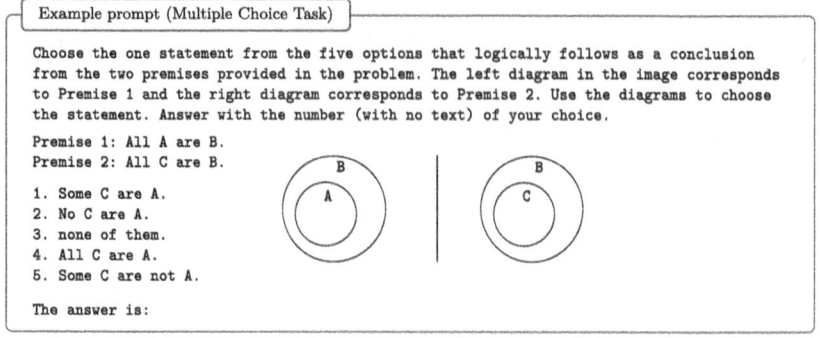

Fig. 4. Example Extended prompt for the MC task (correct answer: 3)

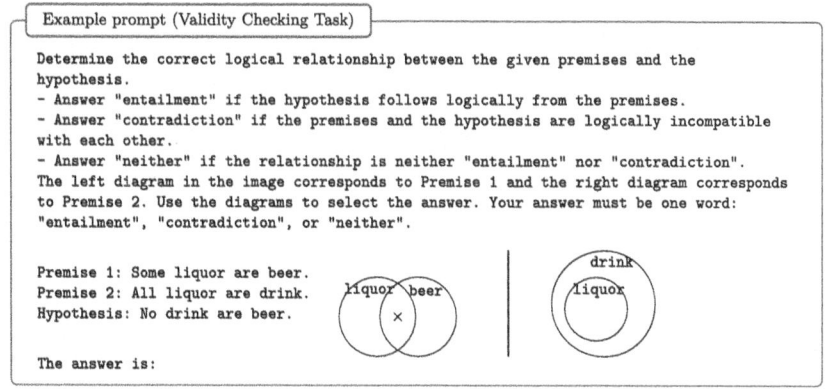

Fig. 5. Example Extended prompt for the VC task (correct answer: entailment)

4.3 Results and Analysis

The evaluation results for the Multiple Choice and Validity Checking tasks are shown in Tables 9 and 10, respectively.

Results for the MC Task. The overall accuracy is highest in GPT-4 without diagrams (66.49%), in comparison to which GPT-4V without diagrams has a lower accuracy (53.09%). Therefore, to compare the model's performances with and without diagrams, we use GPT-4V without diagrams as the baseline.

While GPT-4V with diagrams shows an increase in accuracy of about 5% compared to GPT-4V without diagrams, GPT-4V$_{Ext}$ and GPT-4V$_{CoT}$ show little improvement in accuracy. For Valid problems, the accuracy is higher with diagrams, reaching about 93% in GPT-4V$_{CoT}$.

In the case of NoValid problems, providing diagrams alone increases the accuracy from 20% to 28%. This suggests that diagrams are effective on *Conversion* problems because, in the MC task, all NoValid problems are also *Conversion* problems. Figure 6 shows an example output of GPT-4V$_{CoT}$ for a NoValid problem in the MC task with Euler diagrams. In this example, accurate paraphrasing is provided for the Euler diagrams corresponding to the premises in the first

Table 9. Accuracy (%) on the MC task ($n = 194$).

Condition	Model	Overall	Valid	NoValid	Symbolic	Cong	Incong	Conv
w/o diagram	GPT-3.5	40.72	79.80	0.00	34.38	46.91	37.04	0.00
	GPT-4	66.49	91.92	40.00	59.38	69.14	66.67	40.00
	GPT-4V	53.09	84.85	20.00	50.00	49.38	58.02	20.00
w/ diagram	GPT-4V	58.25	86.87	28.42	65.62	54.32	59.26	28.42
	GPT-4V$_{Ext}$	52.58	87.88	15.79	50.00	53.09	53.09	15.79
	GPT-4V$_{CoT}$	54.12	92.93	13.68	56.25	51.85	55.56	13.68

Table 10. Accuracy (%) on the VC task ($n = 285$). **E** = entailment, **C** = contradiction, **N** = neutral, **Incong** = Incongruent, **Cong** = Congruent, **Conv** = Conversion.

Condition	Model	Overall	E	C	N	Symbolic	Cong	Incong	Conv
w/o diagram	GPT-3.5	47.02	96.84	20.00	24.21	37.78	50.00	47.50	21.05
	GPT-4	77.54	93.68	100	38.95	80.00	83.33	70.83	42.11
	GPT-4V	66.67	87.37	100	12.63	73.33	72.50	58.33	21.05
w/ diagram	GPT-4V	70.53	89.47	100	22.11	88.89	73.33	60.83	26.32
	GPT-4V$_{Ext}$	71.93	90.53	100	25.26	80.00	76.67	64.17	31.58
	GPT-4V$_{CoT}$	76.84	97.89	100	32.63	86.67	75.83	74.17	36.84

paragraph, while in the second paragraph, when synthesizing the two diagrams, the relationship between A and C within B is undetermined, leading to selecting 'none of them' as the correct answer. Note, however, that GPT-4V$_{Ext}$ and GPT-4V$_{CoT}$ show a decrease in accuracy for NoValid problems (about 16% and 14%, respectively). This could be attributed to their tendency to attempt to synthesize information from the two diagrams but fail to determine the position of A and C when 'none of them' is the correct answer. This leads to further errors in inference chains. Interestingly, when explanations are not required, there is an increase in selecting 'none of them' as the correct answer.

In the case of *Incongruent* problems, the effect of diagrams is hardly observed. Detailed results for each problem in the MC Task are provided in Appendix Table 11. Regarding figural effects, the average accuracy for the models without diagrams shows no difference in accuracy between EI1 and EI4 problems (both 77.78%), whereas, for the models with diagrams, the accuracy for EI1 problems (88.89%) is lower than that for EI4 problems (100%). This result is contrary to the human case, where without diagrams, the accuracy for EI4 problems is lower than that for EI1 problems, but providing diagrams eliminates the difference in accuracy between EI1 and EI4 problems.

The diagrams depict two categorical relationships. In the left diagram, the circle labeled "A" is fully contained within the larger circle labeled "B," representing that all members of category A are also members of category B (Premise 1: All A are B). In the right diagram, the circle labeled "C" is fully contained within the same larger circle labeled "B," representing that all members of category C are also members of category B (Premise 2: All C are B).

However, the diagrams do not indicate any direct relationship between A and C; they could either be completely separate, partially overlapping, or one could be fully contained within the other within the confines of B.

The answer is: 3

Fig. 6. Example output of GPT-4V$_{CoT}$ for the MC problem with Euler diagrams shown in Fig. 4.

Results for the VC Task. Similar to the results on the MC task, GPT-4 without diagrams achieved higher accuracy than GPT-4V without diagrams (77.54% vs. 66.67%). We choose GPT-4V without diagrams as the baseline for comparing the models' performances with and without diagrams.

Comparing GPT-4V without diagrams to models with diagrams (GPT-4V, GPT-4V$_{\text{Ext}}$, GPT-4V$_{\text{CoT}}$), all models with diagrams surpass the accuracy of GPT-4V without diagrams, with GPT-4V$_{\text{CoT}}$ reaching 76.84%. Note that GPT-4V without diagrams exhibits considerably lower accuracy (12.63%) for *neutral* problems, compared to *entailment* and *contradiction* problems. However, under conditions with diagrams, the accuracy of *neutral* problems increases by around 10% to 20%.

Regarding the problems involving reasoning biases, the accuracy of models with diagrams increases for *Incongruent* and *Conversion* problems (GPT-4V$_{\text{CoT}}$ reaching approximately 74% and 37%, respectively). In addition, for *Symbolic* problems, the accuracy improves under conditions with diagrams, particularly with GPT-4V achieving a high accuracy of 88%.

5 Conclusion and Future Work

We have investigated how accurately current LLMs can perform syllogistic reasoning when provided with Euler diagrams as auxiliary input. Overall, the experimental results showed that using diagrams as auxiliary input is effective for LLMs, although the effect varies under different conditions. In the VC task, where a single specific conclusion is presented, *neutral* problems are particularly challenging; however, diagrams have been shown to slightly improve the models' performance. In the MC task, selecting the "No valid conclusion" option ('none of them') is notably difficult, and it was observed that providing diagrams in a Chain-of-Thought setting decreased the accuracy. In both tasks, the diagrams improved the accuracy for *Conversion* problems, but the improvement in accuracy was not as high as that seen in humans. Regarding *Incongruent* problems, the inclusion of diagrams has been seen to slightly improve performance.

Future work will explore logical inferences beyond syllogisms, such as those in propositional logic augmented with Venn diagrams. Additionally, conducting more detailed qualitative analyses of the models' outputs to compare the explainability of models provided with sentences and those with diagrams would be interesting. Our results suggest a tendency for the paraphrasing of diagrams to be accurate, while the models make mistakes in the subsequent reasoning process concerning diagrams. A more detailed analysis will shed light on aspects that cannot be adequately evaluated by overall accuracy alone.

Acknowledgements. We thank the anonymous reviewers for their insightful comments and suggestions, which have greatly improved the paper. This work is partially supported by JST, CREST Grant Number JPMJCR2114, and JSPS Kakenhi Grant Numbers JP24K00004, JP21K00016, JP21H00467, JP23K20416, and JP21K18339.

A Appendix

Table 11 presents detailed results on the response distributions of LLMs for 32 syllogisms in the Multiple Choice task. The human results (N = 45) are those reported in Sato et al. [22] for the group presented only with sentences and the group presented with both sentences and Euler diagrams.

Table 11. Response distributions for 32 syllogisms. Each conclusion has the form: A for *all(C, A)*, E for *no(C, A)*, I for *some(C, A)*, O for *some-not(C, A)*, N for *no-valid*. The rows shaded in gray indicate the syllogisms labeled as *Conversion*, all being *neutral* problems. Items in bold indicate the correct answers.

Code/	Premises	LLMs						Human: w/o diagrams					Human: w/ diagrams				
		w/o diagrams			w/ diagrams			Conclusion					Conclusion				
Figure	P1, P2	G3.5	G4	4V	4V	4Ve	4Vc	A	E	I	O	N	A	E	I	O	N
AA1	all(B,A), all(C,B)	A	A	A	A	A	A	–	–	–	–	–	–	–	–	–	–
AA2	all(A,B), all(C,B)	I	N	N	N	N	N	55.6	6.7	4.4	2.2	**31.1**	4.4	0.0	0.0	0.0	**95.6**
AA3	all(B,A), all(B,C)	E	I	I	I	I	I	60.0	0.0	26.7	0.0	**13.3**	0.0	0.0	13.3	0.0	**86.7**
AA4	all(A,B), all(B,C)	I	I	I	I	I	I	60.0	0.0	28.9	0.0	**11.1**	8.9	0.0	11.1	2.2	**77.8**
AI1	all(B,A), some(C,B)	I	I	I	I	I	I	2.2	2.2	**88.9**	2.2	4.4	0.0	0.0	**100.0**	0.0	0.0
AI2	all(A,B), some(C,B)	I	I	O	O	O	O	0.0	0.0	55.6	17.8	**26.7**	0.0	0.0	6.7	2.2	**88.9**
AI3	all(B,A), some(B,C)	I	I	I	I	I	I	4.4	2.2	**80.0**	6.7	6.7	0.0	0.0	**84.4**	0.0	15.6
AI4	all(A,B), some(B,C)	I	I	I	N	I	I	4.4	0.0	57.8	6.7	**31.1**	0.0	0.0	15.6	0.0	**84.4**
IA1	some(B,A), all(C,B)	I	I	I	I	I	I	2.2	2.2	60.0	8.9	**26.7**	0.0	0.0	8.9	2.2	**84.4**
IA2	some(A,B), all(C,B)	I	N	I	N	I	N	6.7	0.0	51.1	11.1	**31.1**	0.0	0.0	13.3	2.2	**84.4**
IA3	some(B,A), all(B,C)	I	I	I	I	I	I	0.0	2.2	**93.3**	0.0	4.4	0.0	0.0	**75.6**	0.0	24.4
IA4	some(A,B), all(B,C)	I	I	I	I	I	I	11.1	0.0	**73.3**	6.7	6.7	0.0	0.0	**68.9**	0.0	31.1
AE1	all(B,A), no(C,B)	E	O	E	N	N	O	0.0	64.4	0.0	6.7	**26.7**	0.0	6.7	0.0	4.4	**88.9**
AE2	all(A,B), no(C,B)	E	E	E	E	E	E	2.2	**93.3**	0.0	2.2	2.2	0.0	**97.8**	0.0	0.0	2.2
AE3	all(B,A), no(B,C)	E	O	N	E	E	E	0.0	64.4	2.2	11.1	**20.0**	0.0	15.6	0.0	4.4	**80.0**
AE4	all(A,B), no(B,C)	E	E	E	E	E	E	0.0	**77.8**	4.4	6.7	11.1	0.0	**95.6**	0.0	2.2	2.2
EA1	no(B,A), all(C,B)	E	E	E	E	E	E	0.0	**91.1**	0.0	2.2	2.2	0.0	**97.8**	0.0	0.0	2.2
EA2	no(A,B), all(C,B)	I	E	E	E	E	E	2.2	**88.9**	0.0	4.4	4.4	0.0	**100.0**	0.0	0.0	0.0
EA3	no(B,A), all(B,C)	E	O	O	O	O	O	0.0	62.2	0.0	20.0	**17.8**	0.0	11.1	0.0	4.4	**84.4**
EA4	no(A,B), all(B,C)	O	O	O	O	O	O	2.2	60.0	2.2	13.3	**17.8**	0.0	6.7	0.0	11.1	**82.2**
AO1	all(B,A), some-not(C,B)	O	O	O	O	O	O	0.0	4.4	8.9	66.7	**17.8**	0.0	4.4	0.0	20.0	**75.6**
AO2	all(A,B), some-not(C,B)	O	O	O	O	O	O	0.0	4.4	4.4	**75.6**	15.6	0.0	0.0	2.2	**91.1**	6.7
AO3	all(B,A), some-not(B,C)	O	N	O	N	O	N	0.0	2.2	17.8	53.3	**26.7**	0.0	0.0	8.9	6.7	**77.8**
AO4	all(A,B), some-not(B,C)	O	N	O	O	O	O	0.0	4.4	11.1	55.6	**28.9**	0.0	2.2	0.0	6.7	**91.1**
OA1	some-not(B,A), all(C,B)	O	O	O	O	O	N	2.2	2.2	4.4	66.7	**24.4**	0.0	4.4	0.0	13.3	**82.2**
OA2	some-not(A,B), all(C,B)	O	O	O	N	O	O	2.2	2.2	4.4	64.4	**26.7**	0.0	6.7	2.2	8.9	**82.2**
OA3	some-not(B,A), all(B,C)	I	O	I	O	I	O	0.0	4.4	11.1	**80.0**	4.4	0.0	4.4	4.4	**66.7**	24.4
OA4	some-not(A,B), all(B,C)	O	I	O	I	I	I	0.0	11.1	20.0	42.2	**26.7**	0.0	2.2	4.4	2.2	**91.1**
EI1	no(B,A), some(C,B)	E	O	O	O	O	E	0.0	22.2	2.2	**62.2**	13.3	0.0	8.9	0.0	**84.4**	6.7
EI2	no(A,B), some(C,B)	O	O	O	O	O	O	0.0	26.7	4.4	**46.7**	22.2	0.0	8.9	0.0	**84.4**	6.7
EI3	no(B,A), some(B,C)	O	O	O	O	O	O	0.0	20.0	2.2	**53.3**	17.8	0.0	4.4	0.0	**84.4**	11.1
EI4	no(A,B), some(B,C)	E	O	O	O	O	O	0.0	28.9	6.7	**35.6**	28.9	0.0	6.7	0.0	**75.6**	15.6

References

1. Ando, R., Morishita, T., Abe, H., Mineshima, K., Okada, M.: Evaluating large language models with NeuBAROCO: Syllogistic reasoning ability and human-like biases. In: Proceedings of the 4th NALOMA Workshop, pp. 1–11 (2023)

2. Barwise, J., Shimojima, A.: Surrogate reasoning. Cogn. Stud.: Bull. Jpn. Cogn. Sci. Soc. **2**(4), 7–27 (1995)
3. Brown, T., et al.: Language models are few-shot learners. Adv. Neural. Inf. Process. Syst. **33**, 1877–1901 (2020)
4. Chater, N., Oaksford, M.: The probability heuristics model of syllogistic reasoning. Cogn. Psychol. **38**(2), 191–258 (1999)
5. Dagan, I., Roth, D., Zanzotto, F., Sammons, M.: Recognizing Textual Entailment: Models and Applications. Springer, Heidelberg (2022). https://doi.org/10.1007/978-3-031-02151-0
6. Dasgupta, I., et al.: Language models show human-like content effects on reasoning tasks. arXiv:2207.07051 (2023)
7. Devlin, J., Chang, M.W., Lee, K., Toutanova, K.: BERT: pre-training of deep bidirectional transformers for language understanding. In: Proceedings of NAACL 2019, pp. 4171–4186 (2019)
8. Eisape, T., Tessler, M., Dasgupta, I., Sha, F., van Steenkiste, S., Linzen, T.: A systematic comparison of syllogistic reasoning in humans and language models. arXiv preprint arXiv:2311.00445 (2023)
9. Evans, J.S., Barston, J.L., Pollard, P.: On the conflict between logic and belief in syllogistic reasoning. Mem. Cogn. **11**(3), 295–306 (1983)
10. Evans, J.S., Newstead, S.E., Byrne, R.M.J.: Human Reasoning: The Psychology of Deduction. Psychology Press (1993)
11. Geurts, B.: Reasoning with quantifiers. Cognition **86**(3), 223–251 (2003)
12. Gurr, C., Lee, J., Stenning, K.: Theories of diagrammatic reasoning: distinguishing component problems. Mind. Mach. **8**, 533–557 (1998)
13. Johnson-Laird, P.N., Steedman, M.: The psychology of syllogisms. Cogn. Psychol. **10**(1), 64–99 (1978)
14. Khemlani, S., Johnson-Laird, P.N.: Theories of the syllogism: a meta-analysis. Psychol. Bull. **138**(3), 427–457 (2012)
15. Manktelow, K.: Reasoning and Thinking. Psychology Press (1999)
16. OpenAI: GPT-4V (ision) System Card
17. OpenAI: GPT-4 technical report. arXiv preprint arXiv:2303.08774 (2023)
18. Ouyang, L., et al.: Training language models to follow instructions with human feedback. Adv. Neural. Inf. Process. Syst. **35**, 27730–27744 (2022)
19. Ozeki, K., Ando, R., Morishita, T., Abe, H., Mineshima, K., Okada, M.: Exploring reasoning biases in large language models through syllogism: insights from the NeuBAROCO dataset. In: Findings of the Association for Computational Linguistics: ACL 2024 (2024)
20. Richardson, K., Hu, H., Moss, L., Sabharwal, A.: Probing natural language inference models through semantic fragments. In: Proceedings of the AAAI Conference on Artificial Intelligence, vol. 34, pp. 8713–8721 (2020)
21. Sato, Y., Mineshima, K.: How diagrams can support syllogistic reasoning: an experimental study. J. Logic Lang. Inform. **24**, 409–455 (2015)
22. Sato, Y., Mineshima, K., Takemura, R.: The efficacy of Euler and Venn diagrams in deductive reasoning: empirical findings. In: Goel, A.K., Jamnik, M., Narayanan, N.H. (eds.) Diagrams 2010. LNCS (LNAI), vol. 6170, pp. 6–22. Springer, Heidelberg (2010). https://doi.org/10.1007/978-3-642-14600-8_6
23. Shimojima, A.: Semantic Properties of Diagrams and Their Cognitive Potentials. CSLI Publications (2015)
24. Stenning, K., van Lambalgen, M.: Human Reasoning and Cognitive Science. MIT Press (2012)

25. Wang, D.: Neural diagrammatic reasoning. Ph.D. thesis, University of Cambridge (2020)
26. Wang, D., Jamnik, M., Liò, P.: Investigating diagrammatic reasoning with deep neural networks. In: Chapman, P., Stapleton, G., Moktefi, A., Perez-Kriz, S., Bellucci, F. (eds.) Diagrams 2018. LNCS (LNAI), vol. 10871, pp. 390–398. Springer, Cham (2018). https://doi.org/10.1007/978-3-319-91376-6_36
27. Wei, J., et al.: Chain-of-thought prompting elicits reasoning in large language models. Adv. Neural. Inf. Process. Syst. **35**, 24824–24837 (2022)
28. Williams, A., Nangia, N., Bowman, S.: A broad-coverage challenge corpus for sentence understanding through inference. In: Proceedings of NAACL 2018, pp. 1112–1122 (2018)
29. Yanaka, H., Mineshima, K., Bekki, D., Inui, K., Sekine, S., Abzianidze, L., Bos, J.: Can neural networks understand monotonicity reasoning? In: Proceedings of BlackboxNLP 2019, pp. 31–40 (2019)

Open Access This chapter is licensed under the terms of the Creative Commons Attribution 4.0 International License (http://creativecommons.org/licenses/by/4.0/), which permits use, sharing, adaptation, distribution and reproduction in any medium or format, as long as you give appropriate credit to the original author(s) and the source, provide a link to the Creative Commons license and indicate if changes were made.

The images or other third party material in this chapter are included in the chapter's Creative Commons license, unless indicated otherwise in a credit line to the material. If material is not included in the chapter's Creative Commons license and your intended use is not permitted by statutory regulation or exceeds the permitted use, you will need to obtain permission directly from the copyright holder.

Diagrams in Logic

Mozi's Square of Opposition and Logemes as New Logical Approach

Andrew Schumann[✉] [ID]

University of Information Technology and Management in Rzeszow, Sucharskiego 2,
35-225 Rzeszow, Poland
andrew.schumann@gmail.com
https://wsiz.edu.pl/kadra-akademicka/aschumann/

Abstract. The first logical treatise in Chinese was written by Mozi and entitled the *Dialectical Chapters* (*Mòbiàn*). In this work we cannot detect a single logical system, but it contains many valuable logemes (logical fragments that are consistent). In particular, we find a non-Aristotelian square of opposition in which the existential quantifier is understood in the narrow sense as some and only some. We remember that Aristotle treats the existential quantifier in a broad sense as some, or perhaps all. Mozi was the first logician who proposed considering logic not as a single system, but as a set of different logemes, realized on some partially ordered sets (posets). This is a semantic approach to logic, which was then further developed in the Chinese school of names. In this paper, I logically formalize this approach, proposing to use simplicial complexes from topology as logical diagrams. The point is that these complexes uniquely represent finite posets up to homeomorphisms. Therefore, within the framework of simplicial complexes we can consider different logemes realized on different finite posets at once.

Keywords: square of opposition · Mozi · Aristotle · univalent foundations of mathematics

1 Introduction

In the ancient world, complex logical reasoning techniques were first developed in Mesopotamia. The judicial reform of Ur-Nammu in Ur III (approximately 4000 years ago) unified not only the judicial system in the empire, but also judicial acts, which affected the unification of forensic evidence techniques. Subsequently, we observe complex logical inference techniques in three types of texts in Akkadian [11]: (1) divination (prediction) based on logical inference (usually *modus ponens*) from positive or negative signs (omens); (2) hermeneutics of signs using logical inferences (for example, the use of the transitivity property of implication); (3) judicial acts recording logical conclusions from articles of the code (which are implied but not directly mentioned) and established facts.

The development of logical thinking in China can also be associated with the judicial reform carried out during the Qin dynasty and reflected in the motto "[all]

is under Heaven" (*tiānxià*) said in the *Edict of the Twenty-Sixth Year* (*Niànliù nián zhào*; *èrshíliù nián zhào*):

Niàn liù nián huángdì jǐn bìng tiānxià zhūhóu qiánshǒu dà ānlì hào wèi huángdì nǎi zhào chéngxiàng zhuàng wǎn fǎ dùliàng zé bù yī qiàn yí zhě jiē míng yī zhī.

In the twenty-sixth year (221 B.C.), the August Thearch completed the annexation of all-under-Heaven. The various lords and commoners were greatly pacified. [He] inaugurated the title of August Thearch. Thereupon, [he] instructed the grand councilors Zhuang and Wan: "Those laws and measurements that are not standard or are ambiguous should all be distinguished and standardized" [2, p. 62], [17, pp. 4–19].

Following the judicial reforms instituted by the Qin dynasty and continued by the Han dynasty, court verdicts were exclusively rendered by officials in ascending rank order, detailed in [8, pp. 31–32]: (a) county-level government officers (*xiànguān*); (b) district heads, that is, the leading officials of the respective districts (*xiànlìng* or *xiànzhǎng*); (c) county governors (*jùnshǒu*). In instances of uncertainty, cases were escalated to superior authorities. This structure resulted in a judiciary system that was not autonomous from the administrative body, with officials directly overseeing trial proceedings. Consequently, the court setting did not facilitate a critical dialogue between the plaintiff and defendant, focusing instead purely on determining punishments based on predefined guidelines, such as laws and legal codes inscribed on specific bamboo strips (*biǎnshū*) and disseminated extensively from the central government to the farthest regions, ensuring the widespread communication of the state's legal standards. This practice is documented in writings from the Western Han dynasty period (206 B.C. – 25 A.D.):

Míngbái dà biǎnshū xiāng tíng shì lǐ ménwài yè shě xiǎnjiàn chù lìng bǎixìng jǐn zhīzhī.

Prominently display [the edict/law] on a large *biǎnshū* in conspicuous locations at townships, stations, markets, outer gates, and hostels causing the hundred names to fully comprehend it (57 2000ES7S: 4A; [2, p. 35]).

As a consequence, the legal framework operated on directives for officials on governance and disciplinary actions, bypassing the need for propositional logic to conduct critical discussions. Instead, a few analogical syllogisms sufficed to underpin the officials' judgment processes, involving particularization (*sī*), generalization (*dá*), and the identification of analogous cases (*lèi*), which served the function of deduction and induction in their reasoning. While no logical textbook has yet been found in Akkadian that could be used by lawyers to study logical techniques, there was such a textbook in China. Its very first source was ascribed to Mozi (Mò Dí; from the 5th to the 4th century B.C.) who authored the *Mohist Canon* (*Mòjīng* or *Mòzi*), [6], which laid the foundation for the philosophical movement known as Mohism, named after him. This work includes what are referred to as the *Dialectical Chapters* (*Mòbiàn*), specifically chapters 40 to 45, which explore various semantic and logical dimensions of argumentation. Notably, the 45th chapter is considered one of the earliest proto-logical texts in Chinese history, known as the *Minor Illustrations* (*Xiǎoqǔ*). This text was a significant contribution to traditional Chinese hermeneutics, highlighting how Mozi's insights into the nature of reasoning and naming underpinned the foundations of Chinese legal tradition.

Mozi's most important insight was the idea that there cannot be one logical theory that covers all specific cases of reasoning. He classified some basic reasoning that can contradict each other, so they cannot be applied simultaneously. At the same time, he tested the judgments themselves on some specific examples that show stable semantic connections, quite intuitive. Thus, this approach can be formalized as the existence of different logemes (reasoning schemes that do not yield a system) that are realized on different partially ordered sets (posets). Each logeme can be represented as a logical diagram. To consider different diagrams at once, I propose to use simplicial complexes as logical diagrams, since simplicial complexes uniquely represent posets up to homeomorphisms.

In Sect. 2, I will analyze Mozi's reasoning rules. In Sect. 3, I will show the logical significance of Mozi's work in the context of simplicial complexes.

2 Analogical Syllogisms of the *Mohist Canon*

The first principle of the Mohist hermeneutics from his *Minor Illustrations* (*Xiǎoqǔ*) is known as the principle of distinction, which "serves to clarify the demarcation between correct and incorrect" (*fū biàn zhě, jiāng yǐ míng shìfēi zhī fēn*; *Mohist Canon* 45:1). This principle establishes the aim of argumentation (*biàn*) as the precise separation of truth from falsehood (*shìfēi*), essentially discerning between what is correct and what is not.

The second principle addresses the strict limitations on drawing similarities (analogies): "Being partially so doesn't equate to being wholly so, and hypothesizing something doesn't confirm its current existence" (*huò yě zhě, bù jìn yě. jiǎ zhě, jīn bùrán yě*; *Mohist Canon* 45:2). It introduces four inference techniques based on similarities, each with a method for verification (*xiào*) to establish norms or laws (*fǎ*): (i) *analogy* (*pì*), justifying one aspect through its resemblance to another; (ii) *concurrency* (*móu*), observing when two events happen simultaneously; (iii) *authority* (*yuán*), adhering to a leader or an authoritative opinion; (iv) *extension* (*tuī*), expanding from one finding to another based on similarity.

The third hermeneutic principle cautions against one-dimensional analysis, stating, "Given the multitude of methods [like analogy, concurrency, authority, extension], diverse scenarios, and various rationales, one should not examine them from a singular viewpoint" (*yán duōfāng, shū lèi, yì gù, zé bùkě piān guān yě*; *Mohist Canon* 45:3). This principle is very important for understanding the specifics of Chinese logical discourse. Initially, it is assumed that there are not and cannot be universal logical techniques for all occasions. All these techniques are purely contextual. This is a completely different understanding of logic (reasoning) than Aristotle had in his apodictics.

In the *Xiǎoqǔ*, we encounter several examples of analogical reasoning, expressed as: "If a specific instance belongs to a broader category, then interacting with the specific instance equates to interacting with the broader category" ("If B is a consequence of A, then engaging with B is a consequence of engaging with A"; formally: $(A \Rightarrow B) \Rightarrow$ ["acting with" $A \Rightarrow$ "acting with" B]). For example, the text illustrates, "A white horse is a horse; thus, to ride a white horse is to ride a horse" (*báimǎ, mǎ yě; chéng báimǎ, chéng mǎ yě*; *Mohist Canon* 45:4); "Zang is a human; to love Zang is to love a human" (*zāng, rén yě; ài zāng, àirén yě*; *Mohist Canon* 45:4). These examples are categorized under positive premises leading to positive conclusions (*shì ér rán zhě yě*; *Mohist Canon* 45:4).

Conversely, the chapter also presents examples of a contradictory form of analogical reasoning: "If a particular instance belongs to a broader category, interaction with the specific instance does **not** result in interaction with the broader category" ("If B is a consequence of A, then engaging with B is not a consequence of engaging with A"; formally: $(A \Rightarrow B) \Rightarrow \neg$ ["acting with" $A \Rightarrow$ "acting with" B]). An instance provided is, "His younger brother is handsome; to love his younger brother does not mean to love all handsome people" (*qí dì, měirén yě; ài dì, fēi ài měirén yě*; *Mohist Canon* 45:5); "A boat is made of wood; boarding a boat is not the same as boarding wood" (*chuán, mù yě; rùchuán, fēi rù mù yě*; *Mohist Canon* 45:5). These examples fall under the category of positive premises with negative conclusions (*cǐ nǎi shì ér bùrán zhě yě*; *Mohist Canon* 45:5).

Thus, we are presented with two opposing forms of inference: "If B follows from A, then using/acting with B follows from using/acting with A" and "If B follows from A, then using/acting with B does not follow from using/acting with A". From the same first premise $A \Rightarrow B$ by applying these inference rules, we obtain a contradiction: ["using/acting with" $A \Rightarrow$ "using/acting with" B] and $\neg$["using/acting with" $A \Rightarrow$ "using/acting with" B]. This contradiction highlights the third Mohist principle of interpretation, suggesting that relying solely on one method or inference pattern is insufficient for a comprehensive analysis. It advocates for the integration of diverse, even seemingly contradictory, methods. For example, the transitivity of "acting with" can both apply and not apply to the implication "If A implies B, then acting with A implies acting with B." Its applicability depends solely on the semantic relations between concepts A and B which can be different. And this is the main feature of Mozi's approach, according to which logical reasoning is always contextual for some semantic dependencies. There are no logical relationships in themselves. This understanding greatly distinguishes Mozi from Aristotle with his apodictics. In the next section I will find out that Mozi was right about many things.

Further examples showcase analogical syllogisms with nuanced implications: "Engaging with A does not imply A, yet appreciating engaging with A implies appreciating A": $\neg$("acting with" $A \Rightarrow A$) $\Rightarrow$ ["appreciating acting with" $A \Rightarrow$ "appreciating" A]. For instance, "Reading books is distinct from books themselves; however, enjoying reading books implies enjoying books" (*qiě fū dúshū, fēi shū yě; hǎo dúshū, hǎo shū yě*; *Mohist Canon* 45:6); "The act of preparing to engage with A does not imply engaging with A, yet preventing the act of preparing to engage with A implies preventing the engagement with A": ¬("preparing to acting with" $A \Rightarrow$ "acting with" A) $\Rightarrow$ ["preventing the act of preparing to acting with" $A \Rightarrow$ "preventing the act with" A]. For example, "Being on the verge of falling into a well is not the same as falling in; preventing someone from being on the verge of falling into a well is the same as preventing them from falling in" (qiě rùjǐng, fēi rùjǐng yě; zhǐ qiě rùjǐng, zhǐ rùjǐng yě; Mohist Canon 45:6). These cases are categorized under negative premises leading to positive conclusions (cǐ nǎi shì ér rán zhě yě; Mohist Canon 45:6).

In Xiǎoqǔ, the author endeavors to delineate a universal quantifier 'all' ('universally') through the following analogical syllogism: "If engaging with A, then engaging with all instances of A; consequently, if not engaging with A, then it is not the case that engaging with not all instances of A." Formally: ["acting with" $A \Rightarrow$ "acting with" $\forall A$] $\Rightarrow$ [¬ "acting with" $A \Rightarrow$ ¬ "acting with" ¬$\forall A$]. Put more plainly: "If A, then all A; therefore, if not A, then not all not A": $(A \Rightarrow \forall A) \Rightarrow (\neg A \Rightarrow \neg \forall \neg A)$. For example, "If one loves people, then universally loving people encompasses loving everyone; if one does not love people, it is not universally not loving people. Not universally [not loving people], because not loving people" (àirén, dài zhōu àirén érhòu wèi àirén. bù àirén, bùdài zhōu bù àirén; bù zhōu ài, yīn wéi bù àirén yǐ; Mohist Canon 45:7). To phrase it more idiomatically: "Loving people entails loving all people, including strangers; not loving people does not mean not loving all people. If one does not love people, then [one does not love] not all people." In essence, if you love people (i.e., you are compassionate), you love all people (e.g., including strangers); if you do not love people (i.e., you lack kindness), then you love only some people (e.g., excluding strangers) – you do not love not all people. Semantically, this holds true. In terms of categorical statements: "If A is B, then A is true for all instances of B; hence, if A is not B, then it is not true that A is true for not all instances of B": $[(A \Rightarrow B) \Rightarrow \forall B(A \Rightarrow B)] \Rightarrow [\neg(A \Rightarrow B) \Rightarrow \neg\forall B\neg(A \Rightarrow B)]$. Consequently, in the Mohist Canon 45:7, the author concludes: "If not [bù] loving people [àirén], it does not imply [bù] loving [àirén] not all people [bùdài zhōu]. If not loving people [bù àirén yǐ], then [not loving] not all people [bùdài zhōu]."

In these arguments, the interpretation of quantifiers is clearly not Aristotelian. Namely, the quantifier of generality and the quantifier of existence are presented as opposing. A categorical statement "A is B" ($A \Rightarrow B$) means a quantifier of generality "A is B for all B" ($\forall B(A \Rightarrow B)$), and a categorical negation "A is not B" ($\neg(A \Rightarrow B)$) means a quantifier of existence "A is B for some B" ($\exists B(A \Rightarrow B)$) or "$A$ is not B for not all B" ($\neg\forall B\neg(A \Rightarrow B)$). As a result, $\forall B(A \Rightarrow B)$ opposes $\exists B(A \Rightarrow B)$. If we take this reasoning very shortly: $(A \Rightarrow \forall A) \Rightarrow (\neg A \Rightarrow \neg\forall\neg A)$, then $\forall A$ opposes $\neg\forall\neg A$ (e.g. $\exists A$). How can this be? Is this a logical fallacy? But no. The point is that existence can be understood in two senses: (1) as existing or for everyone and (2) as existing but not for everyone. The first sense is Aristotelian and the second is not. And Mozi gives a good example for the second sense. I cannot love all people and only some at the same time. This non-Aristotelian undersdaning of the existential quantifer was first examined in the European logic by Nicolai Vasiliev (1880 – 1940) in his syllogistic [12].

Additionally, according to Mozi, another analogical syllogism holds true: "If engaging with A, then engaging with not all instances of A; consequently, if not engaging with A, then not engaging with all instances of A." Formally: ["acting with" $A \Rightarrow$ "acting with" $\neg \forall A] \Rightarrow [\neg$ "acting with" $A \Rightarrow \neg$ "acting with" $\forall A]$. In simpler terms: "If A, then not all A; hence, if not A, then it is all not A": $(A \Rightarrow \neg \forall A) \Rightarrow (\neg A \Rightarrow \forall \neg A)$. For instance, "If riding horses, then not universally riding horses counts as riding horses; if riding horses, then having ridden upon a horse. Conversely, if not riding horses, then universally not riding horses counts as not riding horses" (chéng mǎ, bùdài zhōu chéng mǎ ránhòu wèi chéng mǎ yě; yǒu chéng yú mǎ, yīnwèi chéng mǎ yǐ. dǎi zhì bù chéng mǎ, dài zhōu bù chéng mǎ érhòu bù chéng mǎ; Mohist Canon 45:7). In other words, if you ride a horse, then you ride not all horses (only one); therefore, if you do not ride a horse, then you do not ride all horses. Semantically, this is accurate, too. But this is again a non-Aristotelian understanding of quantifiers. Indeed, the quantifier $\neg \forall A$ opposes the quantifier $\forall \neg A$. We can also formulate the last reasoning in terms of categorical judgments: "If A is B, then A is true not for all instances of B; therefore, if A is not B, then it is not true that A is true for all instances of B": $[(A \Rightarrow B) \Rightarrow \neg \forall B(A \Rightarrow B)] \Rightarrow [\neg(A \Rightarrow B) \Rightarrow \forall B \neg(A \Rightarrow B)]$. Then the quantifier $\neg \forall B(A \Rightarrow B)$ opposes the quantifier $\forall B \neg(A \Rightarrow B)$.

Thus, for Mozi, $\forall B(A \Rightarrow B)$ opposes $\exists B(A \Rightarrow B)$, and $\neg \forall B(A \Rightarrow B)$ opposes $\forall B \neg(A \Rightarrow B)$, although for Aristotle, both pairs form subalternation. These two pairs of opposition are examples of a scenario where "one necessitates the use of a universal quantifier, while the other does not" (cǐ yīzhōu ér yī bù zhōu zhě yě; Mohist Canon 45:7).

Mozi offered examples for which the general quantifier and the existential quantifier are opposites. Indeed, the phrases "loving all people" (in predicate logic: "All A such as $P(A)$") and "loving some people" (in predicate logic: "Some A such as $P(A)$") cannot be true together, but they can be false together; so they are contraries. At the same time, "riding not all horses" (in predicate logic: "Some A such as not $P(A)$") and "not riding any horses" (in predicate logic: "All A such as not $P(A)$") cannot be false together, but they can be true together; so they are subcontraries. As a consequence, we get the non-Aristotelian square of opposition, first formally considered in [13], see Fig. 1.

In the next section of the Xiǎoqǔ, its author presents examples of syllogisms that contrast general principles with specific instances: "If a characteristic of A does not belong to A, then a characteristic of a specific instance of A belongs to that specific instance of A." For instance, "The ghost of a person is not the person themselves; however, the ghost of one's elder brother is considered as one's elder brother" (rén zhī guǐ, fēi rén yě; xiōng zhī guǐ, xiōng yě; Mohist Canon 45:8). "If engaging with a characteristic of A does not imply engaging with A, then engaging with a characteristic of a specific instance of A implies engaging with that specific instance of A." For example, "Offering sacrifices to the ghost of a person is not the same as offering sacrifices to the person; offering sacrifices to

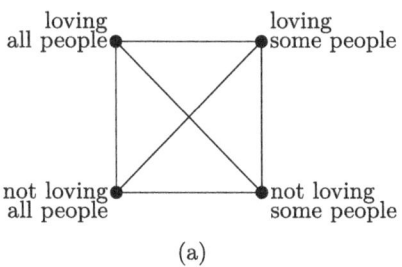 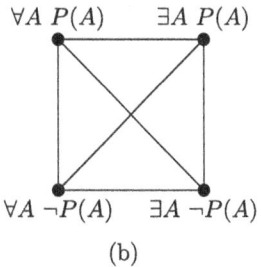

Fig. 1. (a) Mozi's square of opposition, based on his examples in *Mohist Canon* 45:7; (b) Mozi's square of opposition, formulated in predicate logic. In this square, the existential quantifier $\exists A$ means that only some A exist, but surely not all. For more details about this square, please see [13].

the ghost of one's elder brother is considered as offering sacrifices to one's elder brother" (*rén zhī guǐ, fēi jì rén yě; jì xiōng zhī guǐ, nǎi jì xiōng yě*; *Mohist Canon* 45:8). All examples provided in the *Mohist Canon* 45:8 are referred to as "it is the case of one positive and one negative" (*cǐ nǎi yī shì ér yī fēi zhě yě*; *Mohist Canon* 45:8).

There were several attempts to present Mozi's reasoning as some kind of system of logic, but all of them were unsuccessful. The main problem with Mozi is that we can sometimes draw contradictory conclusions from the same premises. So, some Chinese scholars have identified strong logical concepts within the 45th chapter of the *Mohist Canon*, suggesting that the analogical syllogisms presented can be categorized into four distinct patterns based on the combination of affirmative and negative statements in their premises (*bǐ*) and conclusions (*cǐ*). These patterns are: (a) affirmative premise leading to an affirmative conclusion (*shì ér rán*); (b) affirmative premise leading to a negative conclusion (*shì ér bùrán*); (c) negative premise leading to an affirmative conclusion (*bùshì ér rán*); and (d) negative premise leading to a negative conclusion (*bùshì ér bùrán*), see [5, pp. 44–45]. In addition, see [10, p. 433–434], an alternative interpretation for pattern (d) suggests it can be understood as instances where one element is sufficient while another is not (*yīzhōu ér yī bù zhōu*), or one is accurate while another is not (*yī shì ér yī fēi*). Shaokui Mo [10, p. 433–436] labels pattern (a) as a "normal phenomenon," represented by the formula "$A = B$, and $CA = CB$," where C represents an additional attribute. Patterns (b) through (d) are termed "abnormal phenomena," with each having its unique formula that deviates from the norm. Despite this attempt at formalization, the logic is considered flawed since patterns (a) and (b) can lead to contradictions.

The main problem with all these formalizations was that there was a search for a single general logical system for Mozi's reasoning, although he clearly expressed that such a system cannot exist: "Given the multitude of methods [like analogy, concurrency, authority, extension], diverse scenarios, and various rationales, one should not examine them from a singular viewpoint" (*yán duōfāng, shū lèi, yì gù, zé bùkě piān guān yě*; *Mohist Canon* 45:3).

3 Formalization of Mozi's Reasoning as a Family of Logemes

In the 44th chapter, titled the *Major Illustrations* (*Dàqǔ*), within the *Dialectical Chapters* of the *Mohist Canon*, Mozi delineated a framework for understanding propositions that resembles the modern theory of situation semantics [3]. Rather than focusing on factual occurrences, Mozi emphasized classes of situations as the meanings behind propositions: "A proposition (*cí*) is made on the basis of a relevant condition (*gù*), developed in accordance with a relevant idea (*lǐ*) and held in accordance with an appropriate class [of situations] (*lèi*)" (*fū cí yǐ gù shēng, yǐ lǐ zhǎng, yǐ lèi xíng yě zhě*; *Mohist Canon* 44:25). To express our idea (*lǐ*), we must concentrate exclusively on the semantic relationships in the sentence and identify classes of situations (*lèi*) of its concepts. These classes of situations can be generalized (*dá*) into broader classes or limited into narrower classes (*sī*). Thus, logical reasoning appears that reflects generalization (*dá*) and specification (*sī*). At the same time, Mozi understood well that the general follows from the particular, for example, from a "white horse" a "horse" logically follows.

According to Mozi, each term may undergo a process of generalization (*dá*), classification into a class of situations (*lèi*), and specification into particular cases (*sī*):

Míng, dá, lèi, sī.
Míng: wù, dáyě. yǒu shí bì dài zhī míng yě. míng zhī mǎ, lèi yě. ruò shí yě zhě, bì yǐ shì míng yě. míng zhī zāng, sī yě. shì míng yě zhī yúshì shí yě. shēng chūkǒu, jù yǒumíng, ruò xìng zì sǎ.

A name [may be] generalized (*dá*), attributed to the class of similar items (*lèi*), and detalized (*sī*).

'Thing' is a generalization. If there is an entity, it necessarily gets this name. Its class of situations is 'horse.' If there is an entity like this, it is necessarily named by this [word of 'horse']. Naming someone Zāng is a [final] detalization. This name stops at this entity. The sounds issued from the mouth all have names – like the pairing of surname and given name (*Mohist Canon* 40: 79).

As we can see, Mozi believed that logical reasoning is correct if and only if it holds true on the semantic relationships between appropriate classes of situations. Moreover, the reasoning itself has the form of implications $A \Rightarrow B$, and then the judgment $A \Rightarrow B$ is true if and only if the class of situations $||A||$ of concept A is contained in the class of situations $||B||$ of concept B: $||A|| \subseteq ||B||$. Thereby, each reasoning is contextual to classes of situations. Suppose that we have a partially ordered set (poset) P, and for this set there is a set of implications F that are true in P. Let us call this set of implications a logeme. The notion of logeme was first introduced in [11].

Mozi's square of opposition (see Fig. 1) is a logeme verified on an appropriate poset. Let us formally restore this logeme, as well as its poset. Let us take the logeme F as the fragment of propositional logic consisting of the following implications:

- subalternation: $(p \wedge q) \Rightarrow ((p \wedge q) \vee (\neg p \wedge \neg q))$ and $((p \vee q) \wedge \neg (p \wedge q)) \Rightarrow (\neg p \vee \neg q)$;
- contrary: $(p \wedge q) \Rightarrow \neg ((p \vee q) \wedge \neg (p \wedge q))$;
- subcontrary: $\neg ((p \wedge q) \vee (\neg p \wedge \neg q)) \Rightarrow (\neg p \vee \neg q)$;
- contradictory: $(p \wedge q) \Rightarrow \neg (\neg p \vee \neg q)$; $\neg (p \wedge q) \Rightarrow (\neg p \vee \neg q)$; $((p \vee q) \wedge \neg (p \wedge q)) \Rightarrow \neg ((p \wedge q) \vee (\neg p \wedge \neg q))$; and $\neg ((p \vee q) \wedge \neg (p \wedge q)) \Rightarrow ((p \wedge q) \vee (\neg p \wedge \neg q))$.

Then these implications form Mozi's square of opposition, see Fig. 2(a). Let us compare it with Aristotle's square of opposition presented by the logeme F', consisting of the following implications of propositional logic:

- subalternation: $(p \wedge q) \Rightarrow (p \vee q)$ and $(\neg p \wedge \neg q) \Rightarrow (\neg p \vee \neg q)$;
- contrary: $(p \wedge q) \Rightarrow \neg (\neg p \wedge \neg q)$;
- subcontrary: $\neg (p \vee q) \Rightarrow (\neg p \vee \neg q)$;
- contradictory: $(p \wedge q) \Rightarrow \neg (\neg p \vee \neg q)$; $\neg (p \wedge q) \Rightarrow (\neg p \vee \neg q)$; $(p \vee q) \Rightarrow \neg (\neg p \wedge \neg q)$; and $\neg (p \vee q) \Rightarrow (\neg p \wedge \neg q)$.

It is depicted in Fig. 2(b).

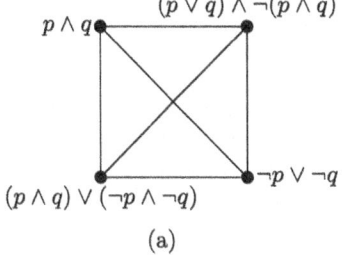
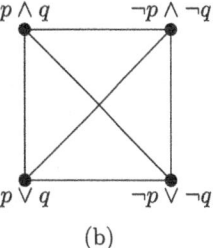

Fig. 2. (a) Mozi's square of opposition for the logeme F; (b) Aristotle's square of opposition for the logeme F'.

As we see, graphically both logemes (F and F') are the same. Nevertheless, they are verified on very different posets. Let us show it. Assume that P is a poset for F and P' is a poset for F'. To construct these posets, we should find their atoms. Let $1 = ||p \wedge q||$; $2 = ||p \wedge \neg q||$; $3 = ||\neg p \wedge q||$; $4 = ||\neg p \wedge \neg q||$. Then they are atoms for both P and P'. As a consequence, $P = \{1, 2, 3, 4, 12, 13, 14, 24, 34, 23, 123, 234\}$ and $P' = \{1, 2, 3, 4, 12, 13, 24, 34, 123, 234\}$, where $12 = 1 \vee 2 = ||p||$; $13 = 1 \vee 3 = ||q||$; $24 = 2 \vee 4 = ||\neg q||$; $34 = 3 \vee 4 = ||\neg p||$;

$14 = 1 \vee 4 = ||(p \wedge q) \vee (\neg p \wedge \neg q)||$; $23 = 2 \vee 3 = ||(p \vee q) \wedge (\neg p \vee \neg q)||$; $123 = 1 \vee 2 \vee 3 = ||p \vee q||$; $234 = 2 \vee 3 \vee 4 = ||\neg p \vee \neg q||$. We see that $P' \subset P$.

Posets may be studied from the point of view of homotopy type theory [4]. In this framework, starting with a poset P, a simplicial complex $\Delta(P)$ is constructed from all P's chains. The geometric realization of $\Delta(P)$, denoted $||\Delta(P)||$, systematically represents these chains: a single-element chain becomes a vertex, a two-element chain forms an edge, a three-element chain creates a coloured triangle, and a four-element chain constructs a coloured tetrahedron in $\Delta(P)$, etc. Therefore, in short, it is called a triangulation of a poset. Furthermore, the geometric realization of $\Delta(P)$ ensures a unique representation up to homeomorphisms.

For any finite poset P, its triangulation $||\Delta(P)||$ may be regarded as a logical diagram that is unique up to homeomorphisms. For standard logical diagrams (such as a square of opposition), there are specific challenges related to their logical function. For instance, consider a square of Fig. 2(b) with four vertices: $p \wedge q$, $\neg p \wedge \neg q$, $p \vee q$, $\neg p \vee \neg q$. This diagram represents a poset, yet it omits six elements: p, q, $\neg p \wedge q$, $p \wedge \neg q$, $\neg p \vee q$, $p \vee \neg q$. Although these elements are implied, the diagram does not explicitly depict all components of the corresponding poset.

With more complex diagrams, such as the U4 hexagon, we encounter more significant issues. Both propositional logic and Aristotle's syllogistic share the same hexagon structure, yet they differ in the set of atoms within their posets. Propositional logic uses four atoms ($p \wedge q$, $\neg p \wedge \neg q$, $\neg p \wedge q$, $p \wedge \neg q$), while Aristotle's syllogistic uses five ($A(S, P) \wedge A(P, S)$, $A(S, P) \wedge O(P, S)$, $O(S, P) \wedge A(P, S)$, $E(S, P)$, $I(S, P) \wedge O(S, P) \wedge O(P, S)$). Consequently, the algebras that produce the same hexagon are not isomorphic [9]. Thus, we have a single diagram representing different algebras!

In this context, Venn diagrams provide a more reliable method for representing posets. However, their drawback is that they can depict the same poset differently when dealing with multiple elements. To address this issue, Ueberweg's diagrams were introduced. While these diagrams are more effective, the fundamental problem persists: we lack a unique representation of posets up to homeomorphisms. This indicates the absence of a mathematical language for uniquely representing posets graphically. However, such a language does exist. It is homotopy, with its triangulation of finite posets. Triangulation of a poset means that the poset is divided into chains: points (there is only one element in the chain), lines (only two) and filled triangles (there are three elements in the chain). After triangulation, the poset becomes a simplicial complex. Each such complex is the only possible graphical representation of a poset. In fact, this is a unique logical diagram up to homeomorphisms. The space of all simplicial complexes is the space of all possible logical diagrams with a unique representation. Here, for example, the U4 hexagon for propositional logic and the U4 hexagon for syllogistic are homotopically equivalent. They have non-isomorphic algebras, but they are equivalent in topology!

Let $P = \{1, 2, 3, 4, 12, 13, 14, 24, 34, 23, 123, 234\}$. Then its simplicial complex $\Delta(P) = \{\{1, 12, 123\}, \{1, 13, 123\}, \{1, 14\}, \{2, 12, 123\}, \{2, 23, 234\}, \{2, 23, 123\},$

{2, 24, 234}, {3, 13, 123}, {3, 23, 123}, {3, 34, 234}, {3, 23, 234}, {4, 14}, {4, 24, 234}, {4, 34, 234}}. It contains two two-element chains and 12 triangles, see Fig. 3(a). In the universe with the top member 1234, Mozi's square of opposition is depicted in Fig. 3(b). But the simplicial complex $\Delta(P)$ of Fig. 3(a) represents this figure in a unique way up to homeomorphisms. Let us take $P' = \{1, 2, 3, 4, 12, 13, 24, 34, 123, 234\}$. Then its simplicial complex $\Delta(P') = \{\{1, 12, 123\}, \{1, 13, 123\}, \{2, 12, 123\}, \{2, 24, 234\}, \{3, 13, 123\}, \{3, 34, 234\}, \{4, 24, 234\}, \{4, 34, 234\}\}$. It contains 8 triangles, see Fig. 3(c). In the universe with the maximal element 1234, Aristotle's square of opposition is depicted in Fig. 3(d). Its simplicial complex $\Delta(P')$ of Fig. 3(c) represents this figure in a unique way up to homeomorphisms.

Simplicial complexes $\Delta(P)$ and $\Delta(P')$ are homotopy equivalent, although $||\Delta(P)|| \neq ||\Delta(P')||$. So, we can distinguish different logemes on the basis of the simplicial complexes of their posets, whether their geometrizations are the same or they are homotopy equivalent. If they are, this allows us to identify logemes, and if they are not, then there are different logemes. We see that the squares of opposition in the standard scheme (Fig. 3(b) and Fig. 3(d)) are the same, but their simplicial complexes are different (Fig. 3(a) and Fig. 3(c)). Hence, we can identify two logemes F and F', based on their posets P and P' respectively, if and only if their simplicial complexes $\Delta(P)$ and $\Delta(P')$ are the same:

$$F = F' \text{ if and only if } ||\Delta(P)|| = ||\Delta(P')||. \qquad (1)$$

We can generalize it to the requirement that their simplicial complexes $\Delta(P)$ and $\Delta(P')$ are homotopy equivalent:

$$F \simeq F' \text{ if and only if } ||\Delta(P)|| \simeq ||\Delta(P')||. \qquad (2)$$

In the case of two squares of opposition from Fig. 3, $F \neq F'$ in the meaning of Axiom (1), but $F \simeq F'$ in the meaning of Axiom (2).

The approach expressed through Axiom (1) allows us to consider numerous families of logemes without dealing with only one logic. This is precisely Mozi's main intuition. We cannot go beyond specific semantic examples in our reasoning. We always deal with specific logemes and their posets (simplicial complexes). For example, the relations, mentioned in the *Mohist Canon* 45:7, form the logeme, consisting of the following implications:

- *subalternation*: ["loving all people" $\Rightarrow$ "not loving all people"] and ["loving some people" $\Rightarrow$ "not loving some people"];
- *contrary*: ["loving all people" $\Rightarrow$ ¬ ("loving some people")];
- *subcontrary*: [¬ ("not loving all people") $\Rightarrow$ "not loving some people"];
- *contradictory*: ["loving all people" $\Rightarrow$ ¬ ("not loving some people")]; [¬ ("loving all people") $\Rightarrow$ "not loving some people"]; ["not loving all people" $\Rightarrow$ ¬ ("loving some people")]; and [¬ ("not loving all people") $\Rightarrow$ "loving some people"].

And its simplicial complex is depicted in Fig. 3(a). In this logeme, "not loving all people" ∨ "not loving some people" is always true. Meanwhile, [||"not loving all people"|| ∨ ||"not loving some people"||] = 14 ∨ 234, where 1 means "loving all

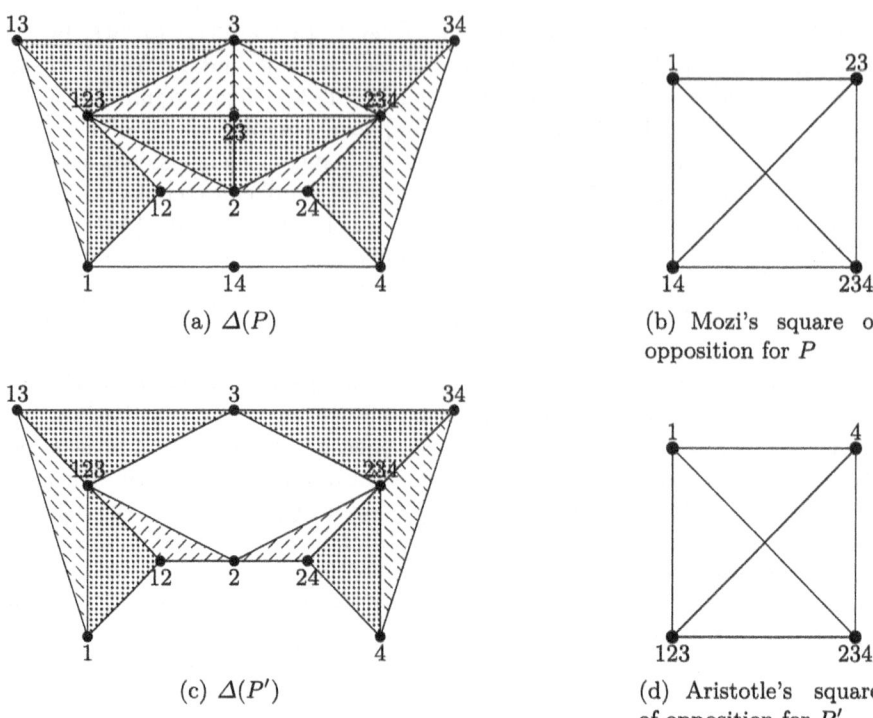

Fig. 3. (a) The simplicial complex $\Delta(P)$ for Mozi's square of opposition for the logeme F. (b) Within the framework of the universe with the maximum 1234, Mozi's square of opposition delineates the following relationships between vertices: 1 and 23 stand as contraries, 14 and 234 as subcontraries, the pairs 1 and 14, as well as 23 and 234, are related through subalternation, and the pairs 1 and 234, along with 14 and 23, are identified as contradictories. (c) The simplicial complex $\Delta(P')$ for Aristotle's square of opposition for the logeme F'. (d) Aristotle's square of opposition, similarly laid out in the universe with the maximum 1234, outlines that 1 and 4 are contraries, 123 and 234 function as subcontraries, 1 and 123, as well as 4 and 234, share a subalternation relationship, and finally, 1 and 234, along with 4 and 123, are established as contradictories.

people," 4 means "hating all people," and 23 means "loving only some people," see Fig. 3(b).

Taking into account that the quantifier of existence is an unbounded disjunction, and the quantifier of generality is an unbounded conjunction, we can generalize the logeme F to the logeme F_Q, and the logeme F' to the logeme F'_Q, where F_Q includes the following implications:

– *subalternation*:
  - $\forall A\ P(A) \Rightarrow (\forall A\ P(A) \vee \forall A\ \neg P(A))$;
  - $(\exists A\ P(A) \wedge \neg(\forall A\ P(A))) \Rightarrow \exists A\ \neg P(A)$;
– *contrary*: $\forall A\ P(A) \Rightarrow \neg(\exists A\ P(A) \wedge \neg(\forall A\ P(A)))$;

- subcontrary: $\neg(\forall A\ P(A) \lor \forall A\ \neg P(A)) \Rightarrow \exists A\ \neg P(A)$;
- contradictory:
  - $\forall A\ P(A) \Rightarrow \neg(\exists A\ \neg P(A))$;
  - $\neg(\forall A\ P(A)) \Rightarrow (\exists A\ \neg P(A))$;
  - $(\forall A\ P(A) \lor \forall A\ \neg P(A)) \Rightarrow \neg(\exists A\ P(A) \land \neg(\forall A\ P(A)))$;
  - $\neg(\exists A\ P(A) \land \neg(\forall A\ P(A))) \Rightarrow (\forall A\ P(A) \lor \forall A\ \neg P(A))$.

And F'_Q includes the following implications:

- subalternation: $\forall A\ P(A) \Rightarrow \exists A\ P(A);\ \forall A\ \neg P(A) \Rightarrow \exists A\ \neg P(A)$;
- contrary: $\forall A\ P(A) \Rightarrow \neg(\forall A\ \neg P(A))$;
- subcontrary: $\neg(\exists A\ P(A)) \Rightarrow (\exists A\ \neg P(A))$;
- contradictory: $\forall A\ P(A) \Rightarrow \neg(\exists A\ \neg P(A));\ \neg(\forall A\ P(A)) \Rightarrow \exists A\ \neg P(A)$;
$\exists A\ P(A) \Rightarrow \neg(\forall A\ \neg P(A));\ \neg(\exists A\ P(A)) \Rightarrow \forall A\ \neg P(A)$.

Let F_Q hold true on the poset P_Q and F'_Q on the poset P'_Q, where $1 = ||\forall A\ P(A)||;\ 4 = ||\forall A\ \neg P(A)||;\ 14 = 1 \lor 4 = ||\forall A\ P(A) \lor \forall A\ \neg P(A)||;\ 23 = 2 \lor 3 = ||\exists A\ P(A) \land \exists A\ \neg P(A)||;\ 123 = 1 \lor 2 \lor 3 = ||\exists A\ P(A)||;\ 234 = 2 \lor 3 \lor 4 = ||\exists A\ \neg P(A)||$. Then F_Q forms Mozi's square of opposition and F'_Q gives Aristotle's square of opposition, see Fig. 4.

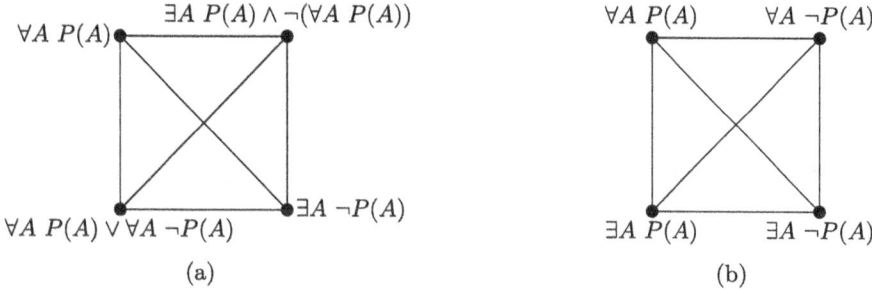

Fig. 4. (a) Mozi's square of opposition for the logeme F_Q; (b) Aristotle's square of opposition for the logeme F'_Q.

We can prove that $||\Delta(P)|| = ||\Delta(P_Q)||$ and $||\Delta(P')|| = ||\Delta(P'_Q)||$. This means that we can identify the logemes according to Axiom (1): $F = F_Q$ and $F' = F'_Q$, see Fig. 3.

In turn, we can generalize the quantifier of existence to the modal operator of possibility $\Diamond$, and the quantifier of universality to the modal operator of necessity $\Box$. Let $F_\Box$ contain the implications:

- subalternation: $\Box A \Rightarrow (\Box A \lor \Box \neg A);\ (\Diamond A \land \neg \Box A) \Rightarrow \Diamond \neg A$;
- contrary: $\Box A \Rightarrow \neg(\Diamond A \land \neg \Box A)$;
- subcontrary: $\neg(\Box A \lor \Box \neg A) \Rightarrow \Diamond \neg A$;
- contradictory:
  - $\Box A \Rightarrow \neg(\Diamond \neg A)$;

- $\neg(\Box A) \Rightarrow \Diamond\neg A$;
- $(\Box A \vee \Box\neg A) \Rightarrow \neg(\Diamond A \wedge \neg\Box A)$;
- $\neg(\Box A \vee \Box\neg A) \Rightarrow (\Diamond A \wedge \neg\Box A)$.

Let $F'_\Box$ consist of the following implications:

- subalternation: $\Box A \Rightarrow \Diamond A$; $\Box\neg A \Rightarrow \Diamond\neg A$;
- contrary: $\Box A \Rightarrow \neg(\Box\neg A)$;
- subcontrary: $\neg(\Diamond A) \Rightarrow \Box\neg A$;
- contradictory:
  - $\Box A \Rightarrow \neg(\Diamond\neg A)$;
  - $\neg(\Box A) \Rightarrow \Diamond\neg A$;
  - $\Box\neg A \Rightarrow \neg(\Diamond A)$;
  - $\neg(\Box\neg A) \Rightarrow \Diamond A$.

Assume that $F_\Box$ is true on the poset $P_\Box$ and $F'_\Box$ on the poset $P'_\Box$, where $1 = ||\Box A||$; $4 = ||\Box\neg A||$; $14 = 1 \vee 4 = ||\Box A \vee \Box\neg A||$; $23 = 2 \vee 3 = ||\Diamond A \wedge \Diamond\neg A||$; $123 = 1 \vee 2 \vee 3 = ||\Diamond A||$; $234 = 2 \vee 3 \vee 4 = ||\Diamond\neg A||$. Consequently, $F_\Box$ represents Mozi's square of opposition and $F'_\Box$ Aristotle's square of opposition, see Fig. 5.

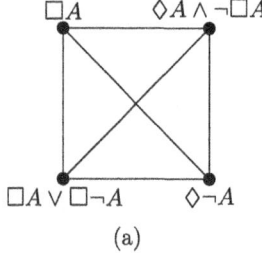
(a)

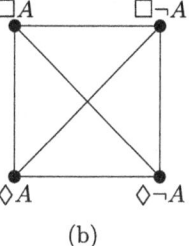
(b)

Fig. 5. (a) Mozi's square of opposition for the logeme $F_\Box$; (b) Aristotle's square of opposition for the logeme $F'_\Box$.

As a result, it is provable that $||\Delta(P)|| = ||\Delta(P_\Box)||$ and $||\Delta(P')|| = ||\Delta(P'_\Box)||$. This means that we can identify the logemes according to Axiom (1): $F = F_\Box$ and $F' = F'_\Box$, see Fig. 3.

Summing up, homotopy type theory [16] allows logical diagrams to be considered through simplicial complexes. This makes it possible to typologize all diagrams in an effective way, and together with diagrams also logemes – inference rules that are reflected in a particular diagram. The study of logic through diagrammatic representations [1,15], enhanced by the use of simplicial complexes, can evolve into an efficient mathematical method grounded in pure mathematics (univalent foundations).

4 Conclusion and Discussion

The principles of Mozi's hermeneutics, along with its logemes and situation semantics, found their way into the Confucian hermeneutics of Xunzi (Xún Kuàng; Xúnzi; 3rd century B.C.), who authored a Confucian text named after himself, the *Xúnzi*, which features the 22nd chapter titled the *Rectification of Names* (*Zhèngmíng*). This chapter delves into hermeneutics rooted in the concepts and principles of Mozi, focusing on the proper usage of 'names' (*míng*) and elucidating their semantic connections to 'entities,' 'objects,' 'occurrences,' or 'contexts' (*shí*). Additionally, the Legalist canon *Hánfēizi*, attributed to Han Feizi (Hán Fēizi; 3rd century B.C.), also reflects the Mohist approach to handling "names and reasoning" (*míng biàn*).

According to Mozi, we cannot deal with a single logic, but with a plenty of different logemes: "Items are sometimes positive and positive, sometimes positive and negative, sometimes one requires a universal quantifier and one does not, sometimes one positive and one negative" (*fū wù huò nǎi shì ér rán, huò shì ér bùrán, huò yīzhōu ér yī bù zhōu, huò yī shì ér yī fēi yě*; *Mohist Canon 45:2*). In particular, we can deal with the non-Aristotelian square of opposition in which the existential quantifier is treated as some, but surely not all. This approach can be realized through the application of the univalent foundations of mathematics [16], according to Axiom (1) or (2). This involves treating individual logemes and their posets as logical diagrams, representing as simplicial complexes, rather than considering a singular logic and its models.

The insight that it's possible to engage with a wide variety of logemes simultaneously, rather than focusing on a single logic system, was also shared by the Tanaim, who were the earliest sages of the Talmud. This perspective acknowledges that the Talmud does not adhere to a single, unified logic. Instead, it encompasses a multitude of logemes, encompassing areas such as modal, temporal, mereological, and relational logic, among others. Each logeme operates within a specific poset, which is articulated through natural language as consistent semantic relationships among certain concepts. As a result, the complexity and richness of Talmudic logic emphasizes that it cannot be confined to a single logical system. Instead, it embraces a diverse array of logical structures, each grounded in legal concepts and relationships as understood through natural language. In Judaism this approach was laid down by Rabbi Ishmael (Rabbî Yišmā'ē'l; ca. 70–ca. 135 A.D.), one of the founders of Judaic hermeneutics. In the treatise **Bāraytā' Darabbî Yišmā'ē'l**, he collected 13 inference rules (*middôṯ*, 'measures') that are applied in parallel, but do not form a closed logical system. In fact, these are logemes which are consistent [14]. The first mathematical formalization of the hermeneutical rules of the Talmud was introduced by Hirschfeld in 1840 [7], marking the pioneering recognition that these rules are applicable and meaningful specifically within certain posets, rather than universally.

Thus, it is entirely feasible to revisit and rejuvenate the distinct approach to logic as practiced by Mozi and Rabbi Ishmael. This view emphasizes working with individual logemes rather than logic as a single entity. Moreover, the use of homotopy types provides a unique representation of posets up to homeomorphisms, which facilitates this approach. Logemes can also enrich the diagram logic [1,15] with new methods. In particular, there is a way to unify any logical diagrams as a simplicial complex due to Axiom (1) or its generalization (2).

References

1. Anger, C., Berwe, T., Olszok, A., Reichenberger, A., Lemanski, J.: Five dogmas of logic diagrams and how to escape them. Lang. Commun. **87**, 258–270 (2022)
2. Caldwell, E.: Writing Chinese Laws. The Form and Function of Legal Statutes Found in the Qin Shuihudi Corpus. Routledge (2018)
3. Cooper, R., Mukai, K., Perry, J. (eds.): Situation Theory and its Applications, vol. 1. CSLI Publications, Stanford (1990)
4. Folkman, J.: The homology groups of a lattice. J. Math. Mech. **15**(4), 631–636 (1966)
5. Fung, Y. (ed.): Dao Companion to Chinese Philosophy of Logic. Springer, Cham (2020)
6. Johnston, I. (tr.): Book of Master Mo. The Chinese University Press (2010)
7. Hirschfeld, H.S.: Der Geist der talmudischen Auslegung der Bibel, vol. 1. Die Halagische Exegese Midot U-derashot Ha-halakhah. M. Simion (1840)
8. Korolkov, M.V.: Zòuyànshū ("Sbornik sudebnykh zaprosov"): paleograficheskiye dokumenty drevnego Kitaya. Nauka, Moscow (2013). (in Russian)
9. Demey, L., Erbas, A.: Boolean subtypes of the U4 hexagon of opposition. Axioms **13**(2) (2024)
10. Mo, Sh.: Mòzi xiǎoqu piān luójí de tǐxì (Logic System in Mozi Xiǎoqu). In Zhōngguó luójíshǐ lùnwénxuǎn (Collection of Theses on History of Chinese Logic) (1949-1979). Sanlian Publishing House, Beijing (1981)
11. Schumann, A.: Archaeology of Logic. Taylor & Francis Group, Oxon (2023)
12. Schumann, A.: A lattice for the language of Aristotle's syllogistic and a lattice for the language of **Vasiľév's** syllogistic. Logic Log. Philos. **15**(1), 17–37 (2006)
13. Schumann, A.: On two squares of opposition: the Leśniewski's style formalization of synthetic propositions. Acta Analytica **28**, 71–93 (2013)
14. Schumann, A.: Rabbi Ishmael's thirteen hermeneutic rules as a kind of logic. J. Appl. Logics **10**(1), 37–56 (2023)
15. Schumann, A., Lemanski, J.: Logic, spatial algorithms and visual reasoning. Log. Univers. **16**, 535–543 (2022)
16. The Univalent Foundations Program. Homotopy type theory: Univalent foundations of mathematics. Technical report, Institute for Advanced Study (2013)
17. Weizu, S., Gufu, G. (eds.): Qín Hàn jīnwén huìbiān. Shanghai shudian, Shanghai (1997)

Implicational Existential Graphs

Arnold Oostra[✉][iD]

Universidad del Tolima, Ibagué, Colombia
noostra@ut.edu.co

Abstract. We present a diagrammatic logical system, inspired by and very similar to C.S. Peirce's existential graphs Alpha, which corresponds to non-classical implicational propositional calculus and can be extended to various related logics.

Keywords: Existential graphs · Implicational propositional calculus · Scroll

1 Introduction

Existential graphs were invented by C.S. Peirce [16,17] and include a completely diagrammatic version of classical logic, whose Alpha part corresponds to propositional calculus [20–22]. In Peirce's original graphs Alpha, conjunction is represented simply by juxtaposition and the only additional sign is a simple closed curve, called a cut, which expresses the negation of its interior. All other propositional connectives are then obtained from conjunction and negation.

Recently, Oostra proposed a variant system of existential graphs that corresponds to intuitionistic logic [10,11,13–15]. Since intuitionistic connectives are independent of each other, new graphical signs were required to represent implication and disjunction. The sign chosen for implication is called a scroll and is composed of two simple closed curves, one of them inside the other and both intersecting at a single point, while the sign for disjunction has two interior curves. A quite natural development of this system consisted of considering the scroll as the only additional sign, which led to a system of existential graphs for non-classical implicational propositional logic with conjunction [3,5,12].

In a subtle additional step we can completely eliminate juxtaposition within graphs constructed solely of letters and scrolls. In this way we obtain a system of existential graphs for implicational logic that does not satisfy all the classical laws of this connective. It does validate the basic axioms of implication but it does not verify Peirce's law. Therefore, this propositional logic is a subsystem of intuitionistic propositional calculus.

In the following Sect. 2, we present the system of existential graphs for non-classical implicational logic. Section 3 contains the axioms and rules of this propositional calculus, in addition to its algebraic semantics known as Hilbert algebras. There we also point out the difference with classical implicational logic. Finally, in Sect. 4 we discuss the equivalence between the graphical and algebraic systems presented and we point out graphical systems for some extensions.

2 Implicational Existential Graphs Alpha

The components from which the implicational existential graphs Alpha are built are:

- The plane surface without border upon which we draw all graphs, called the *sheet of assertion*;
- Propositions, symbolized by capital letters;
- Curves called *scrolls*, which are composed of two simple closed curves, one of them inside the other and the two intersecting at only one point, as shown in Fig. 1.

Fig. 1. The scroll.

The outer curve of a scroll is called cut and the inner curve is called loop. The region bounded by the cut and the loop is the outer area of the scroll, and the interior of the loop is the inner area. An implicational existential graph Alpha is defined inductively by the following clauses.

1. The sheet of assertion, or any part homeomorphic to it, without letters or scrolls drawn on it, is a graph.
2. A letter written on the sheet of assertion is a graph.
3. A scroll containing a graph in each of its areas is a graph.
4. That is all.

Thus, more descriptively, a graph is an empty area, or a single letter, or a single scroll. In the latter case, each of its areas is empty, or contains only one letter, or contains only a single scroll, for which the same conditions apply. We highlight the fact that the juxtaposition of two or more graphs cannot appear in any internal area: neither two letters, nor a letter and a scroll, and so forth. On the other hand, it is permissible to consider two or more graphs drawn simultaneously on the sheet of assertion, but in that case they together do not constitute a single graph. Like the formulas of a propositional calculus, that would be just a set of different graphs. These are the fundamental differences from the usual existential graphs. Otherwise, these new graphs share most of the features with traditional graphs: there may be repeated letters in a graph, but they now all occupy different areas; different scrolls do not touch the letters nor do they touch each other; two graphs that can be continuously deformed into each other are equal.

We derive the interpretation of the implicational existential graphs Alpha from the following clauses.

- The sheet of assertion is the universe of possibilities of truth.
- To draw a graph on the sheet means to assert its interpretation. To write a letter means to assert the proposition it stands for.
- To draw two graphs on the sheet of assertion means to assert both.
- To draw a scroll means to assert the implication whose antecedent is the graph in the outer area and whose consequent is the graph into the loop.

From this we obtain unique graphs for all propositional formulas whose only connective is implication. Figure 2 shows some examples.

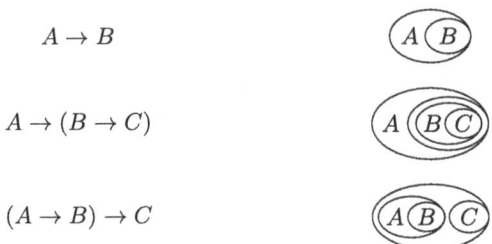

Fig. 2. Implicational existential graphs Alpha for some formulas.

In general, an area is defined as a region of the sheet of assertion limited by curves, both cuts and loops. Topologically, this is a connected component of the complement of the curves. An area is odd or even if there is an odd or even number of curves around it, here we count both cuts and loops alike.

These are the rules of transformation allowed for intuitionistic Alpha graphs:

1 *Erasure.* In an even area, any graph may be erased.
2 *Insertion.* In an odd empty area, any graph may be scribed.
3 *Iteration.* Any graph drawn on the sheet of assertion may be iterated on the sheet or in any empty area of another graph on the sheet. Any graph in the outer area of a scroll may be iterated in an empty area contained in its loop.
4 *Deiteration.* Any graph may be erased if a copy of it persists on the sheet of assertion. Any graph contained in a loop may be erased if a copy of it persists in the outer area of that scroll.
5 *Scrolling.* A scroll with empty outer area may be drawn around or removed from any graph on any area.

Some clarification may be in place here. The rule **1** means that any graph may be erased from the sheet of assertion, and that the contents of any interior even area may be deleted altogether. The rule **2** differs only from the usual permission in that the area must be empty. In rule **3** the empty area contained in the loop can be enclosed in other scrolls. The deiteration rule **4** means that any graph that might have been drawn by iteration may be deleted. Finally, rule **5** implies, in particular, that a scroll with both empty areas may be drawn or erased on any empty area, or on the sheet of assertion.

270   A. Oostra

The intuition behind these rules is that odd areas act as the antecedent of a valid implication, and even areas as its consequent. Then, for instance, the consequent may be ignored (rule **1**) and an antecedent may be added (rule **2**).

A finite set $\mathcal{G} = \{G_1, \ldots, G_k\}$ of implicational existential graphs Alpha entails a graph H, which we denote $G_1, \ldots, G_k \Rightarrow H$, if there exists a finite sequence of sets of implicational graphs $\mathcal{A}_1, \ldots, \mathcal{A}_n$ with $\mathcal{A}_1 = \mathcal{G}$, $\mathcal{A}_n = \{H\}$ and such that each set $\mathcal{A}_{i+1}$ follows from the previous $\mathcal{A}_i$ by applying any of the above rules, say **r**, to one of the graphs in it, in which case we write $\mathcal{A}_i \stackrel{\mathbf{r}}{\Rightarrow} \mathcal{A}_{i+1}$. Note that it suffices to require $H \in \mathcal{A}_n$, but in graphical proofs it is customary to delete all remaining graphs except H.

For instance, Fig. 3 shows a proof of $A, A \to B \Rightarrow B$, which is the traditional rule of inference *modus ponens*. Figure 4 is a proof in this system of the hypothetical syllogism $A \to B, B \to C \Rightarrow A \to C$. This diagrammatic argument is quite different from the original one given by Peirce, in which juxtaposition is used in the first step.

Fig. 3. Proof of *modus ponens*: $A, A \to B \Rightarrow B$.

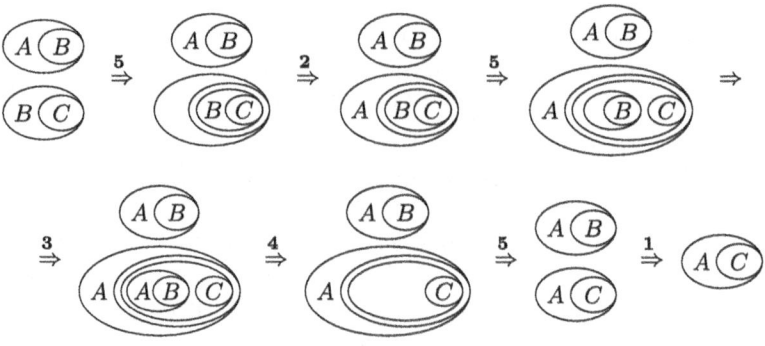

Fig. 4. Proof of the hypothetical syllogism $A \to B, B \to C \Rightarrow A \to C$.

3   Implicational Propositional Calculi

The only connective for non-classical implicational propositional calculus (IMP) is the implication $\to$. We take the following formulas as axioms:

1. $A \to (B \to A)$
2. $(A \to (B \to C)) \to ((A \to B) \to (A \to C))$

The only inference rule is *modus ponens*: from formulas $A \to B$ and A we may proceed to B. A set of formulas Φ entails a formula F, which we denote $\Phi \vdash F$, if there exists a finite sequence of formulas $A_1, \ldots, A_n$ with $A_n = F$ and such that each formula A_i has the form of an axiom, or belongs to Φ, or follows from two previous formulas by *modus ponens*. A formula T entailed by the empty set, that is $\vdash T$, is called a theorem of IMP.

All formal proofs that depend solely on axioms 1 and 2 and the fact that *modus ponens* is the only rule of inference are valid in IMP. In particular this includes the deduction theorem, *i.e.*, $\Phi \vdash A \to B$ if and only if $\Phi, A \vdash B$ [1,9].

Since the first Hilbert-type axiomatization of propositional logic, its authors suggested studying connectives separately [7]. Consequently, the algebraic semantics of IMP was called a Hilbert algebra by Diego [4], although Henkin earlier studied its dual [6] and Rasiowa later called them positive implication algebras [19]. A Hilbert algebra, which acts as a model for IMP, is a set H with a binary operation $\to$ and a constant 1, which satisfy the following axioms.

H1. $a \to (b \to a) = 1$.
H2. $(a \to (b \to c)) \to ((a \to b) \to (a \to c)) = 1$.
H3. If $a \to b = 1$ and $b \to a = 1$ then $a = b$.

In [19] the next identity is added as an axiom, but this is not necessary because it follows from H1 and H3.

Proposition 1. *In a Hilbert algebra we have* $a \to 1 = 1$ *for any element* a.

Proof. First, $1 \to (a \to 1) = 1$ by H1. Substituting we obtain $(a \to 1) \to 1 = (a \to 1) \to (1 \to (a \to 1))$, but $(a \to 1) \to (1 \to (a \to 1)) = 1$ again by H1, hence $(a \to 1) \to 1 = 1$. Applying H3 to the first and the last identities obtained, we conclude $a \to 1 = 1$.

Every Hilbert algebra is partially ordered by the relation $a \leq b$ if $a \to b = 1$. Now Proposition 1 means that the constant 1 is the maximum for this order. In the other way, any partially ordered set with maximum 1 is a Hilbert algebra defining the arrow as follows:

$$a \to b = \begin{cases} 1 & \text{if } a \leq b; \\ b & \text{otherwise.} \end{cases}$$

However, this correspondence between Hilbert algebras and partially ordered sets is not bijective.

The soundness and completeness theorems of this logic express that a formula is a theorem of IMP if and only if any of its valuations is equal to 1 in every Hilbert algebra. For instance, if in a partially ordered set with maximum 1 certain elements satisfy $1 > a > b$, then in the associated Hilbert algebra as above we have $((a \to b) \to a) \to a = (b \to a) \to a = 1 \to a = a \neq 1$. Thus, the formula $((A \to B) \to A) \to A$, known as Peirce's law, is not a theorem of IMP.

Non-classical implicational propositional calculus has various fascinating extensions. For instance, (classical) implicational propositional calculus is defined as IMP plus Peirce's law:

3. $((A \to B) \to A) \to A$

Łukasiewicz [8] proved that this logic is obtained from a single axiom, namely:

$$((A \to B) \to C) \to ((C \to A) \to (D \to A)).$$

In classical implicational propositional calculus we can define the disjunction as $A \vee B = (A \to B) \to B$ because this formula satisfies the usual axioms. To attain completeness it is enough to add a constant $\bot$ (absurd) that satisfies the principle of explosion:

4. $\bot \to A$

Now negation is defined as $\neg A = A \to \bot$, and conjunction in the usual De Morgan way as $A \wedge B = \neg(\neg A \vee \neg B)$. This is a complete axiomatization of classical propositional calculus, very similar to Peirce's 1885 system [16]. In this system all classical identities are valid, including e.g. $A \to B = \neg A \vee B$.

Another way to attain conjunction, without the other connectives, is to simply add the symbol $A \wedge B$ subject to the usual Hilbert-type axioms. The result is non-classical implicational logic with conjunction, whose algebraic semantics is a special class of Hilbert algebras [3,12].

4 Equivalence and Extensions

The basic interpretation of the scroll allows all implicational formulas to be translated into graphs. Transfer of the deduction relation is derived from two facts. On the one hand, we can deduce the translation of both axioms from the empty sheet of assertion, for instance Fig. 5 shows the diagrammatic proof of axiom 1. On the other hand, *modus ponens* corresponds to a graphical entailment as shown in Fig. 3. Now we can reconstruct every step of a deduction $E_1, \ldots, E_k \vdash F$ graphically: if an axiom is needed, we construct its graph on the empty sheet; if *modus ponens* is applied, we follow the steps of Fig. 3. Always retaining the set of all graphs, and deleting at the end the resulting graphs one by one except the translation of the conclusion F, we obtain a graphical proof of the algebraic result.

Fig. 5. Graphical proof of axiom 1: $\Rightarrow A \to (B \to A)$.

By assigning a fixed theorem of IMP to each empty area, all implicational graphs Alpha are directly translated into formulas. This double translation is bijective modulo equivalent graphs on the sheet of assertion. To algebraically interpret a graphical proof, we first observe that the odd and even areas translate into well-defined odd and even positions in implicational formulas. Next, with careful inductions it is possible to establish contraposition theorems both in implicational existential graphs Alpha and in implicational formulas: if one graph can be transformed into another, the same transformation is possible in any even area, and the reverse transformation can be achieved in any odd area. Exactly the same result applies to the deduction of formulas in even and odd positions. Then, by verifying each rule of transformation we can translate its application in one step of a graphical transformation into a valid deduction of IMP. Finally, if $G_1, \ldots, G_k \Rightarrow H$ then each transformation of sets of graphs $\mathcal{A}_i \stackrel{r}{\Rightarrow} \mathcal{A}_{i+1}$ translates into a proof in IMP, and from there we can always construct an algebraic deduction in this logic of the complete transformation.

Thus we establish the mathematical equivalence between the system of implicational existential graphs Alpha and non-classical implicational propositional calculus. More details, especially the comprehensive proofs of the different inductions, can be found in [2].

We can extend this new system of existential graphs in several ways. To include axiom 3 (Peirce's law), it suffices to add its graph as a diagrammatic axiom, see Fig. 6. That is, agree that we may draw any instance of this graph on the sheet of assertion or in any empty area, and also that we may delete any instance of it. Of course, we could alternatively include the graphical version of Lukasiewicz's axiom, as suggested in [18].

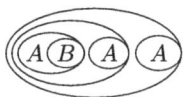

Fig. 6. Peirce's law in implicational existential graphs Alpha.

On the other hand, we could maintain the same graphs but dropping the juxtaposition restriction, that is, allowing two or more letters or cuts to be drawn together in any area. The result is a system of existential graphs with exactly the same rules of transformation as Peirce's original graphs Alpha, only changing the cuts for scrolls and the double cut rule for scrolling. Then the corresponding logic is non-classical implicational propositional calculus with conjunction. For more details and a proof of equivalence, see [5,12].

Furthermore, the three proposed systems of existential graphs can be enriched to include negation in the associated logics. By adding a constant graph which represents the absurd $\bot$, even without any additional axiom we can define the negation of a graph as the scroll that expresses that this graph implies the absurd. Many characteristics of negation follow then from the properties of implication, for example the rule of inference *modus tollendo tollens*. By adding the principle of explosion as an additional axiom to the graph of $\bot$ we achieve a full-fledged negation. In particular, this process applied to the existential graphs for (classical) implicational propositional calculus leads to an alternative system of existential graphs Alpha for classical logic.

References

1. Caicedo, X.: Elementos de lógica y calculabilidad. Una empresa docente, Bogotá (1990)
2. Calderón, F.J., Calderón, M.Y.: Gráficos Alfa para la lógica implicativa. Undergraduate thesis, Universidad del Tolima, Ibagué (Colombia) (2022)
3. Castillo, M., Oostra, A.: Álgebras para la lógica implicativa con conjunción. Matemáticas: Enseñanza Universitaria **18**(2), 31–50 (2010)
4. Diego, A.: Sobre álgebras de Hilbert. Ph.D. thesis, Universidad de Buenos Aires, Buenos Aires (1961)
5. Gómez, A.Y.: Gráficos Alfa para la lógica implicativa con conjunción. Undergraduate thesis, Universidad del Tolima, Ibagué (Colombia) (2013)
6. Henkin, L.: An algebraic characterization of quantifiers. Fundam. Math. **37**, 63–74 (1950)
7. Hilbert, D., Bernays, P.: Grundlagen der Mathematik, vol. I. Springer, Berlin (1934)
8. Lukasiewicz, J.: The shortest axiom of the implicational calculus of propositions. Proc. R. Irish Acad. **52**, 25–33 (1948)
9. Mendelson, E.: Introduction to Mathematical Logic, 6th edn. CRC Press, Boca Raton (2015)
10. Oostra, A.: Los gráficos Alfa de Peirce aplicados a la lógica intuicionista. Cuadernos de Sistemática Peirceana **2**, 25–60 (2010)
11. Oostra, A.: Gráficos existenciales Beta intuicionistas. Cuadernos de Sistemática Peirceana **3**, 53–78 (2011)
12. Oostra, A.: Representación compleja de los gráficos Alfa para la lógica implicativa con conjunción. Boletín de Matemáticas **26**(1), 31–50 (2019)
13. Oostra, A.: Equivalence proof for intuitionistic existential alpha graphs. In: Basu, A., Stapleton, G., Linker, S., Legg, C., Manalo, E., Viana, P. (eds.) Diagrams 2021. LNCS (LNAI), vol. 12909, pp. 188–195. Springer, Cham (2021). https://doi.org/10.1007/978-3-030-86062-2_16
14. Oostra, A.: Intuitionistic and geometrical extensions of Peirce's existential graphs. In: Zalamea, F. (ed.) Advances in Peircean Mathematics. The Colombian School, Peirceana, vol. 7, pp. 103–178. De Gruyter, Berlin/Boston (2022)
15. Oostra, A.: Existential graphs as a visual tool of abductive cognition in intuitionistic logic and various sublogics. In: Magnani, L. (ed.) Handbook of Abductive Cognition, pp. 647–668. Springer, Cham (2023). https://doi.org/10.1007/978-3-031-10135-9_39
16. Peirce, C.S.: Collected Papers of Charles Sanders Peirce, 8 vols. Harvard University Press, Cambridge (1931–1958)
17. Peirce, C.S.: Logic of the future. In: Pietarinen, A.V. (ed.) Logic of the Future, Peirceana, vol. 1–3. De Gruyter, Berlin/Boston (2020–2024)
18. Pietarinen, A.V.: On the supreme beauty of logical graphs. Cuadernos de Sistemática Peirceana **8**, 5–40 (2016)
19. Rasiowa, H.: An Algebraic Approach to Non-classical Logics. North-Holland, Amsterdam (1974)
20. Roberts, D.D.: The Existential Graphs of Charles S. Peirce. Mouton, The Hague (1973)
21. Zalamea, F.: Peirce's Logic of Continuity. A Conceptual and Mathematical Approach. Docent Press, Boston (2012)
22. Zeman, J.J.: The Graphical Logic of C. S. Peirce. Ph.D. dissertation, University of Chicago, Chicago (1964)

Aristotelian Diagrams as Logic Diagrams

Stef Frijters[✉] and Atahan Erbas

Center for Logic and Philosophy of Science, KU Leuven, Leuven, Belgium
{stef.frijters,atahan.erbas}@kuleuven.be

Abstract. Recently, diagrams are more seen as logical systems. Curiously, Aristotelian diagrams, such as the square of opposition, have mostly been excluded from this. In this paper we challenge this prejudice and show that Aristotelian diagrams can be considered as proper logic diagrams that can be used to draw inferences. We do so by describing a logical system of Aristotelian diagrams: we provide a diagrammatic vocabulary and syntax, a set of axioms and (transformation) rules, and a formal semantics. We show how this diagrammatic logic system can be used to make inferences and argue that the crucial step here is to not only consider finished diagrams, but also unfinished diagrams; it is in the process of (re)constructing the diagram that the reasoning steps are made. We finish the paper by commenting on the soundness and completeness of the system.

Keywords: Logical Geometry · Square of Opposition · Aristotelian Diagrams · Logic Diagrams

1 Introduction

Aristotelian diagrams such as the square of opposition have been drawn since at least the second century CE [1] and have been used ever since in a variety of disciplines [4,12]. However, until now Aristotelian diagrams have not been taken seriously as logic diagrams, and are often easily dismissed as 'merely illustrative'. In their influential overview article, Moktefi and Shin define a logic diagram (which they also call a diagrammatic logic system) as "a logical system (i) which has a list of transformation rules, (ii) which has a formal semantics, but (iii) whose vocabulary is diagrammatic" [8, p. 612]. They state that Aristotelian diagrams are not logic diagrams in this sense, but merely "illustrative diagrams" used as "visual aids in order to make some immediate inferences" [8, p. 614].

In this paper we show that Aristotelian diagrams *are* logic diagrams. We do so by constructing a logical system based on Aristotelian diagrams that fulfills all three of Moktefi and Shin's requirements: it has (i) a list of transformation rules, (ii) a formal semantics, and (iii) a diagrammatic vocabulary. After some preliminaries about logic diagrams and Aristotelian diagrams in Sect. 2, we present the vocabulary and syntax of the logical system in Sect. 3. This vocabulary is diagrammatic, thus showing that the logical system fulfills requirement (iii).

In Sect. 4 we give an axiomatization consisting of axioms and transformation rules, thus showing that we satisfy requirement (i). Section 5 presents a formal semantics for the system, showing that it satisfies requirement (ii). In Sect. 6 we show how the system (and thus Aristotelian diagrams in general) can be used to draw inferences. We comment on soundness and completeness in Sect. 7. Finally, Sect. 8 contains some concluding thoughts and a short discussion of possible future research paths.

2 Preliminaries

Aristotelian diagrams graphically represent the four Aristotelian relations of contradiction, contrariety, subcontrariety and subalternation holding among a set of formulas or sentences. These four Aristotelian relations are defined in Definition 1. In what follows we use the abbreviations $CD_\mathbf{S}(\varphi,\psi)$, $C_\mathbf{S}(\varphi,\psi)$, $SC_\mathbf{S}(\varphi,\psi)$ and $SA_\mathbf{S}(\varphi,\psi)$. These days the contradiction relation is typically represented by a solid line, contrariety by a dashed line, subcontrariety by a dotted line and subalternation by an arrow. Some examples of Aristotelian diagrams following this convention are given in Fig. 1.

Definition 1. *Let* **S** *be a logical system based on a language* $\mathcal{L}_\mathbf{S}$, *which is assumed to have Boolean operators and a model-theoretic semantics* $\models_\mathbf{S}$. *The formulas* $\varphi, \psi \in \mathcal{L}_\mathbf{S}$ *are*

S-*contradictory*	*iff* $\models_\mathbf{S} \neg(\varphi \wedge \psi)$	*and*	$\models_\mathbf{S} \varphi \vee \psi$
S-*contrary*	*iff* $\models_\mathbf{S} \neg(\varphi \wedge \psi)$	*and*	$\not\models_\mathbf{S} \varphi \vee \psi$
S-*subcontrary*	*iff* $\not\models_\mathbf{S} \neg(\varphi \wedge \psi)$	*and*	$\models_\mathbf{S} \varphi \vee \psi$
in **S**-*subalternation*	*iff* $\models_\mathbf{S} \varphi \to \psi$	*and*	$\not\models_\mathbf{S} \psi \to \varphi$

In this paper we are only concerned with diagrams that contain no equivalent formulas, tautologies or contradictions. Furthermore, we demand that for every formula φ, the diagram contains exactly one formula that is equivalent to the negation of φ. Most of the Aristotelian diagrams found in the literature adhere to these constraints, and there are also some theoretical considerations motivating these constraints [4,12].

Even though Aristotelian diagrams represent logical relations, they have generally not been considered logic diagrams. Shin [11, p. 1] complains of a "general prejudice against diagrams", namely the view "that diagrams can be only heuristic tools but not valid proofs". She argues against this prejudice by showing that Venn diagrams can be viewed as a logical system in its own right, with a syntax and a semantics. This shows that Venn diagrams are real logic diagrams, and not merely illustrative diagrams. Since then, this approach has been used to show that other diagrammatic methods also need to be taken seriously as logical systems (see e.g. [6] for a recent example).

However, as was already indicated in the introduction, even if one is sympathetic to the general idea of viewing diagrams as logical systems, remainders of the "prejudice against diagrams" reported by Shin still continue to surround particular cases, in this case Aristotelian diagrams. For example, Moktefi and

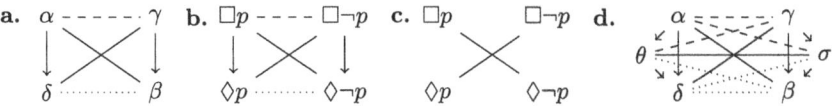

Fig. 1. (a) the traditional square of opposition (with formula variables), (b) one instantiation for the modal logic **S5**, (c) a degenerate square for the modal logic **K**, and (d) a Sherwood-Czeżowski hexagon (with formula variables).

Shin claim that, in contrast to e.g. Venn diagrams, Aristotelian diagrams are *not* logic diagrams, but merely "illustrative diagrams" and "visual aids" [8, p. 614]. And they are by no means alone in this prejudice. For example, already in 1967 MacQueen claimed that the square of opposition "is not a true logic diagram" [7, p. 105]. More recently, Lemanski mentions in passing that three texts "do not contain analytical logic diagrams in the strict sense, but rather some squares of oppositions" [5, p. 9], thus confirming that he does not consider Aristotelian diagrams to be logic diagrams. In what follows, we challenge this remaining prejudice by presenting a diagrammatic logic system for Aristotelian diagrams that satisfies the three requirements set out by Moktefi and Shin in [8]: it has (i) a list of transformation rules, (ii) a formal semantics, and (iii) a diagrammatic vocabulary.

3 Vocabulary and Syntax

In this section we focus on fulfilling requirement (iii). We define a diagrammatic vocabulary, as well as a syntax. For now, we forget the meaning of the diagrams set out in Sect. 2, and look at the diagrams as purely syntactical objects. In Definition 2 we give the vocabulary and in Definition 3 we give the syntactical rules that determine whether a diagram is well-formed. By Definition 2 the wffs of the system are diagrams, thus the vocabulary is diagrammatic. Hence, requirement (iii) has been fulfilled.

Definition 2 (Vocabulary). *The vocabulary consists of (1) a set of formula-constants $\mathcal{C}$, (2) solid, dotted and dashed lines, and (3) arrows.*

We use the formula-variables $\alpha, \beta, \gamma, \ldots$ as meta-variables ranging over $\mathcal{C}$, i.e. to refer to the formula-constants in the vocabulary.

Definition 3 (Syntax). *A well-formed Aristotelian diagram, WFAD, is defined as follows:*

1. *Every non-zero, but finite number of formula-constants written on a page is a WFAD.*
2. *If D is a WFAD containing the formula-constants α and β, then the diagram obtained by connecting α and β with a line (solid, dotted, dashed) or an arrow is also a WFAD.*
3. *Only diagrams obtained in this way are WFAD's.*

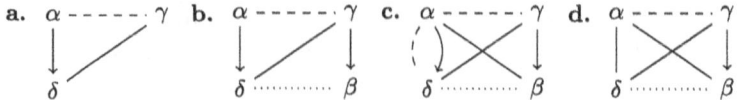

Fig. 2. Four well-formed Aristotelian diagrams contradicting the axioms

Note that according to these definitions, the diagrams in Figs. 1, 2, and 3 are all well-formed Aristotelian diagrams. A diagram that is not a WFAD would for example be a Venn diagram, a diagram with a dashed arrow, or a diagram where one of the lines does not end at a formula-constant. It might strike some readers as odd that the diagrams in Figs. 2 and 3 are well-formed. We will comment on this at the end of the next section.

4 Axiomatization

In this section, we focus on fulfilling requirement (i). We define transformation rules, which are in effect introduction rules, in Definition 5, as well as axioms, that deal with consistency, (Definition 4), structural rules, among which is an elimination rule, (Definition 6) and a consequence relation (Definition 7). The transformation rules are based on earlier work, in particular: the properties of the relations in AD-frames from [4], the theorems in Section. 3.3 of [12], the Aristotelian subdiagrams in [13], and Proposition 19 in [9].

Definition 4 (Axioms). *Within a given WFAD:*

1. *For every α, there is no solid, dashed or dotted line between α and α, and no arrow from α to α.*
2. *For any α and β there is at most one line or arrow between them.*
3. *For any α, β and γ, if $\alpha \neq \beta$ and there is a solid line between α and γ, then there is no solid line between β and γ.*
4. *For every α, there is at least one β such that there is a solid line between α and β.*

Definition 5 (Transformation rules). *Apply the transformation rules only if the consequent hasn't been fulfilled. For every α, β, γ and δ in a WFAD:*

- $-I$ *If there is a solid line between α and β, and an arrow from γ to β, then draw a dashed line between α and γ.*
- $\bullet I$ *If there is a solid line between α and β, a solid line between γ and δ and a dashed line between α and γ, then draw a dotted line between δ and β.*
- $\rightarrow I$ *If there is a solid line between α and β, and a dotted line between δ and β, then draw an arrow from α to δ.*
- Tr. *If there is an arrow from α to β and an arrow from β to γ, draw an arrow from α to γ.*

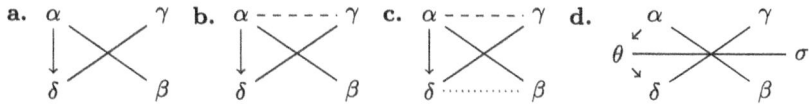

Fig. 3. Four well-formed but unfinished diagrams.

Definition 6 (Structural Rules).

EFSQ If any of the axioms does not hold, then draw any new WFAD.
Weak Remove any dashed line, dotted line, or arrow.

Definition 7 (Consequence Relation). *A Diagram D follows from a diagram D', written as $D' \vdash D$, iff D can be obtained by applying a finite number of transformation and/or structural rules to D'.*

Figure 3 illustrates the transformation rules. We can apply −I (dashed line introduction) to diagram (a) to obtain diagram (b), hence Diagram 3(b) follows from Diagram 3(a). Applying a transformation rule to obtain a new diagram is exactly like deriving a new wff from another wff in any other logic. So we can in turn use •I (dotted line introduction) to obtain diagram (c) from (b), and use →I (arrow introduction) to obtain diagram 1(a) from 3(c). Similarly, Diagram 1(d) follows from Diagram 3(d).

As long as it is still possible to apply such transformation rules, we call the diagram *unfinished*. More precisely: a diagram is unfinished iff it is still possible to apply transformation rules, i.e. there is at least one transformation rule whose antecedent is fulfilled while the consequent has not been fulfilled. A diagram is finished iff it is not unfinished.

The idea of working with unfinished diagrams instead of only finished diagrams is crucial for our approach. If one only looks at finished diagrams, then it is indeed tempting to view Aristotelian diagrams as purely illustrative visual aids. However, it is in the process of constructing a diagram that important reasoning steps are made.[1]

At the end of the previous section we mentioned that some readers might feel uneasy about classifying not only the diagrams in Fig. 1, but also those in Figs. 2 and 3 as well-formed diagrams. This unease can now be explained. Firstly, all of the diagrams in Fig. 2 contradict at least one of the axioms, just like in classical propositional logic, CPL, $p \land \neg p$ is well-formed, but contradicts the axioms of CPL. This should explain the unease about these diagrams. Secondly, the diagrams in Fig. 3 are unfinished diagrams, while those in Fig. 1 are finished diagrams. Thus (of the diagrams in this paper) only the diagrams in Fig. 1 are well-formed, finished, and satisfy all of the axioms of Definition 4.

---

[1] This might explain why in logic courses students are asked to (re)construct the diagrams, instead of simply being given the diagrams to read off.

5 Semantics

In this section we focus on fulfilling requirement (ii). We provide a formal semantics for our logical system. This is mostly analogous to the formal semantics that one is used to. Definition 8 defines models on which WFAD's are interpreted, Definition 9 provides the semantic clauses, and Definition 10 gives the semantic consequence relation. Just as in any other logic, a diagram can be satisfied by multiple models, and one model can satisfy multiple diagrams.

Note however that one of the elements of a model is an *underlying logic* $\mathbf{S}$. Such an underlying logic can for example be CPL, first-order logic, or a normal modal logic. This underlying logic should not be confused with the diagrammatic logic system developed in this paper. In most practical applications of Aristotelian diagrams, the diagram is interpreted on a model $\langle \mathcal{F}, \mathbf{S}, I \rangle$ such that the formula-constants used in the diagram will be the elements of $\mathcal{F} \subseteq \mathcal{L}^S$. In that case $\mathcal{C}$ is a subset of the wff's of $\mathbf{S}$ and $I(\varphi) = \varphi$. However, to avoid confusion we have clearly distinguished the two languages in this paper (cf. also [4], where a similar distinction is made).

Definition 8 (Models). *A model is a triple $\langle \mathcal{F}, \mathbf{S}, I \rangle$ such that $\mathbf{S}$ is a logical system that has a language $\mathcal{L}_\mathbf{S}$, Boolean operators and a model-theoretic semantics $\models_\mathbf{S}$, $I : \mathcal{C} \to \mathcal{L}_\mathbf{S}$ is a bijection, and $\mathcal{F} \subseteq \mathcal{L}_\mathbf{S}$ is such that:*

1. *for every $\varphi \in \mathcal{F}$ there is a $\psi \in \mathcal{F} \setminus \{\varphi\}$ such that $\models_\mathbf{S} \psi \leftrightarrow \neg\varphi$,*
2. *for every $\varphi \in \mathcal{F}$ there is no $\psi \in \mathcal{F} \setminus \{\varphi\}$ such that $\models_\mathbf{S} \varphi \leftrightarrow \psi$,*
3. *there is no $\varphi \in \mathcal{F}$ such that $\models_\mathbf{S} \varphi$ or $\models_\mathbf{S} \neg\varphi$.*

Definition 9 (Semantic Clauses). *Let $M = \langle \mathcal{F}, \mathbf{S}, I \rangle$ be a model, then M satisfies an Aristotelian diagram D iff all of the following conditions are met:*

1. *For every α occurring in D, $I(\alpha) \in \mathcal{F}$.*
2. *For every α and β occurring in D such that α and β are connected by a solid line, $CD_\mathbf{S}(I(\alpha), I(\beta))$.*
3. *For every α and β occurring in D such that α and β are connected by a dashed line, $C_\mathbf{S}(I(\alpha), I(\beta))$.*
4. *For every α and β occurring in D such that α and β are connected by a dotted line, $SC_\mathbf{S}(I(\alpha), I(\beta))$.*
5. *For every α and β occurring in D such that there is an arrow from α to β, $SA_\mathbf{S}(I(\alpha), I(\beta))$.*

Definition 10 (Semantic Consequence). *D is a semantic consequence of D', written as $D' \models D$, iff every model that satisfies D' also satisfies D.*

6 Reasoning with Aristotelian Diagrams

One of the canonical uses of logic is to facilitate and evaluate reasoning. The logical system presented in this paper can be used for this as well. Suppose that we know that for some logical system S, $SA_\mathbf{S}(\alpha, \delta)$, $CD_\mathbf{S}(\alpha, \beta)$, and $CD_\mathbf{S}(\delta, \gamma)$.

We formalise this symbolic information as the well-formed Aristotelian diagram in Fig. 3(a), which will act as a premise in our diagrammatic reasoning system. By applying transformation rules $-I$, $\bullet I$ and $\to I$, we obtain as a conclusion the diagram in Fig. 1(a). Finally, we can return from the diagrammatic to the symbolic realm, for example by reading off from Fig. 1(a) that $SA_\mathbf{S}(\gamma, \beta)$. (And thus one can also conclude e.g. that $\models_\mathbf{S} \gamma \to \beta$.)

One big advantage of this approach is that one does not need to understand all the ins and outs of the underlying logic $\mathbf{S}$ in order to make the derivation. Thus, Aristotelian diagrams are especially useful for very complex underlying logics, or for people not yet familiar with the underlying logic. These include students in an introductory logic course, but also researchers who encounter a logical system for the first time – because the logic is new altogether, or because it belongs to another research tradition, e.g. consider a philosophical logician reading for the first time about the AI formalism of Sugeno integrals [3].

7 Soundness and Completeness

For reasons of space, we cannot provide detailed soundness and completeness proofs in this paper. However, we comment on both in this section. To prove soundness, it suffices to check for every axiom and rule that they hold in every model. As an example, consider part of Axiom 1: there is no solid line between α and α. Towards a contradiction, suppose there is such a line in a diagram D satisfied by a model $M = \langle \mathcal{F}, \mathbf{S}, I \rangle$. By the semantic clauses, $CD_\mathbf{S}(I(\alpha), I(\alpha))$, which contradicts that the operators of $\mathbf{S}$ are Boolean (Definitions 1 and 8).

Proving (strong) completeness would take up even more space than proving soundness. However, results in [4] (especially Theorem 2) suggest a promising proof strategy. In [4] a number of structural properties of Aristotelian diagrams are given, and it is proven that all other structural properties follow from those. Furthermore, the properties in [4] can all be derived from the axioms and transformation rules in this paper. Thus, it is reasonable to suspect that the axiomatization provided in this paper is complete, but this still remains to be proven.

8 Concluding Thoughts

We have shown that there is a persistent prejudice against Aristotelian diagrams even among those who are generally favorable about diagrammatic reasoning. To argue against this prejudice, we constructed a diagrammatic logic system for Aristotelian diagrams that fulfills all three requirements set out by Moktefi and Shin: it has (i) a list of transformation rules (Sect. 4), (ii) a formal semantics (Sect. 5), and (iii) a diagrammatic vocabulary (Sect. 3). In addition, we illustrated that this system can be used to draw inferences (Sect. 6), and we commented on the soundness and completeness (Sect. 7).

This study opens up multiple possibilities for future research. We have only looked at diagrammatic arguments with a single premise. We can also look at deriving a diagram from combining the information in two or more diagrams, or we might investigate the rules for adding new formulas to a diagram.

There are also connections with other diagrams that are worth exploring. For example, one can investigate whether the Diagrammatic system introduced here falls within the general framework of Single Feature Indicator Systems introduced in [10]. One could also look at the relation with Euler diagrams. In [2] it is shown that there is a close relation between Euler diagrams and Aristotelian diagrams. Is there e.g. also a relation between diagrammatic logic systems for Euler diagrams and the system in this paper?

Finally, it is worth pointing out that we have focussed on the description of logic diagrams given by Moktefi and Shin in [8]. However, the definition of logic diagrams differs slightly amongst different authors. For example, in [11] Shin seems to require that the diagrammatic logic system is complete as well. Exploring these other definitions for Aristotelian diagrams is left for future work.

Acknowledgments. The authors would like to thank Lorenz Demey and Hans Smessaert for valuable comments on earlier versions of this paper. This research was funded by the European Union (ERC, STARTDIALOG, 101040049). Views and opinions expressed are however those of the author(s) only and do not necessarily reflect those of the European Union or the European Research Council Executive Agency. Neither the European Union nor the granting authority can be held responsible for them.

References

1. Correia, M.: Boethius on the square of opposition. Around and beyond the square of opposition, pp. 41–52 (2012)
2. Demey, L., Smessaert, H.: From Euler diagrams to Aristotelian diagrams. In: Giardino, V., Linker, S., Burns, R., Bellucci, F., Boucheix, J.M., Viana, P. (eds.) Diagrams 2022. LNCS, vol. 13462, pp. 279–295. Springer, Cham (2022). https://doi.org/10.1007/978-3-031-15146-0_24
3. Dubois, D., Prade, H., Rico, A.: The cube of opposition and the complete appraisal of situations by means of Sugeno integrals. In: Esposito, F., Pivert, O., Hacid, M.-S., Raś, Z.W., Ferilli, S. (eds.) ISMIS 2015. LNCS (LNAI), vol. 9384, pp. 197–207. Springer, Cham (2015). https://doi.org/10.1007/978-3-319-25252-0_21
4. Frijters, S., Demey, L.: The modal logic of Aristotelian diagrams. Axioms **12**(5), 471 (2023)
5. Lemanski, J.: Logic diagrams in the Weigel and Weise circles. Hist. Philos. Logic **39**(1), 3–28 (2018)
6. Lemanski, J., Jansen, L.: Calculus CL as a formal system. In: Pietarinen, A.-V., Chapman, P., Bosveld-de Smet, L., Giardino, V., Corter, J., Linker, S. (eds.) Diagrams 2020. LNCS (LNAI), vol. 12169, pp. 445–460. Springer, Cham (2020). https://doi.org/10.1007/978-3-030-54249-8_35
7. MacQueen, G.W.: The logic diagram. Master's thesis, McMaster University (1967)
8. Moktefi, A., Shin, S.J.: A history of logic diagrams. In: Handbook of the History of Logic, vol. 11, pp. 611–682. Elsevier (2012)
9. Schang, F.: Logic in opposition. Studia Humana **2**, 31–45 (2013)

10. Shimojima, A., Barker-Plummer, D.: A generic approach to diagrammatic representation: the case of single feature indicator systems. In: Jamnik, M., Uesaka, Y., Elzer Schwartz, S. (eds.) Diagrams 2016. LNCS (LNAI), vol. 9781, pp. 83–97. Springer, Cham (2016). https://doi.org/10.1007/978-3-319-42333-3_7
11. Shin, S.J.: The Logical Status of Diagrams. Cambridge University Press, Cambridge (1994)
12. Smessaert, H., Demey, L.: Logical geometries and information in the square of oppositions. J. Logic Lang. Inform. **23**, 527–565 (2014)
13. Smessaert, H., Shimojima, A., Demey, L.: Free rides in logical space diagrams versus Aristotelian diagrams. In: Pietarinen, A.-V., Chapman, P., Bosveld-de Smet, L., Giardino, V., Corter, J., Linker, S. (eds.) Diagrams 2020. LNCS (LNAI), vol. 12169, pp. 419–435. Springer, Cham (2020). https://doi.org/10.1007/978-3-030-54249-8_33

Open Access This chapter is licensed under the terms of the Creative Commons Attribution 4.0 International License (http://creativecommons.org/licenses/by/4.0/), which permits use, sharing, adaptation, distribution and reproduction in any medium or format, as long as you give appropriate credit to the original author(s) and the source, provide a link to the Creative Commons license and indicate if changes were made.

The images or other third party material in this chapter are included in the chapter's Creative Commons license, unless indicated otherwise in a credit line to the material. If material is not included in the chapter's Creative Commons license and your intended use is not permitted by statutory regulation or exceeds the permitted use, you will need to obtain permission directly from the copyright holder.

Sentence Negation and Term Negation as Syntactic Operations in Diagram Logic

Sohail Hossain[✉] and Mihir Kumar Chakrobarty

School of Cognitive Science, Jadavpur University, Kolkata 700032, India
hsohail.t96@gmail.com, mihirc4@gmail.com

Abstract. A formal treatment of negation operation in diagram logic is presented. Syntactic definitions of sentence negation and term negation are given in the Venn$_i$ diagram system. The difference between sentence negation and term negation is established in the language of diagrams. The syntactic way of negating a well-formed diagram does not increase the expressiveness power of the logical system. However, it can be very useful for quickly constructing diagrams that are in opposition relation to a given diagram.

Keywords: Diagram Logic · Sentence Negation · Term Negation

1 Introduction

The study of diagram logic may be traced back to Euler (1707–1783), Venn (1834–1923), and Peirce (1839–1914) along with others. Based on the work of Venn and Pierce diagrammatic logical systems with proper syntax and semantics have been developed by Shin [1] and Hammer [2]. Their works have been extended by incorporating names of individuals in [3]. Parallelly, spider diagrams are introduced with existential and constant spiders in [4,5]. Based on [3] a non-classical diagrammatic system Venn$_i$ has been developed in [6].

The notion of negation in the spider diagram system has been introduced in [8]. In [3] the absence of an individual was introduced in diagram logic which is similar (though not equivalent) to the negation of a singular proposition in predicate logic [7]. Stepleton and Mastoff also talked about the negation operation in a case study on Euler diagrams in [9]. One of the drawbacks of the Euler diagram system lies in solving some logic problems involving negative terms. A historical and explanatory survey on various techniques used by subsequent researchers to eliminate this difficulty is mentioned in a recent paper [10].

In the present work, a syntactic way of negating a diagram in Venn$_i$ is presented [3]. We shall talk about two different types of negation operations, sentence-negation, and term-negation in the system Venn$_i$. Section 2 mentions previous works on negation in diagram logic. In Sect. 3, a brief account of the system Venn$_i$ is given. In Sect. 4, a result similar to the simplification rule in natural deduction is formulated, and using the results distribution of disjunction and conjunction over each other is shown. A new operation called separation operation is defined which is the inverse of the unification operation in Venn$_i$. By applying the separation operation on a diagram having n different diagrammatic objects, a set of n diagrams having exactly one diagrammatic object in respective counterpart regions in each diagram is obtained. Then the rules of sentence negation operation are defined by the manipulation of diagrammatic objects. Semantic justification of the sentence negation operation is established by proving some theorems. Using the sentence negation operation, De Morgan's laws and the double negation Law are proved. In Sect. 5, the rules of term-negation operation in Venn$_i$ are formulated. The double negation law for term negation operation and the commutativity relationship between sentence negation and term negation operation are established. Finally, in Sect. 6, the difference between the sentence negation and the term negation is shown through a modified cube of opposition. Section 7 contains some concluding remarks.

2 Previous Works on Negation in Diagram Logic

In [9] Gem Stapleton and Judith Masthoff extended the Euler diagram with notations to include the '¬' operator as well as '∧' and '∨'. In the Stapleton-Masthoff Euler system, the negation of a unitary diagram D is syntactically represented by putting a bar above D, as $\bar{D}$ and the negation of a compound diagram is defined using De Morgan's Law (Rule 9 in [9]). The definition of the negation of a diagram in this system [9] mimics its definition in propositional logic. In [8] a constructive way of negating a spider diagram is mentioned. Despite not possessing the notion of negation in Venn-II [1] and Venn$_i$ [3], they are as expressive as the language of monadic predicate logic. As the authors mentioned in [9], "Even when expressiveness is not increased, it is likely that excluding '¬' and '∧' will impact usability because to make a negated sentence without using '¬', one has to find an equivalent non-negated statement." Similar to the construction of the negation of a spider diagram in [8], our aim here is to define the negation operation in Venn$_i$ more rigorously. As mentioned in the introduction various treatments for solving logic problems with negative terms are given by many scholars of diagram logic [10]. However, no syntactic way of obtaining the term negation of a diagram with several closed curves and diagrammatic objects is mentioned explicitly in any previous works. The language of diagrams has special characteristics; here we are to explore them logically and witness some important results. The proofs of some lemma and theorems are skipped due to the shortage of space.

3 A Brief Account of the System Venn$_i$

Before going into the definition of the negation operations, a brief account of the system Venn$_i$ [3] is given below. In the language, primitive symbols include

1. ☐ : Rectangle; representing the universe,
2. ○ : Closed curve; representing monadic predicate,

3. ▬ : Shading; representing emptiness,
4. x: cross; representing non-emptiness,
5. $a, b, c,...$: Names of individuals,
6. $\bar{a}, \bar{b}, \bar{c},...$: Absence of individuals,
7. $A, B, C,...$: Names of closed curves,
8. —: line connecting cross and rectangle, and
9. - - -: Broken line connecting individuals.

There are six types of diagrammatic objects.

1. shading (▬),
2. cross (x),
3. names of individuals ($a, b, c,...$),
4. names of individuals with bars ($\bar{a}, \bar{b}, \bar{c},...$),
5. sequence of crosses (x's) connected by—[line connecting crosses, in short lcc (x—x—x—) represent inclusive disjunction]
6. sequence of individuals (a's) connected by - - - [line connecting individuals, in short lci (a - - a - - a - -) represent exclusive disjunction]

The region surrounded by a rectangle or a closed curve is called a basic region. The rectangle and the closed curves together divide the space within the rectangle into disjoint spaces, each space is called a minimal region. A region is the union of minimal regions.

Two basic regions enclosed by curves having the same label are in counterparts(cp) relation. If r_1 and r_2 are regions of diagram D and s_1 and s_2 are regions of E such that r_1 and r_2 are counterparts of s_1 and s_2 respectively i.e. $<r_1, s_1>$, $<r_2, s_2> \in cp$, then $r_1 + r_2$, $r_1 \cdot r_2$ and $r_1 \setminus r_2$ are counterparts of $s_1 + s_2$, $s_1 \cdot s_2$ and $s_1 \setminus s_2$ respectively, where $+, \cdot$ and $\setminus$ denotes a union, intersection and relative complement of a region respectively.

There are three types of well-formed diagrams.

1. Type-I
   Diagrams having a single closed curve within the rectangle.
2. Type-II
   Diagrams having more than one closed curve within the rectangle. Each closed curve divides each region obtained by the collection of other closed curves into exactly two minimal regions.
3. Type-III
   Diagrams resulting from connecting more than one Type-I/II diagram by lines. Each Type-I/II diagrams are called a component of the actual diagram.

The rules of transformations are listed below.

1. Introduction Rules (for Closed curves, a, $\bar{a}$, and x).
2. Extension Rules (for Lcc, Lci, and Components).
3. Elimination Rules (for Lcc, Lci, Shading, $\bar{a}$, and Closed curves).
4. Unification Rule.
5. Rule of Splitting Sequence.

6. Rule of Excluded Middle.
7. Rule of Construction.
8. Inconsistency Rule.

A diagram D_2 is a consequence of D_1 (denoted by $D_1 \vdash D_2$) iff D_2 can be obtained from D_1 by using these above rules of transformation. Two diagrams D_1 and D_2 are said to be ρ-equivalent iff $D_1 \vdash D_2$ and $D_2 \vdash D_1$.

A Model of a diagram is defined as a triple (U, I, h), where

1. U is a non-empty set,
2. I is a function from the set of regions of well-formed diagrams to the power set of U, such that
   (a) $I(r) = U$ and $I(-r) = U \setminus I(r)$ if r is a basic region enclosed by a rectangle,
   (b) $I(r) = I(s)$, if r and s are two counterpart regions,
   (c) $I(r + s) = I(r) \cup I(s)$, $I(r \cdot s) = I(r) \cap I(s)$, and $I(r - s) = I(r) \setminus I(s)$ if r and s are two regions of a diagram D,
3. h is a function from the set of names of individuals to U.

Let $M = (U, I, h)$ be a model.

1. A Type-I/II diagram D is True in M (denoted by $M \models D$) iff it satisfies the following conditions.
   (a) $I(r) = \emptyset$ (nullset) if r is shaded.
   (b) $I(r) \neq \emptyset$ if r ($= m_1 + m_2 + \ldots + m_n$) has x-sequence.
   (c) $h(a) \in I(m_i)$ for exactly one i if 'a' is in r ($= m_1 + m_2 + \ldots + m_n$).
   (d) $h(a) \notin r$ i.e. $h(a) \in U \setminus r$ if $\bar{a}$ is in r.
2. A type-III diagram $D = D_1 - D_2 - D_3 - \cdots - D_n$ is True in M iff $M \models D_i$ for at least one component D_i.

Let Δ be a set of diagrams and D be a diagram. We say that D is a semantic consequence of Δ and write $\Delta \models D$ iff for any Model M. if $M \models \Delta$, then $M \models D$. Let S be a set of diagrams, say $\{D_1, \ldots, D_k\}$. For a model M, $M \models S$ iff $M \models D_i$ for all i.

The system Venn$_i$ is shown to be a sound and complete system.

4   Sentence Negation as Operator in Venn$_i$

We define the negation of a diagram as a syntactic rule in the Venn$_i$ system. To obtain a unique negated diagram, a diagram D is considered to be a representative of an equivalence class produced by the ρ-equivalence relation as mentioned in the previous section.

Definition 1 (Unification Operation). *In [6] the Unification Rule is one of the rules from the proof system and it applies to two diagrams. Here for our purpose, the rule is considered as an operation and also modified and extended for any number of diagrams recursively. The unification operation is denoted by Uni and is defined as below.*

1. For only one Type-I/II/III diagram
   $Uni(D) \equiv D$. (Identity)
2. For more than one Type-I/II/III diagram.
   (a) For the Type-I/II diagram,
       i. $Uni(D_1, D_2)$ can be obtained by introducing all closed curves of D_2 to which there was no counterpart in D_1 and introducing all diagrammatic objects of D_2 in D_1 in the respective region obeying the rule of the introduction of a closed curve (Fig. 1).

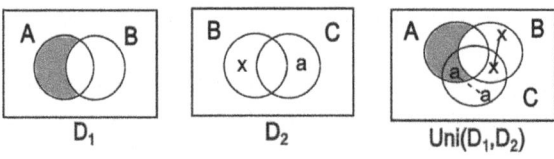

Fig. 1. Example of unification operation.

   ii. $Uni(D_1, \ldots, D_n) \equiv Uni(D_1, Uni(D_2, Uni(D_3, \ldots Uni(D_{n-1}, D_n))))$
   (b) For the Type-III diagram,
       i. $Uni(D_{11} - D_{12}, D_{21} - D_{22}) \equiv Uni(D_{11}, D_{21}) - Uni(D_{11}, D_{22}) - Uni(D_{12}, D_{21}) - Uni(D_{12}, D_{22})$.
       ii. The unification of two diagrams having m_1 and m_2 components respectively can be defined recursively.

A few results that follow from the definition are listed below,

1. $Uni(D, D) \equiv D$. (Idempotency)
2. $Uni(D_1, D_2) \equiv Uni(D_2, D_1)$. (Commutativity)

In propositional logic, we have seen the simplification rule and the distribution of conjunction and the disjunction over each other. Here we shall show similar results.

Lemma 1. *For any two Type-I/II diagrams D_1 and D_2 having the same set of closed curves, $Uni(D_1, D_2) \vdash D_1$.*

Proof. The Elimination Rule (for Lcc, Lci, Shading, and $\bar{a}$) is used in this proof.

Lemma 2. *Let D and D^1 be two Type-I/II diagrams such that $D \vdash D^1$ (Introduction Rule for Closed Curves) then, $D^1 \vdash D$.*
In language, For Any diagram D, if a closed curve C is introduced by the Introduction Rule then, C can be eliminated by the Elimination Rule for a closed curve.

Proof. The Elimination Rule for closed curves is used suitably in this proof.

Lemma 3. *For any two Type-I/II diagrams D_1 and D_2, $Uni(D_1, D_2) \vdash D_1$.*

Proof. Lemma 1 and Lemma 2 are used in this proof.

Theorem 1. *(Distribution of conjunction over disjunction)*
$Uni(D_1, D_2 - D_3) \equiv Uni(D_1, D_2) - Uni(D_1, D_3)$.

Proof. It is a direct implication of the Unification operation.

Theorem 2. *(Distribution of disjunction over conjunction)*
$D_1 - Uni(D_2, D_3) \equiv Uni(D_1 - D_2, D_1 - D_3)$.

Proof. Extension rule for components, Rule of construction, Lemma 3, and Theorem 1 are used in this proof.

4.1 Separation Operation

Definition 2 (Zero-Counter Operation). $Z(D)$ is the diagram obtained by erasing(removing) all the diagrammatic objects from D (Fig. 3).

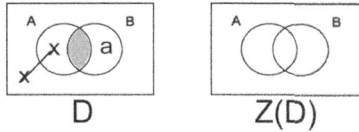

Fig. 2. Example of zero-counter operation.

Definition 3 (Separation Operation). *(for Type-I/II diagrams only)*

1. *Case(I)*
   If D is a Type-I/II diagram having no diagrammatic object, then define $S(D) = \{D\}$.
2. *Case(II)*
   If D be a Type-I/II diagram having k diagrammatic object, say $O_1, O_2 \ldots O_k$, where $k \geq 1$. Define $S(D) = \{D_1, D_2, \ldots, D_k\}$ where each D_i is $Z(D)$ with exactly one diagrammatic object, say O_i in the corresponding counterpart region. This means if we eliminate all diagrammatic objects of the set $\{O_j \mid j \neq i\}$ the obtained diagram would be identical to D_i. Every region of $Z(D)$ has a counterpart region in D, so every region of D_i has a counterpart in D (Fig. 2).

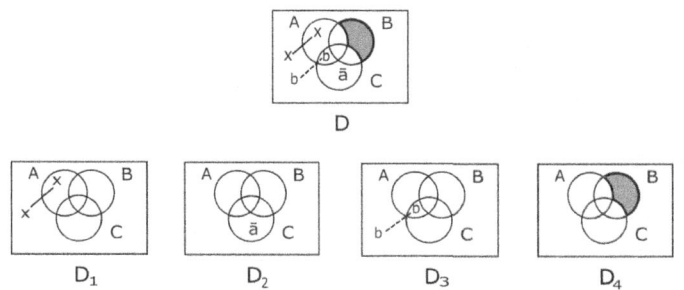

Fig. 3. Example of separation operation.

A few results related to the separation operation are shown below.

Lemma 4. *For any Type-I/II diagram D, $Uni(S(D)) \equiv D$.*

Lemma 5. *For any two Type-I/II diagrams D_1 and D_2, $Uni(D_1, D_2) \equiv Uni(S(D_1), S(D_2))$.*

4.2 The Rules of Sentence Negation Operation

Definition 4. *Sentence negation of a diagram D is denoted by $\neg D$. The Rules of Operation of the sentence negation of a well-formed diagram D are defined below.*

1. *Case(I) (for Type-I/II diagram having no diagrammatic object)*
   Let D be a well-formed Type-I/II diagram having no diagrammatic Object. The rule (R1): shade every counterpart region of D. The obtained diagram is $\neg D$, which is the negation of diagram D (Fig. 4).

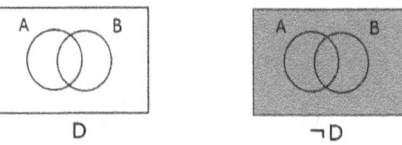

Fig. 4. Example of R1.

2. *Case(II) (for Type-I/II diagram having at least one diagrammatic object)*
   (a) *Subcase(I) (exactly one diagrammatic object in exactly one minimal region)*
   Let D be a well-formed Type-I/II diagram having exactly one type of diagrammatic object in exactly one minimal region, say r. There are four possible cases.
   The rules (R2):
   i. (R2.1) if r is shaded, then $\neg D$ is obtained by inserting a cross in s, the counterpart region of r in $Z(D)$.
   ii. (R2.2) if r includes a cross, then $\neg D$ is obtained by shading s, the counterpart region of r in $Z(D)$ (Fig. 5).

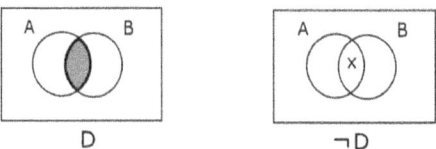

Fig. 5. Example of R2.1.

   iii. (R2.3) if r includes a name of individual (say a), then $\neg D$ is obtained by inserting an '$\bar{a}$' in s, the counterpart region of r in $Z(D)$.
   iv. (R2.4) if r includes a name of individual with a bar (say $\bar{a}$), then $\neg D$ is obtained by inserting an 'a' in s, the counterpart region of r in $Z(D)$.

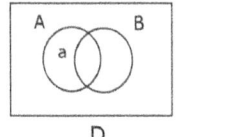
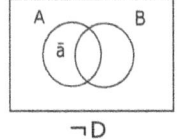

Fig. 6. Example of R2.3.

(b) *Subcase(II) (exactly one diagrammatic object in more than one minimal region)*
Let D be a well-formed Type-I/II diagram having exactly one diagrammatic object in exactly m minimal regions, say $r_1, r_2, \ldots, r_m$, where $m \geq 2$, and $r = \sum r_i$.
The rules (R3):
  i. (R3.1) If r has a x-sequence, then we obtain a Type-I/II diagram $\neg D$ by shading s, the counterpart region of r in $Z(D)$, where $<r, s> \in cp$ (Fig. 7).

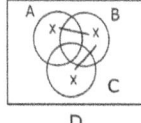
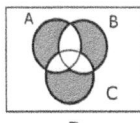
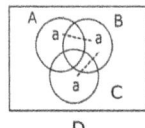
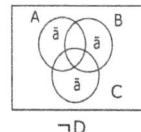

Fig. 7. Example of R3.1.   Fig. 8. Example of R3.2.

  ii. (R3.2) If r has a a − sequence, then we obtain a Type-I/II diagram $\neg D$ by inserting '$\bar{a}$' in s_i, the counterpart region of r in $Z(D)$, where $<r_i, s_i> \in \bar{cp}$ for all i (Fig. 8).

(c) *Subcase(III) (more than one diagrammatic object in one or more than one minimal region)*
Let D be a well-formed Type-I/II diagram having m diagrammatic objects. The rule (R4): first apply the separation function to D, we get a set of m diagrams having exactly one diagrammatic object, $S(D) = \{D_1, D_2, \ldots, D_m\}$ then take the negation of all diagrams of $S(D)$ using rules R2 and R3, we obtain a Type-III diagram $\neg D$ by connecting m diagrams, say $\neg D_1 - \neg D_2 - \cdots - \neg D_m$ (Fig. 9).

3. Case(III) (for Type-III diagram)
Let D be a well-formed Type-III diagram having m components.
$D = D_1 - D_2 - D_3 - \cdots - D_m$, without loss of generality let D_i's are all Type-I/II diagrams.
The Rule (R5): We obtain a Type-I, Type-II, or Type-III diagram $\neg D$ by applying the Unification rule after negating all of its components by using rules R2-R4 i.e. $\neg D = Uni(\neg D_1, \neg D_2, \neg D_3, \ldots, \neg D_m)$ (Fig. 10).

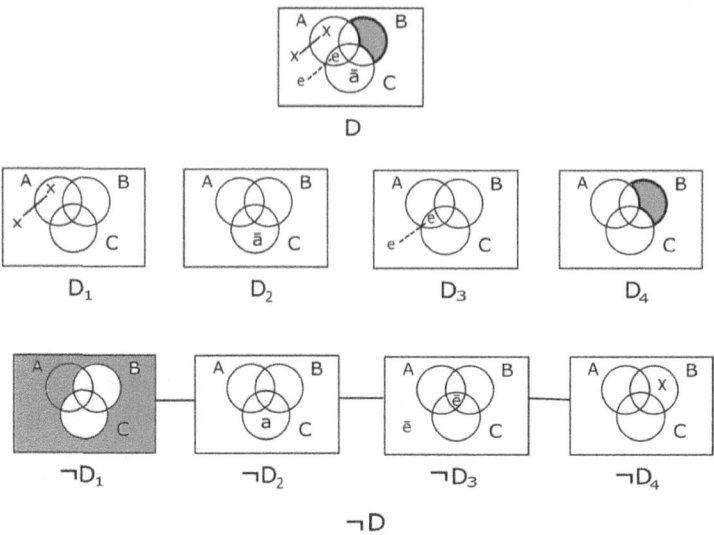

Fig. 9. Example of R4.

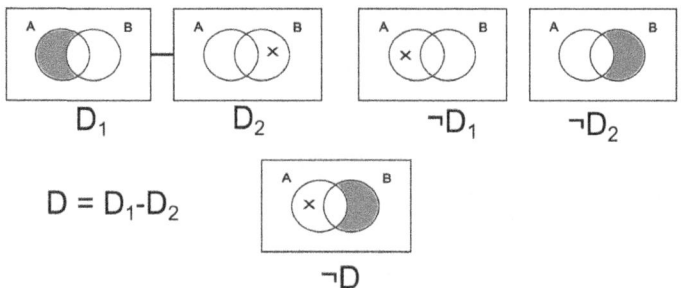

Fig. 10. Example of R5.

4.3 Semantic Justification

The system Venn$_i$ is shown sound and complete [6]. Since the negation operation is syntactically defined, we have to give the semantic justification of the negation operations and other related operations. The results are shown below.

The class of all Type-I/II diagrams can be divided into three categories tautological, consistent, and inconsistent.

Theorem 3. *For any Type-I/II inconsistent diagram D and for any model M, $M \not\models D$ and $M \not\models S(D)$ i.e. $M \not\models D_p$ for some $D_p \in S(D)$.*

Proof. Lemma 4 is used in this proof.

Theorem 4. *For any Type-I/II consistent diagram that is not a tautology D and for any model M, $M \models D$ iff $M \models S(D)$.*

Proof. The principle of mathematical induction on the number of diagrammatic objects in a diagram is used in this proof along with Lemma 4 and Lemma 5.

Theorem 5. *Let $M = (U, I, h)$ be a model and D be a well-formed diagram. $M \models D$ iff $M \not\models \neg D$.*

Proof. The proof is divided into three cases.

1. Case(I) (for Type-I/II diagram having no diagrammatic object).
   A diagram D having no diagrammatic object is a tautology. A tautology is true in every Model M i.e. $M \models D$ (Lemma 3.21 in [6]). As we have defined $\neg D$ is entirely shaded (R1). $\neg D$ has become an inconsistent diagram. An inconsistent diagram is false in every model (Proposition 3.20 in [6]). Hence $M \not\models \neg D$.
2. Case(II) (for Type-I/II diagram having at least one diagrammatic object).

   (a) Subcase(I) (exactly one diagrammatic object in exactly one minimal region).
   Let D be a Type-I/II diagram having exactly one diagrammatic object in exactly one minimal region, say r. Let s be the counterpart region of r in $\neg D$. r have shading or a cross or a name of an individual, say a or a name of an individual with a bar, say $\bar{a}$.
   Now, Let $M \models D$.
   $\iff I(r) = \emptyset$ or $I(r) \neq \emptyset$ or $h(a) \in r$ or $h(a) \notin r$ respectively.
   $\iff I(s) \neq \emptyset$ (R2.1) or $I(s) = \emptyset$ (R2.2) or $h(a) \notin s$ (R2.3) or $h(a) \in s$ (R2.4) respectively.
   $\iff M \not\models \neg D$.

   (b) Subcase(II) (exactly one diagrammatic object in more than one minimal region).
   Let D be a Type-I/II diagram having exactly one diagrammatic object in exactly m minimal region, say r_i for all i. Let s_i be the counterpart region of r_i for all i in $\neg D$. Each r_i has a node of the $x - sequence$ or the $a - sequence$ and no other region of D has a node.
   Let $M \models D$.
   $\iff I(r_i) \neq \emptyset$ for some i or $h(a) \in r_i$ for exactly one i.
   $\iff s_i$ is shaded for all i (R3.1) and $h(a) \notin I(s_i)$ for all i(R3.2)
   $\iff M \not\models \neg D$.

   (c) Subcase(III) (more than one diagrammatic object in more than one minimal region).
      i. Let D be a Type-I/II inconsistent diagram having k diagrammatic object. $S(D) = \{D_1, D_2, \ldots, D_k\}$.
         To show $M \models \neg D$ for any model M.
         $M \not\models D$ and $M \not\models S(D)$ for any model M (Theorem 3).
         $\iff M \not\models D_p$ for some p
         $\iff M \models \neg D_p$ for some p (from Subcase(II)).
         $\iff M \models \neg D_1 - \cdots - \neg D_p - \ldots - \neg D_k$ (Semantics of disjunction).
         $\iff M \models \neg D$.

ii. Let D be a Type-I/II consistent diagram and not a tautology having k diagrammatic object, say $O_1, O_2, \ldots, O_k$, where $k > 1$. $S(D) = \{D_1, D_2, \ldots, D_k\}$. Let M be a model of D.
   Let, $M \models D$.
   $\iff M \models S(D)$ (Theorem 4).
   $\iff M \models D_i$ for all i (Definition of Semantics).
   $\iff M \not\models \neg D_i$ for all i (from Subcase(II)).
   $\iff M \not\models \neg D_1 - \neg D_2 - \cdots - \neg D_k$ (Semantics of disjunction).
   $\iff M \not\models \neg D$. (Definition of Negation Operation).
3. Case(III) (for Type-III diagram).
   Let D be a Type-III diagram having m components, $D = D_1 - D_2 - \cdots - D_m$. We know $\neg D = Uni(\neg D_1, \neg D_2, \ldots, \neg D_m)$ from R5. Let M be a model of D.
   Let, $M \models D$.
   $\iff M \models D_1 - D_2 - \cdots - D_m$ (Substitution).
   $\iff M \models D_q$ for some q.
   $\iff M \not\models \neg D_q$. (Subcase(II))
   $\iff M \not\models Uni(\neg D_1, \ldots, \neg D_q, \ldots, \neg D_m)$ (Soundness of the Uni operation).
   $\iff M \not\models \neg D$ (Definition of Negation Operation).

Hence, $M \models D$ iff $M \not\models \neg D$ □.

De Morgan's laws come along with the definition of negation, disjunction, and conjunction operators. Now it is time to formulate the De Morgan's Law in diagrammatic language.

Theorem 6 (De Morgan's Law 1). *The negation of a disjunction is the conjunction of the negations.* $\neg(D_1 - D_2) = Uni(\neg D_1, \neg D_2)$.

Proof. The proof follows from the definition of the negation operation.

Theorem 7 (De Morgan's Law 2). *The negation of a conjunction is the disjunction of the negations.* $\neg Uni(D_1, D_2) = \neg D_1 - \neg D_2$.

Proof. Lemma 5, Theorem 1, Theorem 2, and Theorem 6 (De Morgan's Law 1) are used in this proof.

Theorem 8 (Double Negation). *For any diagram D, $\neg \neg D \equiv D$.*

Proof. Introduction Rule for lcc, Rule of Extension, Rule of Construction, Elimination Rule of lcc, Lemma 4, and Theorem 7 (De Morgan's Law 2) are used in this proof.

Theorem 9. *For any diagram D, $D - \neg D$ is a tautology.*

Proof. Extension Rule for components, Rule of Excluded Middle, Theorem 2, Lemma 4, and Theorem 6 (De Morgan's Law 1) are used in this proof.

Corollary 1. *For any diagram D, $Uni(D, \neg D)$ is an inconsistent diagram.*

Proof. It is a consequence of Theorem 9.

5 Term Negation as Operation in Venn$_i$

It is well known that in Aristotle's term logic, two kinds of negation are mentioned - predicate denial consists of a predicate being denied of a subject, e.g. 'Stone is not ill', and term negation consists of a negative predicate being affirmed of a subject, e.g. 'Stone is not-ill' [11]. Negative predicates are obtained by applying a term-negation operation (in English usually 'not-', 'non-', 'in-', or 'un-' etc.) before a positive predicate. Denying a predicate is linguistically equivalent to putting 'It is not the case that' before a sentence e.g. 'It is not the case that Stone is ill'. Thus predicate denial is equivalent to sentence negation. Though semantically, 'Stone is not ill' and 'Stone is not-ill' are the same but it is argued in [12] that the latter entails the existence of the subject while the former does not.

We first show the diagrammatic representation of categorical propositions and the relations between them.

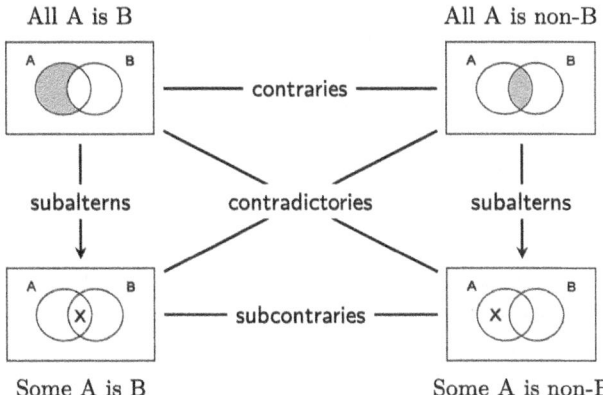

Fig. 11. Traditional square of opposition.

Only the predicate term has been negated in Fig. 11. But one can have the subject term negated also. Thus obtaining the following sentence: All non-A is B, All non-A is non-B, Some non-A is B, and Some non-A is non-B. Pictorial representations of the above sentences are shown in Sect. 6. A special case of 'All non-A is B' is 'All non-a is B', where 'a' is a singular term. In Venn$_i$ 'non-a' is really 'absence of a' and has been denoted by $\bar{a}$ (see Fig. 6) [3].

It may be mentioned here that though in Venn$_i$ which picturized classical monadic predicate logic, '$\bar{a}$ is A' and 'a is non-A' are intentionally equivalent, they are not so in other systems, e.g. Venn$_{in}$ [6]. Even in Venn$_i$, intentionally they are different as mentioned in [12]. The representation of the absence is more useful in minimizing clutter in a diagram having several terms or lines connecting individuals [13]. In this paper, we shall restrict to Venn$_i$ and proceed to a set of rules for term negation. Similar ideas are loosely embedded in [10] but not developed rigorously.

The logical form of the sentence negation and the term negation of a given sentence must be different and here we are to establish this. In this section, the term negation operation in Venn$_i$ is syntactically defined, and the difference between the sentence negation and the term negation is established. Using the term negation operation any term of an arbitrary diagram in Venn$_i$ can be negated.

5.1 The Rules of Term Negation Operation

Definition 5. *Let D be well-formed Type-I/II diagram having k closed curves say $C_1, C_2, \ldots, C_k$. As we know $C_1, C_2, \ldots, C_{k-1}$ divides the rectangle into 2^{k-1} regions say $r_1, r_2, \ldots, r_{2^{k-1}}$. C_k divides every region thus obtained into two minimal regions one is inside C_k and the other is outside C_k. Let D have m diagrammatic objects. Now C_k term negation of the diagram D denoted by $D_{\sim C_k}$ is defined by the following rules.*

1. *(R6.1) If $r_i \cdot C_k$ is shaded in D, then $r_i \setminus C_k$ is shaded in $D_{\sim C_k}$ and vice-versa for all i.*
2. *(R6.2) If $r_i \cdot C_k$ has $\bar{a}$ in D, then $r_i \setminus C_k$ has $\bar{a}$ in $D_{\sim C_k}$ and vice-versa for all i.*
3. *(R6.3) If $r_i \cdot C_k$ has a cross in D, then $r_i \setminus C_k$ has a cross in $D_{\sim C_k}$ and vice-versa for all i. If D has cross connecting lines i.e. $x-$ sequence, then in $D_{\sim C_k}$ the corresponding crosses are also connected and vice-versa.*
4. *(R6.4) If $r_i \cdot C_k$ has a in D, then $r_i \setminus C_k$ has a in $D_{\sim C_k}$ and vice-versa for all i. If D has a's connecting lines i.e. $a-$ sequence, then in $D_{\sim C_k}$ the corresponding a's are also connected and vice-versa.*
5. *(R6.5) For a Type-III diagram $D = D_1 - \cdots - D_m$,*
   $D_{\sim C} = (D_1 - \cdots - D_m)_{\sim C} = D_{1 \sim C} - \cdots - D_{m \sim C}.$

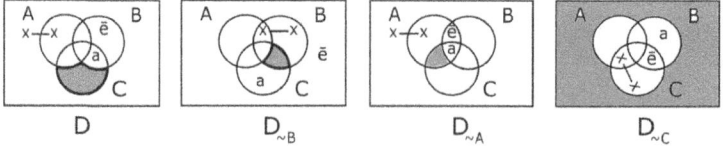

Fig. 12. Example of term-negation operation.

In Fig. 12, the diagram D consists of three closed curves A, B, C, and four diagrammatic objects. The B term negation of D is denoted by $D_{\sim B}$. Using the defined rules, we will explain how we obtain $D_{\sim B}$ from D. At first, we identify the minimal regions where there is at least one diagrammatic object, (i) In D the region $A^c \cap B^c \cap C$ is shaded, from R6.1, in $D_{\sim B}$ the region $A^c \cap B \cap C$ is also shaded. (ii) In D the region $A^c \cap B \cap C^c$ has a $\bar{e}$, from R6.2, in $D_{\sim B}$ the region $A^c \cap B^c \cap C^c$ also has a $\bar{e}$. (iii) In D the regions $A^c \cap B^c \cap C^c$ and $A \cap B^c \cap C^c$ has an $x - sequence$, from R6.3, in $D_{\sim B}$ the regions $A^c \cap B \cap C^c$ and $A \cap B \cap C^c$ also has a $x - sequence$. (iv) In D the region $A^c \cap B \cap C$ has a a, from R6.4, in $D_{\sim B}$ the region $A^c \cap B^c \cap C$ also has a a. Similarly, $D_{\sim A}$ and $D_{\sim C}$ can be obtained.

A few mathematical relations related to term negation operation are shown below.

Lemma 6. *Let D be a diagram having m diagrammatic objects. Let $S(D) = \{D_1, \ldots, D_m\}$. Then, $S(D_{\sim C}) = \{D_{1 \sim C}, \ldots, D_{m \sim C}\}$.*

Proof. The proof only uses the definition of the separation operation and the term negation operation.

Lemma 7. *Let D be a diagram having m diagrammatic objects. Let $S(D) = \{D_1, \ldots, D_m\}$. Then, $(Uni(D_1, \ldots, D_m))_{\sim C} \equiv Uni(D_{1 \sim C}, \ldots, D_{m \sim C})$.*

Proof. Lemma 4 is used in this proof.

Definition 6. *For any diagram D, $D_{\sim \sim C} \equiv (D_{\sim C})_{\sim C}$.*

Theorem 10 (Double Negation). *For any diagram D, $D_{\sim \sim C} \equiv D$.*

Proof. Only the definition of the term negation operation is used in this proof.

Theorem 11. *For any diagram D, $D_{\sim A \sim B} \equiv D_{\sim B \sim A}$.*

Proof. Only the definition of the term negation operation is used in this proof.

Theorem 12. *For any diagram D, $\neg(D_{\sim C}) \equiv (\neg D)_{\sim C}$.*
The operation Sentence-Negation and Term-Negation commute with each other.

Proof. De Morgan's Law 1, De Morgan's Law 2, Lemma 6, and Lemma 7 are used in this proof.

6 Difference Between Sentence Negation and Term Negation and the Cube of Oppositions

Let us consider the diagram D corresponding to the proposition 'All A is B'. Taking all its term negations the diagrams $D_{\sim A}$, $D_{\sim B}$, and $D_{\sim A \sim B}$ indicating 'All non-A is B', 'All A is non-B', and 'All non-A is non-B' respectively are obtained. Again consider the diagram $\neg D$ corresponding to the proposition 'Some A is non-B'. Taking all its term negations the diagrams $\neg D_{\sim A}$, $\neg D_{\sim B}$, and $\neg D_{\sim A \sim B}$ indicating 'Some non-A is non-B', 'Some A is B', and 'Some non-A is B' respectively are obtained. Assuming the non-emptiness of all four terms A, non-A, B, and non-B, a square representing contrary relations consisting of all four universal propositions and a square representing subcontrary relations of all four particular propositions can be achieved. A cube of oppositions can be

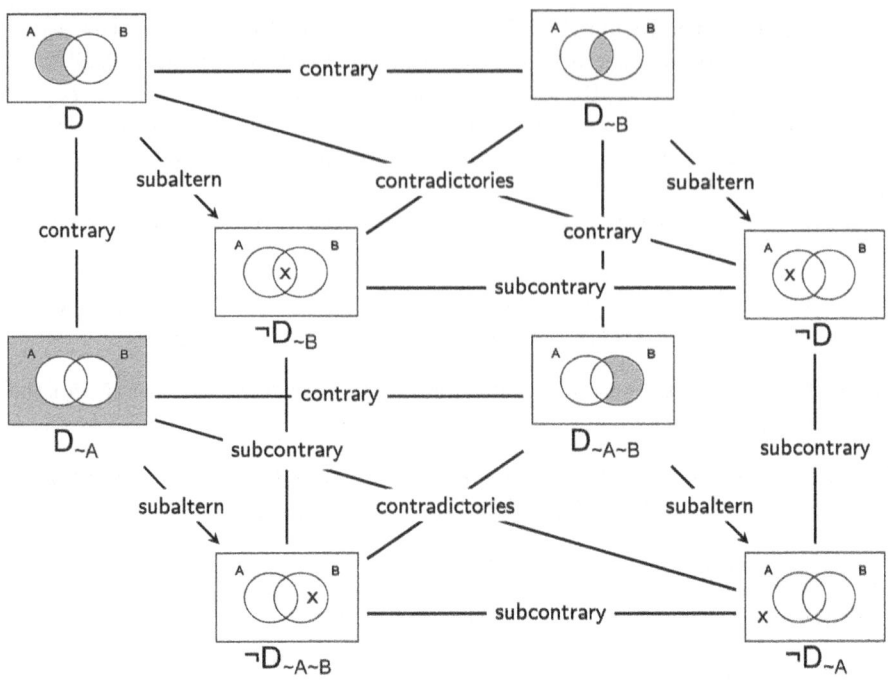

Fig. 13. Cube of opposition.

obtained by connecting these two squares as shown in Fig. 13. The cube shows contradictory, contrary, subcontrary, and subaltern relationships among these eight propositions. The upper face of the cube is the same as the traditional square of oppositions, while the subject term is negated in the lower face. The Keneys-Johnson octagon of opposition is considered to be a natural extension of the square of opposition [14,15]. The octagon as a two-dimensional diagram shows all the respective relationships among the propositions. However, in our case, a three-dimensional extension would be more natural to visualize the term negation relationships. The cube is a collection of squares having different characteristics. The back and front faces of the cube represent the relationship between the universal and particular propositions respectively.

It is very clear from the cube and the definitions of negation operations that the notions of sentence negation and term negation are different. In sentence negation operation the diagrammatic objects are changed but stay in the same minimal regions. In term negation operation the diagrammatic objects remain the same but move to another suitable minimal region. The difference between these two negation operations is thus visually established.

7 Conclusion

The syntactic way of negating a well-formed diagram does not increase the expressiveness of the system as we have already mentioned. But using these operations we can easily formulate the sentence negation and the term negation of a diagram without paying attention to its meaning. The definition of the negation operations also present an algorithm to negate a diagram. It would be very easy for a machine to perform the negation operation. Besides the mathematical relationship that we have found here, there might be more interesting results that could be found in the structure. These will be explored in future work.

References

1. Shin, S.-J.: The Logical Status of Diagram. Cambridge University Press, Cambridge (1994)
2. Hammer, E.: Logic and Visual Information. CSLI Publications (1995)
3. Choudhury, L., Chakraborty, M.K.: On extending Venn diagram by augmenting names of individuals. In: Blackwell, A.F., Marriott, K., Shimojima, A. (eds.) Diagrams 2004. LNCS (LNAI), vol. 2980, pp. 142–146. Springer, Heidelberg (2004). https://doi.org/10.1007/978-3-540-25931-2_14
4. Gil, J., Howse, J., Kent, S.: Formalizing spider diagrams. In: Proceedings of the IEEE Symposium on Visual Languages (VL 1999), Tokyo, pp. 130–137 (1999)
5. Howse, J., Stapleton, G., Taylor, J.: Spider diagrams. LMS J. Comput. Math. **8**, 145–194 (2005)
6. Bhattacharya, R., Chakraborty, M.K., Choudhary, L.: Venn Diagram with names of individuals and their absence: a non-classical diagram logic. Logica Universalia **12**(1–2), 141–206 (2018)
7. Choudhury, L., Chakraborty, M.K.: Singular propositions and their negations in diagrams. In: Proceedings of the First International Workshop on Diagrams, Logic and Cognition. CEUR Workshop Proceedings, vol. 1132, pp. 43–48 (2014)
8. Howse, J., Molina, F., Taylor, J.: On the completeness and expressiveness of spider diagram systems. In: Anderson, M., Cheng, P., Haarslev, V. (eds.) Diagrams 2000. LNCS (LNAI), vol. 1889, pp. 26–41. Springer, Heidelberg (2000). https://doi.org/10.1007/3-540-44590-0_8
9. Stapleton, G., Masthoff, J.: Incorporating negation into visual logics: a case study using Euler diagrams. Vis. Lang. Comput. 187–194 (2007)
10. Bhattacharjee, R., Moktefi, A., Pietarinen, A.V.: The representation of negative terms with Euler diagrams. In: Béziau, J.Y., Desclés, J.P., Moktefi, A., Pascu, A.C. (eds.) Logic in Question, pp. 43–58. Springer, Cham (2022). https://doi.org/10.1007/978-3-030-94452-0_3
11. Horn, L.R.: A natural history of negation (1989)
12. Slater, B.H.: Internal and external negations. Mind **88**(352), 588–591 (1979)
13. Burton, J., Chakraborty, M., Choudhury, L., Stapleton, G.: Minimizing clutter using absence in Venn-i^e. In: Jamnik, M., Uesaka, Y., Elzer Schwartz, S. (eds.) Diagrams 2016. LNCS (LNAI), vol. 9781, pp. 107–122. Springer, Cham (2016). https://doi.org/10.1007/978-3-319-42333-3_9
14. Demey, L., Smessaert, H.: Aristotelian and duality relations beyond the square of opposition. In: Chapman, P., Stapleton, G., Moktefi, A., Perez-Kriz, S., Bellucci, F. (eds.) Diagrams 2018. LNCS (LNAI), vol. 10871, pp. 640–656. Springer, Cham (2018). https://doi.org/10.1007/978-3-319-91376-6_57
15. Moktefi, A., Schang, F.: Another side of categorical propositions: the Keynes-Johnson octagon of oppositions. Hist. Philos. Logic **44**(4), 459–475 (2023)

Playing Games with Diagrams: Truth Diagrams and Game Semantics

Can Başkent[✉]

Department of Computer Science, Middlesex University, London, UK
c.baskent@mdx.ac.uk
https://canbaskent.net/logic

Abstract. In this paper we discuss the connection between truth diagrams and game semantics. Truth diagrams offer a diagrammatic way to represent truth in propositional logic. Game semantics, on the other hand, offers a strategic and game theoretical way to establish the truth value of a given formula. By establishing a relation between the two, we offer another diagrammatic reasoning for game semantics, beyond game trees; and characterise various operations on truth diagrams game theoretically.

Keywords: Truth diagrams · Game Semantics · Propositional Logic · Iterated elimination of strictly dominated strategies

1 Introduction

Logic has always enjoyed a variety of different semantics. Semantic tools have included naive sets, topologies, graphs, diagrams and games. In this paper, we establish a connection between a diagrammatic and a game theoretical semantics for propositional logic. Particularly, for the former we focus on the truth diagrams of Cheng and for the latter the game semantics of Henkin and Hintikka.

Cheng's truth diagrams (TDs, for short) suggest a diagrammatic way to understand truth in propositional logic [7]. Diagrammatic reasoning in logic, as Cheng underlined, is not recent, and can be traced back to Frege and Wittgenstein.

Game semantics (GS, for short), on the other hand, suggests an intuitive and strategic methodology to compute the truth value of a given formula in a given model. In a game semantic model, the truth values of the simplest formulas, that is propositions, are known, yet the truth value of the given formula is *computed* by playing a game. The *semantic verification game* for propositional logic is a two-player, sequential (players take turns to make moves), zero-sum (one player wins, the other loses), pure strategy (players do not roll a die), determined (one player always has a winning strategy) and competitive (players do not cooperate) game with rationality (players make moves and strategise with the purpose of winning). It is played by two players, the Verifier and the Falsifier. Their goal is to win the game and establish the truth value of the given formula. If the

Verifier wins, then the formula is true. If the Falsifier wins, on the other hand, the formula is false. Moreover, the converses of these theorems also hold. True formulas force the Verifier to have a winning strategy, whereas false formulas force the Falsifier to have a winning strategy.

In this paper, we demonstrate the relationship between game semantics and truth diagrams. What we achieve is two-fold: (i) we present a novel domain for truth diagrams, laden with strategic reasoning, (ii) we develop a new research direction for game semantics. As a result, we *gamify* Cheng's truth diagrams. This establishes a straight-forward way to identify TDs with semantic games, opening up further possibilities between different TDs and game semantics for various other logical systems, including non-classical and modal logics.

2 Brief Review of Truth Diagrams and Game Semantics

We start by reviewing the fundamentals of truth diagrams and game semantics.

2.1 Truth Diagrams

Truth diagrams have three elements: (i) *letters* are used to represent countably-many propositional variables $p, q \ldots$, (ii) *nodes* identify the positions around the letters, and (iii) *connectors* link the nodes to express truth or falsity. In classical logic, there are two nodes for each letter: high and low. The high node is for truth and the low one for falsity. Connectors represent the truth values and are colour-coded: black represents truth (T) whereas grey represents falsity (F).

Figure 1 summarises the TDs for the basic logical connectives using the propositional variables p and q: negation, conjunction, disjunction and implication, respectively. The formula $p \rightarrow q$, for example, fails in classical logic when p is true and q is false. Therefore, there is a grey connector, representing F, between the high node of p and the low node of q. And all the other connectors are black as the formula $p \rightarrow q$ is true for all other truth values of p and q.

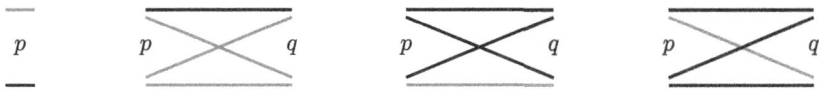

Fig. 1. Truth diagrams for negation, conjunction, disjunction and implication, respectively.

As another example, let us consider the formula $(p \wedge q) \vee p$, given in Fig. 2. The truth table for the aforementioned formula suggests that it is true only when p is true, and false otherwise. The TD reflects this fact diagrammatically.

It is important to note that in a TD, the number of letters and connectors do not depend on the number of subformulas of the given formula, but rather depend on the number of letters (that is propositional variables) appearing in

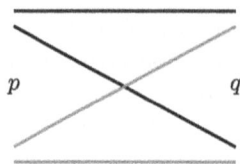

Fig. 2. Truth Diagram for $(p \wedge q) \vee p$ in propositional logic.

the formula. As such, the TD given in Fig. 2 also represents the formula $(p \wedge q) \vee p \vee p \vee p$.

Unlike truth tables, TDs are not necessarily constructed recursively. In truth tables, every subformula needs to be considered by assigning them a column, and the truth values of the given formula depend on the said subformulas. TDs are not recursive, they simply demonstrate the final result of the computation in the truth table by skipping the columns for the subformulas.[1]

We can now define TDs formally, first the elements of TDs, and then their semantics, based on [7].

Definition 1 (Elements of Truth Diagrams). *Elements of truth diagrams for classical logic are given as follows.*

- *A TD is composed of* letters, nodes *and* connectors.
- *Letters are arranged horizontally (with regular spacing for readability).*
- *Nodes are small areas, one above and one below the letters.*
- *Connectors are lines linking nodes. Each connector intersects just one node at each letter and has straight segments that span pairs of immediately adjacent letters.*
- *One connector for each possible combination of high or low nodes of each (type of) letter is permitted: the shape of each connector in a TD is unique.*
- *The style of the connectors is solid and either black or grey.*
- *A TD can contain more than one instance of a letter.*
- *The horizontal order of the letters is arbitrary.*
- *Letters in separate TDs are not linked by connectors.*
- *A connector intersects the nodes at the same level for each instance of the same letter.*

Definition 2 (Semantics for Truth Diagrams). *Semantics of truth diagrams for classical logic is defined as follows.*

- *Letters are propositional variables.*
- *Each node represents a truth-value for the variable: high-node T, low-node F.*
- *The number of the distinct types of variables is the arity of the TD.*

---

[1] This is an important point from an intuitionistic perspective. If truth is proof, and if proofs are computations, and if computations are strategies, then said proofs, computations and strategies are not visible in TDs.

– A *connector* is a *case*: it constitutes a unique set of truth-value assignments to the variables.
– *Connector style* represents the overall truth-value assigned to its case. A black connector assigns T, grey connector assigns F.

TDs reflect the fundamental properties of classical logic directly. Connectors are functional as propositional logic is. They cannot skip a letter, they cannot double a letter. We must have a connector between every possible truth value combinations of letters, and we cannot have more than one connector between each combination, as truth value gaps and gluts, respectively, are not allowed in classical logic. As such, they demonstrate the strengths and limitations of classical propositional logic directly.

2.2 Game Semantics

The semantic verification game is a sequential, zero-sum, pure strategy, determined and competitive game between two players, the Verifier and the Falsifier. The game is played to establish the truth value of a given formula in a given model. The game is played on a model where the truth values of the propositional letters appearing in the formula are known. What is not known is the truth value of the formula in the model. Players take turn and make moves to play the verification game. The goal of the Verifier is to establish the truth of the formula, and the Falsifier aims at establishing the falsity of the formula. We assume that the players are rational, thus forcing the game to an end for their goal.

The rules of the verification game are given based on the form of the given formula. At conjunctions, the Falsifier makes a move and chooses a subformula, and the game continues with the chosen subformula. Similarly, at disjunctions, the Verifier makes a move. The implication $p \to q$ is considered as an abbreviation for $\neg p \lor q$ and treated as such. The negation, however, is tricky. At a negation $\neg \varphi$, the players switch their roles and the game carries on with the new roles at φ.

During the game, the given formula is broken down into subformulas step by step. The game terminates when it reaches the subformulas which cannot be broken further down into subformulas and when the players do not have any further move to make – that is the game terminates when it reaches propositional variables. If the game terminates at a propositional letter which is given true at the model, the Verifier (that is the player with the role of the Verifier) wins. Otherwise, the Falsifier wins. The correctness theorem of game semantics shows that the Verifier has a winning strategy in the verification game *if and only if* the given formula is true in the given model. Similarly, the Falsifier has a winning strategy *if and only if* the formula is false.

Let us now formally describe game semantics for classical logic, following [17]. First, some notation. For a given formula φ, the set $Subf(\varphi)$ denotes the set of subformulas of φ, defined in the usual way. For example, for the formula $(p \land q) \lor p$, formulas q or $p \land q$ belong to $Subf((p \land q) \lor p)$. For a sequence of moves $\sigma = (a_1, \ldots, a_n)$ and a move a, the concatenation of $\sigma + a$ is equal to $(a_1, \ldots, a_n, a)$.

Definition 3. *For a formula φ of propositional logic, and a model M, the semantic game $G(\varphi, M)$ is a two-player, competitive, zero-sum game with perfect information played on the set of histories, defined as follows.*

- *The game has two players: the Verifier and the Falsifier*
- *The set of histories H is defined as $H = \bigcup \{H_\psi : \psi \in Subf(\varphi)\}$ where H_ψ is defined recursively as follows:*
  - *$H_\varphi = \{\varphi\}$,*
  - *If φ is $\neg \psi$, then $H_\psi = \{h + \psi : h \in H_{\neg \psi}\}$,*
  - *If φ is $\psi_1 \vee \psi_2$, then $H_{\psi_i} = \{h + \psi_i : h \in H_{\psi_1 \vee \psi_2}\}$,*
  - *If φ is $\psi_1 \wedge \psi_2$, then $H_{\psi_i} = \{h + \psi_i : h \in H_{\psi_1 \wedge \psi_2}\}$,*
- *Once the game reaches a propositional variable, the game terminates,*
- *If $\varphi = \neg \psi$, then the players switch roles and the game continues with $G(\psi, M)$*
- *If $\varphi = \psi_1 \vee \psi_2$, then the Verifier makes a move and chooses one of ψ_1 or ψ_2, and the game continues with the chosen disjunct $G(\psi_i, M)$*
- *If $\varphi = \psi_1 \wedge \psi_2$, then the Falsifier makes a move and chooses one of ψ_1 or ψ_2, and the game continues with the chosen conjunct $G(\psi_i, M)$*
- *The Verifier wins if the game terminates at a propositional letter p that is true in M, the Falsifier wins if the game terminates at p that is false in M.*

Game trees diagrammatically represent semantic games. Let us consider the formula $(p \wedge q) \vee p$ given in the TD in Fig. 2. Let us further assume that in our model, where we will play the verification game, p is true and q is false. The game tree for this game is given in Fig. 3.

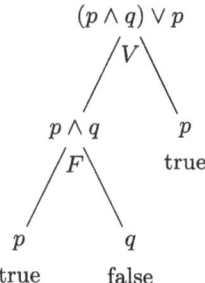

Fig. 3. The game tree for the verification game for the formula $(p \wedge q) \vee p$ where p is true, q is false.

Here, V denotes the player Verifier, and F denotes the player Falsifier. And the truth values of the propositional letters are also specified in the tree. Now, it is easy to see that the Verifier has a winning strategy by choosing p at the very beginning when it is her turn to make a choice. Because otherwise, if the Verifier chooses $p \land q$, then the Falsifier can simply choose a false proposition (that is q) to win the game. This shows that the formula $(p \land q) \lor p$ ends up being true, which can also be verified by a simple truth table.

2.3 A Brief Literature Review

Truth diagrams and game semantics are relatively disjoint research areas.

Diagrams, however, are fundamental elements in game theory in terms of representing epistemic, rational and strategic reasoning, including extensive-form game trees (unlike normal-form games). In GS, extensive-form trees are used widely, as given in Fig. 3.

Earlier work on GS (or dialogical logics) goes back to the 1950s where Lorenzen published his first insights [14]. The work of Paul Lorenzen, later with Lorenz, set the foundations of modern game semantics [15]. Leon Henkin also did foundational work in semantic games [13]. Similarly, an earlier work by Parikh discusses similar ideas (much later published as [19]). The popularity of game semantics, however, is largely due to Hintikka and Helsinki school researchers, including its various extensions [12,17,20]. Recently, a broad variety of logics have been given a game semantical characterisation [2,3,5,10,11].

Another stream of research focuses on game semantics for programming languages and proofs [1]. Arguably, this line of research benefits largely from diagrammatic reasoning due to its wide use of category theory, which is a branch of mathematics that relies heavily on diagrammatic reasoning.

Truth diagrams, on the other hand, have been a recent development by Cheng, published in 2020 [7]. However recent they are, they stand on the shoulders of the earlier work of Frege, Peirce and Wittgenstein. TDs for a variety of non-classical logics were recently given to explore how non-classical logics and TDs interact [4]. Diagrammatic reasoning in mathematics and logic is, most certainly, not new, and can easily be traced back to Euclid. As argued by Macbeth, diagrammatic representation in mathematical disciplines offers a "real extension to our knowledge" [16].

3 From Diagrams to Games, and From Games to Diagrams

What GS achieves is that it starts from the truth values of the propositional variables appearing in the formula, and constructs a way to establish the truth value of the formula in question. Take the simple formula given in Fig. 2. GS suggests that, by playing a verification game, when, for example, p is true and q is false, it is possible to establish the truth value of the given formula. So, this is a bottom-to-top approach. Is it possible to introduce a top-to-bottom approach to game semantics?

By using TDs, it is easy to identify under what conditions a certain player would admit a winning strategy. This becomes easier once we identify the colour of a connector with a player. Black connector is for the Verifier, and the grey connector is for the Falsifier. In short, black connectors identify the winning strategies of the Verifier, grey connectors identify the winning strategies of the Falsifier. This identification ranges over all possible different inputs for the propositional variables.

In this section we describe a natural methodology to identify connectors with strategies. First, we examine the TD to GS direction, then the GS to TD direction.

Given a TD, it is possible to identify what combinations of truth values render the formula true. These combinations are the models where the Verifier has a winning strategy. Therefore, the black connectors identify the verification games where the Verifier has a winning strategy. Similarly, the grey connectors identify the games where the Falsifier has a winning strategy.

Notice that since the TDs do not deal with subformulas (recall the formula in Fig. 2), they do not carry the information of how the players may win the game nor what the winning strategy is for the winning player. Strategic elements of a game tree cannot be retrieved from a TD.

Conversely, given a game tree and a winning strategy for a player, it is possible to identify a connector in the associated TD. For this, one needs to consider truth value distributions of the propositional letters. In those games, where the Verifier has a winning strategy, the given truth value distribution identifies a black connector. Similarly, when the Falsifier has a winning strategy, the distribution identifies a grey connector. Therefore, each different game identifies one connector. In order to construct a complete TD, one needs to consider, at the worst case, all possible 2^n semantic games for possible combinations of truth value distributions over n distinct propositional letters, appearing in the formula.[2]

These observations can easily be extended to multi-valued truth diagrams, where TDs may admit more than two colours for the connectors. This is akin to allowing more than two players in GS, corresponding to multiple truth values in non-classical logics [2].

What is more interesting now is the game theoretical interpretations of certain operations on truth diagrams.

4 Truth Diagram Operations on Game Trees

As van Benthem argued, extensive-form trees of semantic games of logically equivalent formulas are good examples of the "problem of game equivalence" [6]. Given two game trees for the logically equivalent formulas $(p \wedge q) \vee p$ and $(p \vee p) \wedge (q \vee p)$ in Fig. 4, how can we determine if these games are equivalent? In the first game, the Verifer makes the first move and wins. In the second one, however, the Verifier has to wait for the first move by the Falsifier. In each case the Verifier still has a winning strategy, but the strategies are different.

---

[2] Some games can have the same output, therefore may not need to be considered.

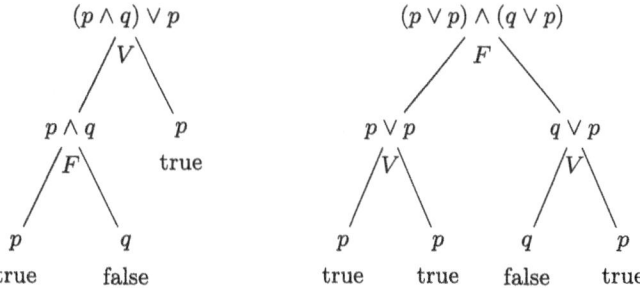

Fig. 4. The game trees for the verification games for the formulas $(p \wedge q) \vee p$ and $(p \vee p) \wedge (q \vee p)$, respectively.

This is a major problem, as it is possible to approach game equivalence from at least two perspectives. The pragmatic approach only focuses on the results, rather than *how* the results are achieved. For pragmatics, the two games above are equivalent. The behaviourist approach, however, considers the way the games are played. For behaviourists, the games are not equivalent as different players take different turns throughout the games and follow different strategies.

We can now ask how TDs can contribute to the understanding of this problem. TDs admit various operations on the diagrams, and by exploring this operations, we can get close to understanding diagrammatic game equivalence. Our goal is to move from TDs to GS by means of diagrammatic operations. In order to achieve this, we start with examining two categories of TD operations: "letter relocation operations" and "composition operations" [7].

Letter Operators. "Letter relocation operator", as Cheng argued, "exploits the arbitrariness of the relative horizontal location of letters in a TD" [ibid]. Using this operator, one can swap, repeat or remove letters as seen in Fig. 5.

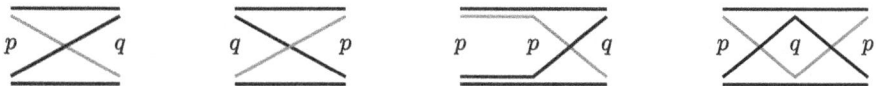

Fig. 5. Relocation operations for the TD for the formula $p \rightarrow q$, as given in [7].

From the viewpoint of game semantics, the first letter relocation operator (switching the positions of p and q for example) does not generate an entirely new game – left moves become right moves, right moves become left moves. Other operators which duplicate the letters introduce new branches to the game as given in Fig. 6. These branches do not introduce any additional strategic power to the game – no player ends up better off after a relocation operation is applied to the game.

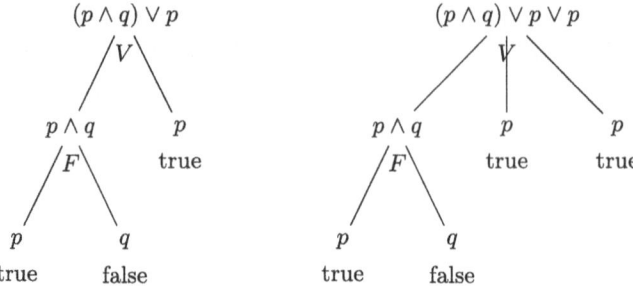

Fig. 6. The game trees for the verification games for the formulas $(p \wedge q) \vee p$ and $(p \wedge q) \vee p \vee p$, respectively, where the latter is obtained after a letter relocation operation is applied to the former one.

Remark 1. Letter relocation operations do not introduce any strategic advantage to any players in semantic games.

Composition Operators. "Composition operators", on the other hand, compose TDs to generate new TDs, as given in Fig. 7 [ibid][3].

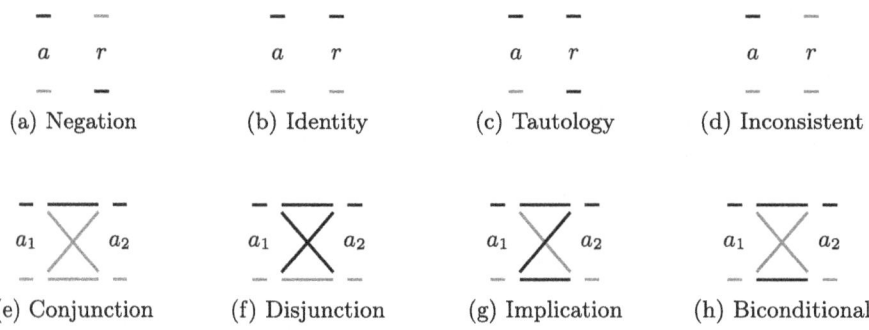

Fig. 7. Unary (top) and binary composition operators, as given in [7].

Composition operators take one (if unary) or more inputs, but produces only one output as they are functional. The unary inputs are represented with a whereas the binary ones are a_1 and a_2. As detailed in [7], composition operators describe how TDs can be "composed" using Boolean connectives. Figure 8 shows the composition of $\neg p$ and q using the conjunction operator, which is given in Fig. 7. In the TD for $\neg p \wedge q$, obtained by composition using the conjunction operator, the only black connector occurs between the black input nodes. In this case, they are the bottom node for $\neg p$ and the top node for q. Hence, the black connector in the output TD for $\neg p \wedge q$ is between these nodes.

---

[3] Cheng places composition operators in dotted-line rectangles which we omit here for simplicity.

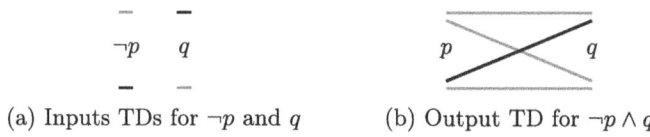

(a) Inputs TDs for ¬p and q   (b) Output TD for ¬p ∧ q

Fig. 8. Composing ¬p and q using the conjunction operator to obtain the TD for ¬p∧q.

It is also possible to compose two binary formulas. Figure 9 shows the composition of $(p \vee q) \wedge (p \wedge q)$.

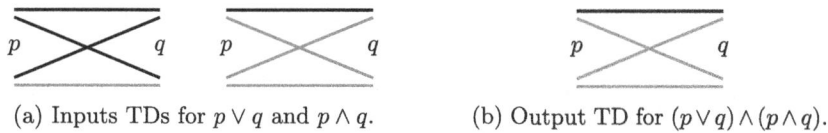

(a) Inputs TDs for p ∨ q and p ∧ q.   (b) Output TD for (p∨q)∧(p∧q).

Fig. 9. Composing the TDs for $p \vee q$ and $p \wedge q$ using the conjunction operator to obtain the TD for $(p \vee q) \wedge (p \wedge q)$.

The composition operators are relevant from the view point of semantic games. In the sequel, we explore how composition operators can be interpreted game semantically.

Let us first consider the semantic games of $p \vee q$ and $p \wedge q$, and see how we can then generate the semantic game of $(p \vee q) \wedge (p \wedge q)$ using the TD-based composition operations. Semantic game trees for these formulas are given in Fig. 10 in a model where p is true and q is false.

In these game trees, the model is fixed: p is true and q is false. Therefore, one only needs to consider the connectors between the top node of p and the bottom node of q in Fig. 9a. The conjunction operator rule given in Fig. 7e suggests that only black connectors produce a black connector. Therefore, in the given model, under conjunction, the output formula $(p \vee q) \wedge (p \wedge q)$ for top-node of p and the bottom-node for q admits a grey connector, hence false, suggesting the Falsifier has a winning strategy as expected.

Let us now systematically explain this process.

First, composition rules give the first move to certain players: game trees composed under the conjunction operator yield the first move to the Falsifier, those composed under the disjunction operator yield the first move to the Verifier.

Second, as argued in Sect. 3, a global look at the input TDs explain under what conditions and presuppositions, players admit winning strategies. As such, determining whether a player has a winning strategy for the composed game directly depends on who has a winning strategy for the composed games.

Now, in this case, for the formula $(p \vee q) \wedge (p \wedge q)$ in Fig. 9, the conjunction operator gives the first move to the Falsifier. He makes the choice and chooses

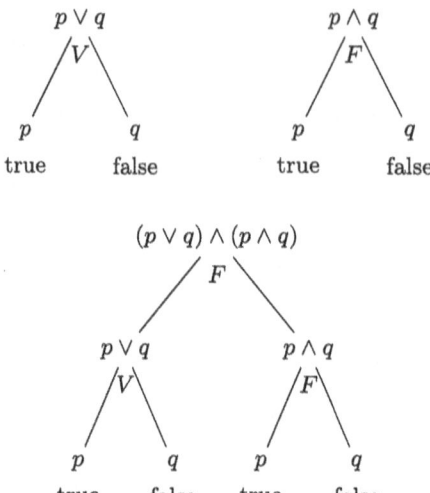

Fig. 10. The game trees for the verification games for the formulas $p \vee q$ and $p \wedge q$, and $(p \vee q) \wedge (p \wedge q)$ respectively.

one of the sub-games, that is one of the inputs. He needs to be able to choose a grey connector to guarantee a win in the subgame.

Let us now consider all four possible input types in order to see whether the Falsifier can win the game under these circumstances. In order for the Falsifier to win, he needs to pick the grey connector.

- If p and q are both true: In both input TDs, the connectors for these cases are black. The Falsifier cannot win, the Verifier has a winning strategy in this case.
- If p and q are both false: In both input TDs, now the connectors are both grey. Therefore, whichever choice the Falsifier makes, he still has a guaranteed win. Hence, in the composed game, the Falsifier has a winning strategy.
- If p is true and q is false: In one of the input TDs (the one for $p \wedge q$), we have a grey connector that the Falsifier can choose. This forms his winning strategy in the composed game.
- If p is false and q is true: In one of the input TDs (the one for $p \wedge q$), we have a grey connector that the Falsifier can choose. This forms his winning strategy in the composed game.

We can now generalise this methodology recursively in order to play semantic games on truth diagrams with connector operations. By doing so, we can game theoretically characterise the TD operations.

Let us start with the rules.

- **Propositional letters**: For a propositional letter p, the Verifier wins for the black connector, the Falsifier wins for the grey connector.

- **Negation**: Players switch colours.
- **Conjunction**: The falsifier makes a choice and selects one of the input games.
- **Disjunction**: The verifier makes a choice and selects one of the input games.

As before, we consider implication as an abbreviation: $p \to q \equiv \neg p \vee q$. And, similarly, the game terminates when a player reaches a propositional letter and has no more moves to make.

This is how we define GS for operations on TDs. For a given formula φ in a model M, the semantic game on truth diagrams is denoted by $G_{TD}(\varphi, M)$.

Theorem 1. *For a given formula φ in a model M, in the truth diagram semantic game $G_{TD}(\varphi, M)$, $M \models \varphi$ if and only if the Verifier has a winning strategy. Consequently, $M \not\models \varphi$ if and only if the Falsifier has a winning strategy.*

Proof. The proof is by induction on the complexity of φ. We will only show it for the base case, negation and conjunction as the case for disjunction is similar.

For φ is a propositional letter p: If $\varphi = p$, then the game terminates. The Verifier wins if and only if $M \models p$. Similarly for the Falsifier.

For φ is a negation $\varphi = \neg \psi$: For $\varphi = \neg \psi$, players switch colours. The composition operator given in Fig. 7a switches the colours, too. By the induction assumption, then, the Falsifier has a winning strategy in the game $G_{TD}(\psi, M)$. Then, the Verifier has a winning strategy in $G_{TD}(\neg \psi, M)$; hence in $G_{TD}(\varphi, M)$.

The argument for the Falsifier is similar.

For φ is a conjunction $\varphi = \psi_1 \wedge \psi_2$: Let $M \models \psi_1 \wedge \psi_2$. Thus, Let $M \models \psi_1$ and $M \models \psi_2$. By the induction hypothesis, the Verifier has winning strategies for both $G_{TD}(\psi_1, M)$ and $G_{TD}(\psi_2, M)$. For the game, $G_{TD}(\psi_1 \wedge \psi_2, M)$, according to the game rules, the Falsifier makes a choice. Whatever choice he makes, the Verifier ends up having a winning strategy at that choice. Thus, the Verifier has a winning strategy for the game $G_{TD}(\psi_1 \wedge \psi_2, M)$ and consequently for $G_{TD}(\varphi, M)$.

Conversely, let the Verifier have a winning strategy for $G_{TD}(\psi_1 \wedge \psi_2, M)$. It is, however, the Falsifier turn to make a move at a conjunction. Therefore, whatever move the Falsifier makes, the Verifier can still win. Therefore, the Verifier has winning strategies for both $G_{TD}(\psi_1, M)$ and $G_{TD}(\psi_2, M)$. By the induction hypothesis, this means that $M \models \psi_1$ and $M \models \psi_2$, suggesting that $M \models \psi_1 \wedge \psi_2$. Consequently, $M \models \varphi$.

The argument for the Falsifier is similar.

This concludes the proof. □

Note that the statement of Theorem 1 benefits from well-defined TD operations given in Fig. 7. What about the converse? Can we make use of game semantical methods to define new composition operations for TDs?

This question pushes us back to the original question regarding game tree equivalences which we mentioned at the beginning of this section. For the example given in Fig. 4, another way to approach the problem is to determine the operations which preserve the strategic power of the players, even before deciding what makes game trees equivalent. This question is deep and challenging. However, we have one answer.

Iterated Elimination of Strictly Dominated Strategies. In the sequel, we explore a well-known solution method in game theory: *iterated elimination of strictly dominated strategies*.

Iterated elimination of strictly dominated strategies (IESDS, for short) is a solution method in games where those strategies which are strictly dominated by other strategies are removed step by step from the game [18]. A strictly dominant strategy is the one that brings higher pay-off to a player, irrespective of how the opponents play. In short, what IESDS does is that once there are alternatives, the strategies which are *strictly* dominated by the others are not the rational moves to make as the *dominant* strategy would bring a higher pay-off. Thus, the strictly dominated ones can be eliminated.

Consequently, IESDS bears significance for the following reason.

Remark 2 ([18]). An action of a player in a finite strategic game is a never-best response if and only if it is strictly dominated.

The same idea applies to logic, particularly to non-classical logics [3]. For example, in classical logic, under disjunction, the truth value True dominates the truth value False. Therefore, the strategy for the sub-game for the False can be eliminated. This idea can be extended to multi-valued logics, too [3,8,9]. Various non-classical logics contain "a non-classical truth value that is 'contaminating' in the sense that a formula must be assigned that value whenever any of its subformulae are assigned the contaminating value" [8]. The game semantic interpretation of this observation is that the contaminating non-classical truth-value is forced by a player whose strategy becomes *strictly dominant* [3]. And this allows us to take advantage of IESDS in semantic games. TDs for such non-classical logical systems fall outside the scope of the current paper, and are thus left for future work.

What is then the equivalent of IESDS in TD games? Can we eliminate some connectors when we are composing TDs?

Let us start with reconsidering the example of the conjunction operator given in Fig. 9. Under the binary conjunction composition rule (Fig. 7e), the Falsifier can eliminate strategies as follows.

- If connectors between same nodes (such as between high p node and low q node) are of different colour, then the Falsifier eliminates the dominated strategy of the Verifier. In the new TD, the grey connectors dominate for the Falsifier.
- If both connectors that are being composed are grey, the Falsifier maintains his winning strategies and does not eliminate any.
- If both connectors that are being composed are black, the Falsifier has no winning strategy under these circumstances.

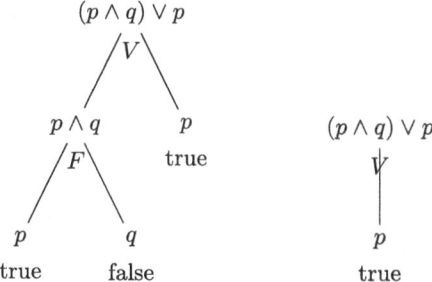

Fig. 11. Iterated elimination of strictly dominated strategies applied to the winning strategy of the Verifier in the game tree of $(p \land q) \lor p$ in the given model. The tree on the left turns to the tree on the right after the strategy elimination.

The same methodology applies to the Verifier for disjunction.

- If connectors between same nodes (such as between high p node and low q node) are of different colour, then the Verifier eliminates the dominated strategy of the Falsifier. In the new TD, the black connectors dominate for the Verifier.
- If both connectors that are being composed are black, the Verifier maintains his winning strategies and does not eliminate any.
- If both connectors that are being composed are grey, the Verifier has no winning strategy under these circumstances.

Notice that the Verifier cannot eliminate any strictly dominated strategy under conjunction, and the Falsifier cannot eliminate any strictly dominated strategy under disjunction. The reason for this is that it is not their turn to make a move at those connectives.

Theorem 2. *Game semantics for truth diagrams for classical logic admits the solution method of iterated elimination of strictly dominated strategies as follows.*

- *If connectors between same nodes are of different colour, then the Falsifier eliminates the dominated strategy of the Verifier. In the new TD, the grey connectors dominate for the Falsifier.*
- *If connectors between same nodes are of different colour, then the Verifier eliminates the dominated strategy of the Falsifier. In the new TD, the black connectors dominate for the Verifier.*

Other cases are given as above.

The proof of this argument follows immediately from the discussion above.

The process of IESDS relates to "decomposition" operators in TDs. The decomposition operators "split apart" the arguments of TDs [7]. Classical logic, however, is not mathematically complex enough to generate composition and decomposition rules for classical TDs. We leave it to future work to examine how various multi-valued logics and their TDs may be characterised by some solution methods in game theory, including the iterated eliminated of *weakly* dominated strategies.

5 Conclusion

In this paper, we established a connection between semantic games and truth diagrams for propositional logic, particularly by focusing on operations on truth diagrams. The immediate extensions of this work is to consider (classical) modal and first-order logics. However portable this methodology is, it is far from trivial to extend truth diagrams to modal or first-order cases by means of game semantics. However, there are benefits. Considering the extensions of propositional logic by means of game semantics provides an alternative to model theoretical methodology to extend truth diagrams to higher order or modal logics. One can ask how the propositional semantic games would change, what additional game theoretical tools would be required, and how strategies need to be advanced to cover modal and higher-order cases.

Non-classical logics is another direction to pursue for truth diagram games. Non-classical logics are often motivated by various philosophical arguments [21]. Diagrammatic reasoning and diagrams themselves are relatively new to non-classical logics, and consequently this is a direction that requires more work.

A direct extension of game semantics is a logical system called "independence-friendly logic" (IF logic, for short) [17]. In IF logic variable dependency of first-order logic can be characterised game theoretically. This is yet another methodology to address the question of truth diagrams for first-order logic by means of focusing on variable dependence (or independence [22]).

Modal and non-classical logics have direct applications in computer science, which can also benefit from diagrammatic reasoning, particularly from truth diagrams. Dynamic epistemic logics (and its power in solving puzzles), knowledge representation (and its power in reasoning with multi-agent systems), and distributed computing (and its power in resource allocation) are some of the areas which can directly benefit from truth-diagrammatic reasoning and representation.

We leave such work for the future.

Disclosure of Interests. The author has no competing interests to declare that are relevant to the content of this article.

References

1. Abramsky, S., McCusker, G.: Game semantics. In: Berger, U., Schwichtenberg, H. (eds.) Computational Logic. NATO ASI Series, vol. 165, pp. 1–55. Springer, Cham (1999). https://doi.org/10.1007/978-3-642-58622-4_1
2. Başkent, C.: Game theoretical semantics for some non-classical logics. J. Appl. Non-Classical Logics **26**(3), 208–39 (2016). https://doi.org/10.1080/11663081.2016.1225488

3. Başkent, C.: A game theoretical semantics for a logic of nonsense. In: Raskin, J.F., Bresolin, D. (eds.) Proceedings of the 11th International Symposium on Games, Automata, Logics, and Formal Verification (GandALF 2020). Electronic Proceedings in Theoretical Computer Science, vol. 326, pp. 66–81 (2020)
4. Başkent, C.: Truth diagrams for some non-classical and modal logics. J. Appl. Non-Classical Logics (to appear)
5. Başkent, C., Henrique Carrasqueira, P.: A game theoretical semantics for a logic of formal inconsistency. Logic J. IGPL **28**(5), 936–952 (2020). https://doi.org/10.1093/jigpal/jzy068
6. van Benthem, J.: Logic in Games. MIT Press, Cambridge (2014)
7. Cheng, P.C.H.: Truth diagrams versus extant notations for propositional logic. J. Logic Lang. Inform. **29**, 121–161 (2020)
8. Ciuni, R., Ferguson, T.M., Szmuc, D.: Logics based on linear orders of contaminating values. J. Log. Comput. **29**(5), 631–663 (2019). https://doi.org/10.1093/logcom/exz009
9. Ferguson, T.M.: Logics of nonsense and parry systems. J. Philos. Log. **44**(1), 65–80 (2015). https://doi.org/10.1007/s10992-014-9321-y
10. Fermüller, C.G.: Semantic games for fuzzy logics. In: Cintula, P., Fermüller, C.G., Noguera, C. (eds.) Handbook of Mathematical Fuzzy Logic, vol. 3, pp. 969–1029. College Publications (2016)
11. Fermüller, C.G., Majer, O.: On semantic games for Łukasiewicz logic. In: van Ditmarsch, H., Sandu, G. (eds.) Jaakko Hintikka on Knowledge and Game-Theoretical Semantics. OCL, vol. 12, pp. 263–278. Springer, Cham (2018). https://doi.org/10.1007/978-3-319-62864-6_10
12. Hintikka, J., Sandu, G.: Game-theoretical semantics. In: van Benthem, J., ter Meulen, A. (eds.) Handbook of Logic and Language, pp. 361–410. Elsevier (1997). https://doi.org/10.1016/B978-044481714-3/50009-6
13. Hodges, W.: Logic and games. In: Zalta, E.N. (ed.) The Stanford Encyclopedia of Philosophy (2013). http://plato.stanford.edu/archives/spr2009/entries/logic-games
14. Lorenzen, P.: Einführung in die operative Logik und Mathematik. Springer, Cham (1955)
15. Lorenzen, P., Lorenz, K.: Dialogische Logik. WBG (1978)
16. Macbeth, D.: Realizing Reason. Oxford University Press, Oxford (2014)
17. Mann, A.L., Sandu, G., Sevenster, M.: Independence-Friendly Logic. Cambridge University Press, Cambridge (2011)
18. Osborne, M.J., Rubinstein, A.: A Course in Game Theory. MIT Press, Cambridge (1994)
19. Parikh, R.: D structures and their semantics. In: Gerbrandy, J., Marx, M., de Rijke, M., Venema, Y. (eds.) JFAK, ILLC, UvA (1999). http://www.illc.uva.nl/j50/
20. Pietarinen, A., Sandu, G.: Games in philosophical logic. Nord. J. Philos. Log. **4**(2), 143–173 (2000)
21. Priest, G.: An Introdiction to Non-Classical Logic. Cambridge University Press, Cambridge (2008)
22. Väänänen, J.: Dependence Logic. Cambridge University Press, Cambridge (2007)

Peirce's Extended Euler Diagrams and the System Atl Based on Ladd-Franklin's Exclusion Relations

Fangzhou Xu[1] and Ahti-Veikko Pietarinen[2](✉)

[1] University of Chinese Academy of Social Sciences, Beijing, China
xufangzhou@ucass.edu.cn
[2] Hong Kong Baptist University, Kowloon, Hong Kong SAR
pietarinen@hkbu.edu.hk

Abstract. We explore Charles Peirce's innovative extension of Euler diagrams to incorporate negative terms through novel star-shaped curves, representing negation and dualizing the relation of inclusion to express exclusion. The study demonstrates the equivalence of Peirce's extended Euler diagrams (**PED**) with the axiomatic system **Atl**, which is based on Christine Ladd-Franklin's work on exclusion relations. The **PED** system is shown to be capable of expressing all fifteen valid syllogisms, thereby streamlining logical reasoning. We also explore the historical development of these diagrams and the implications of considering exclusion as a fundamental relation. By taking exclusion as primary, we connect Peirce's advancements to Ladd-Franklin's earlier work, highlighting the simplification of syllogistic reasoning and the historical evolution of Eulerian diagrammatic logic.

Keywords: Extended Euler Diagrams · Charles Peirce · Copula of Exclusion · Christine Ladd-Franklin · Antilogism

1 Introduction

Representing Negative Terms in Euler Diagrams. Various Euler-style representations have been proposed for syllogisms that involve negative terms or classes [4]. Typically, Euler diagrams represent positive terms within circles and their negative counterparts outside these circles. Space outside a closed shape such as a circle or oval is only one, non-optimal method to denote a negation of a positive term, deriving from representations of positive terms as the regions inside the closed curve. This representation becomes problematic when attempting to express the non-existence of a region or dealing with logical problems that involve negative terms. When inferences are at issue, or when everything must lie within finite, closed regions, standard Eulerian diagrams are inadequate.

Peirce's Eulerian Diagrams. Peirce's contribution to this issue is noteworthy. In 1901, Peirce proposed novel Euler diagrams by reshaping the curves to convey the sign of the terms [15,16]. In Peirce's extension, positive terms are placed on the concave side of the curve and negative terms on the convex side (Figs. 1, 2, 3 and 4).

This method differs from usual Euler diagrams where the curve shape has no logical meaning, straightforwardly representing propositions with negative terms. For example, to represent the proposition "No not-A is not-B" (which means "Everything is A or B"), Peirce's method would use shapes that either include not-A within B or not-B within A, without requiring a space for the outer region that represents neither A nor B (Fig. 3; [14]). In 1903, Peirce presented several other extensions to increase expressivity [3,12,20,22].

2 Peirce's System of Extended Eulerian Diagrams (PED)

The Exclusion Relation in Euler Diagrams. Peirce approached the problem of negative terms by defining that by "'enclosed' we mean being on the concave side" (R 1147, [14,16]), adding to Euler diagrams novel star and triangle-shaped closed non-continuous curves slightly infolded in themselves. Negative terms are represented as their imagined insides. Negative terms resting inside them are "on the concave side" of stars (Figs. 1, 2, 3 and 4).[1]

In standard Euler diagrams, inclusion is represented by an oval within another oval. In Peirce's extended Euler diagrams (**PED**), the relation of inclusion is represented via the relation of *exclusion*. That is, Peirce resorts to two types of shapes: namely 'stars', the inside of which is convex ("curves convex inwards", R 481; [15], [19, p. 282]), and ovals, the inside of which is concave. See, e.g. Fig. 1(b), in which the oval of A rests within the star of B and is thus excluded by B. In this way, Peirce dualized the geometrical relation of inclusion.

The relation of inclusion may be represented either by two different disjoint shapes or by the inclusion of the shape within the same shape. For instance, "Every A is B" is represented by the **PED**s depicted in Figs. 2(b)(c), while the relation of exclusion may be represented either by the two disjoint same shapes, or inclusion of an oval within a negative space, as in "Nothing is A and B", represented as the **PED**s depicted in Figs. 1(b)(c)). Two disjoint stars represent "Not-A is excluded by not-B and not-B exclude not-A", that is, "Not-A and not-B mutually exclude each other", which means "Everything is either A or B" (Fig. 3(a)). This is equivalent to the inclusion of star within an oval.

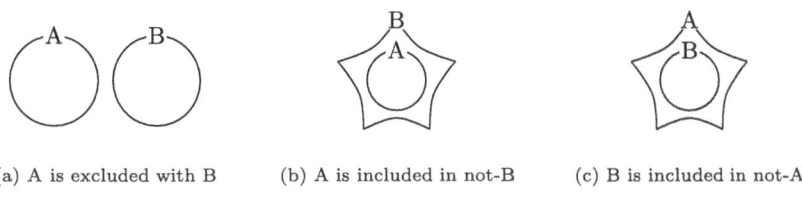

(a) A is excluded with B   (b) A is included in not-B   (c) B is included in not-A

Fig. 1. Nothing is A and B.

---

[1] As a convenient way to think of negations, these curves may be imagined as hyperbolic spaces and hence void of any space in their zero-dimensional interior regions.

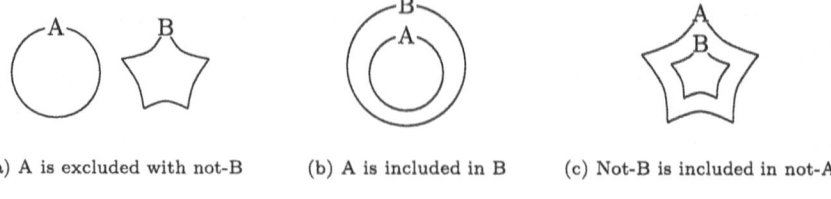

(a) A is excluded with not-B   (b) A is included in B   (c) Not-B is included in not-A

Fig. 2. Every A is B.

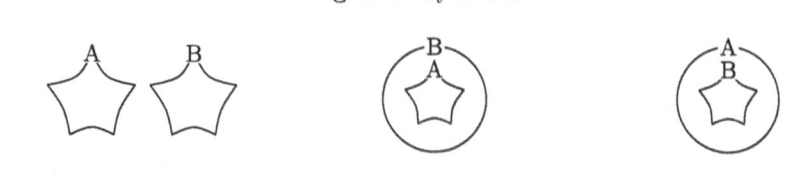

(a) Not-A is excluded with not-B   (b) Not-A is included in B   (c) Not-B is included in A

Fig. 3. Everything is either A or B.

To represent non-universal terms, Fig. 4 depicts the remaining three cases: "Something is both A and B", "Something is both A and not-B", and "Something is both not-A and not-B" derive from the non-exclusion of the two terms which can be either ovals or stars or both. Figures 4(a)(b)(c) are the negations of Figs. 1, 2 and 3, respectively. The converse also holds.

In this way, the inclusion relations are just another version of the exclusion relations. Although the converse also holds, and we have a symmetric system of representing positive and negative terms in **PED**, we reduce inclusions to exclusions rather than the converse, following the arguments given in [27].

Remarks on Peirce's Extended Euler Diagrams. Peirce presented this system in 1901 in the unpublished parts of his article "Logical Diagram" (R 1147; [16]) that appeared in the *Dictionary of Philosophy and Psychology* [17]. He desired also to publish the omitted part on Euler diagrams but the editor James Mark Baldwin did not accept new diagrams or graphs to be printed in this popular dictionary. Peirce wrote in the manuscript margin "This I can't find room for B[aldwin]". By that time, Peirce had also produced another manuscript [15,19],

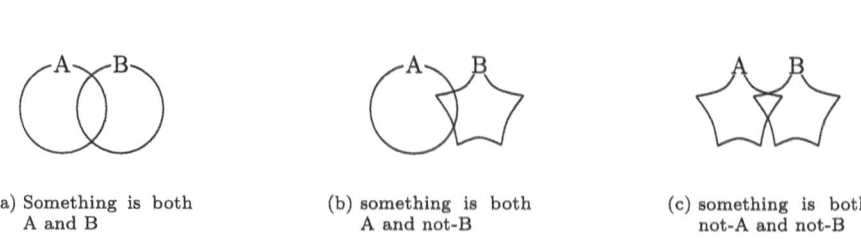

(a) Something is both A and B   (b) something is both A and not-B   (c) something is both not-A and not-B

Fig. 4. One **PED** intersecting with another.

which presents the same idea of **PED** as the excluded parts of the dictionary entry did, including all fifteen syllogisms and the four spurious ones.[2]

Peirce's explicit addition of the representation of the relation of exclusion has some notable repercussions. For one, we get all fifteen syllogisms from that representation. To do that, Peirce added a large dot "to represent some existing individual enclosed by one oval but not by the other" (R 1147).[3] With the dot placed either inside or outside on an oval, one may in this notation present all fifteen syllogisms plus the spurious ones [2].

We take the relation of exclusion as primary and show that from **PED**, we get a single-syllogism representation of exactly the antilogism of Ladd-Franklin, now expressible in **PED**. Beginning with exclusion, we relate **PED** to Ladd-Franklin's earlier work in which the relation of exclusion was equally primitive.

3 Ladd-Franklin's Antilogism and Peirce's Contribution

The development of **PED** was inspired by Peirce's correspondence with Ladd-Franklin [12]. The idea of adding the exclusion relation to the algebra of logic likely originated from her supervisor. Their exchange was most intense in 1901 when the collaborative dictionary entries were being produced [11,12].

Earlier, Ladd-Franklin's algebra of logic [5] was based on the relations of *wholly-exclusion* and *not-wholly-exclusion*. She found that all fifteen valid syllogisms can be reduced into one triad of propositions, termed *antilogism*, such that any two of them imply the negation of the remaining one [6,8–10,13].

The formula of antilogism—the reduction of the Aristotelian syllogistics into a single formula [1], [21, p. 142]—was termed by Ladd-Franklin's student Shen Youngding (Shen Yu-Ting, 1908–1992) "the Ladd-Franklin formula" [25]:

> The Antilogism was at first called by Dr. Ladd-Franklin the 'inconsistent triad'; apropos of it, the late Josiah Royce of Harvard was in the habit of saying to his classes: 'There is no reason why this should not be accepted as the definitive solution of the problem of the reduction of syllogisms'.

It has been remarked that Ladd-Franklin did not prove the "Rule of the Syllogism" [24, p. 463]. What Ladd-Franklin was after was to "extricate the term to be eliminated" [6,26, p. 545] while keeping the term in the subject or predicate position in syllogisms. For this, the dual of exclusion was needed to make the relation of the copula symmetric.

This idea was presented to her colleagues at Johns Hopkins University [23] and worked out in her dissertation during 1881-2 and published in 1883 [5]. The idea of adding the symmetric copula may have originated from Peirce's comments in the Metaphysical Club meeting on January 18, 1881, in which Ladd-Franklin's preliminary work on the algebra of logic was presented as the principal paper to

---

[2] R 481 [15,19] was probably appended to a letter drafted for Christine Ladd-Franklin in late 1900 [12]. In the two surviving manuscripts in which these fifteen syllogisms appear, four figures are represented in slightly different but equivalent ways.

[3] The remark also appears in the published dictionary entry [17, p. 28].

an audience of sixteen colleagues. *The Minute Book* records the talk with the note by the secretary Allan Marquand: "Mr Peirce suggested an extension of the algebra by the addition of a new copula" [18].

Ladd-Franklin later presented a definite statement of the antilogism in terms of "the impossibility of concurrence of the premises and the denial of the conclusion of the ordinary syllogism" [7, p. 575]. In the published dictionary entry "Syllogism" Ladd-Franklin's part expressed it as follows:

> (Syllogism) In this argument the implication contained in the word *but* is that the statements made cannot be all three true together; if the last two are true, the first is not so; if the first and either of the others are true, the remaining one is not true. In other words, the three propositions taken together constitute an inconsistency, or an incompatibility, or, as it may perhaps be called, to distinguish it from the syllogism, an antilogism. Expressed in letters for terms, it affirms that no a is b, no c is non-b, and some a is c cannot be all three true at once; if any two of them are known to be true (it matters not which two) the remaining one is known to be false—that is as much as to say that its contradictory is known to be true, and to be, therefore, the conclusion of a valid syllogism of which the other two are the premises. [13, p. 632]

We will next describe the system based on Ladd-Franklin's symmetric copula(s).

4 Axiomatic System Atl Based on Exclusion and Not-Exclusion

Referring to Ladd-Franklin's and her followers' work, [28] proposed the axiomatic system **Atl** based on Ladd-Franklin's exclusion and not-exclusion relations. Conjointly they express fifteen valid syllogisms, nine special syllogisms, and the square of opposition. For the sake of length, we only focus on the formula implying the fifteen valid syllogisms.

Let us review only the essential parts of **Atl**. See [28] for full definitions.

Definition 1. *A model is a structure* $\mathcal{M} = (M, \bar{\vee}^{\mathcal{M}}, \bar{\bar{\vee}}^{\mathcal{M}}, \bar{}^{\mathcal{M}}, a^{\mathcal{M}}, b^{\mathcal{M}}, \cdots)$ *with equality, where*

- M *is a non-empty collection of classes;*
- $\bar{\vee}^{\mathcal{M}}$ *is the binary exclusion relation;*
- $\bar{\bar{\vee}}^{\mathcal{M}}$ *is the binary not-wholly-exclusion relation, i.e., the negation of* $\bar{\vee}^{\mathcal{M}}$;
- $\bar{}^{\mathcal{M}}$ *is the unary function for a class to represent its complement;*
- $a^{\mathcal{M}}, b^{\mathcal{M}}, \cdots$ *are finitely many classes for all atomic class symbols correspondingly.*

Definition 2. *The satisfaction clauses of the atomic formulas* $t_1 \bar{\vee} t_2$ *and* $t_1 \bar{\bar{\vee}} t_2$ *are the following: for arbitrary terms* t_1 *and* t_2,

$$\mathcal{M} \models t_1 \bar{\vee} t_2 \text{ iff } t_1^{\mathcal{M}} \text{ is (wholly) excluded with } t_2^{\mathcal{M}};$$

$$\mathcal{M} \models t_1 \bar{\bar{\vee}} t_2 \text{ iff } t_1^{\mathcal{M}} \text{ is not (wholly) excluded, i.e., intersects, with } t_2^{\mathcal{M}}.$$

Definition 3. *The axioms are the following:*

1. *First-Order Logic Axioms with equality*
2. *Axioms of Exclusion*
   - *Axiom of Extension of Equality:* $\forall v_1 \forall v_2 (v_1 \bar{\vee} \bar{v_2} \wedge \bar{v_1} \bar{\vee} v_2 \leftrightarrow v_1 \simeq v_2)$.
   - *Axiom of Symmetry:* $\forall v_1 \forall v_2 (v_1 \bar{\vee} v_2 \rightarrow v_2 \bar{\vee} v_1)$.
   - *Axiom of Conversion:* $\forall v_1 \forall v_2 (v_1 \bar{\vee} v_2 \leftrightarrow \neg v_1 \bar{\bar{\vee}} v_2)$.
   - *Axiom of Excluding All:* $\forall v_1 (\exists v_2 (v_1 \bar{\vee} v_2 \wedge v_1 \bar{\vee} \bar{v_2}) \leftrightarrow \forall v_3 (v_1 \bar{\vee} v_3))$.
   - *Axiom of Syllogism:* $\forall v_1 \forall v_2 \forall v_3 (v_1 \bar{\vee} \bar{v_2} \wedge v_2 \bar{\vee} \bar{v_3} \rightarrow v_1 \bar{\vee} \bar{v_3})$.
3. *Axioms of Calculus of Classes*
   - *Axiom of Existence of Complement:* $\forall v_1 \exists v_2 (\bar{v_1} \simeq v_2)$.
   - *Axiom of Double Complement:* $\forall v (\bar{\bar{v}} \simeq v)$.

Let $n \geq 2$ and $t_1, \cdots, t_n$ be arbitrary closed terms. In **Atl**, any antilogism $\{t_1 \bar{\vee} \bar{t_2}, t_2 \bar{\vee} \bar{t_3}, \cdots, t_{n-1} \bar{\vee} \bar{t_n}, t_1 \bar{\bar{\vee}} \bar{t_n}\}$ is minimally inconsistent. That is, any $n-1$ formulas of it imply the negation of the remaining one formula of it. When $n = 3$, it exactly implies all fifteen valid syllogisms.

5 Equivalence Between PED and Atl

Next, we show the equivalence between **PED** and **Atl**. Because the fundamental relations between circles and concaves in **PED** only include the inclusion, the exclusion, and the intersection, **Atl** expresses inclusions and intersections by exclusions and negations of exclusion. **PED** and **Atl** are atomically equivalent to each other through exclusions.

By combining the models of **Atl** with Figs. 1, 2, 3 and 4, we can easily see the equivalence between **PED** and **Atl**, in terms of the following construction:

- If $F = t_1 \bar{\vee} t_2$, then
  - When t_1 and t_2 are both positive,[4]
    F holds *iff* any subfigure of Fig. 1 holds;
  - When one of t_1 and t_2 is positive and another is negative,
    F holds *iff* Fig. 2 holds;
  - When t_1 and t_2 are both negative,
    F holds *iff* Fig. 3 holds;
- If $F = t_1 \bar{\bar{\vee}} t_2$, then
  - When t_1 and t_2 are both positive,
    F holds *iff* Fig. 4(a) holds;
  - When one of t_1 and t_2 is positive and another is negative,
    F holds *iff* Fig. 4(b) holds;
  - When t_1 and t_2 are both negative,
    F holds *iff* Fig. 4(c) holds.

---

[4] That the number of complement functions of a term is even (incl. zero). In contrast, negative term means that the number of its complement functions is odd.

Thus, the minimally inconsistent antilogism $\{t_1 \bar{\vee} \bar{t_2}, t_2 \vee \bar{t_3}, t_1 \bar{\vee} \bar{t_3}\}$ is equivalent to the situation in **PED** that: any two subfigures of Fig. 5 imply the negation of the remaining single subfigure of Fig. 5.

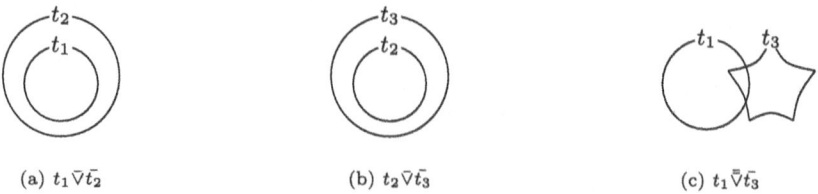

(a) $t_1 \bar{\vee} \bar{t_2}$ (b) $t_2 \vee \bar{t_3}$ (c) $t_1 \bar{\vee} \bar{t_3}$

Fig. 5. Any two imply the negation of the third.

Example 1. Consider the syllogism: $\dfrac{\text{Some M is not P.} \quad \text{Any M is S.}}{\text{Some S is not P.}}$
It is equivalent to the formula $M \bar{\vee} \bar{P} \wedge M \bar{\vee} \bar{S} \rightarrow S \bar{\vee} \bar{P}$ in **Atl**, which takes $M \bar{\vee} \bar{P}$ and $M \bar{\vee} \bar{S}$ true, and $S \bar{\vee} \bar{P}$ false. It is also equivalent to the statement that: Figs. 5(c) and 5(b) hold but Fig. 5(a) does not. This is shown in Fig. 6 in the language of **PED**.

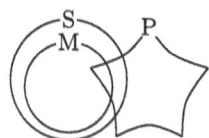

Fig. 6. (**PED** Antilogism) M intersects with not-P, M is included in S, and S intersects with not-P.

6 Conclusions

Peirce's extension of Euler diagrams (**PED**) [15,16], simplifies the representation of negative terms. It is characterized by the innovative use of curvature to indicate the sign of terms and his careful treatment of existence, which requires explicit marking rather than implicit assumption. This approach provides a flexible way to handle negative terms in logical reasoning.

In the present paper, we have taken exclusion relation as primary and shown that by this change, **PED** does what Ladd-Franklin meant to accomplish with her 1883 system, later axiomatically reconstructed as **Atl** [28]. Hence, **PED** amounts to the simplification of syllogistic reasoning in terms of diagrammatic representation of antilogism. Moreover, **PED** does not automatically ascribe existence to a space just because it is represented; instead, if one wishes to express existence, a syntactic device such as a heavy dot is used.

PED was inspired by Peirce's correspondence with Ladd-Franklin, especially via their exchange over the articles they were authoring and co-authoring for the *Dictionary of Philosophy and Psychology* in 1900–1902 [12]. Two decades earlier, the idea of adding the exclusion relation to the system of algebra of logic was vividly discussed by Peirce and Ladd-Franklin at the Johns Hopkins University Metaphysical Club seminar.

References

1. Abeles, F.F.: Christine Ladd-Franklin's antilogism. In: Verbugt, L., Cosci, M. (eds.) Aristotle's Syllogism and the Creation of Modern Logic. Bloomsbury (2023)
2. Bhattacharjee, R., Moktefi, A.: Peirce's inclusion diagrams, with application to syllogisms. In: Pietarinen, A.-V., Chapman, P., Bosveld-de Smet, L., Giardino, V., Corter, J., Linker, S. (eds.) Diagrams 2020. LNCS (LNAI), vol. 12169, pp. 530–533. Springer, Cham (2020). https://doi.org/10.1007/978-3-030-54249-8_50
3. Bhattacharjee, R., Moktefi, A.: Revisiting Peirce's rules of transformation for Euler-Venn diagrams. In: Basu, A., Stapleton, G., Linker, S., Legg, C., Manalo, E., Viana, P. (eds.) Diagrams 2021. LNCS (LNAI), vol. 12909, pp. 166–182. Springer, Cham (2021). https://doi.org/10.1007/978-3-030-86062-2_14
4. Bhattacharjee, R., Moktefi, A., Pietarinen, A.-V.: Representation of negative terms with Euler diagrams. In: Béziau, J.Y., Desclés, J.P., Moktefi, A., Pascu, A.C. (eds.) Logic in Question. Studies in Universal Logic, pp. 43–58. Springer, Heidelberg (2023). https://doi.org/10.1007/978-3-030-94452-0_3
5. Ladd-Franklin, C.: On the algebra of logic. In: Peirce, C.S. (ed.) Studies in Logic by Members of the Johns Hopkins University, pp. 17–71. Little, Brown & Co. (1883)
6. Ladd-Franklin, C.: On some characteristics of symbolic logic. Am. J. Psych. **2**(4), 543–576 (1889)
7. Ladd-Franklin, C.: The reduction to absurdity of the ordinary treatment of the syllogism. Science **13**(328), 574–576 (1901)
8. Ladd-Franklin, C.: Explicit primitives again. J. Phil. Psych. Sci. Met. **9**, 580–583 (1912)
9. Ladd-Franklin, C.: Implication and existence in logic. Phil. Rev. **21**, 641–665 (1912)
10. Ladd-Franklin, C.: The antilogism. Mind **37**, 532–534 (1928)
11. Ladd-Franklin, C.: Christine Ladd-Franklin and Fabian Franklin Papers, ca. 1900–1939. Columbia University, Rare Book and Manuscript Library, New York City
12. Ladd-Franklin, C., Peirce, C.S.: Correspondence 1881–1904. 54 letters. In: Pietarinen, A.-V., Nikulainen, J. (eds.) (2006). Unpublished ms
13. Ladd-Franklin, C., Peirce, C.S.: Syllogism. In: Baldwin, J.M. (ed.) Dictionary of Philosophy and Psychology, vol. 2. Macmillan, New York (1902)
14. Moktefi, A., Pietarinen, A.-V.: Negative terms in Euler diagrams: Peirce's solution. In: Jamnik, M., Uesaka, Y., Elzer Schwartz, S. (eds.) Diagrams 2016. LNCS (LNAI), vol. 9781, pp. 286–288. Springer, Cham (2016). https://doi.org/10.1007/978-3-319-42333-3_25
15. Peirce, C.S.: On Logical Graphs. C.S. Peirce papers at the Houghton library of Harvard University, R 481 (1901)
16. Peirce, C.S.: Logical Graphs. C.S. Peirce papers at the Houghton library of Harvard University, R 1147 (1901)
17. Peirce, C.S.: Logical diagram. In: Baldwin, J.M. (ed.) Dictionary of Philosophy and Psychology, vol. 2, p. 28. Macmillan, New York (1902)

18. Peirce, C.S.: The Metaphysical Club Minute Book. Johns Hopkins University Rare Books and Manuscripts Library (1879–1884)
19. Peirce, C.S.: Writings on Existential Graphs. Volume 1: History and Applications. De Gruyter, Boston (2020). Pietarinen, A.-V. (ed.)
20. Peirce, C.S.: Writings on Existential Graphs. Volume 2/1: The Logical Tracts. De Gruyter, Boston (2021). Pietarinen, A.-V. (ed.)
21. Pietarinen, A.-V.: Christine Ladd-Franklin's and Victoria Welby's correspondence with Charles Peirce. Semiotica **2013**(196), 139–161 (2013)
22. Pietarinen, A.-V.: Extensions of Euler diagrams in Peirce's four manuscripts on logical graphs. In: Jamnik, M., Uesaka, Y., Elzer Schwartz, S. (eds.) Diagrams 2016. LNCS (LNAI), vol. 9781, pp. 139–154. Springer, Cham (2016). https://doi.org/10.1007/978-3-319-42333-3_11
23. Pietarinen, A.-V., Chevalier, J.-M.: The Johns Hopkins Metaphysical Club and its impact in the United States, pp. 1–36. The Commens Working Papers (2015)
24. Russinoff, I.S.: The syllogism's final solution. Bull. Symb. Log. **5**, 451–469 (1999)
25. Shen, E.: Ladd-Franklin formula in logic: the antilogism. Mind **36**, 54–60 (1927)
26. Uckelman, S.L.: What problem did Ladd-Franklin (think she) solve(d)? Notre Dame J. Formal Logic **62**(3), 527–552 (2021)
27. Xu, F.: Antilogism–an axiomatic system based on Ladd-Franklin's relations. University of Chinese Academy of Social Sciences (2023)
28. Xu, F.: Axiomatic syst. on Ladd-Franklin's antilogism. Hist. Phil. Logic (2023)

Diagrams and Applications

CRITICISM AND FICTION

Anxiety Moderates the Effects of Drawing Support on Drawing Accuracy in Mathematical Modeling

Johanna Schoenherr[1](✉) [iD] and Richard E. Mayer[2] [iD]

[1] Department of Mathematics, Paderborn University, Paderborn, Germany
`johanna.schoenherr@uni-paderborn.de`
[2] Department of Psychological and Brain Sciences, University of California, Santa Barbara, USA

Abstract. Many students struggle to make an accurate drawing for a geometry modeling problem, which limits the benefits of drawing. Providing them with drawing support in form of provided elements and a background template can help them make more accurate drawings, but it is an open question whether the effects of drawing support vary based on learner characteristics. We investigate whether enjoyment and anxiety related to the drawing strategy moderate the effects of drawing support, influencing drawing accuracy and modeling performance. In a between-subjects experiment, we randomly assigned 112 undergraduates to either the supported or unsupported drawing condition, assessing their emotions related to the drawing strategy before they solved modeling problems, with or without support. Anxiety moderated the effects of drawing support on drawing accuracy and modeling performance, but not in the expected direction: when provided with support, students with higher anxiety made less accurate drawings. Enjoyment did not contribute significantly to drawing accuracy and modeling performance when students were explicitly asked to make a drawing.

Keywords: Learner-Generated Drawing · Problem Solving · Emotions

1 Introduction

1.1 Objective and Rationale

Imagine two students, Alex and Jordan, who are both good at math, but while Alex enjoys making a drawing for a modeling problem, Jordan experiences drawing as an additional burden that he can barely handle, leading to feelings of anxiety. If you ask both of them to make a drawing when solving a modeling problem, who is more likely to make an accurate drawing and solve the problem adequately?

Previous research has predominantly focused on cognitive learner characteristics, such as working memory and prior knowledge, when investigating the conditions under which drawing benefits learning and problem solving [1]. However, non-cognitive learner characteristics, such as the learner's emotions regarding the drawing strategy

[2], might influence how they implement the strategy, the extent to which they benefit from it, and the impact of instructional methods on outcomes. In this study, which is based on the same sample as Schoenherr and Mayer [3], we investigate how enjoyment and anxiety interact with visual drawing support, drawing accuracy, and modeling performance.

1.2 Literature Review

Drawing for Modeling Problems. Modeling problems are ill-structured real-world problems (for an exemplary problem see Fig. 1), that often include vague conditions, require the learner to make assumptions, and allow for different mathematical operations, tools, and solutions. Solving a modeling problem is difficult for many learners, with building an appropriate problem representation being a major step [4], which might be supported by making a drawing.

> **Fire brigade**
> The fire brigade has a total of 17 locations in the city. A central component of the firefighting trains is a 36 ft long, 8.3 ft wide and 10 ft high turntable ladder vehicle. With the help of the 100 ft long turntable ladder, the fire brigade can rescue people from the upper floors of burning houses. In a rescue operation, the turntable ladder vehicle must maintain a safety distance of 25 ft from the burning house. From what height can the fire brigade rescue people with a turntable ladder vehicle?

Fig. 1. The fire brigade problem [4]

Making a drawing, i.e., creating a visual representation of the problem, has long been recognized as a powerful problem-solving strategy in mathematics education [5, 6]. In modeling problems, two types of drawings are distinguished [7]: situational drawings depict the problem pictorially (as in the right side of Fig. 2), while mathematical drawings represent the problem structure schematically (as in the left side of Fig. 2). Previous research suggests that the benefits of drawing for students, aiding in the selection, organization, and integration of relevant information into a coherent representation and facilitating inferences [8], depend on the student's ability to focus on the mathematical problem structure and to make an accurate drawing [9].

Drawing accuracy refers to how well the drawing aligns with the problem structure as described in the modeling problem [10]. An accurate drawing provides an adequate

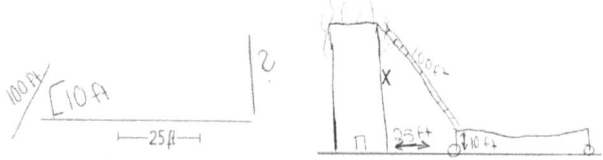

Fig. 2. Drawings of low (left) and high (right) accuracy for the fire brigade problem

representation of relevant elements, their relationships, and numerical values; see Fig. 2 [7].

One instructional technique to assist students in making an accurate drawing is to provide them with visual drawing support in the form of a partially finished drawing to complete; see Fig. 3 [11]. Visual drawing support aims at directing attention to the important elements and facilitating their selection, organization, and integration, thereby enhancing drawing accuracy [12, 13]. Previous studies yield mixed findings on the effects of visual drawing support on drawing accuracy. Results range from an increased representation of relevant elements in drawings without an improvement in organization [14] to enhanced overall drawing accuracy in a previous analysis of the same data as this report [3] and, conversely, a decrease in drawing accuracy [15]. Explanations encompass individual learner characteristics, including their emotions related to the drawing strategy, serving as moderators for the effects of drawing support on drawing accuracy and modeling performance.

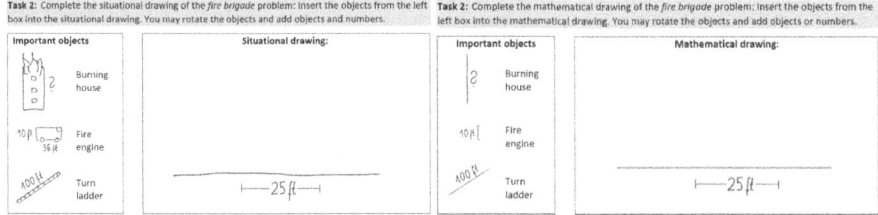

Fig. 3. Visual support in situational (left) and mathematical (right) drawing

Enjoyment and Anxiety. Emotions, such as enjoyment and anxiety, are multifaceted phenomena, having affective, cognitive, physiological, motivational, and expressive parts, that shape students' cognitive responses in a learning environment [16]. Enjoyment is characterized by a positive emotional state or a feeling of pleasure, associated with the sense of being in control and finding value in an activity, i.e., when students are confident of success and perceive the activity as important [17]. Anxiety is characterized by a state of heightened apprehension, worry, or fear often accompanied by physiological arousal. Anxiety is associated with a perceived lack of control combined with high subjective value (i.e., students expect to fail, but the activity is important to them) and can enhance learning when experienced at low anxiety levels, or be detrimental when experienced at high levels [17].

Emotions influence how students process information and the strategies they employ [17]. Recently, Schukajlow et al. [2] found that learners who reported that they enjoyed making a drawing made more drawings than peers who enjoyed drawing to a lesser extent. Whereas anxiety negatively affected the use of the drawing strategy in students with low topic-specific knowledge, this negative emotion positively affected the use of the drawing strategy in students with high topic-specific knowledge. Thus, emotions, including enjoyment and anxiety, can influence the use of the drawing strategy and subsequent performance. How emotions interact with instructional techniques to scaffold drawing is an open question to date.

1.3 Theory and Expectations

The Cognitive Theory of Multimedia Learning (CTML) [16] explains how people learn from multimedia materials, emphasizing generative processing. Drawing, as a generative activity, involves selecting relevant information to represent visually, organizing it into coherent structures, and integrating it with existing knowledge, thereby promoting a deeper understanding [12]. Drawing often enhances learning and performance, especially with instructional support [13]. Understanding which instructional supports benefit specific learners remains unclear to date, prompting a call for systematic investigation [12]. While past research mainly addressed cognitive factors, our study expands into non-cognitive factors, particularly emotions related to drawing.

Control-Value Theory (CVT) [16] can help explain how emotions interact with drawing support. CVT posits that enjoyment and low levels of anxiety are activating emotions that can enhance deep processing, strategy use, and academic achievement [16]. Providing students with drawing support may enhance their perceived control in making an accurate drawing and solving the problem, particularly benefiting those who experience a lack of control, such as students with low levels of enjoyment and high levels of anxiety.

In this study, we examined whether enjoyment and anxiety moderate the impact of drawing support on drawing accuracy and modeling performance, testing the model presented in Fig. 4. The relationships among drawing support, drawing accuracy, and modeling performance were previously analyzed using the same sample [3].

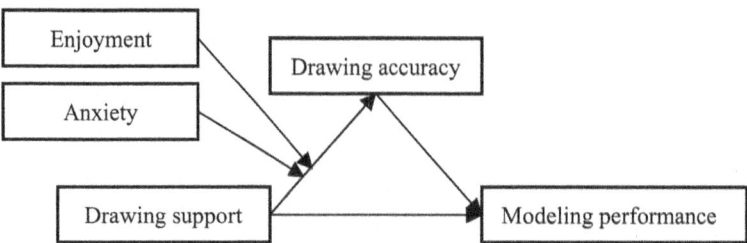

Fig. 4. Hypothesized path model

We predicted that higher enjoyment would be related to higher drawing accuracy (H1) and moderate the positive impact of drawing support, suggesting that drawing support would have a larger positive effect for students with lower rather than higher levels of enjoyment (H2). Additionally, we expected an indirect positive effect of enjoyment on modeling performance through drawing accuracy (H3). For anxiety, we predicted that higher, but still low absolute levels of anxiety, would be related to higher drawing accuracy (H4) and moderate the positive impact of drawing support, with a larger positive effect for higher anxiety levels (H5). The direction of the indirect effect of anxiety on performance via drawing accuracy was unclear, contingent on the experienced anxiety levels.

2 Method

2.1 Participants and Design

The participants were 112 first-year psychology undergraduates ($M = 18.95$, $SD = 1.06$ years old, 69 female, 41 male, 2 non-binary) at a public university in the United States. The randomized experiment followed a 2 × 2 between-subjects design with drawing support (supported drawing or unsupported drawing) and drawing instructions (situational drawing instructions or mathematical drawing instructions) as factors. As we did not find differences in situational and mathematical drawing instructions, we formed two groups for the following analyses: drawing with support ($n = 58$) and drawing without support ($n = 54$). The two groups did not differ on age, gender, English as first language, and topic-specific knowledge.

2.2 Procedure and Instruments

Procedure. Students participated in an individual lab setting of about 45 min. After giving informed consent to participate, students completed a questionnaire on demographics and their emotions related to the drawing strategy, among others. Then, they solved four modeling problems (such as in Fig. 1) with paper and pencil according to their condition.

Drawing Support. Students in the group with drawing support were asked to complete a partially provided drawing by coping the provided elements into the provided background template (as shown in Fig. 3). Students in the group without drawing support were asked to make a free-hand drawing. Assignment to the groups was scored dichotomously with no drawing support (code 0) and drawing support (code 1).

Enjoyment and Anxiety. To assess enjoyment and anxiety, we used Schukajlow et al.'s [2] questionnaire. Sample items included, "I am in a good mood when I make a drawing for a difficult word problem" (emotion scale) and "I often worry that making a drawing for a word problem will be too difficult for me" (anxiety scale). Responses were on a Likert scale from 1 (not at all true) to 5 (completely true). The reliabilities (Cronbach's α) of the enjoyment scale (four items) and the anxiety scale (four items) were good with 0.71 and 0.96, respectively.

Drawing Accuracy. Drawing accuracy was coded on a three-point scale [7]: Code 0 for inadequate representation of elements and relationships, code 1 for adequate representation of elements and relationships with missing numbers, and code 2 for adequate representation of relevant elements, their relationships, and numbers. Intercoder reliability for drawing accuracy was good (Cohen's $\varkappa \geq .69$). Cronbach's α was low with .56, indicating substantial between-task variation in students' drawing accuracy.

Modelling Performance. Modelling performance was coded on a three-point scale across four modeling problems [7]: Full credit (code 2) for an adequate model and numerical result, partial credit (code 1) for an adequate model but incorrect numerical result due to computational errors, and zero credit (code 0) for an inadequate model. Intercoder reliability and scale reliability was satisfactory (Cohen's $\varkappa \geq .91$, Cronbach's $\alpha = .69$).

3 Results

Descriptive statistics are displayed in Table 1. We used the software MPlus (Version 8.7) to analyze the mediational model with maximum likelihood (ML) estimation. Model fit indices indicate that the theoretical model fits the data well: The chi-square goodness of fit, $\chi2(1) = 0.254$, $p = .88$, the Comparative Fit index CFI = 1.0, the 90 Percent Confidence Interval of Root Mean Square Error of Approximation RMSEA = [0.00; 0.09], and the Standardized Root Mean Square Residual SRMR = .007. The model explained 22% of the variance in drawing accuracy, and 50% of the variance in modeling performance. In line with prior analyses of the same sample (Schoenherr & Mayer, under review), drawing support had a positive direct effect on drawing accuracy ($\beta = .245$, $SE = .082$, $p < .01$), and drawing accuracy had a positive direct effect on modeling performance ($\beta = .703$, $SE = .055$, $p < .01$). Drawing support did not have a significant direct effect on modeling performance ($\beta = -.074$, $SE = .070$, $p = .288$).

Table 1. Descriptive statistics

Variable	M (SD)	1	2	3	4	5
1. Enjoyment	2.82 (0.82)	1	−.17	.09	.26**	.20*
2. Anxiety	2.21 (0.82)		1	.01	−.04	−.16
3. Drawing support	0.5 (0.5)			1	.25**	.09
4. Drawing accuracy	1.43 (0.47)				1	.69**
5. Modelling performance	0.98 (0.62)					1

Note. * $p < .05$, ** $p < .01$, $n = 112$

Contrary to our predictions, enjoyment did not have direct effects on drawing accuracy ($\beta = .153$, $SE = .111$, $p = .173$), and did not moderate the effect of drawing support on drawing accuracy ($\beta = .075$, $SE = .114$, $p = .511$). Enjoyment did not have a direct effect ($\beta = .011$, $SE = .071$, $p = .871$), a total effect ($\beta = .090$, $SE = .080$, $p = .258$), or an indirect effect ($\beta = .081$, $SE = .060$, $p = .179$) on modeling performance.

In line with our predictions, anxiety had a positive direct effect on drawing accuracy ($\beta = .262$, $SE = .112$, $p = .019$). Anxiety moderated the effect of drawing support on drawing accuracy, exhibiting a direction opposite to our predictions ($\beta = -.397$, $SE = .112$, $p < .01$). That is, students with higher anxiety levels produced more accurate drawings in the absence of drawing support but less accurate drawings when drawing support was provided. Anxiety did not have a direct effect ($\beta = -.091$, $SE = .052$, $p = .078$) or a total effect ($\beta = .064$, $SE = .106$, $p = .548$), but anxiety had an indirect effect on modeling performance via drawing accuracy ($\beta = .184$, $SE = .081$, $p = .024$).

4 Discussion

Findings of this study have empirical contributions, theoretical implications, and practical implications. First, this study enriches the broader literature on the interplay between emotions, using a drawing strategy, and academic performance. Extending a previous study on the relationships between emotions, spontaneous drawing use, and modeling performance [2], our findings indicate that emotions not only influence the decision to employ the drawing strategy but also impact the quality of its use in terms of drawing accuracy, subsequently influencing modeling performance. For example, anxiety had a positive effect on drawing accuracy in the unsupported drawing condition, suggesting that lower anxiety levels may increase effort to avoid failure [16]. However, surprisingly, when drawing support was provided, higher anxiety levels were associated with less accurate drawings. One explanation is that the additional demands of using the provided elements induce intrusive thoughts and worries in learners, who may tend to interpret drawing as threatening. Cognitive worries reduce the cognitive resources available to accurately make the drawing and solve the problem. While previous research has shown that enjoyment contributes to spontaneous drawing use [2], our findings suggest that enjoyment does not significantly contribute to drawing accuracy and modeling performance when explicitly asking students to draw. Importantly, more research is needed on the role of emotions in other mathematical problems types, such as word problems or intra-mathematical problems.

Second, this study contributes to CTML [1] and CVT [16]. Our findings introduce emotions, particularly anxiety related to the drawing strategy, as a non-cognitive learner characteristic. These emotions not only influence decisions to make a drawing but also impact the quality of implementation, reflected in drawing accuracy, and moderate the effectiveness of instructional techniques, such as visual drawing support. Thus, emotions related to the drawing strategy complement previously identified cognitive learner characteristics as boundary conditions for learning through drawing [11, 12]. Moreover, our findings contribute to CVT by demonstrating that emotions regarding a specific strategy contribute to academic outcomes by moderating the effectiveness of instructional techniques and influencing the quality of strategy use. To better understand how anxiety moderates the effects of drawing support on drawing accuracy, future research should include measures of perceived control and subjective value [16].

Third, considering that instructional techniques, such as drawing support, may yield diverse outcomes for learners with varying emotions, teachers should exercise caution when selecting instructional approaches. As illustrated in this study, providing drawing support can lead to poorer drawing use and modeling performance for learners with higher levels of anxiety. Therefore, teachers should offer drawing support in situations where students are not experiencing higher levels of anxiety.

Acknowledgments. This study was funded by the Deutsche Forschungsgemeinschaft (DFG, German Research Foundation) – RE 4973/2-1.

Disclosure of Interests. The authors have no competing interests to declare that are relevant to the content of this article.

References

1. Mayer, R.E.: The past, present, and future of the Cognitive Theory of Multimedia Learning. Educ. Psychol. Rev. **36**(8) (2024)

2. Schukajlow, S., Blomberg, J., Rellensmann, J., Leopold, C.: Do emotions and prior performance facilitate the use of the learner-generated drawing strategy? Effects of enjoyment, anxiety, and intramathematical performance on the use of the drawing strategy and modelling performance. Contemp. Educ. Psychol. **65**, 101967 (2021)
3. Schoenherr, J., Mayer, R.E.: Maximizing the benefits of student-generated drawing for real world mathematical problem solving. Contemp. Educ. Psychol. (under review)
4. Blum, W., Borromeo Ferri, R.: Mathematical modelling: can it be taught and learnt? J. Math. Model. Appl. **1**(1), 45–58 (2009)
5. Arcavi, A.: The role of visual representations in the learning of mathematics. Educ. Stud. Math. **52**(3), 215–241 (2003)
6. Schoenfeld, A.H.: Mathematical Problem Solving. Academic Press, Cambridge (1985)
7. Rellensmann, J., Schukajlow, S., Leopold, C.: Make a drawing. Effects of strategic knowledge, drawing accuracy, and type of drawing on students' mathematical modelling performance. Educ. Stud. Math. **95**(1), 53–78 (2017)
8. van Meter, P., Garner, J.: The promise and practice of learner-generated drawing: Literature review and synthesis. Educ. Psychol. Rev. **17**(4), 285–325 (2005)
9. Rellensmann, J., Schukajlow, S., Blomberg, J., Leopold, C.: Effects of drawing instructions and strategic knowledge on mathematical modeling performance: mediated by the use of the drawing strategy. Appl. Cogn. Psychol. ac3930 (2022)
10. Ott, B.: Learner-generated graphic representations for word problems: an intervention and evaluation study in grade 3. Educ. Stud. Math. **105**(1), 91–113 (2020)
11. Fiorella, L., Zhang, Q.: Drawing boundary conditions for learning by drawing. Educ. Psychol. Rev. **30**(3), 1115–1137 (2018)
12. Leutner, D., Schmeck, A.: The drawing principle in multimedia learning. In: Mayer, R.E., Fiorella, L. (eds.) The Cambridge Handbook of Multimedia Learning, 3rd edn., pp. 360–369. Cambridge University Press, Cambridge (2022)
13. Schwamborn, A., Mayer, R.E., Thillmann, H., Leopold, C., Leutner, D.: Drawing as a generative activity and drawing as a prognostic activity. J. Educ. Psychol. **102**(4), 872–879 (2010)
14. Jamet, E., Michinov, E.: Effects of verbal and visual support on learning by tablet-based drawing. Comput. Educ. **181**, 104460 (2022)
15. Schmidgall, S.P., Scheiter, K., Eitel, A.: Can we further improve tablet-based drawing to enhance learning? An empirical test of two types of support. Instr. Sci. **48**, 453–474 (2020)
16. Pekrun, R.: The control-value theory of achievement emotions: assumptions, corollaries, and implications for educational research and practice. Educ. Psychol. Rev. **18**, 315–341 (2006)
17. Pekrun, R., et al.: A three-dimensional taxonomy of achievement emotions. J. Pers. Soc. Psychol. **124**(1), 145–178 (2023)

Learning Magnitudes of Energy Consumption with Symbolic or Iconic Representations

Erica de Vries[1]([✉])[iD], Neil Schwartz[2], and Martin Galilée[1]

[1] Univ. Grenoble Alpes, LaRAC, 38000 Grenoble, France
Erica.deVries@univ-grenoble-alpes.fr
[2] California State University Chico, Chico, CA 95929, USA

Abstract. Learning about the magnitude of energy consumption is crucial for responsible energy management in today's society. This study investigates the use of symbolic and iconic representations for learning and mentally comparing the energy consumption of (sets of) appliances. We conducted an experiment in which participants learned about the magnitude of energy consumption either with symbolic representations (Arabic numerals) or with iconic representations (bars of different heights). Participants then performed a recall and a mental comparison task varying in the number of appliances involved and in the distance in energy consumption on both sides. The findings indicate that both types of representations allow learning magnitudes. Moreover, the results showed a distance effect indicating that mental comparisons are conducted with analog internal representations. Finally, compared to symbolic representations, mental comparisons of magnitudes learned with iconic representations, despite being slightly less accurate, were much faster. These results suggest that iconic representations can be an effective tool in environmental education and communication.

Keywords: Symbolic and iconic representations · mental comparisons · magnitudes · learning

1 Introduction

In the context of climate change, citizens will be held more and more responsible for their energy consumption. In order to keep energy consumption low and in tune with the fluctuating energy production of renewable sources, citizens may resort to various strategies, such as identifying both efficient and wasteful appliances, keeping consumption below certain set goals or in the range of their neighbors', or adapting consumption to fluctuating local production [1]. These energy management tasks and strategies involve learning and memorizing the energy consumption of a variety of appliances, performing mental operations such as estimating sums and differences, and mentally comparing energy consumption of (sets of) appliances. Such energy management tasks may become as ubiquitous as electric appliances themselves and are executed in a distributed cognitive environment [2, 3] simultaneously involving internal mental and external representations.

Energy is a topic in STEM learning, but in contrast to physical objects, such as pumps, gears, springs, inclined planes, and the like, energy is an abstract concept, more like force, mass, voltage, and current. Physical objects and their external representations, such as a text, a sketch, a photo, or a diagram, can be juxtaposed and compared as in Joseph Kosuth's work "One and Three Chairs". However, abstract concepts, such as energy and energy consumption, are intangible, invisible, silent, and weightless. In other words, being intangible, neither energy nor energy consumption are available to the senses for comparison with their external representations. Selecting a particular external representation for communicating about energy consumption is important because different isomorphic representations of a common structure can have differential effects on task difficulty, problem solving efficiency and behavioral outcomes, i.e. the representational effect [2–4]. Regarding energy management systems, a multitude of external representations have been used such as bars and pie charts, horizontal lines, bell curves [5], color hue [6], and glow intensity [7, 8]. In the current study, we aim to study the effect of symbolic and iconic external representations on learning, recalling, and mentally comparing magnitudes of energy consumption.

1.1 Symbolic and Iconic External Representations

Going back to Peirce [9, 10], two types of external representations are distinguished: symbolic and iconic. Symbolic representations, such as digits or characters, represent by virtue of convention [10]. In other words, the content of the representation cannot be predicted nor inferred from its form; there is an arbitrary relation between the form and the content of a symbolic representation. By contrast, iconic representations, such as pictures, diagrams, bar charts, represent by similitude, visual or otherwise, to what is represented. There may be different types of iconicity, but these are all characterized by natural and non-arbitrary relations, i.e. a "resemblance" or likeness between the representing and the represented [11]. The distinction between convention and resemblance underlies oppositions in related fields of study: textual and pictorial [12], linguistic and graphical [13], sentential and diagrammatical [14], and descriptive and depictive representations [15]. It also generally is used for distinguishing two systems of internal mental representation [16–18].

It follows that external representations cannot be categorized as symbolic or iconic *without* knowledge of the context or the intended meaning [19]. For example, in a picture of a road sign, a red circle signals a prohibition or a restriction by convention amongst drivers and legislators. In a picture of a car or of a bicycle, a red circle stands for a red car or bicycle tire based on resemblance. According to Peirce, an icon represents its object by virtue of a character or quality which it would equally possess in the absence of the object and an interpreting mind. Thus, in a country without traffic restrictions or in the absence of a road user, a red circle is not of itself subject to convention. To the contrary, even in the absence of cars, bicycles, and tires, or road users, a red circle still possesses the quality of roundness at the core of the resemblance relation. In conclusion, the same red circle is symbolic in the first and iconic in the second pictorial representation.

1.2 External Representations of Magnitude

Arabic numerals are widely used symbolic representations. In the design of visualizations of energy consumption, selecting an iconic representation is problematic because there is no straightforward choice on the grounds of perceptual resemblance. Energy consumption is a magnitude: a quantity that can be represented as a number and subjected to "greater than" comparisons. Thus, iconic representations of magnitude should allow "greater then" comparisons based on transitivity. Palmer speaks of intrinsic representation [20] when the represented and representing world share the same inherent property. In the case of representing magnitudes, the shared inherent property for intrinsic representations is greater-than transitivity (if $A < B$ and $B < C$, then $A < C$). Indeed, in reference to Peirce, transitivity is a quality that does not depend on an object nor interpreting mind, and thus is the basis of iconicity when representing magnitudes.

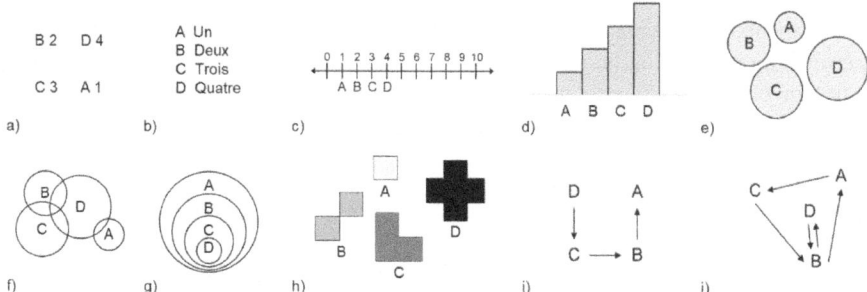

Fig. 1. Candidate representations for four different magnitudes labelled A, B, C and D.

In Fig. 1, a) and b) are symbolic representations ruled by convention. At least one interpreting agent with knowledge of either Arabic numerals or French respectively is required to interpret them as numbers of magnitudes. Figure 1 c), d) and e) are iconic since the property of transitivity holds irrespective of the fact that the line segments, bars, or circles represent different magnitudes and even in the absence of an interpreting mind. Transitivity also holds in f), g) and h), but these have additional features such as layout and shape to be ignored when interpreting them as magnitudes. Figure 1 i) is a case of extrinsic representations in Palmer's definition, because greater-than transitivity is not an intrinsic property of node-link graphs. In Fig. 1 i), greater-than transitivity holds solely due to fact that the node-link graph represents a world of magnitudes. By contrast, Fig. 1 j) shows a node-link graph representing a network of social relations. It possesses points-to transitivity, but is an erroneous representation of greater-than relations in a world of four *different* magnitudes.

1.3 The Triple Code Model of Mental Representations of Magnitude

In number cognition, three mental systems [21] are conjectured for representing numbers and magnitudes: (1) a visual symbolic representation, i.e. Arabic numerals such as "15", see Fig. 1 a), (2) an auditory verbal word representation, i.e. the sound /fifteen/ or the (sound of the) words in Fig. 1 b), and (3) an analog non-symbolic representation, i.e. a number line or bars or dots of different sizes as in Fig. 1 c), d), e). A number line is a spatial representation used in mathematics and consists of a graduated straight line with every point representing a real number. The number line is iconic as it possesses transitivity.

Note that "analog" is still another term for iconic, intrinsic, resembling, or likeness. The theorization of the number line as a spatial analog mental representation followed the discovery of the Spatial-Numerical Association of Response Codes (SNARC) in parity judgments [22]. The SNARC effect is characterized by faster responses to relatively large numbers with the right hand and to relatively small numbers with the left hand, suggesting a spatial and lateralized mental representation of number.

According to the triple-code model [21], the use of a particular internal representation depends on task demands. The visual Arabic number representation is used in multi-digit operations. The auditory verbal word representation is used to communicate through language and to recall arithmetic facts learned by verbal rote. And finally, the analog magnitude representation is used in estimations and comparisons. In theory, different mental representations specifically fit different tasks. In practice, however, all three mental systems can be used jointly. For example, a complex mental calculation can first involve mentally structuring the calculation with visual Arabic number representations, then accessing addition and multiplication tables with auditory verbal word representations, and finally comparing the end result to a reasonable estimate with the analog magnitude representation.

1.4 The Distance Effect in Magnitude and Number Comparisons

In perception, the more similar two entities are on a certain dimension, the longer it takes to decide which entity is larger on that dimension. The same observation was made in number comparisons. Comparing two numbers is faster when their numerical distance is larger, i.e., comparing 2 to 9 is faster than comparing 5 to 6 [23]. Moyer & Bayer coined the term *distance effect* for the relation between distance and time to perform a comparison [24]. The presence of such a distance effect in both stimuli and number comparisons is best explained by common mechanisms: the perceptual processes described in psychophysics are also used, at least partially, in number comparisons, and the representations used must be analog, sharing properties of visuo-spatial stimuli, rather than discrete, like symbols. Using analog mental representations in number processing implies a conversion of Arabic numerals into comparable analog representations of magnitudes [25].

1.5 The Current Study

The present study involved learning and recalling magnitudes of energy consumption learnt with either symbolic or iconic representations and mentally comparing energy consumption of (sets of) fictitious appliances in the absence of the external representations of the magnitudes of energy consumption.

The first task involved learning the energy consumption of eight appliances with either symbolic or iconic external representations. The symbolic external representation consisted of Arabic numerals, the iconic external representation consisted of bars of different heights with no scale. Therefore, two of three mental codes from Dehaene's model [21] were solicited: the visual Arabic number representation in the symbolic condition and the analog non-symbolic representation in the iconic condition. However, in both cases, spontaneous translation to another code is feasible. For instance, Arabic numerals may trigger an analog mental representation of magnitude, and inversely bars of different heights may be mentally attributed a number. Furthermore, both Arabic numerals and heights of bars may be converted into auditory verbal representations.

The second task involved recall of energy consumption of the eight appliances using the same representation as in the learning task (typing digits or sizing a bar). Recall was expected to be slightly higher in the symbolic condition because Arabic numerals are discrete representations of magnitude allowing precise recall, whereas the height of a bar is intrinsically continuous.

The third task involved mental comparisons of the energy consumption of a single appliance to another single one (one-by-one comparisons). The silhouettes of the appliances were presented and energy consumption had to be compared in the absence of either iconic or symbolic representations of magnitude. Such mental comparisons involve retrieving energy consumption from memory, obtaining a mental analog representation [21], and conducting the comparison. The difference in energy consumption of the two appliances was manipulated to be at a large, a medium, or a small distance from each other. We expected a distance effect on both accuracy and response time in both the symbolic and the iconic condition. More precisely, we expected faster and more accurate responses for large distance comparisons. We also expected mental comparisons to take longer in the symbolic condition as compared to the iconic condition because it involves a symbolic-iconic translation step at the time of retrieval. Regarding a possible interaction, the distance effect might be slightly stronger in the symbolic condition, because appliances with a large distance in energy consumption require less precise analog representations to be generated from the symbolic representations (Arabic numerals). Small distance comparisons on the contrary require more precise analog representations which, in the symbolic condition, might take longer to generate, and could be less accurate.

In order to study a more ecological and complex situation, we designed a second type of comparison task between two sets of three appliances. These three-by-three comparisons involved retrieval of energy consumption of six appliances, mentally estimating, and finally comparing energy consumption on both sides. Again, according to Dehaene's [21] triple-code model, estimation and comparison are conducted using the analog magnitude representation. We expected three-by-three comparisons to take longer than one-by-one comparisons, but not necessarily to be less accurate. Furthermore, we expected comparisons to take longer in the symbolic condition as compared to the iconic condition, as well as longer for small distances as compared to large distances. In addition, we expected the distance effect on response time to be stronger in the symbolic condition as compared to the iconic condition because small distance comparisons may require several iterations of a convert-add-compare cycle of Arabic numerals into analog mental representations in order to achieve the required level of precision. Numerical and spatial skills were measured as potential covariates.

2 Method

The two conditions differed only in the materials for learning and recalling energy consumption: symbolic (with digits) or iconic (with bars of varying height). The comparison task was identical in the two conditions and involved mentally comparing magnitudes of energy consumption either between two appliances (two silhouettes presented one-by-one) or between two sets of appliances (six silhouettes presented three-by-three). Mental comparisons involved a large, medium, or small distance between total magnitudes of energy consumption on each side.

Table 1. Eight fictitious appliances and their energy consumption in Arabic numerals (digits) in the symbolic condition and in the height of the bar (in mm) in the iconic condition

Description	Silhouette	Digits	Height
Name: Sock-and-Roll Used in: Bedroom The Sock-and-Roll folds, organizes, and dispenses the inventor's socks according to the forecasted weather.		147	24
Name: Absolute Ruler Used in: Workshop The Absolute Ruler measures and controls the size and weight of objects and parts used in the inventor's work.		184	30
Name: Epic Bubbler Used in: Bathroom The Epic Bubbler generates a deluge of soap bubbles to make every shower unforgettable.		230	37
Name: Pantry Pilot Used in: Pantry The Pantry Pilot deftly rotates and organizes the inventor's goods so that food is never forgotten and never goes bad.		287	46
Name: Smart Spice Used in: Kitchen The Smart Spice automatically grinds and mixes spices according to the smell and flavor of cooking meals.		359	58
Name: Rainbow Maker Used in: Kid's Room The Rainbow Maker makes crayons of any color imaginable, with or without glitter.		449	72
Name: Document Den Used in: Office Room The Document Den scans, stores, and organizes paper documents to make them easily available on demand.		561	90
Name: Tool Trapper Used in: Garage The Tool Trapper snatches any tool forgotten on the floor when the inventor repairs his vehicles.		701	113

2.1 Participants

A hundred and four undergraduate students were recruited via a sign-up sheet in a Californian State University and rewarded with extra credit. Six participants failed to complete the tasks in time and their data were discarded. Of the 98 remaining participants, 78 percent were female. Female and male participants were equally distributed across conditions. Age ranged from 18 to 52 years, $M = 22.2$, $SD = 4.09$.

2.2 Materials and Apparatus

Eight appliances with a silhouette, a name, and a description were invented (Table 1). The silhouettes were carefully drawn to be equally different from one another, so that no silhouette was more recognizable than the others. All silhouettes were in black and white and contained the same number of black pixels (within 1 percent). The names and descriptions were chosen not to evoke particularly low or high energy consumption. Pilot studies showed that a linear relation between the eight appliances and their energy consumption induced memorizing order rather than magnitude and therefore should be avoided. The values chosen followed a mathematical sequence unlikely to be detected by the untrained eye (increments of 25% of the preceding value, logarithm function to base 1.25, see Table 1).

The study encompassed four tasks. First, in a familiarization task, participants got acquainted with the appliances. Silhouette, name and description of the appliances were presented followed by a questionnaire.

Fig. 2. Magnitude in digits in the symbolic and the height of a bar in the iconic condition

Second, participants learned about the energy consumption of the eight appliances. The learning task involved repeated presentation of eight computer-simulated flashcards. The cards were presented on the screen one at a time, with a silhouette on the front side and the corresponding energy consumption on the back, and the cards flipped when clicked. Energy consumption was expressed either as a number or as a vertical bar of a certain size (see Fig. 2 for an example). The presentation order followed a confidence-based repetition design, or CBR [26], i.e. depending on the learner's confidence on a scale from one to five. In CBR, all cards stay in the stack, but cards with lower confidence level appear more often. The task ended after a total of 80 presentations of cards.

Third, the recall task consisted of recalling energy consumption upon appearance of a flashcard with a silhouette on the screen, either by typing numbers (in the symbolic condition) or by scaling a vertical bar with the mouse (in the iconic condition). Thus, recall of energy consumption was conducted in the same format in which it was learned. Positive feedback in the form of temporary green highlighting and a checkmark was provided for answers within 10% of the correct value. The task ended when the energy consumption of all appliances was recalled three times. A brief familiarization with the procedure preceded the actual recall task. The recall task was repeated after the comparison task in order to establish whether memory faded, stayed the same, or improved during the experimental session.

Fourth, the comparison task involved determining from memory which of two appliances (one-by-one comparison) or which set of three appliances (three-by-three comparison) uses more energy. A total of 48 comparison items were created using the eight appliances displayed in Table 1. Twenty-eight comparison items displayed one silhouette of an appliance on each side of the screen. Twenty comparison items displayed three silhouettes of appliances on each side of the screen. Comparisons varied in the distance (difference) between the total magnitude of energy consumption on each side: large, medium, or small. First, a central focus point appeared on a computer screen for one second. Then, the (sets of three) silhouettes appeared on the left and the right of the screen. Participants indicated higher energy consumption by using the keyboard with "Q" for left and "P" for right. Each comparison item was presented four times, twice in its original form (A) and twice mirrored on the screen (B), to compensate for laterality effects. Furthermore, the four items were presented in reversed order the second time (ABBA) in order to balance for learning and fatigue effects.

2.3 Procedure

Participants entered a computer lab on campus in groups of up to 12 people and were randomly assigned to conditions. Participants first expressed their informed consent via a computer form, then completed the familiarization task, the learning task, the recall task, the comparison task, and again the recall task. Finally, participants were asked to complete the Form Board Test (designed to measure spatial visualization) and the Addition Test (designed to measure the ability to perform basic arithmetic operations with speed and accuracy) both from the Kit of Factor-Referenced Tests [27] and adapted to the computer. After this, participants were thanked and debriefed. The procedure lasted approximately 45 min. The experimental procedure received approval from the department's ethics committee.

2.4 Dependent Variables

Recall error in the recall task before and after the comparison task was calculated as the mean percentage of error (absolute difference between the given and the correct value divided by the correct value).

In the comparison task, each response was recorded as either "correct" or "incorrect". Mean accuracy was calculated as a percentage of correct answers for each comparison type (one-by-one and three-by-three) and distance (large, medium, small) creating six accuracy scores per participant.

Response times (RTs) in the comparison task were measured as the delay (in milliseconds) between the presentation of the item and the response. Response times were screened for outliers caused by other processes than the one under study, such as reflex key presses, fatigue, or loss of attention. We adopted the outliers screening method [28] and removed RTs that were below human perceptual reaction time, taken at 160 ms, or particularly long, with values higher than 2.5 standard deviations above the participant's average RT at this task (160 ms $\leq$ RT $\leq M + 2.5\ SD$). This method removed 255 RTs in the symbolic condition and 280 RTs in the graphical condition (3% of the total number of RTs). Only the response times of correct answers were kept in further analysis. For each comparison type (one-by-one and three-by-three) and distance (large, medium, small), data were averaged creating six mean response times per participant.

3 Results

In order to ensure that only the data from participants who had well succeeded at learning the materials were analyzed, data were filtered according to a dual threshold. Accuracy at the recall and comparison tasks was used as indicator of success at learning the materials. Participants who deviated from the mean by more than one standard deviation in the direction of more error in either task were removed from further analysis. This concerned 11 participants in the symbolic condition and 14 in the iconic condition. Thirty-five participants remained in the symbolic and thirty-eight in the iconic condition (N = 73).

3.1 Numerical and Spatial Skills

Numerical skills and spatial skills scores were assessed through the Addition Test and Form Board Test, respectively. Scores for numerical skills ranged from 4 to 33 ($M = 17.0$, $SD = 6.39$). Inter-item reliability on our data was found at $\alpha = .736$. Scores for spatial skills ranged from 36 to 220 ($M = 115$, $SD = 39.2$). Inter-item reliability between sub-items was found at $\alpha = .932$. Analyses of variance indicated that numerical and skills did not differ significantly across experimental conditions ($F(1, 71) = 0.45$, $p = .506$, $\eta_p^2 = .006$ and $F(1, 71) = 2.57$, $p = .113$, $\eta_p^2 = .035$ respectively). Numerical and spatial skills did not significantly correlate with recall error, accuracy, response time, nor with one another. Therefore, they were not used as covariates in the analyses.

3.2 The SNARC Effect

We first checked for a SNARC effect. Comparisons could have been faster when the larger magnitude was on the right side of the screen and was thus to be selected with the right hand [22]. We tested the effect of congruity (congruent comparisons present the larger magnitude on the right side and incongruent comparisons present the larger magnitude on the left side. Response time was analyzed in a 2 (congruity) × 2 (external representation) ANOVA. No effect of congruity on response time was found in one-by-one comparisons, $F(1, 71) = 0.21$, $p = .652$, $\eta_p^2 = .003$, nor in three-by-three comparisons, $F(1, 71) = 0.08$, $p = .778$, $\eta_p^2 = .001$.

3.3 Recall Error

In the first completion of the recall task, average recall error was found at 10.4% ($SD = 8.98$). We ran a 2 (time of testing) × 2 (external representation) repeated measures ANOVA and found a significant effect of external representation, $F(1, 71) = 45.13$, $p < .001$, $\eta_p^2 = .389$. Figure 3 shows lower recall error in the symbolic condition which could be attributed to an advantage for typing numbers compared to sizing a bar with the mouse. Indeed, a pilot study showed an average error of 6.66% ($SD = 2.99$) when

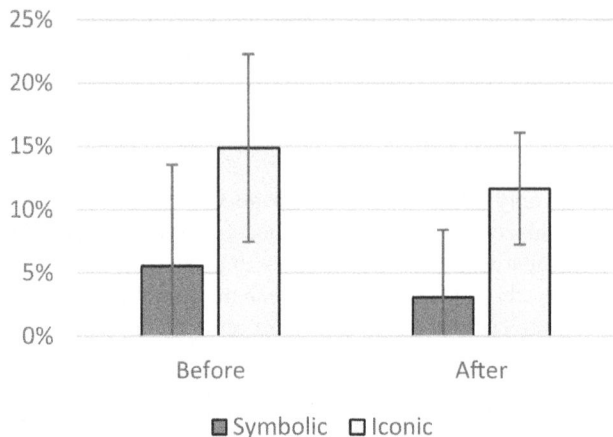

Fig. 3. Mean recall error before and after the mental comparison task

sizing a bar with the mouse immediately after exposure, explaining part of the difference between the two conditions.

We also found a significant main effect of time of testing, $F(1, 71) = 15.9$, $p < .001$, $\eta_p^2 = .183$. Figure 3 shows that recall of energy consumption improved during the experimental session. No significant interaction between external representation and time of testing was found, $F(1, 71) = 0.28$, $p = .601$, $\eta_p^2 = .004$, indicating similar improvement in recall in both conditions. We concluded sufficient learning before the subsequent comparison task with a slight advantage of discreteness in the symbolic condition.

3.4 Accuracy

We ran a 2 (comparison type) × 3 (distance) × 2 (external representation) MANOVA on accuracy scores. Figure 4 shows larger standard deviations for the more difficult small distance comparaisons. Since Mauchly's test indicated a violation of the assumption of sphericity, we used Greenhouse-Geisser corrections.

We found a main effect of distance, $F(1.67, 118.7) = 245.6$, $p < .001$, $\eta_p^2 = .776$ with both linear and quadratic contrasts being significant. Figure 4 shows very high accuracy for comparisons involving a large distance between the (sets of) appliances. Indeed, accuracy approached 100% for one-by-one comparisons and 95% for three-by-three comparisons, dropping to about 85% for one-by-one and to about 70% for three-by-three comparisons.

We found a main effect of type of comparison, $F(1, 71) = 52.62$, $p < .001$, $\eta_p^2 = .426$. One-by-one comparisons resulted in higher accuracy as compared to three-by-three comparisons. No interaction between distance and type of comparison was found, $F(1.62, 115.2) = 25.50$, $p < .264$, $\eta_p^2 = .426$.

Figure 4 shows highest accuracy for the easiest comparisons involving one appliance with another one with a large distance between them, and lowest accuracy for the most difficult comparisons when comparing appliances three by three with a small distance between the total magnitudes on each side. The distance effect occurred in the same manner in the easier one-by-one and the more complex three-by-three comparisons.

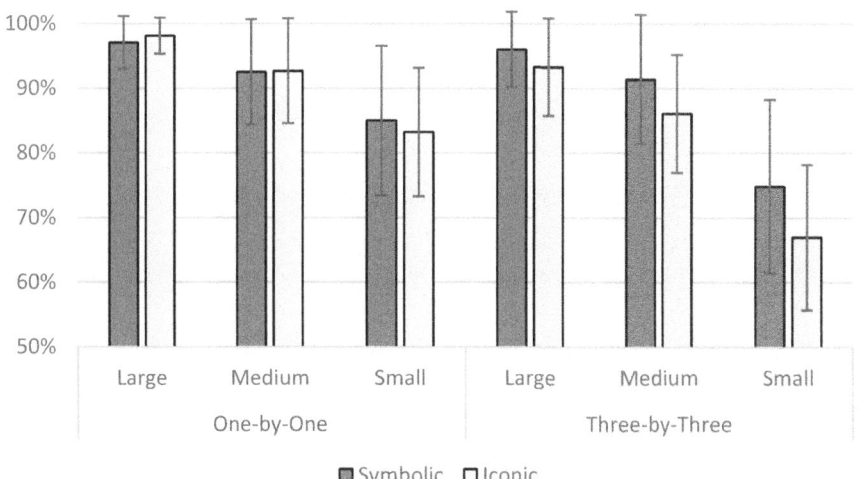

Fig. 4. Mean accuracy as a function of type of comparison, distance in magnitude of energy consumption, and external representation in the learning task

We found a main effect of external representation, $F(1, 71) = 4.38, p = .040, \eta_p^2 = .058$, as well as an interaction effect of comparison type × external representation $F(1, 71) = 7.71, p = .007, \eta_p^2 = .098$. Figure 4 shows an accuracy advantage for learning magnitudes with symbolic representations as compared to iconic representations. The discreteness of the Arabic numerals leads to slightly higher accuracy in mentally comparing magnitudes than learning with the height of bars. The interaction effect indicates that, compared to learning with bars, there was a smaller advantage for learning with Arabic numerals in the one-by-one comparisons, and a larger advantage for three-by-three comparisons. The interaction between distance × external representation did not reach significance, $F(1.67, 118.7) = 2.68, p = .082, \eta_p^2 = .036$, and no three-way interaction was found.

3.5 Response Time

Similarly, we ran a 2 (comparison type) × 3 (distance) × 2 (external representation) MANOVA on response times. Again, Mauchly's test indicated a violation of the assumption of sphericity and Greenhouse-Geisser corrections were used.

We found a main effect of distance, $F(1.50, 106.4) = 78.77, p < .001, \eta_p^2 = .526$, a main effect of type of comparison, $F(1, 71) = 78.47, p < .001, \eta_p^2 = .525$, as well as an interaction between distance and type of comparison, $F(1.48, 104.9) = 7.60, p = .002, \eta_p^2 = .097$. Figure 5 shows faster responses for comparisons involving a large distance between the (sets of) appliances (both linear and quadratic contrasts were

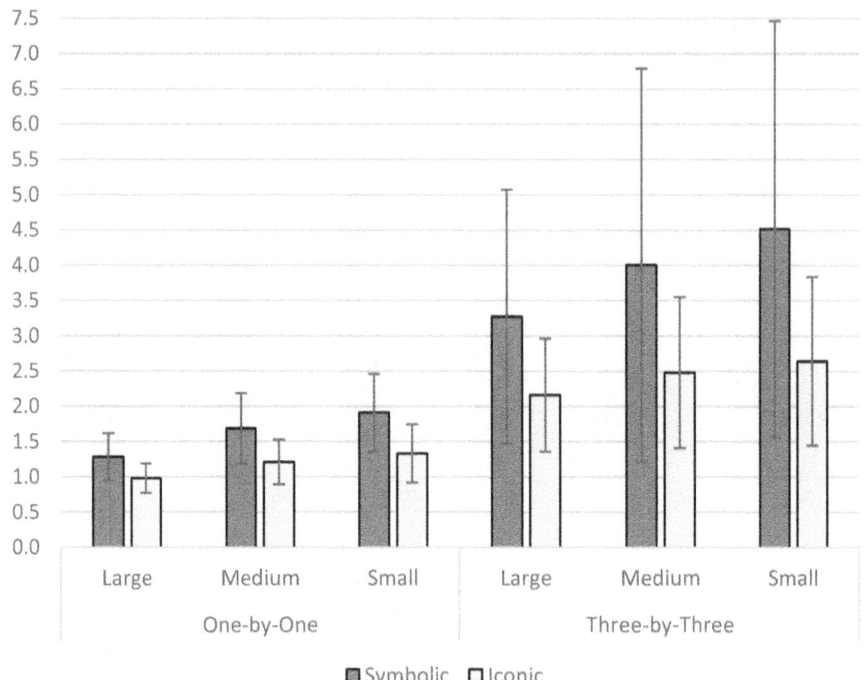

Fig. 5. Mean response time (seconds) as a function of type of comparison, distance in magnitude of energy consumption on each side, and external representation in the learning task

significant) and faster responses for one-by-one comparisons as compared to three-by-three comparisons. Thus, it took much longer to compare whenever there was only a small distance between the energy consumption of (sets of) appliances, and also when comparing sets of three appliances instead of only one by one. Finally, Fig. 5 also allows to interpret the interaction effect between distance and type of comparison. The distance effect is even stronger for the more complex three-by-three comparisons.

We found a main effect of external representation, $F(1, 71) = 16.27, p < .001, \eta_p^2 = .186$. Figure 5 show a speed advantage when mentally comparing magnitudes learned with iconic representations (height of bars) than with symbolic representations (digits).

Both two-way interactions of external representation with comparison type and with distance were also found to be significant ($F(1, 71) = 6.86, p = .011, \eta_p^2 = .088$ and $F(1.50, 106.4) = 11.46, p < .001, \eta_p^2 = .139$ respectively). The advantage of learning with iconic representation emerges for smaller distances and for the more difficult three by three comparisons. Mean response times tripled from about 1 s for large-distance one-by-one comparisons when magnitudes were learned with bars, to about 4,5 s for small distance three-by-three comparisons when magnitudes were learned with digits. The three-way interaction was not significant ($F(1.48, 104.9) = 3.18, p = .060, \eta_p^2 = .043$).

4 Discussion

This study shows that external representation for learning magnitudes of energy consumption affects later performance at recall and mental comparison.

First, results show that learning a magnitude, here a value of energy consumption for fictional household appliances, is possible with both symbolic and iconic external representations. This suggests that learning magnitudes from an iconic external representation is feasible and takes place in the same way as learning magnitudes for entities of a spatial nature, the size of stickmen [16], or of circles [24], or such as learning the relative sizes of animals in everyday life. As expected, due to the relative ease of recalling digits rather than sizing the height of bars, recall error was somewhat lower in the symbolic condition.

Second, a distance effect was found in both accuracy and response time, in both conditions of external representation (symbolic and iconic), and in both types of comparisons (one-by-one and three-by-three). This effect is a manifestation of the analog nature of mental comparisons, where, as in perceptual processes [29], larger differences are easier than smaller differences. The results suggest that mental comparisons of magnitudes of energy consumption are always carried out using an analog internal representation, as shown by Dehaene [21], irrespective of learned external representation or complexity of the comparison.

Third, comparing magnitudes of energy consumption was slightly more accurate in the symbolic condition, but much faster in the iconic condition. The most parsimonious explanation is that higher precision goes at the expense of additional time for the conversion of the symbolic representation into an analog iconic representation. This step is not required for magnitudes learned with iconic representations stored in an analog mental representation. This supports the general hypothesis that mental representations share the properties of initial external representation, as suggested by models of imagery [15, 19]. Furthermore, it shows how mental representations are generated on the fly and processed depending on the task [21].

We also found an interaction between distance and type of external representation. As expected, the distance effect was stronger in the symbolic than in the iconic condition. Comparing magnitudes took the longest time with a small difference in magnitude when learned with a symbolic external representation. It suggests that the generation of a sufficiently precise iconic representation from a symbolic representation takes even longer when differences are smaller. Several retrieve-convert-compare cycles in simple comparisons, or retrieve-convert-add-compare cycles in complex comparisons might be needed for small differences in reaching the required precision.

Finally, the present results are consistent with the grounded cognition's mental simulation hypothesis [30–32]. The process of mental comparison could be summarized like the following. First, on sight of the object to be compared (or its name/silhouette), the associated value is accessed in memory in the format in which it was originally perceived and encoded (digit/height). Then, it is converted into an analog mental representation if needed. Finally, this analog representation is mentally compared via visual-spatial cognitive processes as if it were present before the individual's eyes. These analog processes, which make use of existing perceptual mechanisms such as imaging and visual comparison, explain the present results parsimoniously.

A practical implication is that iconic external representations should be favored in energy management tasks. Although our results indicate that symbolic external representations increase accuracy in recall, iconic external representations may be more conducive to developing magnitude sense [33]. In the case of energy consumption, citizens should not be thought of as rational decision makers using numbers only. They may actually become more efficient decisions makers when acquiring magnitudes of energy consumption with iconic representations. The area of a graphical element, without associated digits, could be used as an iconic representation. For instance, the European Union energy label uses different bar lengths corresponding to different levels of energy efficiency. In our view, the iconic condition resembles the European energy label, since the latter consists of a frame of seven coloured bars A through G without a scale. Longer bars correspond to higher energy consumption, but the lengths of these bars are merely on ordinal level and do not correspond to a specific magnitude. Below it is a number, the annual estimated energy consumption of the appliance, but this is not represented through the area of the bar. Although it would complexify communication and recall, proportional representations of magnitudes would actually benefit rapidity in energy management tasks. In conclusion, the findings underscore the promise of iconic representations as a powerful tool in environmental education and communication. Integrating iconic representations could enhance energy magnitude sense and ultimately play a role in environmental and energy transition.

Acknowledgement. This research was supported by an Academic Research Communities (ARC) doctoral research grant allocated by the Rhone-Alpes Region in France.

References

1. Froehlich, J., Findlater, L., Landay, J.: The design of eco-feedback technology. In: Proceedings of the SIGCHI Conference on Human Factors in Computing Systems, pp. 1999–2008. ACM (2010)
2. Zhang, J.: The nature of external representations in problem solving. Cogn. Sci. **21**, 179–217 (1997)
3. Zhang, J., Norman, D.A.: Representations in distributed cognitive tasks. Cogn. Sci. **18**, 87–122 (1994)
4. Duval, R.: A cognitive analysis of problems of comprehension in a learning of mathematics. Educ. Stud. Math. **61**, 103–131 (2006). https://doi.org/10.1007/s10649-006-0400-z
5. Fischer, C.: Feedback on household electricity consumption: a tool for saving energy? Energ. Effi. **1**, 79–104 (2008). https://doi.org/10.1007/s12053-008-9009-7
6. Wever, R., van Kuijk, J., Boks, C.: User-centred design for sustainable behaviour. Int. J. Sustain. Eng. **1**, 9–20 (2008)
7. Gustafsson, A., Gyllenswärd, M.: The power-aware cord: energy awareness through ambient information display. In: CHI 2005 Extended Abstracts on Human Factors in Computing Systems, pp. 1423–1426. ACM (2005)
8. Gyllensward, M., Gustafsson, A., Bang, M.: Visualizing energy consumption of radiators. In: IJsselsteijn, W.A., de Kort, Y.A.W., Midden, C., Eggen, B., van den Hoven, E. (eds.) PERSUASIVE 2006. LNCS, vol. 3962, pp. 167–170. Springer, Heidelberg (2006). https://doi.org/10.1007/11755494_24

9. Peirce, C.S.: On a new list of categories. In: Proceedings of the American Academy of Arts and Sciences, pp. 287–298 (1868)
10. Peirce, C.S.S.: Prolegomena to an apology for pragmaticism. Monist **16**, 492–546 (1906)
11. Giardino, V., Greenberg, G.: Introduction: varieties of iconicity. Rev. Philos. Psychol. **6**, 1–25 (2015). https://doi.org/10.1007/s13164-014-0210-7
12. Mayer, R.E.: Multimedia Learning. Cambridge University Press, New York (2001)
13. Stenning, K., Oberlander, J.: A cognitive theory of graphical and linguistic reasoning: logic and implementation. Cogn. Sci. **19**, 97–140 (1995)
14. Larkin, J.H., Simon, H.A.: Why a diagram is (sometimes) worth ten thousand words. Cogn. Sci. **11**, 65–100 (1987)
15. Schnotz, W.: Multimedia Comprehension. Cambridge University Press, Cambridge (2023). https://doi.org/10.1017/9781009303255
16. Kosslyn, S.M., Murphy, G.L., Bemesderfer, M.E., Feinstein, K.J.: Category and continuum in mental comparisons. J. Exp. Psychol. Gen. **106**, 341 (1977)
17. Kosslyn, S.M.: Image and Mind. Harvard University Press, Cambridge (1980)
18. Paivio, A.: Mental Representations: A Dual Coding Approach. Oxford University Press, Clarendon Press, New York : Oxford [Oxfordshire] (1986)
19. de Vries, E.: Through the eyes of an archeologist: studying the role of prior knowledge in learning with diagrams. In: Basu, A., Stapleton, G., Linker, S., Legg, C., Manalo, E., Viana, P. (eds.) Diagrammatic Representation and Inference, pp. 315–330. Springer, Cham (2021). https://doi.org/10.1007/978-3-030-86062-2_32
20. Palmer, S.: Fundamental aspects of cognitive representation. In: Rosch, E., Lloyd, B. (eds.) Cognition and Categorization, pp. 259–303. Lawrence Elbaum Associates, Hillsdale (1978)
21. Dehaene, S.: Varieties of numerical abilities. Cognition **44**, 1–42 (1992)
22. Dehaene, S., Bossini, S., Giraux, P.: The mental representation of parity and number magnitude. J. Exp. Psychol. Gen. **122**, 371 (1993)
23. Moyer, R.S., Landauer, T.K.: Time required for judgements of numerical inequality. Nature **215**, 1519–1520 (1967). https://doi.org/10.1038/2151519a0
24. Moyer, R.S., Bayer, R.H.: Mental comparison and the symbolic distance effect. Cogn. Psychol. **8**, 228–246 (1976). https://doi.org/10.1016/0010-0285(76)90025-6
25. Moyer, R.S.: Comparing objects in memory: evidence suggesting an internal psychophysics. Percept. Psychophys. **13**, 180–184 (1973)
26. Cohen, A.S.: Brainscape's "confidence-based repetition" methodology. Brainscape (2008)
27. Ekstrom, R.B., French, J.W., Harman, H.H., Dermen, D.: Kit of Factor-Referenced Cognitive Tests. Educational Testing Service, Princeton (1976)
28. Cousineau, D., Chartier, S.: Outliers detection and treatment: a review. Int. J. Psychol. Res. **3**, 58–67 (2010)
29. Henmon, V.A.C.: The time of perception as a measure of differences in sensations (1906)
30. Barsalou, L.W.: Grounded cognition. Annu. Rev. Psychol. **59**, 617–645 (2008). https://doi.org/10.1146/annurev.psych.59.103006.093639
31. Barsalou, L.W.: Perceptual symbol systems. Behav. Brain Sci. **22**, 577–660 (1999)
32. Wilson, M.: Six views of embodied cognition. Psychon. Bull. Rev. **9**, 625–636 (2002)
33. Dehaene, S.: The Number Sense: How the Mind Creates Mathematics. Oxford University Press, New York (1997)

Designing a Mind-Mapping-Assisted Comparative Literature Course in Chinese Academic Settings

Binfeng Chen[1,2], Jing Zhao[1,2](✉), and Lin He[1]

[1] Fuzhou University, 2 North Wulongjiang Road, Minhou University Town, Fuzhou 350108, China
[2] Fuzhou University Zhicheng College, 50 West Yangqiao Road, Gulou District, Fuzhou 350002, China
zhaojing@fzu.edu.cn

Abstract. Comparative Literature, a course for junior English majors in China, demands advanced cognitive skills and critical thinking to navigate diverse literary works. However, English majors in China traditionally receive intensive foundational language training in the first two years, which often hinders their seamless transition to the analytical, synthesizing, and evaluative skills required in the latter stages of their studies. Therefore, within the Chinese academic context, we design a literature-oriented Comparative Literature Course for English majors assisted by mind mapping tools. Commencing with an orientation on course objectives and the integration of mind mapping for literature review assistance, the 16-week course entails students attending 8-week theoretical lectures and delivering 8-week group presentations. The posted comments and interviews after the implementation of the course indicate that mind mapping acts as a scaffold for the enhancement of higher-order cognitive abilities and logical reasoning among junior students. Besides, the visualization feature of mind maps facilitates students' collaborative learning and interactive communication. Additionally, the paper addresses challenges that may arise in the application of this approach and offering insightful suggestions for its potential implementation in other related courses.

Keywords: mind mapping · comparative literature · scaffold · advanced cognitive skills · collaborative learning

1 Background

The release of "National Standards for Teaching Quality of Foreign Languages and Literature Majors" in 2018 and "Teaching Guides for Undergraduate Foreign Language and Literature Majors in Colleges and Universities" in 2020 reflects a paradigmatic change in Chinese academic settings from merely cultivating compound talents to fostering interdisciplinary humanistic qualities [1, 2]. Listed in these guidelines as a mandatory course in some universities, Comparative Literature is tailored for junior English majors specializing in Foreign Literature. Its teaching content is intricate, encompassing abstract theoretical knowledge and demanding advanced cognitive skills and critical thinking to navigate diverse literary works within various contexts [3]. However, as English majors in China usually receive intensive foundational training in basic linguistic skills in the initial two years [4], it often hinders their seamless transition to the analytical, synthesizing, and evaluative skills required in the latter stages of their studies. Hence, it is worthwhile to explore strategies that effectively scaffold students towards higher academic achievement and facilitate their adjustment to higher-order thinking courses, thereby optimizing the overall learning outcomes.

Against this backdrop, we design a literature-oriented Comparative Literature Course assisted by mind mapping tools in hopes of providing a structured and visually intuitive approach. Mind mapping is generally regarded as a powerful graphic technique, serving as a universal key to unlocking the boundless potential of the human brain and harnessing the full range of cognitive skills [5]. Allowing for the Chinese academic settings, this initiative aims to ease the complexity of theoretical knowledge, fostering a more accessible learning experience for English majors. Additionally, the incorporation of mind mapping is anticipated to enhance their analytical, logical and critical thinking skills, making them better prepared to meet challenges posed by higher-order thinking courses.

2 Literature Review

Mind mapping has been widely acknowledged for its effectiveness in education, aiding in the facilitation of knowledge acquisition and creation [6]. Notably, the mapping process guides the mapper to delve deeper into the information, aiding in the recognition of how new information aligns with an already established knowledge base. Therefore, mind mapping is widely deemed useful for K-12 students in the understanding of new concepts and clarifying the relationships [7]. Nowadays, the adoption of mind mapping in higher educational contexts has become popular, given its perceived utility in advancing learners' cognitive knowledge development [8]. It is noteworthy that targeted mind mapping can assist students in identifying and consolidating the relationships among materials from different classes, thereby enhancing their mastery of their field of study [9]. In general, the mind-mapping techniques align with modern constructivist approaches to learning, emphasizing the active involvement of the learner [10]. This involvement utilizes existing knowledge structures to construct new knowledge by inter-relating new content with the knowledge already stored in memory. Most importantly, the integration of mind mapping into higher education assists college students in brainstorming ideas, facilitating training and development, organizing concepts, and aiding in problem-solving [11]. Nevertheless, despite the widespread adoption of mind mapping in various educational contexts, there has been a dearth of attention given to its utilization in higher-order English major courses, including Comparative Literature.

Among the many advantages of mind mapping, a prominent one is its role in enhancing logical reasoning for mappers. This is achieved through the visual representation and structured organization of information in a coherent and interconnected format. Mind maps adopt a hierarchical structure, emanating from a central idea or topic and branching out into subtopics. This visual hierarchy proves invaluable for organizing information based on relationships and varying levels of significance [12]. Furthermore, mind maps serve as visual representations of concepts and ideas, simplifying the understanding of intricate information. The visual nature of mind mapping allows individuals to perceive the overall picture and the interconnections between different components, thereby facilitating logical reasoning with a clear and comprehensive overview [13].

An additional strength lies in the dynamic nature of digital mind maps, easily modifiable and expandable thanks to the development of different computer-assisted tools. This flexibility empowers users to adapt their logical reasoning processes in response to emerging information or evolving understanding. Importantly, this adaptability underscores another crucial dimension of mind mapping, that is, its effectiveness in a group setting. Peer review of mind maps can assist students in honing their ability to support their ideas by anchoring them in scientific thinking and factual information. Consequently, engaging in group mind mapping might also facilitate students' collaborative learning [14].

3 Present Study

Building upon the benefits of mind-mapping tools and taking into account the academic settings of English majors in China, we have designed and implemented a literature-oriented Comparative Literature course for junior English majors assisted by mind mapping. Acknowledging the differences in article types, we refrain from strictly confining students to the use of specific mind-mapping tools. Instead, we provide a recommended list of tools, including Xmind and Chatmind, aiming to enhance students' learning efficiency. Based on the computer-assisted diagrams, students are then encouraged to collaborate in groups, facilitating the completion of mind maps through discussion and knowledge sharing.

To commence, at the start of the semester, we provided a comprehensive overview to 44 junior English majors, elucidating the essence of Comparative Literature and the methodologies employed in its instruction. Notably, we underscored that the course was not devised to enhance their elementary English language skills such as listening, speaking, writing, or translating. Instead, its primary objective was to cultivate higher-order thinking capabilities and research skills. Specifically, we aimed to better their understanding of the research paradigm within comparative literature and to hone their academic discernment and rigorous research style. Besides, students were explicitly directed toward extensive literature studies. With mind-mapping tools, they were tasked with presenting intricate frameworks, delineating research questions, elucidating argumentative processes, and summarizing conclusions from academic articles and monographs featured in the designated reading list. This approach was implemented to elevate students' engagement, motivating them to delve into the intricacies of comparative literature with analytical precision and critical acumen.

The course spanned a duration of 16 weeks and was structured to maximize student engagement and interaction. In half of the semester, that is, altogether 8 weeks, the instructor delivered lectures elucidating the theories and methodologies of Comparative Literature, offering students a comprehensive understanding of the course's significance, purpose, and practical applications. In parallel, 8 weeks were dedicated to student-led presentations across 15 groups, each consisting of about 3 students, except for one group which had 2 students. These presentations were based on topics that students explored through literature, fostering a collaborative environment for sharing and discussion among peers. Overall, the instructional approach comprised a balanced combination of teacher-led lectures and student seminars, ensuring an interactive and dynamic learning experience.

Specifically, the course unfolded in five distinct segments: the first four sessions (1–4) were devoted to the teacher's introduction of comparative literature theories and methods. Following this, the subsequent four sessions (5–8) were allocated for students to present and discuss articles centered around interdisciplinary themes. The instructional focus shifted in the subsequent three teacher-centered sessions (9–11) to explore concepts related to world literature, literary reception, and translation. The final four sessions (12–15) were dedicated to student presentations and discussions on articles of their choice, allowing for the exploration of diverse topics of world literature and cross-cultural studies. The concluding week (16) was designated for the comprehensive wrap-up of the course, providing an opportunity for summarization, reflection, and a holistic review of the key ideas and insights acquired throughout the semester. The course arrangement is presented in Table 1.

Table 1. Comparative Literature Course Arrangement

Sessions	Teaching Content	Features
1–4	General introduction to theories and methods of comparative literature	Teacher-centered
5–8	Student presentation and discussion on interdisciplinary themes	Student-centered
9–11	Theoretical introduction to concepts related to world literature, literary reception, and translation	Teacher-centered
12–15	Student presentation and discussion on topics of world literature and cross-cultural studies	Student-centered
16	Summary and reflection	Teacher-centered

Throughout the lecture weeks, students were required to engage in the comprehensive reading of pertinent literature and conduct an in-depth examination of 1–2 articles from the assigned reading list. To enhance their understanding, they were instructed to create mind maps. During the two separate seminar sessions, each group was tasked with selecting two articles: one on interdisciplinary themes and the other on topics of world literature and cross-cultural studies. The groups then collaboratively prepared mind maps to aid their class presentations. These presentations were followed by feedback and discussions that extended beyond their own group, fostering a broader exchange of ideas. With minor changes based on the in-class feedback, each group uploaded the mind maps onto the learning platform, specifically storing them in the QQ group folder for the benefit of other students' reference. Based on the feedback, they are also required to improve the diagrams guided by the teacher. The three-phase teaching procedure by using mind mapping is shown in Table 2. In the end, 30 mind maps were collected, with 15 mind maps of the interdisciplinary themed articles and 15 on world literature and cross-cultural topics.

Table 2. The Use of Mind Mapping in the Comparative Literature Course

Phases	How the Mind Mapping Tools are used	Features
Before class	Completion of mind maps within the group	Tool-assisted and collaborative within the group
In class	Class presentation and discussions aided by mind maps	Collaborative beyond the group
After class	Further improvement and feedback	Teacher-guided

4 Findings

Following a semester of teaching practice, we observed that students' comments posted on the teaching assessment platform initiated by the University Teaching Affairs Office generally hold positive views. The end-of-the-semester scoring for the course reached as high as 97.925 out of 100, much higher than the average score in the university (81.675). This, to some extent, demonstrates their general high acceptance of mind mapping in the course study. Most notably, based on the comments and post-class interviews, students generally believe that mind mapping effectively facilitates their cognitive skills.

First and foremost, mind mapping proves instrumental in enabling students to engage in the learning process and improve their cognitive skills. As one student comments, "by leveraging visual hierarchy and sequential flows in the process of reading, mind mapping is instrumental in facilitating my understanding and logical thinking by structuring information according to relationships, levels of significance, and logical coherence". For her, mind mapping is learner-controlled, not teacher-centred or technology-driven; it engages learners actively in the creation of knowledge, reflecting their understanding and conception of information, rather than prioritizing the presentation of objective knowledge. Grounded in a constructivist epistemology, cognitive tools usher in a transformative educational focus, shifting from the mere presentation of fundamental knowledge to motivating learners to construct reflective knowledge based on their understanding [15]. This paradigmatic shift, highlighting a learner-centered, active process of creation, contrasts with traditional, teacher-dominated, and passive teaching methods. The shift is imperative for English majors in China, who are often challenged by active participation in predominantly informative lectures. In essence, mind mapping emerges as a powerful tool, not only fostering students' active participation and creative generation but also revolutionizing the conventional teaching paradigm for English majors. By establishing a learning environment compelling learners to engage in thoughtful, graphically represented cognition, mind mapping facilitates holistic and systematic graphic representation, a challenge without the aid of such tools [16]. Consequently, mind mapping acts as a supportive scaffold for higher-order cognitive processes and logical reasoning, providing a pathway for English majors in China to explore their advanced thinking levels that might otherwise be elusive [8].

Secondly, the visual nature of mind maps facilitates students' collaborative learning and interactive communication. "The process can be demanding, as we have repeatedly discussed on the points, but it is worthwhile", as is reported by an interviewee. In the mapping process, mappers must meticulously identify these patterns through scrutiny, potentially collaborating with peers either within or beyond the group. What is more, as the internal logic of the given text is sometimes challenging to discern, refining the mind map requires additional details for enhanced completeness and accuracy. The comprehensive yet ambiguous nature of mind map node relationships requires mappers to reevaluate and identify these relationships when interpreting the diagram. To ensure a coherent and logical structure within the mind map, the mapper needs to focus on establishing relationships during the mapping process. Post-class interviews indicated that mappers must discern the internal logic within the diagrams, swiftly read associated literature, organize information with the aid of classmates' mind maps, and actively engage in discussions. Simultaneously, they might also identify logical inconsistencies within mind maps or statements derived from them. The mind maps created by students underwent revisions through iterative discussions. Refinements were made by incorporating suggestions from both the mappers within the groups and external input from fellow students and the guiding teacher. For example, an interviewee recounted his experience that in some sections of the article where the logic was less rigorous, mind maps could become particularly complex, with various relationships intertwined, adding a greater cognitive load to him. It is evident that the dual aspects of "comprehensiveness" and "ambiguity" not only present a challenge to the mind mappers but also encourage them to leverage the flexibility and versatility of the map [17]. This proactive approach helps mitigate issues such as inherent illogicality. Ultimately, collaborative learning becomes a catalyst for the meaningful construction of knowledge within the group, leveraging the nuanced attributes of mind maps [14].

Hence, incorporating mind mapping into the literature studies for English majors serves as a valuable mechanism for guiding students through the transition from imitative and drill-based language skill learning to more sophisticated, academic constructs centered on problem-finding and problem-solving [11]. Thus, engaging in literature studies becomes an essential step in cultivating rigorous academic thinking. However, initiating the process of reading literature poses some challenges for these English major students in China, which merits our careful consideration.

First of all, to facilitate the initiation of literature studies, teachers must adopt an open and inclusive approach when assigning tasks. As a student mentioned in the comments, what truly matters is the tool's flexibility and the freedom to select topics that genuinely pique their interest. For instance, the initial presentation topic revolves around interdisciplinary studies and theoretical explorations, with students having the liberty to choose such themes as "Posthumanism", "Intersectional Studies" in "Humanities and Technology" and "Literature and Music", allowing students to link comparative literature to their interested topic. The second session of presentation delves into "World Literature" or "Literary Translation and Reception", granting students the freedom to opt for theories of world literature or present case studies on literary reception and translation between languages. This approach empowers students, including those new to literature, to follow their passions and make personal choices, thereby satisfying their innate curiosity to explore uncharted territories.

Moreover, feedback and suggestions from the teacher and peers significantly enhance the role of mind mapping as a scaffold. We observed that in the initial stages of literature reading, when students lack academic awareness and a solid knowledge structure, the mind maps they create may be too concise or exhibit a confusing organizational structure, hindering their effectiveness as a clear scaffold. However, when teachers and classmates identify issues such as missing information or unclear logic, students can then target the specific content of the literature, logically reorganize, and modify the presentation of each node. This iterative process incorporates the collective academic advanced thinking of the team. As a result, learners has not only improved their disciplinary awareness but also gained a deeper insight into the academic research during this collaborative refinement process [6].

5 Conclusion

Our exploration into the integration of mind-mapping tools in the Comparative Literature course for junior English majors has yielded promising outcomes. The traditional challenges faced by English major juniors in China have been addressed through a structured 16-week course that emphasizes advanced cognitive skills and logical thinking assisted by mind-mapping tools. The efficacy of the constructivist mind mapping approach is evidenced by its role as a scaffold, fostering the higher-order cognitive development and logical reasoning among junior students. Moreover, the visualization aspect of mind maps has proven instrumental in promoting students' collaborative learning and interactive communication, adding an extra layer to students' educational experiences. As we look forward, challenges inherent in this approach are acknowledged, and pragmatic suggestions have been provided for its potential implementation in other related courses. By continuing to refine and expand the application of mind mapping in higher education, we anticipate further advancements in enhancing the cognitive abilities and collaborative skills of English majors within the Chinese academic context.

Acknowledgments. This study was funded by Fujian Social Sciences Fund (grant number FJ2022C013).

Disclosure of Interests. The authors have no competing interests to declare that are relevant to the content of this article.

References

1. Ministry of Education Higher Education Teaching Steering Committee: National Standards for the Teaching Quality of Undergraduate Majors in Colleges and Universities. Higher Education Press, Beijing (2018)

2. Foreign Language and Literature Teaching Committee of Ministry of Education: Teaching Guides for Undergraduate Foreign Language and Literature Majors in Colleges and Universities. Foreign Language Teaching and Research Press, Beijing (2020)
3. Zeng, Y.: On the cultivation of talents for comparative literature and cultural studies at undergraduate English major. J. Foreign Lang. World **208**(1), 14–21 (2022)
4. Li, M., Hu, X.: English major education in China: a chronological analysis. Onomazein **60**, 70–87 (2021)
5. Liu, Z., Kong, X., Liu, S., Yang, Z.: Effects of computer-based mind mapping on students' reflection, cognitive presence, and learning outcomes in an online course. Distance Educ. **44**(3), 544–562 (2023)
6. Renfro, C.: The use of visual tools in the academic research process: a literature review. J. Acad. Librarianship **43**(2), 95–99 (2017)
7. Chang, C., Hwang, G., Tu, Y.: Concept mapping in technology-supported K-12 education. J. Educ. Comput. Res. **60**(7), 1637–1662 (2022)
8. Shi, Y., Yang, H., Dou, Y., Zeng, Y.: Effects of mind mapping-based instruction on student cognitive learning outcomes: a meta-analysis. Asia Pac. Educ. Rev. **24**(3), 303–317 (2023)
9. Schwendimann, B.: Concept maps as versatile tools to integrate complex ideas: from kindergarten to higher and professional education. Knowl. Manag. E-Learn. **7**(1), 73–99 (2015)
10. Dhindsa, H., Kasim, M., Anderson, O.: Constructivist-visual mind map teaching approach and the quality of student' cognitive structures. J. Sci. Educ. Technol. **20**(2), 186–200 (2011)
11. Fun, C., Maskat, N.: Teacher-centered mind mapping vs student-centered mind mapping in the teaching of accounting at pre-U level – an action research. In: International Conference on Learner Diversity 2010, vol. 7, pp. 240–246. Elsevier, Bangi (2010)
12. Swestyani, S., Masykuri, M., Prayitno, B.A., Rinanto, Y., Widoretno, S.: An analysis of logical thinking using mind mapping. J. Phys. **1022**, 012020 (2018)
13. Naghmeh-Abbaspoura, B., Rastgoob, V., Fathic, N., Mortazavi, Y.: The use of mindomo software to improve the logical development of EFL learners' writing. Int. J. Innov. Creat. Change **8**(12), 238–252 (2019)
14. Suharto, R., Rahayu, E., Agustina, H.: The use of mind map in collaborative learning activities of a literary reading class. Briliant **8**(3), 543–550 (2023)
15. Zhao, G., Yang, X., Xiong, Y.: On the principles and focal points of applying thinking visualization tools in teaching and learning. E-Educ. Res. **40**(09), 59–66 (2019)
16. Jones, B., Ruff, C., Snyder, J., Petrich, B., Koonce, C.: The effects of mind mapping activities on students' motivation. Int. J. Scholarsh. Teach. Learn. **6**(1), 060105 (2012)
17. Buzan, T.: Mind Mapping. BBC Active, Harlow (2006)

Integration of Learning Through the Use of Self-constructed Diagrams: Opportunities and Challenges

Emmanuel Manalo(✉) and Mari Fukuda

Graduate School of Education, Kyoto University, Kyoto, Japan
{manalo.emmanuel.3z,fukuda.mari.6y}@kyoto-u.ac.jp

Abstract. The integration of what has been learned so that they become coherent is crucial for learning to become genuinely useful. Despite this importance, integration of learning (IOL) is not usually explicitly taught in formal education. Previous studies have examined various methods to facilitate IOL, and one method shown to be effective is the provision of diagrams to use. However, the usefulness of asking students to create their own diagrams to integrate what they have learned appears not to have been examined yet. That is therefore what we set out to explore in this study. We asked university undergraduate students to create a diagram to explain a process that required integration of information that had been covered in their course. The diagrams that 25 of the students produced were all variations of a flow chart, indicating that they were able to select the most appropriate kind of diagram for representing a process. However, the students varied in the quality of their representation of crucial components of the process, including connections between and within those components. Our findings suggest that constructing diagrams for IOL can serve useful learning diagnostic functions, and that it would be helpful to provide some instruction or guidance to some students so that they can effectively construct their own diagrams for IOL.

Keywords: Self-Constructed Diagrams · Integration of Learning · Apprehension of Meaning · Linking Learning · Student Instructional Needs

1 Introduction

It has long been known and generally accepted that learning depends on the establishment of connections [1]: we learn because we connect different components of what we encounter and experience. However, apart from learning based only on a single or limited connections, such as a new word to its meaning or an action taken to its consequence, we usually connect clusters of these learnings together to achieve a more cohesive and comprehensive understanding. Isolated items of learning (e.g., rote memorized foreign vocabulary words and their definitions, without connection to sentence use in relevant contexts) are usually limited in usefulness and tend to be easily forgotten. Integration of learning (IOL) is crucial for learning to become genuinely meaningful and useful to us [2, 3]. However, despite its importance, IOL is not usually explicitly taught or cultivated in formal education.

Many teachers may have an implicit expectation that students will spontaneously integrate the appropriate components of what they are learning and/or that they will develop skills for integration as their knowledge base expands. Furthermore, some tasks given in class – such as writing summaries and working on tasks that require application of learning – may effectively facilitate IOL. However, unlike other thinking skills/competencies such as critical thinking, analysis, and logical thinking, not many teachers explicitly focus on IOL as a skill to develop and use in the process of designing curricula and in deciding instructional approaches to use.

Although not many, there are research studies that have investigated how effective IOL can be facilitated. Instructional approaches that have been found to work well include the provision of principles or normative concepts that would enable students to understand the connections between different components of knowledge they acquire [4, 5]; the use of constructivist learning environments in which the teacher provides a high degree of instructional support [6]; and the provision of opportunities for structured application and reflection so that students would apprehend the meaningful connections in what they learn and experience [7].

The provision of appropriate diagrams which illustrate the mechanisms that need to be learned has also been shown to be effective [8, 9]. This is quite understandable given that one important quality of diagrams is that they group together in the one location all information that needs to be used together [10]. These previous studies had provided the diagrams for students to use. As far as we are aware, the use of self-constructed diagrams as a tool for students to use in integrating various components of what they have learned has not been properly investigated. The present study therefore was an attempt at trying to understand the extent to which students are able to construct diagrams that might serve such integrative purposes.

The questions we were trying to answer were as follows: When asked to create a diagram to integrate several components of what they have learned,

(i) What kinds of diagrams do students construct to represent and integrate the processes involved in those components?
(ii) How well are they able to represent the necessary components?
(iii) How well do they represent the connections *between* components?
(iv) How well do they represent the connections *within* components?

2 Method

This research comprised analysis of one piece of coursework that undergraduate students completed as part of their regular class session task in a national university in Japan. No experimental intervention or manipulation was made in this study. The students provided written permission for use of their work, and the analysis was undertaken 6 months after the course had finished (it had no bearing on grades). The course was an introduction to educational psychology, and it was taught in English.

Participants. There were 33 students in the course, but 8 students did not submit the coursework in question, so we analyzed the work produced by 25 students (female = 10, male = 15; Japanese = 19, international students = 6).

Procedure. Prior to the class task, the students had received instruction over three class sessions about how humans interact with their environments, and how physiological functions (including those of subcortical and cortical regions of the brain) play crucial roles in that interaction. The task given to the students was to: "Draw a diagram that would show the process, mechanisms, and links from a 'state of need' to 'satisfaction of the need' in socially acceptable ways. Consider what happens when a human is in a state of need? Then what happens next?". As this particular process had not been explicitly discussed in class yet, drawing the diagram would have required the students to integrate what they had learned. We determined that such a diagram would have required showing various connections between and within 7 components from the deficit in the person leading to a disequilibrium, to restoration of equilibrium, and including the hypothalamus, the limbic system, and the cortex of the brain and their functions.

About 30 min of class time was spent on this task, during which students could not only consult their notes and other materials, but also talk with each other if they wanted to. If they did not finish their diagram during the class, they could continue working on it later to finish it. Like other class tasks, they were asked to include this work in their portfolio which they had to submit at the end of the semester.

Analysis of the Diagrams. After discussing and deciding what a diagram that satisfactorily meets the requirements of the task ought to show, we drew up three sets of scoring rubrics to score for the presence and quality of components that needed to be included, presence and kinds of connections *between* components, and presence and kinds of connections *within* components. As an example, the rubrics for scoring the presence and quality of components are shown in the Appendix. We conducted the scoring of the diagrams independently, but in stages (i.e., first 5, next 10, and final 10 diagrams) so that we could check and discuss agreements and disagreements, and resolve the disagreements at each stage until we could reach perfect agreement [11].

3 Findings and Discussion

3.1 Kinds of Diagrams the Students Constructed

Even though at first glance many of the diagrams the students constructed did not look alike, they all fit into the broad category of flow charts. Examples are shown in Fig. 1. The construction of flow charts is probably understandable given that the students were asked to show a process that progresses from one state to another. We have in fact found in an earlier study that the kind of diagram most frequently used by students to portray processes was a flow chart [12], so this current finding provides confirmation of that. In both the earlier study and the current one, no explicit instruction was provided to the students on the kind(s) of diagrams to use, so these findings suggest that, by the undergraduate level, many (if not most) students can determine that flow charts are usually the most appropriate to use in representing processes. This would likely stem from having seen such diagrams representing processes in instructional materials, such as textbooks and their teachers' presentation slides and/or board notes.

Integration of Learning Through the Use of Self-constructed Diagrams   361

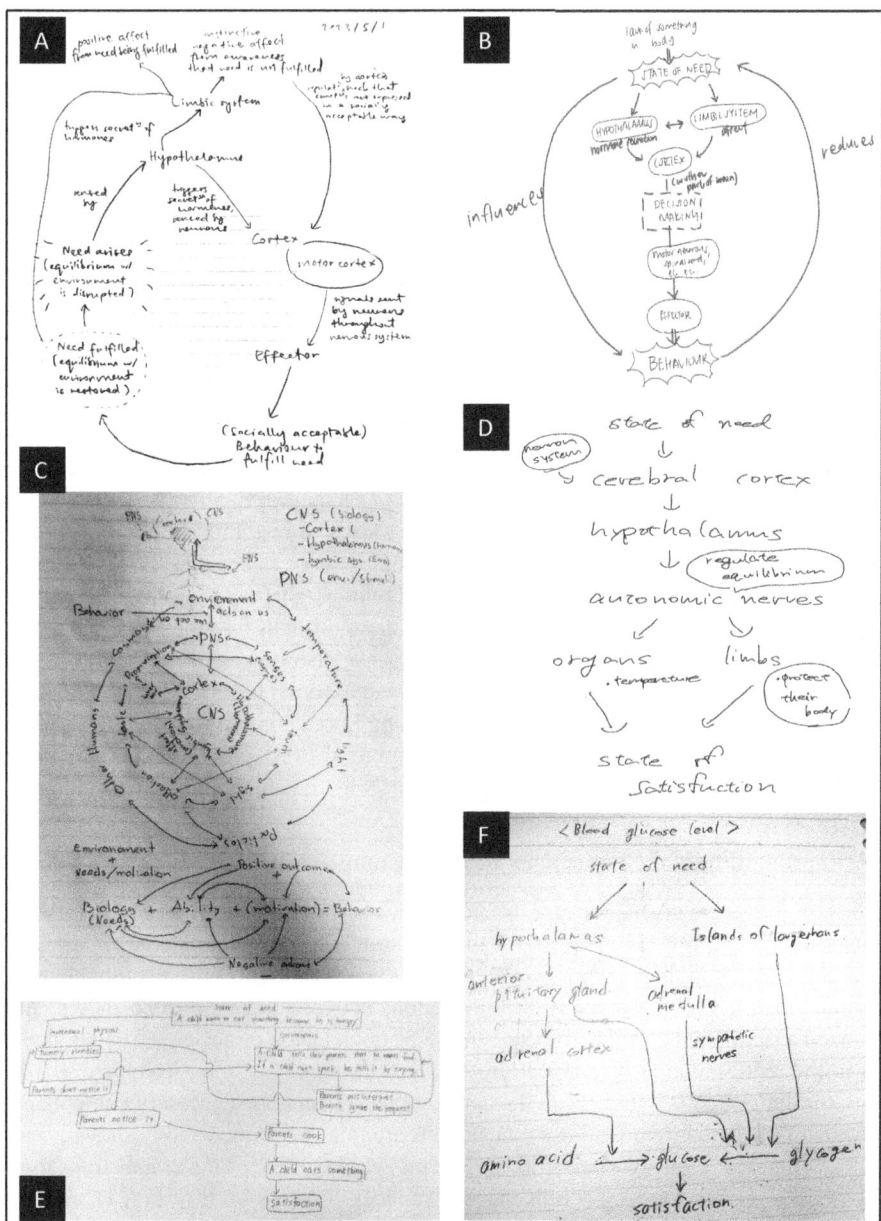

Fig. 1. Examples of diagrams that the students constructed to integrate learning

3.2   Representation of the Necessary Components in the Students' Diagrams

The students' diagrams also varied in the extent to which they represented the components deemed necessary for the task (shown in the Appendix). Figure 1 shows diagrams

A and B which scored the highest in components score (they represented all the components) and in the *quality* of representation of those components. In contrast, E and F scored the lowest in both components represented and in quality of representation.

The correctness of the details portrayed in the diagrams mattered. While in Fig. 1 diagrams A and B show those details correctly, in diagrams E and F the details included are mostly incorrect or irrelevant to the task given. In diagram E, none of the details about the parts of the brain and their functions have been included; and in diagram F, the details contain errors and are mostly different from those covered in the course.

There were also differences in the extent to which the components were represented, and how well the students represented them. While Component 2A ("the hypothalamus enables awareness of the need") was represented by 84% of the students and also scored the highest in quality of representation (mean = 2.40, SD = 1.53), Component 1 ("need leads to disequilibrium") was only represented by 24% of the students and scored the lowest in quality of representation (mean = .68, SD = 1.35), a score which was significantly lower compared to those of Components 2A, 2B, 3 and 4 (range of Cohen's d for the planned comparisons of presence: 1.24–2.08, and of quality: 1.39–1.99). We also found that Component 6 ("reduction of need leads to restoration of equilibrium") was second lowest in representation (52%) and quality score (mean = 1.52, SD = 1.73), supporting previous findings that processes that are more abstract (i.e., in this case, not involving tangible organs/parts of the brain) are more difficult for students to visualize and visually represent [13]. Such processes may have also been more difficult for students to comprehend as the ability to create alternative forms of representation of what has been learned is one indicator of deeper learning [14]. This suggests that instructors may need to pay more attention to explaining more abstract processes, and consider asking students to try to construct diagrams as one way to gauge their understanding.

3.3 Representation of Connections *Between* Components

In the task they were given, because the students were integrating the necessary components of a process, it was important for them to show the correct connections between those components in their diagrams. Of course, it was important for the components to be represented in the first place if the connections between them are to be shown, and we found a significant correlation between scores in inclusion of necessary components and in the representation of connections between components, $r = .766, p < .01$. In diagrams like A and B in Fig. 1, all the connections between the necessary components are shown, but in diagram D, those connections are missing or incorrectly shown.

Regarding the kinds of connections used: on average, 42% of the connections used arrows, 15% used words (e.g., "influences", "reduces" in diagram B of Fig. 1), and 4% used lines. Thus, arrows were the most popular way used to connect components in the flow charts. However, the arrows were rarely labelled, making the actual effect or influence they represent somewhat unclear in some cases (e.g., the numerous two-headed arrows shown in diagram C of Fig. 1). This clarification of what connectors mean in diagrams and how those can be made clearer warrants investigation in future research.

Connections 5 (Cortex to action to reduce need), 6 (Cortex to control to ensure social acceptability), and 7 (Needs reduction to restoration of equilibrium) were the least represented of the *between* components connectors, with only 36% of students including each of those. Where connection 7 is concerned, the low representation may be understandable in terms of the components not being adequately concrete or tangible, as we noted earlier. Where connections 5 and 6 are concerned, it may just be that many students did not fully comprehend the different parts of the cortex, the functions they serve, and how those relate to everyday human actions. Not fully understanding those details would have made it hard, for example, to see the cortex – and more specifically the frontal lobe areas responsible for decision making – being connected to responding in a socially acceptable way to fulfil a need.

3.4 Representation of Connections *Within* Components

For each component of a process to make sense, the information within it needs to be sufficiently cohesive, and that requires depicting any necessary connections. The importance of connections within components of a diagram is confirmed in the present study by our finding of a significant correlation between the scores in quality of the diagram components and in representation of connections *within* components, $r = .934, p < .01$. This means that the clarity of what each component of the diagram conveys depends on whether or not crucial connections between elements within those components are shown or not. Again, diagrams A and B in Fig. 1 are good examples where the connections within each component are all shown. In contrast, diagram F in Fig. 1 portrays none of the necessary connections within components.

Regarding the kinds of connections that the students used to depict connections within the components, 27% used arrows, 18% used words, and 2% used lines. These percentages are lower than those the students used for depicting connections *between* components (see above), but the distribution across the kinds of connectors is more or less the same (i.e., arrows were used the most, while lines were used the least).

Connection 1 (Need leads to disequilibrium) was the least represented of the within component connections with only 21% of students depicting that connection in their diagrams. However, connection 5 (Cortex to motor area, action and/or behavior), 6 (Cortex to control of action for social acceptability), and 7 (Need reduction leads to restoration of equilibrium) were also not well represented in the diagrams with only 36%, 32%, and 28% of the students depicting them, respectively. The reasons for the low representation of these are likely to be similar to those previously explained. Connections 1 and 7 were likely due to their elements being less concrete/tangible in nature (i.e., not including any part of the brain or human physiology), and connections 5 and 6 were likely due to students not fully comprehending those connections to begin with.

3.5 Conclusion: Opportunities and Challenges in Using Diagrams for Integration of Learning

From the analysis of the students' performance in this study, it appears that giving the task of drawing a diagram to integrate various components of what has been learned can potentially serve very useful functions for both students and teachers. For students, drawing the diagram can confirm their understanding and/or enable them to identify aspects that they may not have fully understood (i.e., the parts that they may struggle or even fail to represent). For teachers, the student-constructed diagrams can likewise confirm the students' grasp of the subject material, as well as reveal any gaps and/or misconceptions in their understanding of that material.

There are, however, also challenges. One of those is that some components of learning and their connections – especially those that are more abstract – can be very difficult for students to effectively visualize and represent in diagrammatic form. Hence, developing and evaluating the efficacy of methods for cultivating student skills in visually representing more abstract concepts and processes would be useful to pursue in research. It would also be useful to investigate the use of different kinds of diagrams (other than flow charts) to match variations in the nature of the learning content that needs to be integrated. Finally, it would be valuable in future studies to determine the extent to which successful integration of learning might depend on understanding of the content to be integrated. The opposite should also be investigated: the extent to which IOL through diagram construction might lead to improvements in understanding.

Acknowledgment. This research was supported by grants-in-aid (20K20516, 23H00065) received from the Japan Society for the Promotion of Science.

Appendix

Rubrics for Scoring the Presence and Quality of Diagram Components
Components that need to be shown

[1] Need leads to disequilibrium: Need (lack or deficit in the person) leads to upsetting of equilibrium, or creates an imbalance in equilibrium

[2A] Hypothalamus enables awareness of need: Disequilibrium is detected by the hypothalamus which produces hormones that enable awareness of hunger, thirst, etc.

[2B] Limbic system enables emotional reactions: Limbic system also detects disequilibrium which leads to emotional reactions like unhappiness and motivation to find food

[3] Hypothalamus and limbic system connect to cortex: This enables actual awareness/sensations of need (sensory cortex), and awareness and expression of emotions

[4] Cortex enables movement/action to reduce need: Through the motor cortex movements can be made to find food, water, do what is necessary to reduce need

[5] Cortex controls the response to ensure it is socially acceptable: Through the frontal lobe (responsible for executive functions) thinking and deciding what is appropriate to do becomes possible (e.g., not just grab another person's food or drink)

[6] Reduction of need leads to restoration of equilibrium: Once the need is reduced, the hypothalamus stops producing hormones for hunger, etc., leading to feelings of satiation and pleasure through the limbic system and cortex, and action to reduce need stops.

Quality Scoring of Representation of the Components Above

$5 =$ The relevant part(s) of the brain/process is(are) included, and what happens is clearly shown or stated

4 = The relevant part(s) of the brain/process is(are) included, and what happens is shown or stated, but may be a bit vague
3 = The relevant part(s) of the brain/process is(are) included, and what happens is shown or stated, but either or both of these may be incomplete or contain a minor error
2 = The relevant part(s) of the brain/process is(are) included, OR what happens is adequately shown or stated - but not both (either the part or what happens is missing)
1 = The relevant part(s) of the brain/process is(are) included, OR what happens is shown or stated - but not both (either the part or what happens is missing), AND there is some lack of clarity in what is included or shown
0 = Not included or shown.

References

1. Thorndike, E.L.: Human Learning. Century, New York (1931)
2. Locke, J.: Curriculum integration in secondary schools. Curriculum Matters **4**, 69–84 (2008)
3. Manalo, E.: The need to cultivate more linking in learning to promote more effective thinking. In: Rezaei, N. (ed.) Integrated Education and Learning. IS, vol. 13, pp. 73–94. Springer, Cham (2022). https://doi.org/10.1007/978-3-031-15963-3_5
4. Linn, M.C.: The knowledge integration perspective on learning and instruction. In: Sawyer, R.K. (ed.) The Cambridge Handbook of the Learning Sciences, pp. 243–264. Cambridge University Press, Cambridge (2006)
5. McCrudden, M.T., Kulikowich, J.M., Lyu, B., Huynh, L.: Promoting integration and Learning from multiple complementary texts. J. Educ. Psychol. **114**, 1832–1843 (2022)
6. Flaig, M., Simonsmeier, B.A., Mayer, A.-K., Rosman, T., Gorges, J.: Reprint of "conceptual change and knowledge integration as learning processes in higher education: a latent transition analysis. Learn. Individ. Differ. **66**, 92–104 (2018)
7. Perez, R.J., Barber, J.P.: Intersecting outcomes: promoting intercultural effectiveness and integration of learning for college students. J. Divers. High. Educ. **11**, 418–435 (2018)
8. Ainsworth, S., Th Loizou, A.: The effects of self-explaining when learning with text or diagrams. Cogn. Sci. **27**, 669–681 (2003)
9. Butcher, K.R.: Learning from text with diagrams: promoting mental model development and inference generation. J. Educ. Psychol. **98**, 182–197 (2006)
10. Larkin, J.H., Simon, H.A.: Why a diagram is (sometimes) worth ten thousand words. Cogn. Sci. **11**, 65–99 (1987)
11. Brinkmann, S., Kvale, S.: InterViews: Learning the Craft of Qualitative Research Interviewing. Sage, Los Angeles (2015)
12. Manalo, E., Fukuda, M.: Diagrams in essays: exploring the kinds of diagrams students generate and how well they work. In: Basu, A., Stapleton, G., Linker, S., Legg, C., Manalo, E., Viana, P. (eds.) Diagrams 2021. LNCS, vol. 12909, pp. 553–561. Springer, Cham (2021). https://doi.org/10.1007/978-3-030-86062-2_56
13. Manalo, E., Uesaka, Y.: Students' spontaneous use of diagrams in written communication: understanding variations according to purpose and cognitive cost entailed. In: Dwyer, T., Purchase, H., Delaney, A. (eds.) Diagrams 2014. LNCS, vol. 8578, pp. 78–92. Springer, Berlin (2014). https://doi.org/10.1007/978-3-662-44043-8_13
14. National Research Council: Education for Life and Work: Developing Transferable Knowledge and Skills in the 21st Century. National Academies Press, Washington, DC (2012)

Chinese Children' Drawing in Science Class

Ran Lu(✉) and Emmanuel Manalo

Graduate School of Education, Kyoto University, Kyoto, Japan
lu.ran.65j@st.kyoto-u.ac.jp, manalo.emmanuel.3z@kyoto-u.ac.jp

Abstract. Numerous studies have shown that the use of drawing in primary school classrooms is effective. Drawing helps children understand scientific concepts, express their ideas, and communicate. However, the use of drawing in classrooms in China is limited. In this qualitative exploratory study, we examined the role of drawing in a primary school science classroom in China. The teacher used drawings at all stages of the implementation of the teaching, and provided the 12 students of the class with drawing options when they were asked to complete tasks. Before and after a teaching intervention, we administered questionnaires and conducted interviews with students. Through analysis of the students' drawings and information from a survey and interviews, we found that the students all subsequently chose to complete the tasks by drawing independently and, over time, the teacher reported that the students were able to complete the tasks faster and better. Through drawing, the students also behaved more actively during group communication activities. In addition, we found that the drawing examples provided by the teacher greatly influenced the students' drawings. These findings provide support for the effectiveness of drawing in primary school science classrooms in China. However, we also found some problems in the use of drawing and we propose some ways for solving such problems.

Keywords: Science education · Children' drawing · Chinese science class

1 Introduction and Theoretical Background

Drawings are visual representations that belong in the category of "diagrams", as broadly defined in this forum. According to science education research, drawing can enhance students' conceptual understanding [1], their ability to communicate knowledge [2], and their ability to critically use and interpret visual media [3]. Also, in teaching, the examples teachers provide have a large impact on how children represent science content in their drawings [4]. Apart from enhancing students' participation in learning activities and helping them learn to represent information, drawing can also be used as a learning strategy to help students learn and communicate more effectively [1]. Drawing is not only a way to record and express information, but also a learning method to promote reasoning, design conceptualization, modeling, and other advanced cognitive development [5]. However, drawing is not necessarily appropriate all the time in the science classroom, and if students are not prepared or the teacher does not provide appropriate help, the students can end up performing worse, thus discouraging their learning efforts. Therefore, the use of drawing requires the teacher to choose appropriate topics in which to implement it [6].

Previous work by Areljung and colleagues [4], Westlund [7], and Kress and van Leeuwen [8] have proposed various ways by which children may represent science content, including representations of such content as theory, natural experience, event, art, person, attitude evoking, and cultural heritage. In representing science as *theory*, the science content gets divided and depicted in terms of its analytical units (from the scientific point of view), using minimalist shapes, without much detail and with little if any coloring. As *context*, the content is placed in a setting where it is likely to be found (often based on the child's experience), accompanied by details that integrate the content into that setting. As *event*, the drawings convey movement or processes (what happens), and visual techniques such as alignment, contrast, proximity, and symbols (e.g., arrows, lines) are often – but not always – employed. As *art*, the child represents the content using vibrant colors, decorations, patterns, and various shapes that can make that content visually arresting or interesting. In representing the content as *person*, the drawings may include the face and body or parts of it, and human characteristics – even when the content is non-human. As *attitude evoking*, the drawing of the science content evokes attitude and emotions toward it through the use of emoticons and other symbols that convey meaning of the child's understanding of that content. Finally, as *culture*, the drawing includes visual elements from culture in its broadest sense, such as societal practices, religion, folklore, and pop culture.

Numerous studies have shown the role of drawing in science classrooms. However, the existing studies on the science classrooms of Chinese primary school students have only shown that teachers use drawings and images to help students understand abstract knowledge. As far as we are aware, there are no studies that have examined the current situation of Chinese primary school students' drawing in science classrooms, including how their drawings may change following instructional interventions.

Based on this situation, in the present study, the main question we tried to answer was: How would primary school students respond when a teacher provided more drawing examples during science class? We were particularly interested in findings out any changes in (1) the students' ways of representing science content in class tasks and (2) their engagement in learning tasks and communication with peers; and (3) whether the kinds of examples the teacher used influenced the students' drawings.

2 Method

The participants were 12 students in the 4th grade in Xinghe Experimental School Pingfang Branch, Beijing, China (mean age = 10 years, females = 9). The teacher was a science teacher, 55 years old, female. All necessary permissions from the school administration, the grade teachers, and the students and their parents were sought and obtained prior to the conduct of the study. The drawing intervention was incorporated as part of the regular science instruction to be provided to the class.

The study was conducted over 7 lessons, each of which was 45 min in duration. The topics covered in those 7 lessons were comparable: "sound" in lessons 1–3, and "air" in lessons 4–7. Prior to the start of Lesson 1, the students were given a questionnaire to obtain demographic details and their beliefs and perceptions about the use of drawings. A pre-intervention task was given during the first half of Lesson 1, after which the teacher already provided the intervention (i.e., also during Lesson 1) and gave the first post-intervention task, which was the same as the pre-intervention task (i.e., "Describe the sounds that you hear in your daily life").

The teacher provided the science class instruction using 5 steps. **Step 1.** Introducing topics and asking questions (to establish links between old and new knowledge). **Step 2.** Brainstorming (Assigning class exercises on the science topic. Plus, this step could change into designing an experiment depending on the goal of teaching.). **Step 3.** Exploring (Showing a video; the videos were about the new knowledge being covered in class. After the students watched the video, the teacher would ask relevant questions, requiring the students to answer in their preferred way, through drawing or writing). **Step 4.** Communicating and displaying (students engage in group discussion, exchange of views, improving their work then displaying them). **Step 5.** Branching out (Combining the science knowledge with real life things to promote deeper understanding of the practical application of scientific knowledge in familiar daily life environments).

For the instructional intervention, the teacher simply increased the proportion of drawings included in the above teaching steps. Also, when she assigned tasks to students, she emphasized that they could use drawings in completing those tasks.

We interviewed the students and the teacher to gauge any changes in their views on the use of drawings in the science classroom after the intervention. The data we collected included questionnaires and interview transcripts, the teaching materials, and class tasks and assignments completed by the students. In analyzing the students' drawings, we used the categories from Areljung et al.'s [4] and Westlund's [7] studies, adhering as much as possible to the descriptions of the categories they provided.

3 Results and Discussion

3.1 Drawing Use Increased Following the Teacher's Intervention

In the pre-intervention task, none of the 12 students used drawing: they only used words to describe the sounds they could hear (see left panel of Fig. 1 for a typical task response). In the first post-intervention task, all of the 12 students used drawings (see the right panel of Fig. 1 for a typical response). This increase in drawing was maintained in subsequent lessons, including when the topic changed from "sound" to "air".

The students used drawings (on their own or in combination with words) to express their understanding of the science content, summarize what they had learned, design experiments (including showing both their assumptions and outlining the process), and communicate with their peers. Figure 2 shows an example of a student's drawing for designing an experiment to prove that air takes up space (with the teacher providing materials such as containers, water, water bottles, and small balls for the students to use).

From the interviews, according to the students, drawing was of great use in helping them understand new content and in expressing their ideas. The teacher reported apprehending better the students' actual levels of understanding of the content via the drawings

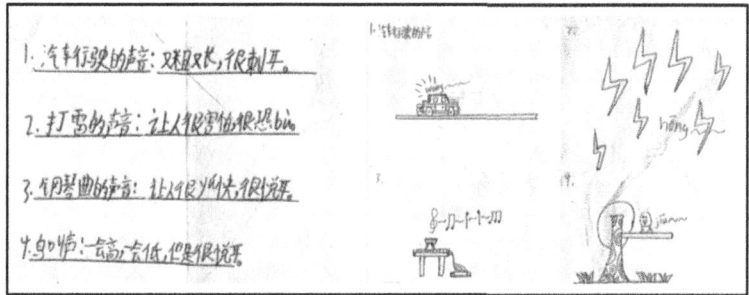

Fig. 1. Task response about sounds we hear in daily life, before and after intervention.

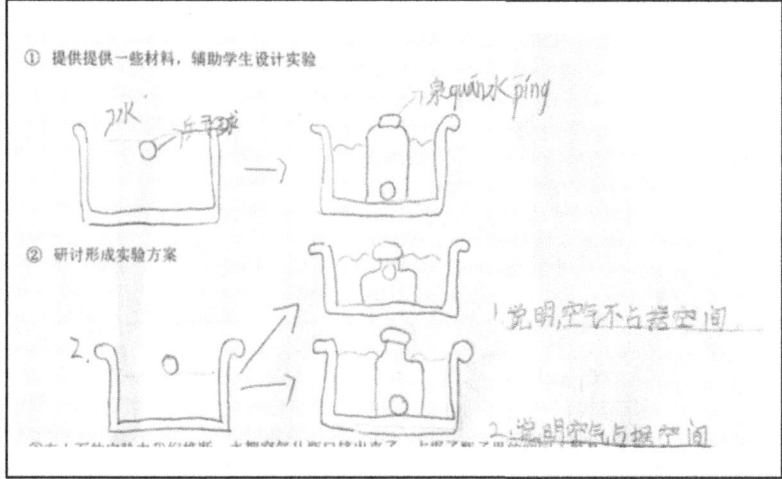

Fig. 2. One student's drawings about an experimental design.

they produced. Also, following the intervention, according to the teacher, many of the students' enthusiasm in class markedly increased. For example, students who were usually quiet became more actively engaged in sharing their drawings and explaining their understanding of the lesson with their classmates during group work. These reported experiences from the students and their teacher – although only anecdotal – suggest that drawings can produce desirable outcomes in Chinese primary science classes.

3.2 The Examples Provided by the Teacher Affected the Students' Drawings

Because the students initially had limited and varied abilities in drawing, the teacher provided example drawings for them. We observed indications that those examples might have affected the students' drawings. For example, when drawing about "sound" before receiving any examples, the students visually represented the propagation of sound in five different ways, such as straight lines, irregular curved lines, elongated irregular shapes, repeated half brackets, and waves (examples of those are shown in Fig. 3, left panel). After the teacher provided examples that used repeated half brackets and waves to show sound travel (Fig. 4), the students produced only two kinds of drawings that were the same as or similar to the teacher's examples (examples of those are shown in Fig. 3, right panel). This change suggests that the students likely imitated the teacher's examples resulting in less variation in their drawings. In hindsight, it might have been better if the teacher had more clearly emphasized beforehand that the examples were not the only ways to draw.

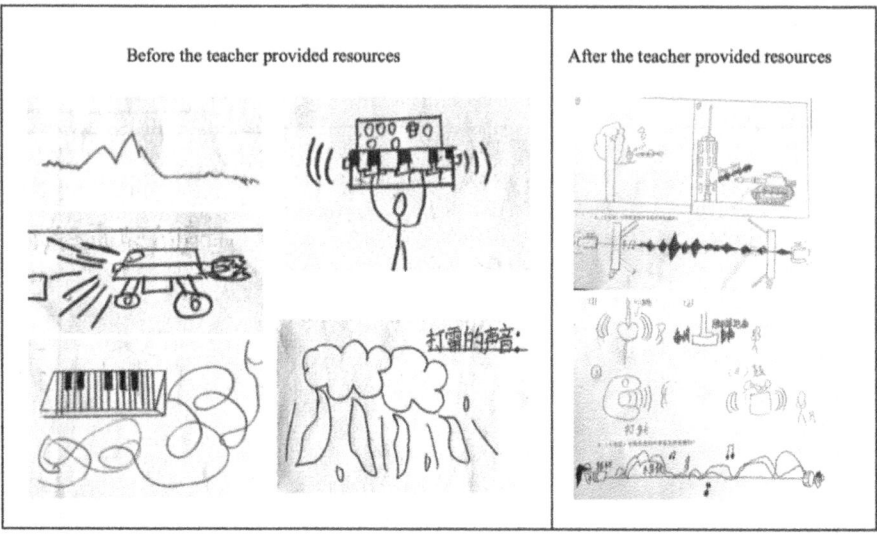

Fig. 3. Examples of students' drawings before (left panel) and after (right panel) the teacher provided the drawing examples.

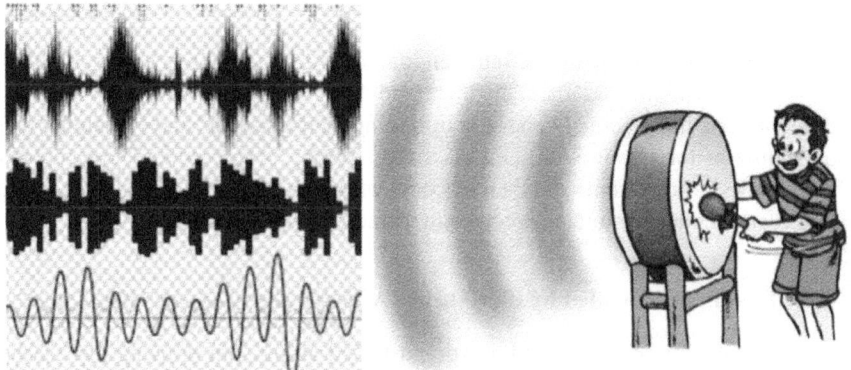

Fig. 4. The drawing examples provided by the teacher for the topic of "sound".

3.3 How the Students Visually Represented Science Content

As previously noted, we used categories from Areljung et al.'s [4] and Westlund's [7] studies to examine the students' representations of the science content they were dealing with. Table 1 shows examples of the students' drawings and which of those categories they corresponded to. We also coded the representations in the 70 drawings that the students produced in the lessons about "sound" using those categories (note that each drawing could contain one or more categories of representations in it). We found that the largest proportion of the representations were in the "Event" category, followed by "Context" and "Person": all 12 students produced drawing representations that belonged to these three categories (see Table 2). Because the topic of "sound" includes movement and processes, the largest proportion being in the "Event" category is probably not surprising. We also found that the students tended to create contexts to show their understanding of science concepts: in other words, they tend to portray contexts that include some 'stories' to make sense of the 'science' they were learning (e.g., the two people under an umbrella during a thunderstorm in Table 1 below). However, none of the students drew anything that was categorizable as "Art". This was probably not because students lacked imagination or creativity. We guessed that one possible reason was that the students

Table 1. Examples of students' drawings content and how they were categorized.

Drawing	Category	Explanation
The sound of thunder	Theory	Three simple symbols to represent lightning.
	Event	Four straight lines below the thunder symbol represent the sound travel.
The sound of thunder	Context	The student gave us a setting: In thunder storm, two people hide under a big umbrella.
	Event	Brackets represent the sound travelling.
	Person	There are two people.
	Attitude evoking	The two people cover their ears, showing a feeling of fear.

were not provided with colored pencils, because the teacher considered that coloring the drawings would take up too much time. A second possible reason was that the content of their drawings was related to the science topic being taught and they tried to draw something more directly related to that content: as a consequence, the students probably did not even consider drawing more creative, but unnecessarily elaborate shapes.

Table 2. Frequency of use of each of the representation categories in the students' drawings.

Category	No. of representations	Percentage of total	No. of students who used this representation category
Event	40	27.59%	12
Context	38	26.21%	12
Person	28	19.31%	12
Theory	19	13.10%	4
Culture	10	6.90%	10
Attitude evoking	10	6.90%	6
Art	0	0%	0

3.4 Additional Findings About What Some of the Students Found Difficult

The responses the students wrote on the questionnaire administered to them suggested that they also experienced some problems in drawing. Many students stated that sometimes they felt it difficult to draw some specific things (especially those that were abstract or that they had never seen before). Therefore, they hoped the teacher could show them how to draw things like sound, air, and so on. Also, many of the students struggled with positioning their drawing on paper. Figure 5 shows an example in which the student drew at the bottom edge of the paper and left a lot of blank space in the top half of the paper.

Due to differences in students' abilities to draw, it took longer for some students to draw and the teacher had to wait until all the students had finished. As a result, the teacher did not accomplish all her teaching objectives in some of the earlier lessons. However, as the students became more proficient in drawing, the teacher reported to the first author that she could achieve all her teaching objectives in later class lessons.

③在上面的实验中我们推断，水把空气从瓶口挤出来了，占据了瓶子里的空间。但是，你看到跑出来的空气了吗？你有办法让我们看到这些空气？

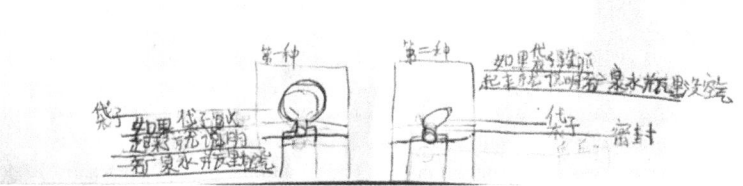

Fig. 5. Example of a poorly positioned drawing created by one of the students.

4 Conclusion

In this study, we found that although drawing is at present rarely used in science classes in China, students are actually willing to draw. From our observations and the class teacher's comments, drawing appeared to help students understand scientific concepts better, express their ideas more clearly, and communicate more effectively – which are all helpful in promoting deeper learning. There were also indications that drawings raised their enthusiasm for learning. In terms of the representations shown in their drawings, the seven categories we used for coding were not equally depicted: three categories (Event, Context, Person) were used the most and by all the students, but one category (Art) was not used at all. These distributions are different from those reported by Areljung et al. [4]: the reasons for such differences warrant further examination.

We also identified some problems with applying drawings in class. Those comprised challenges for both the teacher and the students. To promote the application of drawing in Chinese science classes, the school can build a research group to integrate drawing into the curriculum and put it into the teachers' instruction manual. Furthermore, the group can develop training courses for teachers, which can include how to realistically apply drawing to the teaching process and how to assist students in drawing. Providing some training for the students' drawing would also likely be helpful.

Acknowledgment. This research was supported by grants-in-aid (20K20516, 23H00065) received from the Japan Society for the Promotion of Science.

References

1. Ainsworth, S., Prain, V., Tytler, R.: Drawing to learn in science. Science **333**(6046), 1096–1097 (2011)

2. Danish, J.A., Phelps, D.: Representational practices by the numbers: how kindergarten and first-grade students create, evaluate, and modify their science representations. Int. J. Sci. Educ. **33**(15), 2069–2094 (2011)
3. García Fernández, B., Ruiz-Gallardo, J.R.: Visual literacy in primary science: exploring anatomy cross-section production skills. J. Sci. Educ. Technol. **26**, 161–174 (2017)
4. Areljung, S., Skoog, M., Sundberg, B.: Teaching for emergent disciplinary drawing in science? Comparing teachers' and children's ways of representing science content in early childhood classrooms. Res. Sci. Ed. **52**, 909–926 (2022)
5. Tan, L.-H., Feng, S.-J.: Scientific drawing in science education. Res. Teach. **44**(3), 79–86 (2021). (in Chinese)
6. Heckler, A.F.: Some consequences of promoting novice physics students to construct force diagrams. Int. J. Sci. Educ. **32**(14), 1829–1851 (2010)
7. Westlund, E.: Visual formation of science content in young students' multimodal compositions – seven content representations. J. Vis. Lit. **37**(4), 294–316 (2018)
8. Kress, G., van Leeuwen, T.: Reading Images: The Grammar of Visual Design. Routledge (1996)

Diagram Tools

Hoop Diagrams: A Set Visualization Method

Peter Rodgers[1(✉)], Peter Chapman[2], Andrew Blake[3], Martin Nöllenburg[4], Markus Wallinger[4], and Alexander Dobler[4]

[1] University of Kent, Canterbury, UK
p.j.rodgers@kent.ac.uk
[2] Edinburgh Napier University, Edinburgh, UK
[3] University of Brighton, Brighton, UK
[4] Technische Universität Wien, Vienna, Austria

Abstract. We introduce Hoop Diagrams, a new visualization technique for set data. Hoop Diagrams are a circular visualization with hoops representing sets and sectors representing set intersections. We present an interactive tool for drawing Hoop Diagrams and describe a user study comparing them with Linear Diagrams. The results show only small differences, with users answering questions more quickly with Linear Diagrams, but answering some questions more accurately with Hoop Diagrams. Interaction data indicates that those using set order and intersection highlighting were more successful at answering questions, but those who used other interactions had a slower response. The similarity in usability suggests that the diagram type should be chosen based on the presentation method. Linear Diagrams increase in the horizontal direction with the number of intersections, leading to difficulties fitting on a screen. Hoop Diagrams always have a square aspect ratio.

Keywords: Hoop Diagrams · Linear Diagrams · Set Visualization

1 Introduction

This paper introduces and evaluates a new set visualization technique, Hoop Diagrams. Hoop Diagrams are a circular visualization, where each hoop represents a set, and each sector represents a set intersection. The hoops are broken concentric circular lines, so that the appearance of lines in a sector indicates that the corresponding intersection of sets is non-empty.

To aid our investigation into Hoop Diagrams as a visualization tool, we have designed and implemented an interactive tool for drawing Hoop Diagrams. This allows set and intersection highlighting, reordering of sets and intersections in the diagram, and diagram rotation.

Whilst such set data can also be visualized with Venn and Euler diagrams, the closest current visualization method to Hoop Diagrams are Linear Diagrams. Linear Diagrams have been shown to be an effective visualization method [13, 20] hence we concentrate on an empirical and practical comparison between these two diagram types.

Figure 1 illustrates a Hoop Diagram visualising the relationships between six sets. The legend provides the set names in alphabetical order. Each coloured hoop (which follow a circular path) represents one of the six sets. Colour classifies each curve and corresponds to a name on the legend. A black circle on the outside and a smaller circle on the inside form the limits of the hoops. Concentric grey circles delineate the path of each coloured curve. Circle sectors are divided by spokes emanating from the circle centre. Each sector represents a set intersection, so the presence of hoops in a sector means that there is an non-empty intersection of the corresponding sets in the data.

In the examples in this paper, our sets are interests that people may have and the intersections are combination of interests that at least one person has. So that in Fig. 1, the first sector reading clockwise (at the 1 O'clock position) has the red and blue lines present, indicating that there is someone that has both the interests "Dogs" and "Poker", but no other interests. The sector immediately to the left (11 O'clock position) has only the brown line present, indicating that there is a person who has only the interest "Esport". There is no sector where the red and purple lines are both present, hence there is no person with only the two interests "Dogs" and "Hifi".

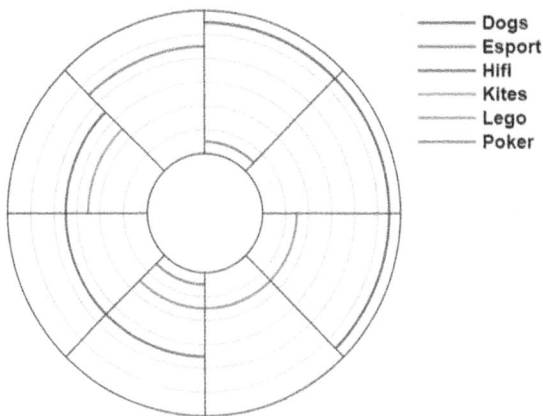

Fig. 1. Hoop Diagram.

Figure 2 shows a Linear Diagram visualising the same data as the Hoop Diagram in Fig. 1. The set names are shown in alphabetical order on the left. Each row contains a coloured line and represents one of the six sets. Horizontal grey lines delineate the path of each coloured line. Columns are divided by vertical lines. Each column represents a set intersection, so the presence of coloured lines in a column means that there is an non-empty intersection of the corresponding sets in the data.

We compared both static and interactive versions of Hoop Diagrams against Linear Diagrams. This empirical study shows only small differences in user performance between Hoop and Linear Diagrams. Users were presented with a mix of two types of question: set oriented (requiring the close examination of entire sets) and intersection oriented (requiring the close examination of entire intersections). They answered questions with Linear Diagrams more quickly, but users answered intersection oriented questions

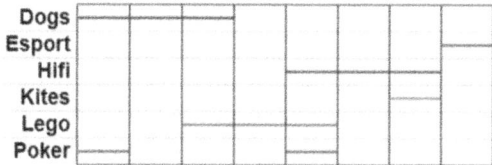

Fig. 2. Linear Diagram showing the same data as Fig. 1.

with Hoop Diagrams more accurately. We present an exploration of interactions which indicate that those using set order and intersection highlighting were more accurate, but those who used other interactions had a slower response. Our study details, as well as access to interactive Hoop Diagram and Linear Diagram software tools can be found at https://www.eulerdiagrams.com/hoop/.

We also explore the usability of these diagrams from a screen real estate perspective. Hoop Diagrams have hoops of different sizes – inner sets use hoops with a smaller radius than outer sets, thus, they use less space and have lower visual prominence in the diagram, which may be seen as a disadvantage compared to Linear Diagrams, which have the same width for all set lines. Hoop diagrams do not change size as the number of intersections increase, instead reducing the space available for each intersection. As the number of sets increases, Hoop Diagrams grow in both horizontal and vertical directions. This means that the diagrams maintain a square aspect ratio. In contrast, Linear Diagrams grow only in the vertical direction as the number of sets increase and grow in the horizontal direction when the number of intersections increase.

The rest of the paper is organized as follows. Section 2 gives a summary of prior efforts to visualize set based data and gives a theoretical explanation of why Hoop Diagrams might be an effective visualization from a cognitive perspective. Section 3 gives a detailed definition of Hoop Diagrams and explains the interactions available in the software tool. Section 4 outlines our empirical approach. Section 5 gives the results of our comparison between Hoop Diagrams and Linear Diagrams. Section 6 explores the data from the log of interactions. Finally, Sect. 7 gives our conclusions and discusses future possible work.

2 Background

2.1 Linear Diagrams

Linear Diagrams can be traced back to Leibniz [8]. Each set is represented as a line drawn in the horizontal direction. The presence of vertically overlapping lines in a column means that the corresponding intersection of sets is not empty. A line may be broken and sets are often represented by multiple line sectors. Set intersections not appearing in the diagram are considered to be empty. See Fig. 2 for a example of a Linear Diagram. The method shares similarities to parallel bargrams [24], double decker plots [15] and UpSet [16].

A design study into Linear Diagrams [20] provided evidence that using guidelines, a minimal number of line sectors (a lower number of line breaks) and thin lines for sets led to significantly improved user performance. Given the similarity in the design of Linear Diagrams to Hoop Diagrams, this study informs the design of Hoop Diagrams used in this paper, which follow these guidelines. A study on interactivity in Linear Diagrams [4], found that interactivity improved participants' accuracy, confidence and speed, a result that motivates our use of interactive Hoop and Linear Diagrams.

Theoretical work exists that shows that satisfying one of the design guidelines of linear diagrams – minimizing the number of line segments – is NP-hard in general [6, 9]. However, there exist algorithms that achieve the minimum number of line segments for real-world data [9]. Furthermore, a recent paper explores reducing the vertical space required by Linear Diagrams by visualizing multiple sets in the same row of the linear diagram [23].

2.2 Other Set Visualization Methods

Venn [22] and Euler [12] diagrams are widely used alternatives to Linear Diagrams. Here, sets are represented by curves enclosing regions. Region overlaps show which set intersections are not empty. Venn and Euler differ in the representation of empty sets. In Venn diagrams, all possible intersections are shown and empty intersections are shaded. In Euler diagrams, empty intersections are not shown. The consequence for Venn diagrams is that the number of intersections is exponential to the number of sets regardless of the number of set intersections present, hence this method has serious scalability issues. The consequence for Euler diagrams of not including empty intersections is that poor wellformedness can often appear. This includes features such as concurrency, triple points and non-simple curves. There is evidence to suggest that breaking such conditions adversely impacts understanding [21]. Limiting the shape of Euler diagrams has been attempted in systems such as RectEuler [19] which represent Euler-like diagrams as rectangles, although studies indicate that simpler shapes such as circles may be more effective for understanding [3]. We note that prior studies indicate Linear Diagrams are more understandable than Euler diagrams [5, 13], hence in this paper, we concentrate our comparison between Hoop and Linear Diagrams.

We also note the use of a variety of other set visualization methods, for a survey, see [2]. Those that use lines or region encoding include LineSets [1] and Bubble Sets [7]. However, prior embedding of items contained in the sets is required for these methods as they place lines or regions over existing data items. Node-link techniques, such as PivotPaths [10] also rely on the existence of items, using links between items to indicate shared set membership. Also related are circular visualization methods, see a survey on radial visualizations [11].

3 Hoop Diagrams

3.1 Hoop Diagram Overview

Hoop Diagrams have sets represented by broken concentric circles ("Hoops"), see Fig. 1. Comparing Hoop Diagrams to Linear Diagrams, the same labels and lines are present, as are the intersections, but intersections are represented as columns in Linear Diagrams, rather than sectors as in Hoop Diagrams. Figure 2 shows an equivalent Linear Diagram to the Hoop Diagram in Fig. 1. Hoop Diagrams can be seen as transforming into Linear Diagrams by separating the circle at 12 O'clock, straightening the set lines and moving the labels to the left of the diagram.

Compared to Linear Diagrams, we consider the benefit of Hoop Diagrams to be:

1. Compact representation. We consider that the consistent use of two dimensions for intersections provides a more desirable aspect ratio and effective use of display real estate than Linear Diagrams, which suffer when the number of intersections is large, as the diagram becomes unreadable as the intersections get very far from the labels in the X direction. Linear Diagrams scale in the Y direction as the number of sets increase. Hoop diagrams do not increase in size as the number of intersections increases, instead the intersections are presented in a smaller sector of the circle. Hoop diagrams scale linearly in both X and Y directions with the number of sets. This means that the labels, although always separated from the hoops they label, do not get more distant from them.
2. No excessive spacing between intersections. The circular nature of the Hoop Diagram leaves all intersections in close proximity. This contrasts with Linear Diagrams, where the leftmost and rightmost intersections may be a considerable distance apart.

Of course, Linear Diagrams may be considered to have advantages over Hoop Diagrams. We regard the visual simplicity of Linear Diagrams to be the representation's major advantage. In particular, the broken straight line of sets and the column structure of intersections are simpler than the broken circles of sets and sector-based structure of intersections in Hoop Diagrams. Linear Diagrams provide the same space for each set in an intersection, whereas the concentric circles of Hoop Diagrams means there is different spacing for each set, with the lines for inner sets having considerably less circumference than outer sets. Whilst Linear Diagrams can disappear off the screen as intersection numbers increase, for Hoop Diagrams, the size of diagram does not change, but the width of intersections decreases, making discerning the sets present in intersections more difficult with a large number of sets.

3.2 Interactions

Here we describe the interactions provided by our web based system. We regard these as a basic set of functionality. More sophisticated interactions, including those that are task dependent, could also be provided by expanding the system.

1. Set Highlighting: Fig. 3 top left shows the diagram with a set highlighted by hovering over the hoop. The mouse is in the second hoop from the outside, "Dogs".
2. Intersection Highlighting. Figure 3 top right shows the diagram with an intersection highlighted by hovering inside or outside the intersection sector. The mouse is just outside the second hoop clockwise, the intersection with "Cars" and "Hifi" only.
3. Bring Set to Outside. Figure 3 middle left shows the diagram after the label "Food" has been clicked by the left mouse button. "Food" is brought to the outside with "Cars" and "Dogs" moved inwards by one hoop.

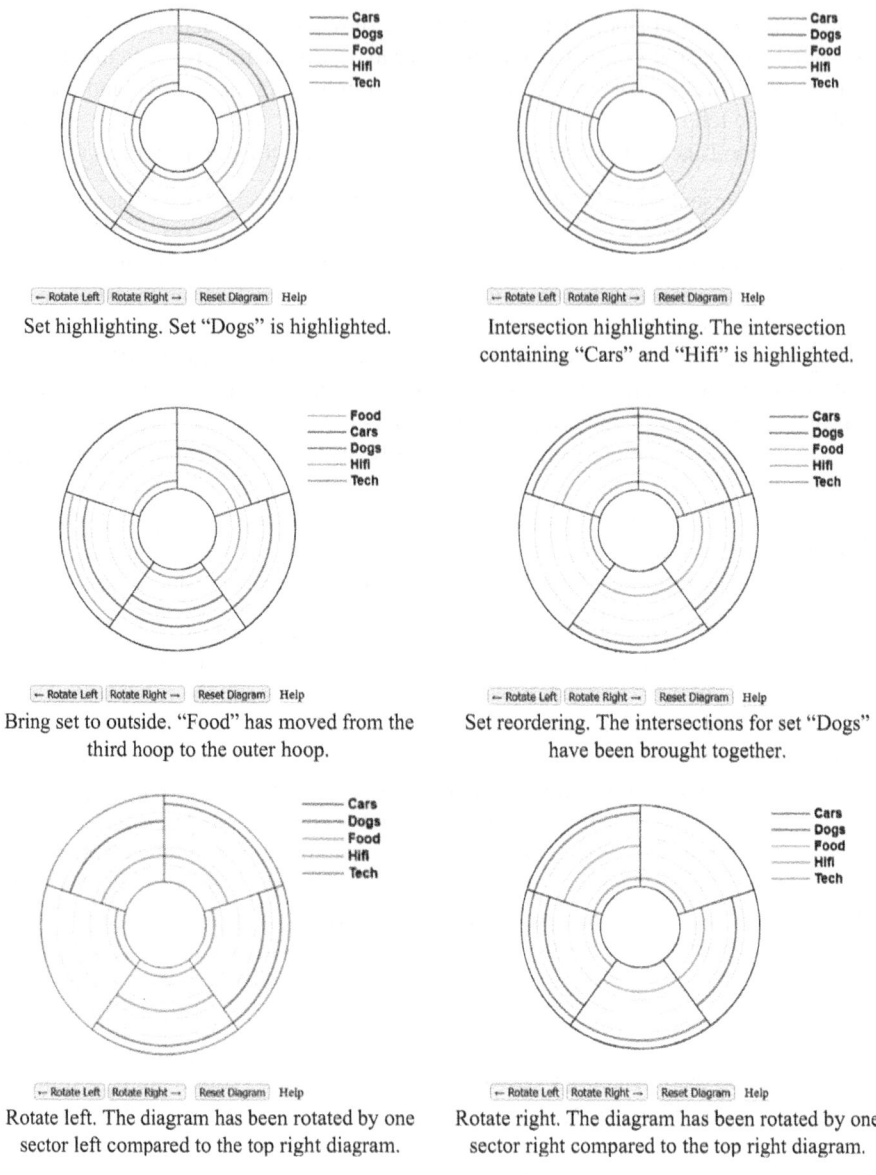

Fig. 3. Layout changes after interaction is applied to the second training diagram.

4. Reorder Set. Figure 3 middle right shows the diagram after the second hoop from the outside, "Dogs" has been clicked by the left mouse button. The sectors are reordered so that all the sectors containing "Dogs" are together, starting from 12 O'clock, meaning the "Dogs" line is continuous. The system attempts to minimize the number of line segments for the other hoops.

5. Rotate. Figure 3 bottom left shows the diagram after the "Rotate Left" button has been pressed once. Figure 3 bottom right shows the diagram after the "Rotate Right" button has been pressed twice after the Rotate Left button was pressed.

The "Reset Diagram" button returns the diagram to its original configuration. The effect on the screen of reordering is animated, using HTML Transitions, with each animation taking 1 s. The reordering interactions are composable, for example, it is possible to have a different set on the outside, along with the set line of choice continuous whilst rotating as many times as desired.

We also developed an interactive tool for Linear Diagrams, to support the studies given in Sect. 4 where each of the above Hoop Diagram interactions has an equivalent in the Linear Diagram tool:

1. Set Highlighting: A set can be highlighted by hovering over a line.
2. Intersection Highlighting. An intersection can be highlighted by hovering above or below a column.
3. Bring Set to Top. If a label on the left is clicked by the left mouse button, it is raised to the top, with the other labels that were above it lowered by one set.
4. Reorder Set. Clicking on a line reorders the columns so that all the columns for the clicked set are together starting from the left, so that the chosen set line is continuous.
5. Rotate. Pressing a Rotate button rotates the diagram by one column in the desired direction, so "Rotate Left" moves all columns left by one, except the leftmost column that becomes the rightmost column. The "Rotate Right" button moves all columns right by one and the rightmost column becomes the leftmost column.

As with interaction in Hoop Diagrams, the "Reset Diagram" button returns the diagram to its original configuration. All changes are animated.

4 Methodology

We defined the following research questions to guide our exploration:

- RQ1: Is data visualized with Hoop Diagrams more understandable than data visualized with Linear Diagrams?
- RQ2: Does interaction in Hoop Diagrams or Linear Diagrams aid understanding?

This led to the following Research Hypotheses related to RQ1:

- H1: Hoop Diagrams are more understandable than Linear Diagrams.
- H2a: Dynamic Linear Diagrams are more understandable than Static Linear Diagrams.
- H2b: Dynamic Hoop Diagrams are more understandable than Static Hoop Diagrams.

RQ2 was to be addressed by a more exploratory approach, where we logged interaction activity to look at:

- Which are the most/least used interactions in Dynamic Hoop/Linear Diagrams?
- Which interactions in Dynamic Hoop/Linear Diagrams are used in successful/unsuccessful tasks?

4.1 Data Capture

The study was aimed at both research questions and aimed to answer H1, H2a and H2b, as well as provide data for the exploration of interaction activity.

Participants were recruited through the Prolific platform and redirected to a web based study vehicle hosted at the University of Kent. On completion they were returned to Prolific. All those completing and correctly answering at least one attention check question were paid, although answering both attention check questions correctly was required for the data to be included in our analysis. We calculated the payment depending on expected time for completion (mostly 20 min, although some early studies were 25 min) and the reward was set at the UK National Living Wage, which at 1 April 2023 was £10.42 an hour, although some studies took place before this for a reward at the UK National Living Wage of £9.50 an hour. The maximum time for completion is assigned by Prolific, and for a 20 min expected completion the maximum time allowed was 67 min. Throughout the studies the median completion time was under the expected time. Participants were distributed through the Prolific standard sample method. We placed prescreening conditions as follows: they must use laptop or desktop computers; participant approval rate must be 99/100; they had a minimum number of 5 prior submissions; and we excluded participants from previous studies on Hoop Diagrams. Participants passed from prolific would be assigned to each condition in turn in an attempt to keep condition completion relatively even. This is a random assignment, as we made no use of participant information when making condition assignments. The completion per condition ratio was not always even as some participants would return the study before completion or would time out. We ran studies in varying size batches of participants to control load on the web servers and make an attempt to even up condition completion ratios towards the end of the study by removing overrepresented conditions from the rotation.

Our diagrams were formed from real-world Twitter Circles data, obtained from the SNAP data set collection [18]. We used the same input data to form the diagrams for each condition of each study. For each Twitter Circle, a set is formed with the users that have those interests. The set intersections that have at least one member with exactly those interests are shown in the diagram. The set intersections that do not have at least one member with those interests are not shown. To avoid any confusion as to context, our diagrams and questions made no reference to Twitter. All set names were changed, whilst keeping the real-world scenario: interests people share. The set names were chosen so that no two sets in any one Linear Diagram started with the same letter to reduce the potential for misreading and so that they were sufficiently short to avoid occlusion when presented as Hoop Diagrams with all our labelling variants. After prepilot experimentation by the project researchers, we settled on a suitable size of 6 sets, with 8 to 16 intersections, providing a variety of difficulties. After piloting this configuration in the first study, this was confirmed as an appropriate range of values to avoid ceiling and floor effects for correctness whilst permitting completion of the study in a reasonable time frame.

The study was performed using a web site developed on a Html/Javascript/Node.js platform. Prolific passed the prolific id, session id and study id to the initial web page. The initial page explained what the participants would do in the study, what information we would be recording, stated that they could stop the study at any time and asked for their consent to begin the study. If they gave their consent, those undertaking static conditions were presented with 4 training examples. Those undertaking the interactive conditions were presented with a short video (1 min 33 s) of the interactions followed by the 4 training examples. The training examples gave details of how to interpret the diagram, and, after a choice was made, some feedback on their answer was presented. See Fig. 4 for an example of a training page with feedback after the choice was made. The training data sets were of varying sizes, starting small with 2 options, and ending with diagrams of the size typically seen in the study. Training Question 4 had 6 sets, 12 intersections and 4 options to choose from.

After the training phase, the study phase began. Questions were presented in a random order. For example, Fig. 5 shows a question presented to a participant which is Question 9 in the list of questions. The order of the 4 options was randomized. Two attention check questions were presented, again randomly in the question sequence and with random answer order. These showed a diagram with the text "This is an attention check question." Three of the 4 options were standard looking answers similar to those in Fig. 5 (although they did not relate to the diagram displayed, so could not be plausible alternative answers). The fourth option was the correct one which had the text "Choose this option". The second to last page was the demographics page. This asked questions

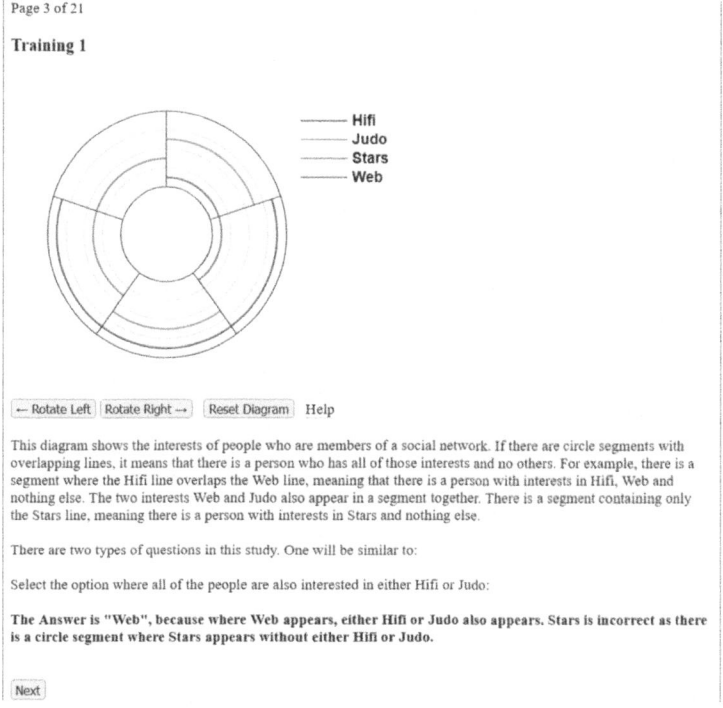

Fig. 4. Screenshot. First training example after the user makes a choice, so feedback on the correct answer is shown in bold. Previous to this, the participants were presented with the radio buttons "Stars" and "Web". This is a dynamic Hoop Diagram.

about the participants Gender, Sight Difficulties, Age and an optional Any Comments free text box. The final page was a return link to Prolific with the completion code in the query string.

All diagrams, questions, anonymized data and code can be found at https://www.eulerdiagrams.com/hoop/. This web site also includes an interactive interface for creating Hoop Diagrams and Linear Diagrams.

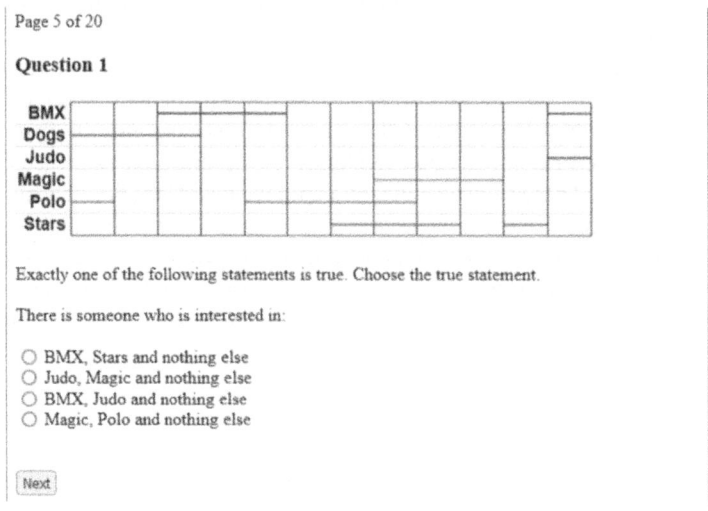

Fig. 5. A question as presented to the participants. This is a Static Linear Diagram. The correct answer is "BMX, Judo and nothing else".

We stored anonymized participant and response data on a local server (University of Kent, UK). For the interactive diagrams we stored hovers where changes in mouse position altered the highlights on the diagram, and clicks on the diagram and buttons.

Questions were of two types, with the same number of each in the study. The training had two of each type. The first question type was a set oriented question that was intended to require the close examination of entire sets to elicit the answer. It had the general form of: "Select the option where all of the people are also interested in X or Y" with single sets as answers. E.g., for Fig. 6 the question was: "Select the option where all of the people are also interested in either Lego or Yoga:" with the options:

- Art
- Dogs
- Web
- Zumba

The correct answer was "Zumba".

The second question type was an intersection oriented question that intended to require the close examination of entire intersections to elicit the correct answer. It had the general form of: "Exactly one of the following statements is true. Choose the true

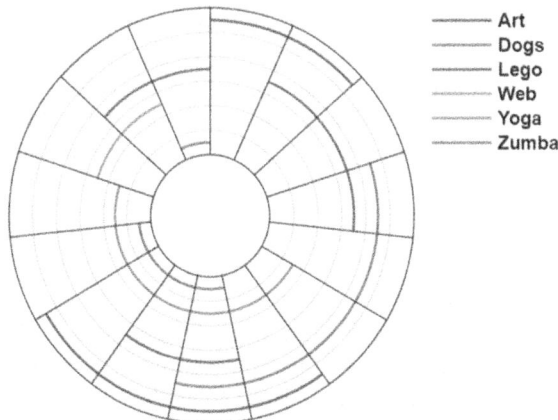

Fig. 6. Static Hoop Diagram for Question 4 from the final study.

statement. There is someone who is interested in:" with groups of sets as answers. For example, see Fig. 5 which includes the question text and answer.

Our pilot had 20 participants. This was used to gauge task difficulty and to spot any technical issues with the studies, of which none were found and so no changes were made for the main study. The data from the pilot was discarded. The free text responses by participants was read immediately after all studies, pilots and main studies. No patterns or problems were identified from the free text entry.

We used Linear Diagrams drawn based on previous guidelines [20]. The design of Hoop Diagrams was consistent with the Linear Diagram design where possible. Labelling was via a legend. The interactive diagrams used the same diagram design as the static diagrams.

4.2 Statistical Methodology

We recorded two response variables: the time taken to answer a question (in seconds), and whether or not a participant correctly answered a question (a binary response). Taken across all participants and question instances, the mean correct responses provide a probability; namely, the probability that a randomly selected participant answers a randomly selected question correctly.

Multiple responses were collected from each participant (i.e. a time and success for each question, of which there were 12). These responses will therefore exhibit clustering, as the twelve responses from each individual are correlated with each other [17]. Use of standard statistical tests, such as ANOVA and chi-squared tests of goodness of fit, are insufficient, as such tests have an assumption of independence of observations. If we violate this assumption, then it can lead to overstated statistical significance, and underestimated standard errors. In other words, we may see a difference between groups where none exists.

To address this problem, a combination of generalized estimating equations (GEE) and generalized linear models (GLM) were used. The statistical software R was used for the analysis, primarily the package geepack [14]. The approach produces p-values, which can be interpreted in the usual way. In this instance, the p-value is the probability of obtaining the observed data given the coefficient of the independent variable in the model is 0. The models used had either a normal response variable (for time), or a binomial response variable (for accuracy). We considered the results to be significant if the p-value was less than 0.05.

5 Study Results

The study compared four conditions: static Hoop Diagrams, static Linear Diagrams, dynamic Hoop Diagrams and dynamic Linear Diagrams in an attempt to answer H1, H2a and H2b. The goal was to provide guidance about the most effective diagram type for those visualizing set-based data.

Table 1. Hoop Diagram vs Linear Diagram. The green shaded cells indicate the value is significantly better than the other groups.

Group		Accuracy			Time (s)		
		Overall	Set	Intersection	Overall	Set	Intersection
Hoop	Overall	0.798	0.6474	0.9025	51	55.4	48.5
	Interactive	0.779	0.6386	0.8825	53	57.8	51.1
	Static	0.808	0.6568	0.9236	48	52.9	45.7
Linear	Overall	0.757	0.6138	0.8662	46	49	44.1
	Interactive	0.762	0.6205	0.8656	48	52	46.2
	Static	0.752	0.6074	0.8668	43	46	42.1

To see which of Hoop Diagrams and Linear Diagrams are most suited to these sort of tasks, two analyses were performed: one with no reference to interaction (i.e. there were two groups, those who saw Hoop Diagrams, and those who saw Linear Diagrams), and a second analysis with four groups (i.e. interactive Hoop Diagrams, static Hoop Diagrams, interactive Linear Diagrams, and static Linear Diagrams). There was no split made on interactivity (i.e. interactive versus non-interactive) for reasons discussed in Sect. 6. There were 208 participants in the hoop group (107 interactive, 101 static), and 205 participants in the linear group (101 interactive, 104 static). The descriptive statistics are given in Table 1.

For the overall results, two generalized linear models were constructed with one independent variable (group) and either a binomial response variable (accuracy) or a normal response variable (time). For accuracy, the p-value is 0.061 (i.e. no significance), whereas for time there was a significant difference (p-value is 0.017), providing evidence that Linear Diagrams can be used more quickly than Hoop Diagrams. The effect size (Cohen's d) is 0.099, which is small.

For the split by interaction as well as diagram type, two generalized linear models were constructed with one independent variable (group) and either a binomial response variable (accuracy) or a normal response variable (time). There were no significant differences between the groups for accuracy, whereas the linear static group were significantly faster than the other three (p-value is 0.0015). This improvement in time for linear static diagrams is evident in both question types (for intersection oriented questions, the p-value is 0.0074, whereas for set oriented questions, the p-value is 0.0046).

When looking at question type, for set oriented questions, there was no difference in accuracy between hoop and Linear Diagrams (p-value 0.253), but there is a difference in time: Linear Diagrams were faster (p-value 0.039). For intersection oriented questions, Hoop Diagrams improved accuracy (p-value 0.034) but were slower (p-value 0.046) than Linear Diagrams.

With regard to accuracy, our primary measure of diagram understandability, there is a significant result for Hoop Diagrams over Linear Diagrams in the case of intersection oriented questions. This leads us to the conclusion that when accurate analysis is the primary concern, and the questions are related to examination of intersections, rather than sets, Hoop Diagrams are preferred. For time, there are significant results favouring Linear Diagrams overall, as well as for static Linear Diagrams. This leads us to the conclusion that when time taken to complete an interpretation is the primary concern static Linear Diagrams are preferred.

With regards to our hypotheses, the differing statistical results for accuracy and time means we conclude that "H1: Hoop Diagrams are more understandable than Linear Diagrams" is false. For "H2a: Dynamic Linear Diagrams are more understandable than Static Linear Diagrams" has a timing result that indicates that this hypothesis is false. Finally, there is no evidence for "H2b: Dynamic Hoop Diagrams are more understandable than Static Hoop Diagrams" which we conclude to be false.

6 Exploration of Interaction Data

In this section, we explore what sort of interactions were used, and whether or not the used interactions were beneficial to participants (in terms of time and accuracy). We make a distinction between necessarily intentional interactions (those which require a mouse click: reordering sets, reordering sectors, rotating diagrams, and resetting the diagram) and those which could be coincidental (highlighting). For the latter, for example, as a participant moves their cursor around the screen, they may highlight different sections of the diagram as an unintended consequence.

In what follows, we look only at the interactive group; that is, participants who had access to the interactive controls. Tables 2, 3 and 4 show the proportion of question instances for which the indicated interactive element was used.

As discussed above, the intentional interactions are the two reorderings and the two rotations. Rotations appear to be used less frequently in Hoop Diagrams than in Linear Diagrams, and in general are used in at most 6.5% of diagram instances. Reordering is more common, with set reordering (i.e. vertical reordering) used in 27.8% of question instances, and sector reordering (i.e. horizontal reordering) used in 24.8% of instances. In general though, these intentional interactions are not particularly common. For (possibly) unintentional highlighting, it is used in around 85% of question instances.

We can also examine whether interactions appear to be helpful. This comparison is different to that given in Sect. 5. There, we were comparing the results of the groups

Table 2. Proportion using diagram highlighting

	Highlight (horizontal)			Highlight (vertical)		
	Overall	Set	Intersection	Overall	Set	Intersection
Hoop	0.89	0.9	0.879	0.783	0.807	0.76
Linear	0.863	0.869	0.857	0.851	0.861	0.84

Table 3. Proportion using set reordering

	Reorder (horizontal)			Reorder (vertical)		
	Overall	Set	Intersection	Overall	Set	Intersection
Hoop	0.27	0.292	0.247	0.264	0.294	0.234
Linear	0.225	0.259	0.191	0.294	0.342	0.246

Table 4. Proportion using rotation

	Rotate (left)			Rotate (right)		
	Overall	Set	Intersection	Overall	Set	Intersection
Hoop	0.011	0.011	0.011	0.026	0.034	0.018
Linear	0.051	0.062	0.04	0.051	0.064	0.038

Table 5. Interaction accuracy. Green shaded cells show the highest accuracy.

Interactive Element	Accuracy whilst using the element	Accuracy whilst not using the element
Highlight - horizontal	0.7871	0.6993
Highlight - vertical	0.7865	0.6599
Reorder - horizontal	0.8395	0.7483
Reorder - vertical	0.8586	0.7371
Rotate - left	0.6438	0.7749
Rotate - right	0.6374	0.7762

against each other. Here, however, we are looking at average rates of success for participants who always had access to interactions, and either chose to use them or not. As the previous discussion on take-up shows (Tables 2, 3 and 4), these rates vary widely amongst interactions, in some cases are very low or very high, and different participants used interactive elements on different questions. We therefore only give descriptive statistics, as the study presented was not designed to test these sorts of data.

As Table 5 shows, questions for which reordering was used had a higher absolute accuracy rate. As noted above, we cannot make any inferences about this apparent difference, but it is certainly suggestive that reordering is a useful interactive element. The same is true of highlighting, with the additional caveat that we cannot determine the participant's intention when using highlighting. By contrast, absolute accuracy rates are lower when using rotations, although rotations were used in only a limited number of cases (around 5% of question instances) so the evidence for any effect is weak.

7 Conclusions

We have introduced Hoop Diagrams, a new set visualization technique using an interactive diagram creating tool and empirical studies for diagram design and comparison against the closest existing set visualization technique, Linear Diagrams. The empirical results show that users answer questions on Linear Diagrams more quickly, but answer some questions more accurately with Hoop Diagrams. From a screen real estate perspective, Hoop Diagrams maintain a square aspect ratio, scaling in both horizontal and vertical directions, whereas Linear Diagrams only scale in the vertical direction with the number of sets and in the horizontal direction with the number of intersections. This horizontal scaling can quickly make Linear Diagrams too large for many displays. Hence, Hoop Diagrams may be preferred when presented on screen or paper.

Regards future possible work, the interaction functionality could be extended. The close relationship between Hoop Diagrams and Linear Diagrams could be exploited as Hoop Diagrams can be seen as Linear Diagrams with the ends of the diagram connected. Hence, it would be possible to show an animated transition between the two diagram types. A user could choose the diagram type depending on their personal preference. This would also allow the benefits of both diagram types for the user. So it would be possible to move from Linear to Hoop Diagram if the size of Linear Diagram means that it exceeds the space available for display, so allowing all the data to be shown at once for appropriate data sets.

Other future work includes developing interactivity aimed at improving users' performance at particular tasks. Being able to hide sets or intersections so that uninteresting areas of the diagram are no longer shown may simplify the display, so making analysis of data of interest easier. Similarly, using automatic highlighting to show potential areas of interest, such as sets or intersections might improve performance significantly.

Acknowledgments. This research has been partially funded by the Vienna Science and Technology Fund (WWTF) [https://doi.org/10.47379/ICT19035]. It was discussed at Dagstuhl Seminar 22462.

References

1. Alper, B., Henry Riche, N., Ramos, G., Czerwinski, M.: Design study of linesets, a novel set visualization technique. IEEE TVCG **17**(12), 2259–2267 (2011)
2. Alsallakh, B., Micallef, L., Aigner, W., Hauser, H., Miksch, S., Rodgers, P.: The state-of-the-art of set visualization. Comput. Graph. Forum **35**(1), 234–260 (2016)

3. Blake, A., Stapleton, G., Rodgers, P., Cheek, L., Howse, J.: The impact of shape on the perception of Euler diagrams. In: Dwyer, T., Purchase, H., Delaney, A. (eds.) Diagrams 2014. LNCS (LNAI), vol. 8578, pp. 123–137. Springer, Heidelberg (2014). https://doi.org/10.1007/978-3-662-44043-8_16
4. Chapman, P.: Interactivity in linear diagrams. In: Basu, A., Stapleton, G., Linker, S., Legg, C., Manalo, E., Viana, P. (eds.) Diagrams 2021. LNCS, vol. 12909, pp. 449–465. Springer, Cham (2021). https://doi.org/10.1007/978-3-030-86062-2_47
5. Chapman, P., Stapleton, G., Rodgers, P., Micallef, L., Blake, A.: Visualizing sets: an empirical comparison of diagram types. In: Dwyer, T., Purchase, H., Delaney, A. (eds.) Diagrams 2014. LNCS (LNAI), vol. 8578, pp. 146–160. Springer, Heidelberg (2014). https://doi.org/10.1007/978-3-662-44043-8_18
6. Chapman, P., Sim, K., Chen, H.: Minimising line segments in linear diagrams is NP-hard. J. Comput. Lang. **71**, 101136 (2022)
7. Collins, C., Penn, G., Sheelagh, M., Carpendale, T.: Bubble sets: revealing set relations with isocontours over existing visualizations. IEEE TVCG **15**(6), 1009–1016 (2009)
8. Couturat, L.: Opuscules et fragments inédits de Leibniz. G. Olms, Hildesheim (1903)
9. Dobler, A., Nöllenburg, M.: On computing optimal linear diagrams. In: Giardino, V., Linker, S., Burns, R., Bellucci, F., Boucheix, J.-M., Viana, P. (eds.) Diagrams 2022. LNCS, vol. 13462, pp. 20–36. Springer, Cham (2022). https://doi.org/10.1007/978-3-031-15146-0_2
10. Dörk, M., Riche, N.H., Ramos, G., Dumais, S.: PivotPaths: strolling through faceted information spaces. IEEE TVCG **18**(12), 2709–2718 (2012)
11. Draper, G.M., Livnat, Y., Riesenfeld, R.F.: A survey of radial methods for information visualization. IEEE TVCG **15**(5), 759–776 (2009)
12. Euler, L.: Lettres a Une Princesse d'Allemagne Sur Divers Sujets de Physique et de Philosophie. Letters **2**, 102–108 (1775)
13. Gottfried, B.: A comparative study of linear and region based diagrams. J. Spat. Inf. Sci. **10**, 3–20 (2015)
14. Højsgaard, S., Halekoh, U., Yan, J.: The R package geepack for generalized estimating equations. J. Stat. Softw. **15**, 1–11 (2006)
15. Hofmann, H., Siebes, A., Wilhelm, A.: Visualizing association rules with interactive mosaic plots. In: Proceedings of the Sixth ACM SIGKDD 2000, pp. 227–235. ACM (2000)
16. Lex, A., Gehlenborg, N., Strobelt, H., Vuillemot, R., Pfister, H.: UpSet: visualization of intersecting sets. IEEE TVCG **20**(12), 1983–1992 (2014)
17. Liang, K., Zeger, S.: Longitudinal data analysis using generalized linear models. Biometrika **73**(1), 13–22 (1986)
18. Leskovec, J.: Stanford large network dataset (2011). http://snap.stanford.edu/data/
19. Paetzold, P., Kehlbeck, R., Strobelt, H., Xue, Y., Storandt, S., Deussen, O.: RectEuler: visualizing intersecting sets using rectangles. Comput. Graph. Forum **42**(3), 87–98 (2023)
20. Rodgers, P., Stapleton, G., Chapman, P.: Visualizing sets with linear diagrams. ACM Trans. Comput. Hum. Interact. **22**(6), 27:1–27:39 (2015)
21. Rodgers, P., Zhang, L., Purchase, H.: Wellformedness properties in Euler diagrams: which should be used? IEEE TVCG **18**(7), 1089–1100 (2012)
22. Venn, J.: On the diagrammatic and mechanical representation of propositions and reasonings. Phil. Mag. (1880)
23. Wallinger, M., Dobler, A., Nöllenburg, M.: LinSets.zip: compressing linear set diagrams. IEEE TVCG **29**(6), 2875–2887 (2023)
24. Wittenburg, K., Lanning, T., Heinrichs, M., Stanton, M.: Parallel bargrams for consumer-based information exploration and choice. In: Proceedings of the 14th Annual ACM Symposium on User Interface Software and Technology, pp. 51–60. ACM (2001)

Building a Large Dataset of Human-Generated Captions for Science Diagrams

Yuri Sato[1](✉) , Ayaka Suzuki[2] , and Koji Mineshima[3]

[1] Ochanomizu University, Tokyo, Japan
sato.yuri@ocha.ac.jp
[2] Chiba University, Chiba, Japan
[3] Keio University, Tokyo, Japan

Abstract. Human-generated captions for photographs, particularly snapshots, have been extensively collected in recent AI research. They play a crucial role in the development of systems capable of multimodal information processing that combines vision and language. Recognizing that diagrams may serve a distinct function in thinking and communication compared to photographs, we shifted our focus from snapshot photographs to diagrams. We provided humans with text-free diagrams and collected data on the captions they generated. The diagrams were sourced from AI2D-RST, a subset of AI2D. This subset annotates the AI2D image dataset of diagrams from elementary school science textbooks with types of diagrams. We mosaicked all textual elements within the diagram images to ensure that human annotators focused solely on the diagram's visual content when writing a sentence about what the image expresses. For the 831 images in our dataset, we obtained caption data from at least three individuals per image. To the best of our knowledge, this dataset is the first collection of caption data specifically for diagrams.

Keywords: diagram caption · human annotation · machine learning dataset

1 Introduction

Human-generated captions for photographs, particularly snapshots, have been extensively collected in recent AI research (e.g., [13,24]). They play a crucial role in the development of machine learning models capable of multimodal information processing that combines vision and language (for a survey, see [3]). In this paper, we shift our focus from snapshot photographs to diagrams.

There is a crucial difference between how people extract information from things drawn by humans, such as diagrams and illustrations, and how they extract information from photographs. The difference can be explained by the distinction between natural and non-natural meanings as made by Grice [8]. When extracting information from illustrations, understanding the intentions of the creator (or user) plays a crucial role, in contrast to photographs where this is not the case. The former is an instance of natural meaning, while the latter is an instance of non-natural meaning. In the same vein, Daston and Galison [5] discuss the distinction between photographs and illustrations as the distinction between the two types of images commonly used in science publications. They argue that photographs aim to capture nature mechanically, with as little human intervention as possible, whereas illustrations attempt to represent the characteristics of the subject as an underlying type rather than individual tokens, thereby requiring human abstraction and generalization. Given this background, we focus on text-free diagrams that are best understood as a medium akin to illustrations—that is, visual representations that require human intervention, such as inferring intent, to derive appropriate information. This perspective suggests that constructing AI models capable of reflecting a broader spectrum of human intelligence is feasible by learning not only from photographic captions but also from diagram captions.

We focus on the diagram image dataset AI2D [10] by the Allen Institute for AI (AI2).[1] It consists of 4,903 diagram images collected from elementary science textbooks. Recently, the AI2D-RST corpus has been published[2], in which 1,000 images from this dataset are annotated with diagram types (table, depiction, graph, etc.) [9]. We build a human-generated caption dataset targeting the same 1,000 images in AI2D as analyzed by the AI2D-RST corpus.

There are at least three types of diagrams used in science publications [16], which are referred to here as science diagrams. The first is the one used among scientific experts (academic paper stage), the second is the one used among students (textbook stage), and the third is the one used among the general public (popularization stage). For example, regarding the first type, [4,7,15] quantitatively analyzed the usage patterns of the diagrams that appear in the science papers. The above AI2D focuses on diagrams in the second type.

Several previous studies have collected annotations for diagram images. Alikhani and Stone [1] collect human-generated annotations for each type of arrow, which is a component of diagrams in AI2D. Hiippala et al. [9] collect annotations on diagram types for human workers, as mentioned above. However, none of these are the diagram captions we are aiming for. Most recently, Zala et al. [25] introduce the AI2D caption dataset and is attempting to generate diagram images from text, but the captions here are generated by other large language models, not by humans. To the best of our knowledge, our dataset is the first collection of human-generated caption data specifically for diagrams.

2 Data Collection

2.1 Method

Participants. Three hundred participants who can read and write in Japanese were recruited through Crowd Works, a Japanese crowdsourcing service. The

---

[1] https://prior.allenai.org/projects/diagram-understanding.
[2] https://korp.csc.fi/download/AI2D-RST/v1.1/.

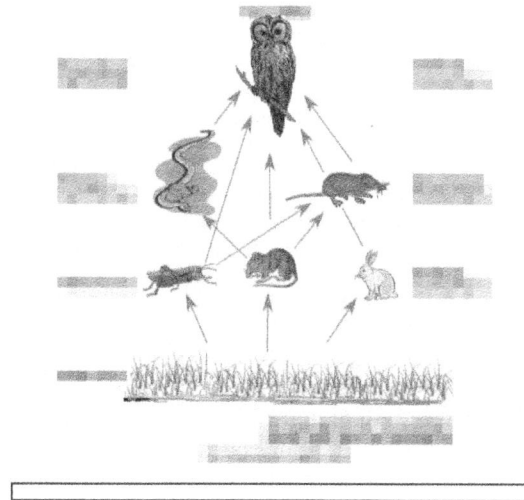

Fig. 1. A trial of image captioning task (AI2D image id#2083, AI2D-RST group #4)

average age of the participants was 40.65 (minimum 18, maximum 66). Participants worked remotely online with informed consent and were paid a fixed honorarium for their participation. The experimental procedures and the release of the collected data were approved by the local ethics committee.

Materials. Figure 1 is a trial of experimental task. We mosaicked all textual elements within the diagrams (1,000 AI2D images) to ensure that human annotators focused solely on the diagram's content when writing a sentence about what the image expresses. As a result, the diagrams consisting only of text became only mosaic images, and their meaning was unintelligible. There were 80 such diagrams. These diagrams were not used in subsequent tasks. Therefore, 920 of the 1,000 diagrams were used in the task. By type, 127 were tabular (group 1), 526 were spatial depiction (group 2), 224 were graphs or networks (group 3), and 43 were mixes (group 4)[3]. The sentences and instructions were given in Japanese. The stimulus presentation program was made in qualtrics.

Procedures. After the general informed consent, the following three specific notes were given and the participants were asked to press the "I understand" or "I do not understand" button for each (only those who pressed the former button could proceed to the main task).

<u>Note 1.</u> You are presented with an image, so please write what it represents in the form of a single sentence (subject + predicate). The subject should be the object(s) depicted in the image(s). Do not use meta-subjects such as "the image" or meta-predicates such as "be depicted" or "represent". The following is a bad example; please do not write sentences like this: "This image depicts∼".

---

[3] The four groups follow the data of diagram type given in [9].

Note 2. If you have no idea what an image represents, you may write "I don't know" or "I can't answer".

Note 3. There is a mosaic in the text that was included in the original image, but please do not mention it.

In the main task, 20 of the 920 questions were given at random. As shown in Fig. 1, the participants were asked to write a sentence describing the displayed diagram image in Japanese in the box provided below the image. Finally, age and gender demographic questions were given.

2.2 Results

Two hundred ninety-one participants answered all 20 trials. Three of them, all of whom answered all the trials with "I don't know" or "I can't answer" (hereinafter called "unknown"), were excluded from the data. There were 26 people who included a meta-subject or meta-predicate (see note 2 in Procedure) in all trials (except for the "unknown" responses)[4]. A typical answer of this type is like "Food Chain Diagram". These 26 people were excluded from the data since they clearly did not follow the instruction (note 1) of the experiment. This is essentially the same as finding and excluding cases that do not respond to seriousness in an online experiment [11]. The data for the remaining 262 people were included in our dataset. Our dataset is available at https://github.com/yuri-ocha/DiagramCaption.

We obtained an average of 5.73 captions per image (min 2, max 9). Subtracting the "unknown" responses from the captions obtained, the images have captions with different numbers from 2 to 9. One hundred eight images remained with three captions, 164 images remained with four captions, 202 images remained with five captions, 212 images remained with six captions, 98 images remained with seven captions, 42 images remained with eight captions, and 5 images remained with nine captions. Of the default 920 items of diagram images, 831 items of diagram images were given 3 or more captions, resulting in our dataset.

The captions that participants gave for the trial in Fig. 1 were as follows.[5]

(1) Fukuroo-ga shokumotsurensa-no chooten-ni i-ru
    owl-NOM food.chain-GEN top-DAT be.at-NON.PST
    'Owls are at the top of the food chain'
(2) Shizenkai-no shokumotsurensa
    nature-GEN food.chain
    'Food chain in nature'

---

[4] This does not include cases where the dependent clause in caption sentences contains a verb other than the meta-predicate in even one case.

[5] We use the following abbreviations: NOM: nominative case, ACC: accusative case, GEN: genitive case, DAT: dative case, TOP: topic, LOC: locative, PASS: passive, PROG: progressive, NON.PST: non-past tense.

1. The surface of the stone has various textures.
2. There are four stones.
3. There are various types of stones, each with different characteristics.
4. These are the names of various types of stones.
5. What types of minerals there are.

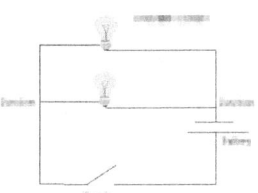
1. There are two types of electrical circuits.
2. Electrical array diagram
3. When you connect a light bulb with a copper wire and turn on the switch, it lights up.
4. The switch is off.
5. That electricity passes through a circuit and lights up a light bulb

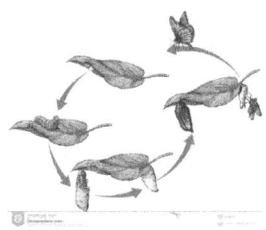
1. A butterfly larva is growing into adults while eating leaves.
2. A caterpillar is born from an egg, the caterpillar grows into a pupa, the pupa becomes a butterfly, the butterfly lays an egg, and the process repeats.
3. A butterfly's larva hatches from an egg, transforms into a pupa, and then becomes an adult.
4. This is the growth process of a caterpillar.
5. It shows the process of a swallowtail butterfly from egg to larva to pupa and finally to adult.

Fig. 2. Diagram images and captions in the cases in which the majority of caption data were collected in the response form following the instruction: Upper: image id#4621, group #1, Middle: image id#1531, group #2, Lower: image id#729, group #3.

(3) Fukuroo-wa kusa-wo tabe-ru mushi-ya shoodoobutsu-wo
owl-TOP grass-ACC eat-NON.PST insect-and small.animal-ACC
hoshoku-su-ru dobunezumi-ya hebi-wo
predation-do-NON.PST brown.rat-and snake-ACC
hoshoku-su-ru node aru-kankyoo-de-wa
predation-do-NON.PST because some.environment-LOC-TOP
shokumotsurensa-no chooten-ni a-ru
food.chain-GEN top-LOC be.at-NON.PST
'Owls are at the top of the food chain in some environments because they prey on grazing insects, brown rats (which themselves prey on small animals), snakes, and more'.

(4) Fukuroo-ga tekii-wo muke-rare-tei-ru
owl-NOM hostility-ACC target-PASS-PROG-NON.PST
'Owls are being treated with hostility'

(5) Shokumotsurensa-ga oki-tei-ru
food.chain-NOM occur-PROG-NON.PST
'Food chain is occurring'

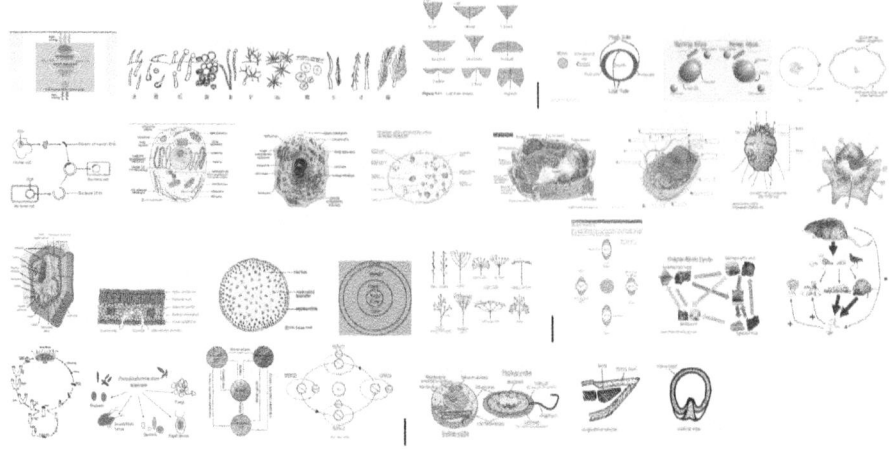

Fig. 3. 28 diagram images with over 70% of participants answering "unknown": AI2D image id#4112, #4504, #4714 for group 1; #2656, #2661, #2695, #2796, #3034, #3060, #3078, #3316, #3323, #3451, #3527, #3536, #3863, #3903, #551, #968 for group 2; #132, #1362, #2048, #2339, #2827, #4174, #757 for group 3; #3048, #3972 for group 4 (from top left to right)

The image of this trial in Fig. 1 is AI2D image #2083 and mix group (#4). As shown in the list above, this is a case where five captions were obtained; the second caption is a noun only, while the other four captions are sentences. Thus this is a case where the majority of caption data was collected in the response form following the instruction (especially, note 1). Similarly, the cases in which the majority of caption data were collected in the form following the instruction are shown in Fig. 2 for each of the other diagram groups: tabular (#1), spatial depiction (#2), and graph/network (#3).

The cases in which five captions were obtained by removing the "unknown" responses are, 40 out of 127 (31.5%) in group #1, 108 out of 526 (20.5 %) in group #2, 54 out of 224 (24.1%) in group #3, and 11 out of 43 (25.6%) in group 4. There were no big differences among diagram types. Figure 3 lists the diagram images of the cases in which more than 70% of the participants answered "unknown". In 20 of the total 28 images, the case did not include an arrow: 71.4%. Reducing the criteria a bit more, there were 54 images in which more than half of the participants (more than 51%) were "unknown". In 35 of those cases, the arrows were not included: 64.8%.

3 Discussion

In total, human-generated caption data for at least three people were obtained for 831 diagram images. The number of captions per image varied from 3 to 9, partly because 15.1% of all captions (796 out of 5271 total) were "unknown" responses and we excluded them from the dataset. The result that there were a substantial number of "unknown" responses is a distinct difference from the photograph caption data. Although the details of the experimental setting (we explicitly instructed participants to be allowed to answer "unknown" if they could not answer the question) should be taken into account, in [13,24], few "unknown" responses were given when asked to describe what the photograph image represented.

What makes the difference? Given the fact that more than half of the "unknown" responses were for diagrams without arrows, we first narrow the scope of the discussion from diagrams to illustrations. As mentioned in the Introduction, illustrations are characterized by the fact that they are drawn by the sender (author) to convey something to the receiver. In other words, the intention of the sender plays a role in the presupposition for understanding the meaning. Therefore, if the receiver cannot grasp the intention, the meaning of the image cannot be grasped either. By contrast, a photograph does not necessarily include such an intention. Indeed, Walton [23] argues that a photograph is a transparent representation of the world, which means that it does not include the author's idea (in terms of which aspect of the subject was emphasized and depicted).

Furthermore, since the diagrams in this study are used in science publications, what the diagrams represent could be used in the form of textual descriptions in the main text. In such cases, partial components of the diagrams may be given names, whereby the reference may be held in a natural way [6]. The importance of this label is more emphasized in the case of rigorous diagrammatic systems used in mathematics and logic. In Euler diagrams and Venn diagrams, naming an ellipse refers to a particular set, while an unnamed ellipse refers to an anonymous set [17,22]. The effect of the labels has also been emphasized in the context of educational psychology. Mayer [14], for example, reported that labeled diagrams are more effective in learning than unlabeled diagrams. However, we collected human-generated caption data by targeting diagram images with the label and tag information erased (by mosaicing). This may have influenced the understanding of the diagrams, especially the increase in "unknown" responses. However, we focused on text-free diagrams as image data that could serve as an alternative to photographs and constructed a human-generated caption dataset for them. We have the intention to use it as training data for machine learning models that address the challenges of image classification, image captioning, and visual question answering, as in [19–21], for example. Therefore, images that contain textual data are not good for this purpose.

Comparison of the newly constructed diagram caption dataset with the existing photograph caption dataset is an issue for the future. In the past, Leibniz [12] and Berkeley [2] argued that pictures on paper only depict particular objects. Indeed, it has been reported that the occurrence of universal quantifiers and individual-level predicates (rather than stage-level predicates) is extremely low in the photograph caption dataset [18,21]. Our shift to diagram captions is a challenge to go beyond this limitation and depict generality. Whether this can be accomplished with diagrams is the next issue to be addressed.

Acknowledgments. This study was supported by Grant-in-Aid for JSPS KAKENHI JP24K15676, JP24K00004, JP19K13172, as well as JST CREST JPMJCR2114.

References

1. Alikhani M, Stone M: Arrows are the verbs of diagrams. In: COLING 2018, pp. 3552–3563. ACL (2018)
2. Berkeley G.: A Treatise Concerning the Principles of Human Knowledge. The Floating Press (1710/2014)
3. Bernardi, R., et al.: Automatic description generation from images: a survey of models, datasets, and evaluation measures. J. Artif. Intell. Res. **55**, 409–442 (2016). https://doi.org/10.1613/jair.4900
4. Best, L.A., Smith, L.D., Stubbs, D.A.: Graph use in psychology and other sciences. Behav. Process. **54**, 155–165 (2001). https://doi.org/10.1016/S0376-6357(01)00156-5
5. Daston L, Galison P: Objectivity. Zone Books (2007)
6. Greenberg, G: Tagging: semantics at the iconic/symbolic interface. In: AC 2019, pp. 11–20. University of Amsterdam (2019)
7. Fanjoy, L.P., MacNeill, A.L., Best, L.A.: The use of diagrams in *science*. In: Cox, P., Plimmer, B., Rodgers, P. (eds.) Diagrams 2012. LNCS (LNAI), vol. 7352, pp. 303–305. Springer, Heidelberg (2012). https://doi.org/10.1007/978-3-642-31223-6_33
8. Grice, P.: Meaning. Philos. Rev. **66**, 377–388 (1957). https://doi.org/10.2307/2182440
9. Hiippala, T., et al.: AI2D-RST: a multimodal corpus of 1000 primary school science diagrams. Lang. Resour. Eval. **55**, 661–688 (2021). https://doi.org/10.1007/s10579-020-09517-1
10. Kembhavi, A., Salvato, M., Kolve, E., Seo, M., Hajishirzi, H., Farhadi, A.: A diagram is worth a dozen images. In: Leibe, B., Matas, J., Sebe, N., Welling, M. (eds.) ECCV 2016. LNCS, vol. 9908, pp. 235–251. Springer, Cham (2016). https://doi.org/10.1007/978-3-319-46493-0_15
11. Krosnick, J.A.: Response strategies for coping with the cognitive demands of attitude measures in surveys. Appl. Cogn. Psychol. **5**, 213–236 (1991). https://doi.org/10.1002/acp.2350050305
12. Leibniz G.: Philosophical Papers and Letters; Dialogue. L.E. Loemker (Trans. & Ed.). University of Chicago Press, Chicago (1677/1956)
13. Lin, T.-Y., et al.: Microsoft COCO: common objects in context. In: Fleet, D., Pajdla, T., Schiele, B., Tuytelaars, T. (eds.) ECCV 2014. LNCS, vol. 8693, pp. 740–755. Springer, Cham (2014). https://doi.org/10.1007/978-3-319-10602-1_48
14. Mayer, R.E.: Systematic thinking fostered by illustrations in scientific text. J. Educ. Psychol. **81**, 240–246 (1989). https://doi.org/10.1037/0022-0663.81.2.240
15. Mogull S A, Stanfield C T: Current use of visuals in scientific communication. In: IPCC 2015, pp. 1–6. IEEE (2015). https://doi.org/10.1109/IPCC.2015.7235818
16. Pauwels, L. (ed.): Visual Cultures of Science: Rethinking Representational Practices in Knowledge Building and Science Communication. Dartmouth College Press (2006)
17. Sato, Y., Stapleton, G., Jamnik, M., Shams, Z.: Human inference beyond syllogisms: an approach using external graphical representations. Cogn. Process. **20**, 103–115 (2019). https://doi.org/10.1007/s10339-018-0877-2

18. Sato, Y., Mineshima, K.: Visually analyzing universal quantifiers in photograph captions. In: Giardino, V., Linker, S., Burns, R., Bellucci, F., Boucheix, J.M., Viana, P. (eds.) Diagrams 2022. LNCS, vol. 13462, pp. 373–377. Springer, Cham (2022). https://doi.org/10.1007/978-3-031-15146-0_34
19. Sato, Y., Mineshima, K., Ueda, K.: Can negation be depicted? Comparing human and machine understanding of visual representations. Cogn. Sci. **47**(3), e13258 (2023). https://doi.org/10.1111/cogs.13258
20. Sato, Y., Mineshima, K.: Can machines and humans use negation when describing images? In: Baratgin, J., Jacquet, B., Yama, H. (eds.) Human and Artificial Rationalities. Lecture Notes in Computer Science, vol. 14522, pp. 39–47. Springer, Cham (2024). https://doi.org/10.1007/978-3-031-55245-8_3
21. Sato, Y., Suzuki, A., Mineshima, K.: Capturing stage-level and individual-level information from photographs: Human-AI comparison. In: CogSci 2024, pp. 803–810. Cognitive Science Society (2024)
22. Shams, Z., Sato, Y., Jamnik, M., Stapleton, G.: Accessible reasoning with diagrams: from cognition to automation. In: Chapman, P., Stapleton, G., Moktefi, A., Perez-Kriz, S., Bellucci, F. (eds.) Diagrams 2018. LNCS (LNAI), vol. 10871, pp. 247–263. Springer, Cham (2018). https://doi.org/10.1007/978-3-319-91376-6_25
23. Walton, K.L.: Transparent pictures. Crit. Inq. **11**, 246–277 (1984). https://doi.org/10.1086/448287
24. Yoshikawa, Y., Shigeto, Y., Takeuchi, A.: STAIR captions: constructing a large-scale Japanese image caption dataset. In: ACL 2017, pp. 417–421 (2017)
25. Zala, A., Lin, H., Cho, J., Bansal, M.: DiagrammerGPT: generating open-domain, open-platform diagrams via LLM planning. arXiv:2310.12128 (2023)

Open Access This chapter is licensed under the terms of the Creative Commons Attribution 4.0 International License (http://creativecommons.org/licenses/by/4.0/), which permits use, sharing, adaptation, distribution and reproduction in any medium or format, as long as you give appropriate credit to the original author(s) and the source, provide a link to the Creative Commons license and indicate if changes were made.

The images or other third party material in this chapter are included in the chapter's Creative Commons license, unless indicated otherwise in a credit line to the material. If material is not included in the chapter's Creative Commons license and your intended use is not permitted by statutory regulation or exceeds the permitted use, you will need to obtain permission directly from the copyright holder.

KIELER: A Text-First Framework for Automatic Diagramming of Complex Systems

Maximilian Kasperowski[✉], Niklas Rentz, Sören Domrös, and Reinhard von Hanxleden

Department of Computer Science, Kiel University, Kiel, Germany
{mka,nre,sdo,rvh}@informatik.uni-kiel.de

Abstract. In Model-Driven Engineering, editing models is typically not merely a purely textual endeavor, but rather a mix between textual and graphical editors and views. Both have their advantages and use cases where either textual or diagrammatic representations are better suited to edit and understand models. Therefore, a modeling framework offering the best of both worlds can be advantageous.

We define the *text-first* approach to combine the textual and diagrammatic representations by automatically synthesizing the textual model into a diagram. We present the KIELER text-first diagramming framework and its take on current challenges for model visualization and compare it to the diagram-first approach, as exemplified by the GLSP framework.

Keywords: Diagramming Framework · Modeling Tools · Automatic Visualization · Diagram Synthesis

1 Introduction

Visualization is a useful tool in Model-Driven Engineering (MDE), especially for communication and documentation purposes during system development. In the context of developing complex, real-world systems, handmade visualizations are often tedious to create and maintain. In the case of ongoing developments and changes they can quickly become a burden [10,16].

We here consider three paradigms for modeling and implementation of complex systems: *text-only*, *diagram-first* and *text-first*. Text-only is the classic process where a system is developed entirely as a collection of textual sources. Diagram-first describes a workflow where a diagram is edited directly and text-first describes a workflow where both a textual source and an automatically generated diagram are available side-by-side. Text-only tools are arguably very established and therefore serve as a good baseline for the requirements that system development tools must provide.

In this paper we introduce the KIELER text-first diagramming framework. We argue that there are significant benefits of a text-first approach, and we compare our approach to the diagram-first approach, as exemplified by the Graphical Language Server Platform (GLSP).

We aim to answer the research question: "How do text-only, diagram-first, and text-first approaches compare in the context of MDE?"

In this paper, we lay a stronger focus on the diagrammatic aspects because we argue that text-first solutions to browse and understand complex models, via e.g. finding definitions or showing documentation on hover, are already widely implemented and used. Our contributions are

- a comparison of the concepts, technology, and motivation behind the text-first KIELER framework and the diagram-first approach, as exemplified by GLSP in the context of our assumptions about the MDE requirements of developing complex systems in Sect. 3,
- an overview of the generic APIs offered by the KIELER framework in Sect. 4.1 including the diagram synthesis, options sidebar, and semantic filtering,
- a summary of modern browsing techniques for large, hierarchical diagrams in KIELER such as off-screen element visualization, semantic zooming, and the novel top-down layout approach in Sect. 4.2, and
- our vision of the future of diagram technologies in Sect. 5.

2 Related Work

We will first discuss diagramming frameworks other than KIELER and GLSP for a broader overview, as well as some users of such diagramming frameworks.

In addition to KIELER and GLSP there are multiple frameworks and toolboxes for MDE with support for diagrams. Sirius [25] and Sprotty[1] are examples of low-level diagramming frameworks. Sirius supports customization of elements through palettes and configuration views within Eclipse, which automatically generate diagrams for different models. However, these diagrams do not support advanced browsing techniques for large hierarchical models that we discuss in this paper but only zooming and panning. Sprotty is a fully customizable framework for building diagramming tools on the basis of SVG that KIELER and GLSP use as their foundation and provide generic solutions for common diagram-related tool improvements.

Ptolemy II [5] is an older tool that uses a window-based diagram-first solution to browse complex diagrams of hierarchical models. In contrast to other diagram-first solutions presented here, nested subgraphs are visualized in separate windows and not via compound graphs, which offloads parts of the layout to the user and the operating system's window management. KIELER and GLSP support the integration of hierarchical layout through interactions such as expand/collapse and with top-down layout (see Sect. 4.2).

The bigER [9] tool is an implementation of a modeling tool for Entity Relationship (ER) diagrams, which offers both text-first editing and diagram-first editing using Sprotty similar to the KIELER approach but without the advanced

---

[1] https://sprotty.org/.

browsing techniques for large hierarchical models presented here. The bigUML tool [14] developed later by the same group employs the diagram-first approach for UML diagrams, based on GLSP.

Petzold et al. [18] developed a tool (PASTA) to aid with System-Theoretic Process Analysis (STPA). The tool uses automatic visualization with a text-first approach, which sets PASTA apart from other STPA tools that are either purely textual or purely graphical. In particular, the purely graphical modeling is noted to be particularly tedious, mainly due to the size of the graphs and the non-linear workflow. Petzold et al. build on Sprotty and heavily utilize filtering, showing multiple different views of the underlying model to deal with large graphs.

The coordination language Lingua Franca (LF) also comes with an automatic visualization using the text-first approach [13]. LF programs are developed as textual code in one or more source files, and the diagram serves mainly as an abstract high-level overview over the program. Code snippets in different target languages, so-called detailed reaction code, is usually filtered out of the diagram as it would only distract from the high-level view. Therefore, LF diagrams are always highly filtered. LF uses KIELER for its diagram synthesis and it is possible to utilize the advanced browsing techniques presented here out of the box.

The variety of frameworks and tools that use KIELER and GLSP highlights the need for diagrammatic representations. However, we also want to highlight the importance of an editable textual source in addition to a diagram.

3 Modeling Paradigms for Model-Driven Engineering

Before discussing editing paradigms for diagrams we must first discuss the context in which we consider the application of diagrams. We are specifically interested in diagram tools that can support diagrams of large, complex models that may include some form hierarchy. We propose the requirement that diagrams should always provide a degree of readability, both in overview and detail, that is similar to the readability of small diagrams.

We define two diagram editing approaches by the way users interact with the model behind the diagram.

The *text-first* diagram approach revolves around editing the source model textually, while a diagram is automatically synthesized in real time. A textual editor and diagram can be used side-by-side, as seen in Fig. 1. Moreover, the editor and diagram can interact with each other. e.g., clicking on a node or an edge in the diagram navigates the textual editor to its definition or vise versa. The addition of a textual editor lets this text-first approach inherit all advantages of the text-only approach, since the editor can support any text-based IDE and editing features. In the KIELER framework, the text remains the ground-truth of each diagram, and any modifications of the model change the textual source. Changing this textual source will automatically generate a new diagram.

Compared to KIELER's text-first approach, the popular GLSP framework [2,21] practices the *diagram-first* approach. In the diagram-first approach, a

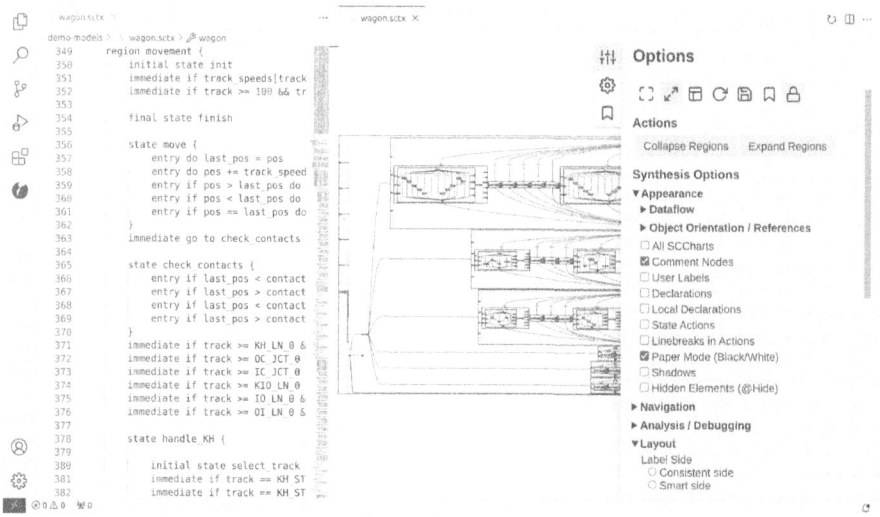

Fig. 1. The KIELER SCCharts VS Code extension shows the textual model and the graphical view of the WAGON model side-by-side with opened synthesis option sidebar in the diagram view.

user mainly edits via diagram interaction, which modifies the typically invisible, underlying source model. This is a fundamentally different approach that compromises many of the text editing advantages of modeling pragmatics [10], as elaborated on in the following.

3.1 Text-First and Diagram-First Editing Paradigms

When developing new tools to support the development of large, complex systems, it is important to keep the strengths of existing technologies and established workflows. Text-only tools are well established and have long been used to model complex systems. Text-first tools employ modeling pragmatics to get the best of a textual and graphical view on the model. e.g., text is easily supported by version management, and it is easy to build tool support in form of content assist, finding references, error and warning markers, and similar standard IDE features. KIELER utilizes all of these features together with a diagram view. The text-first approach utilizes the textual source as a detailed view that can be accurately edited while the diagram can be adjusted to the current needs. GLSP is often configured to have one or more textual models as ground truth. Usually a graphical editor view on the model is built, instead of using a textual editor and a diagram side-by-side, which only allows diagram-first editing. This requires to do all editing operations by interacting with the diagram using palettes, context menus, or input fields.

The text-first framework KIELER was created to relieve the user of the burden of palettes and the need to place everything themselves, by using automatic layout, in the case of KIELER provided by the Eclipse Layout Kernel (ELK) [4]. KIELER is built to use the diagram and text together and is able to map diagram elements to elements in the textual model and vice versa. This enables users to use the view—textual or diagram—most suitable for the given task. Although GLSP integrates elkjs[2] (ELK for JavaScript) for automatic layout, the standard approach involves the user creating and placing everything manually, via diagram interaction either using structure-based editing or palettes.

Diagram-first editing is often argued to be novice-friendly and intuitive. However, this can also be the case for well designed textual languages, as Eumann and Wechselberg reported for railway domain experts than can and do write textual models of SCCharts, a Statecharts dialect, using the KIELER framework without any prior knowledge in programming or SCCharts [6]. Moreover, in real-world models that use diagram-first editing a lot of developer time is spent moving nodes and edges around to manually control the layout [16]. GLSP focuses on such manual control, which is often missing in automatically synthesized diagrams. Automatic layout is rarely able to capture secondary notation [16]. Hence, users are happy with a layout they created themselves and that they can control. However, Domrös et al. [3] show that control over the automatic layout can also be achieved using the order of the textual model. Nevertheless, KIELER partly employs the diagram-first editing paradigm using structure-based editing [11]. Structure-based editing [19] enables users to create a structurally correct model based on diagram interaction using a context menu. In contrast to the WYSIWYG approach employed by GLSP, structure-based editing always employs automatic layout.

Additionally, filtering does not integrate well into diagram-first editing while it is a first-class citizen in the KIELER framework to handle complex models using transient views [22]. e.g., SCCharts provides an *induced data-flow* view that shows a different graph structure than the regular *state-and-transition* view [26]. GLSP can also filter a diagram. However, if the diagram view is filtered to create a smaller model, the usual approach is to use automatic layout to place the remaining elements to close gaps that would appear between elements and to make the filtered view more readable. Additionally, graphical programming has the disadvantage that coordinates for filtered elements cannot easily be inferred from the unfiltered view, for which they would be necessary. This issue is not inherent to the diagram-first approach if an editing approach that continuously employs automatic layout is used. The issue occurs since manual coordinates and coordinates determined by automatic layout cannot trivially be merged.

Generally, GLSP implements many interesting concepts and both KIELER and GLSP use a very similar technology stack, which might make the two approaches compatible, as detailed in Sect. 5.

---

[2] https://github.com/kieler/elkjs.

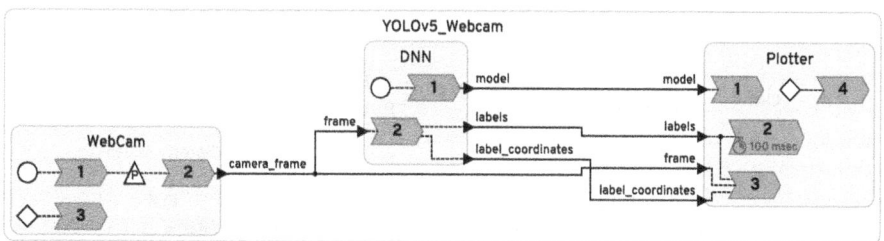

(a) A diagram generated from a Lingua Franca program taken from the Lingua Franca playground repository.

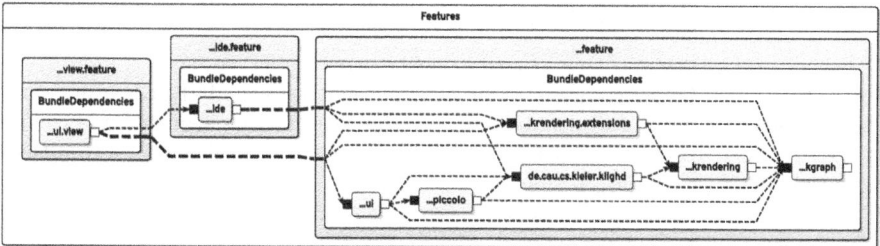

(b) A diagram generated using the Software Project Visualization (SPViz) tool presented by Rentz [20].

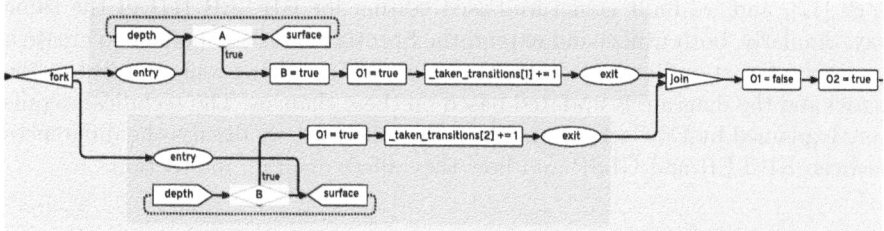

(c) An excerpt from a Sequentially Constructive Graph (SCG) created during the SCCharts compilation process [24].

Fig. 2. Different diagram applications that use the KIELER framework.

3.2 Infrastructure of KIELER and GLSP

KIELER is intended to be a framework mainly for Domain Specific Languages (DSLs) and provides a set of configurable standard features. Figure 2 shows several diagrams from projects that utilize KIELER [20,24]. They demonstrate the versatility of visualizations that KIELER can create as they have very different underlying models and come from very different domains. In comparison, GLSP is more of a sandbox than a framework by extending Sprotty's functionality without offering default implementations as provided by KIELER. Instead, GLSP provides a set of tools that enables adopters to build the features they require.

Both KIELER and GLSP utilize Sprotty and its extension to the Language Server Protocol (LSP) to create interactive diagrams. GLSP additionally pro-

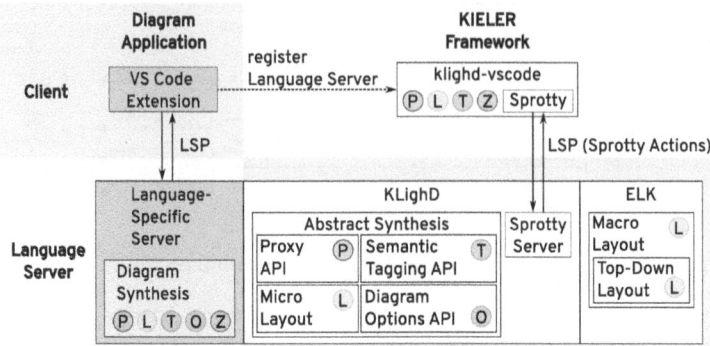

Fig. 3. The general architecture of an application built with the KIELER framework. The icons indicate where features discussed in this paper are implemented and configured: Ⓟ proxies, Ⓛ layout, i.e. positioning of shapes and graph elements, Ⓣ semantic tagging, Ⓞ diagram options, Ⓩ smart zoom.

vides a standard set of diagram interaction mechanisms by extending the LSP. Petzold et al. extended the LSP for their diagram interaction constraint framework [17], and we built structure-based editing for KIELER [11] in the same way. Similarly, both utilize and extend the Sprotty server component to create a one-way interaction between diagram and model, i.e. changes are applied to the model and the diagram is updated based on these changes. The technical details were explained by De Carlo et al. [2]. In the following, we discuss the differences between KIELER and GLSP and how they affect diagram interaction.

4 KIELER Features

Most of KIELER's features are agnostic to the diagram type, meaning they work out of the box for general diagrams. The KIELER API directly includes tooling for a diagram synthesis, diagram options including a sidebar, diagram interaction methods, and advanced browsing techniques as presented below. This enables the designers of domain-specific diagrams to quickly configure the diagram interaction, configuration, and visualization.

4.1 The KIELER Framework

Figure 3 is divided into four different sections by two orthogonal areas and illustrates the typical architecture when working with the framework. The first area, the area of responsibility, divides between the *diagram application*, i.e. the part that the diagram designer implements themselves, and the *KIELER framework*, i.e. the technology and its public API. The second area divides between the *client* represented by the two top boxes, i.e. the parts directly executed in the web environment implemented in TypeScript, and the *language server* represented by the bottom box, i.e. the parts implemented in Java or Xtend, connecting to the client via the LSP. Figure 3 outlines the individual features and highlights the API to be used for individual diagram applications, further discussed below.

Diagram Synthesis. The diagram synthesis is the point of entry for users of the KIELER framework designing their own diagram application. It is located on the server and connects the configuration of all other features. The main goal is to let the users define how their model is translated to a diagram. Specifically, KIELER Lightweight Diagrams (KLighD) [23] defines a graph and rendering model named *KGraph* and *KRendering*, respectively. It also defines an API to synthesize a model to this graph structure with attached rendering and styling information. The diagram application uses this API to write a diagram synthesis that defines the translation from their DSL model to a KGraph. The rendering and styling information define the visual appearance of all graph elements. For example, the simple states from multiple figures in this paper have a rounded rectangle with a certain background color and a centered text inside, and some specific transitions are dashed lines with arrow heads at the end.

Next to the visual appearance of, e.g., lines and boxes, styles can also modify the size of graph elements. Therefore, the diagram generation needs to first estimate the graph element sizes via their styles (called *micro layout*), then do the automatic layout with ELK (*macro layout*), and finally render the diagram. We currently execute both layout steps on the server.

While GLSP configures the micro layout on the server, the absolute element sizes are calculated on the client[3]. This comes at the cost of a larger communication overhead during layouting, as it requires an additional roundtrip between the client and the server because they execute the macro layout on the server. KIELER utilizes KLighD for more advanced server-side positioning [23], while GLSP only supports basic server-side micro layout configuration (padding, horizontal/vertical gap, minimal width/height) [2].

Diagram Options. ⓞ There can be multiple graphical representations of models with varying levels of detail and different focuses. Therefore, we want to give the user options to configure the diagram, so that different views can be shown for each model. As options should be easily accessible to the user, we provide them via an options sidebar in KIELER. Our implementation can be seen in Fig. 1 for SCCharts. The sidebar consists of two parts, the synthesis options depicted in Fig. 1 and the render options, both hosting language-specific options.

The server-side synthesis options filter the model, can configure the layout, or even change what the graphical model representation is in its entirety. The diagram synthesis allows modular configuration of the sidebar for the respective language. e.g., the SCCharts language has buttons to collapse and expand all hierarchical elements, different categories such as **Layout**, and a wide range of check, choice, text, and select boxes to configure the diagram.

---

[3] https://eclipse.dev/glsp/documentation/clientlayouting/.

The client-side render options configure how the client-side view model interaction works. e.g., whether selecting a diagram element selects the corresponding text or vice versa, whether and how movements are animated, the visualization of layout constraints [17], or what the size threshold of smart-zoom (see Sect. 4.2) is configured to be.

The sidebar is implemented as a UI extension for Sprotty, so the concept is not limited to text-first frameworks such as KIELER and could also be provided directly in Sprotty so that any diagram-first framework can use it.

Semantic Tags and Filters. ⓣ The diagram should not only be a static representation of the textual model, but offer the user and the system further information about the model and the semantic elements represented by the diagram. KIELER supports to add such information to the graph by using semantic tags and filters. The KGraph elements that are produced during the diagram synthesis can be tagged to retain semantic information about the original model that would otherwise be lost in the graph and rendering structure. Additional semantic filters can be created and attached to the graph, which may later be used on the diagram client. This allows the KIELER rendering framework to perform client-side diagram-type-specific interactions and overlays, e.g. popups or filtered proxies (see Sect. 4.2). Sprotty and GLSP usually attach such semantic information as properties on the graph elements to be able to work with that information in similar ways.

Semantic tags are user-defined strings that may optionally contain a number value. A tag itself is an atomic filter rule expression. Complex filter rules can be constructed by combining other filter rules with numeric or logical operators.

When applying filters to a set of graph elements, an element is retained if the applied filter rule evaluates to true for that element. The filter rule #someTag returns true for an element if it contains the tag someTag. Using $someTag we can get the number associated with the tag—if the tag does not exist, then 0 is returned instead—and use it for further evaluation. e.g. #state && $children > 4 evaluates to true for nodes tagged with state and children, where the children tag has a value larger than 4.

4.2 Browsing Techniques for Large Diagrams

In the following, we present several browsing techniques that are particularly useful for large, hierarchical diagrams available in KIELER, which are made possible by the infrastructure illustrated in Fig. 3 and outlined in the previous sections. Figure 4 shows a view of a large, hierarchical model with disabled browsing techniques that we use to compare our techniques to.

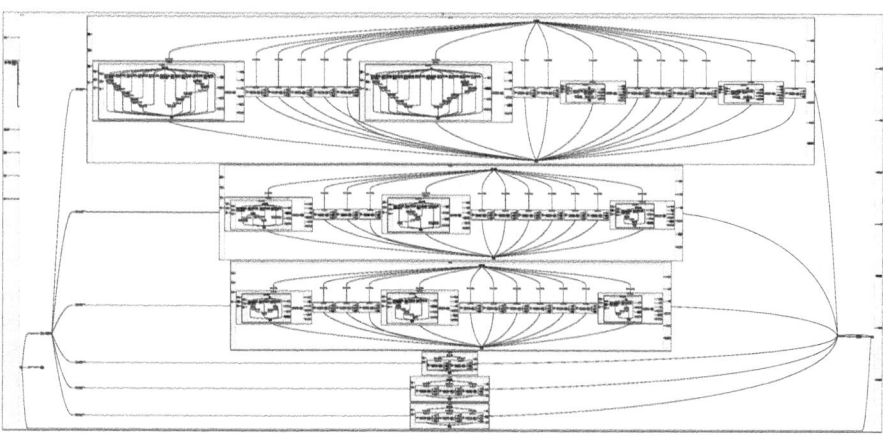

Fig. 4. The WAGON SCChart is a large model that was created as part of a model railway project. The examples shown in this paper use this model.

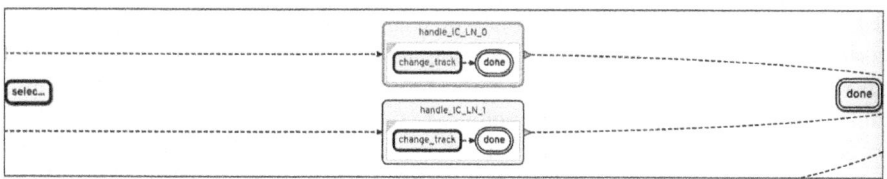

Fig. 5. The WAGON model zoomed in with proxies enabled. The renderings on the left and right side are proxies of off-screen nodes. They may be selected to automatically pan the viewport to focus on their respective nodes.

Proxies for Off-Screen Elements. ⓟ *Proxies* have been used in many settings and different strategies exist to place, merge, and interact with them effectively [1,7]. They are a useful visual aid for diagrams that are too large to be easily viewed in their entirety on a computer screen. Our implementation of proxies works on generic KGraphs.

The proxies are realized as an overlay on top of the diagram rendering. By default, a proxy is created automatically from the existing node rendering so that proxies are always available independent of the diagram type. It is, however, also possible to define a custom proxy rendering for each node. In Fig. 5 we see an example of synthesis-defined proxy renderings used for SCCharts. Instead of showing the entire state with all information, just the typical state shape and styling with a shortened text label is shown.

When there are many nodes in a diagram, decisions must be made about which proxies to show, otherwise the view becomes cluttered and the benefit of the proxies is diminished. We filter proxies based on their nodes' adjacency and hierarchical inclusion in relation to on-screen nodes.

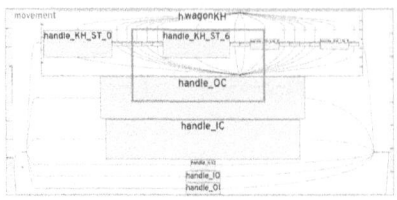
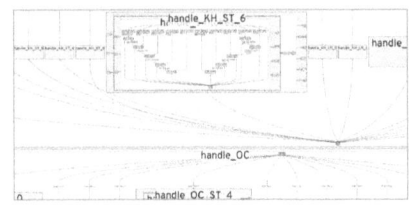

(a) Fully zoomed out model.   (b) Partially zoomed in.

Fig. 6. The WAGON model at different zoom levels with smart zoom enabled. Even when zoomed out, the titles remain readable and connections between elements remain visible, while hiding inner behavior to reduce clutter. In Fig. 6a the view is fully zoomed out and in Fig. 6b the view is partially zoomed in on the state labeled handle_KH_ST_6.

Filtering based purely on the structure of the graph is often insufficient. If a language defines different types of nodes then we may want to create proxies for one type but not the other. Semantic information is necessary to make this distinction and this is one occurrence of where the semantic filtering API, discussed in Sect. 4.1, is used. Filter rules defined by the synthesis are automatically inserted as toggleable options into the sidebar. This gives a synthesis developer control over client-side rendering behavior.

A common challenge with the introduction of proxies for off-screen elements is visual clutter at the edge of the viewport due to many, potentially overlapping proxies [2,7]. The primary information proxies provide is the direction in which their nodes lie or the target node of an edge leaving the viewport. Further useful information is the distance to the nodes. When proxies overlap, we draw them with closer nodes' proxies placed on top. Additionally, we decrease the opacity as the distance increases. This technique aims to reduce visual clutter and add helpful visual cues for navigating complex models.

KIELER improves the utility of proxies with semantic tags and filters, which makes it possible to define proxies and a semantic context in the synthesis for further filtering. De Carlo et al. [2] note that this could be done on the client only, but they want to move it partly to the server. In any case, it is important to utilize semantic information about the model for filtering. If proxies are not sufficiently filtered, one cannot use them at all and this is not a matter of merging or overlaying proxies with clever algorithms but a general problem that occurs in complex models.

Smart Zoom. ② Proxies of off-screen nodes provide context when the view is zoomed in on the details of the diagram. However, they do not provide a good overview over a large diagram in the zoomed out state. When viewing large diagrams as a whole it becomes difficult to discern any details. This makes it challenging to build a mental model of the diagram, and it is difficult to determine where to navigate to.

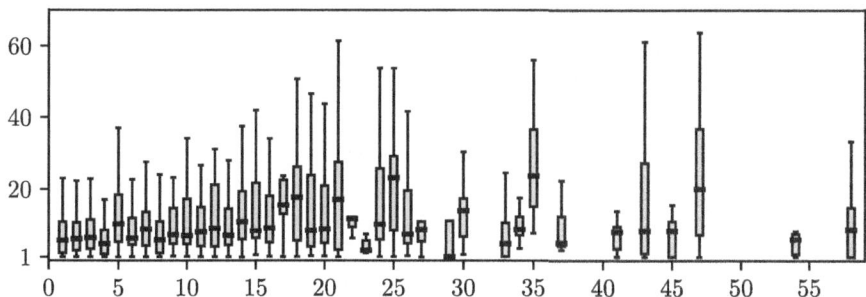

Fig. 7. The y-axis represents the node-key width ratio, i.e. the maximum factor by which smart zoom can upscale a key rendering. The x-axis shows the number of scalable key renderings in a graph. The box plots show the distributions of node-to-key ratios for different graph sizes. This data stems from an analysis of 1250 SCCharts collected over approximately ten years from different projects and teaching. We removed outlier ratios that were larger than 75 in order to not hide the ratios in the denser part of the small to medium-sized graphs, which represents the majority of the graphs.

An approach to help with this is semantic zooming, which is an overview-and-detail technique that has seen many iterations and applications. The key idea is to provide different views depending on the current zoom level according to the required level of detail [2,8,15].

We here propose a variant of semantic zooming that we refer to as *smart zoom*. We approach the readability of diagram elements using only zoom and pan operations based on a *static base diagram*. The diagram application can define a part of the rendering of graph elements as its *key rendering* via a semantic tag. Such a tag will only be used within the framework on the client to make such a rendering more legible, while keeping the layout and overall appearance of all graph elements stable based on the layout of the static base diagram. For example, Fig. 6a shows the large model from Fig. 4 using smart zoom, scaled down by approximately 6000%, yet with legibly represented titles. Here we scale up and overlay the titles as the key renderings within the pre-layouted bounding box of the graph element they represent to a constant 100% scaling, if the space permits, to keep the key renderings legible at many zoom levels. Compared to Fig. 4, this shows better readability.

The framework allows any part of the rendering (names, icons, shapes, etc.) to be the key rendering, for this example it is the text rendering of name of the region. Scaled up key renderings may also overlap other parts of the diagram that are lower down in the hierarchy making the key rendering more prominent. While zooming in, the key rendering will remain at that size until the diagram around it has been scaled up to match its size and any overlaps are then resolved. Furthermore, the client simplifies elements with content that would be too small by hiding the content entirely. Compare the state labeled handle_KH_ST_6 between Figs. 6a and 6b, where zooming in reveals further inner behavior.

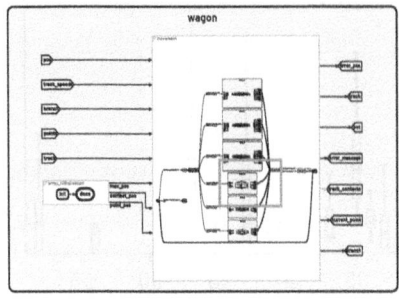
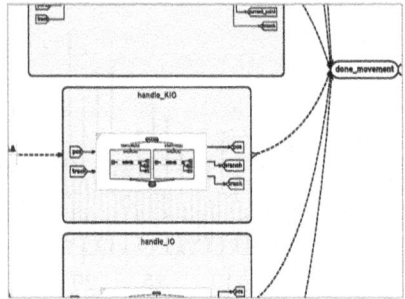

(a) High-level overview. (b) Partially zoomed in view.

Fig. 8. Top-down drawing of the WAGON model used also in the other examples. Figure 8a shows the entire diagram. Node titles are readable similar to the smart zoom effect. The structure can also be seen without zooming or panning. Figure 8b shows a zoomed in excerpt. Details of the current layer are readable and the general structure of the model can easily be recognized.

GLSP follows a different concept. De Carlo et al. [2] point out that their semantic zoom re-computes the layout and, therefore, requires extra roundtrips to the server, limiting the performance of their implementation. In KIELER we decided to always base such zoom and pan operations on the layout of the static base diagram to keep the performance high and even increase it by omitting too small elements, while also maintaining a mental map of the model by having a consistent layout where no elements move around due to a new layout.

Smart zoom can only work effectively when there is sufficient space to upscale the key renderings, i.e. when the node containing the key rendering is significantly larger than the key rendering itself. At the same time smart zoom is only necessary when diagrams reach a certain size, since there is simply no need for upscaling in small diagrams.

In Fig. 7 we show how *node to key rendering* ratios (*node-key* ratios) are typically distributed in graphs of varying sizes. The examples analyzed are all nested graphs, i.e. graphs with hierarchical containment. This means that nodes can be relatively large since they have to accommodate their children. In these types of graphs the node-key ratios generally tend to become larger in larger graphs. Smart zoom is effective in all cases except for very large graphs without any nesting where each node has roughly the same size as its key rendering.

Top-Down Layout. Dynamic adjustments of the displayed diagram are limited in what accessibility features can be realized with them while remaining agnostic to the concrete diagram type. For large diagrams that visualize nested graphs we can also use *top-down layout* [12]. We obtain a similar benefit to smart zoom, i.e. names of nodes are readable when zoomed out but, in contrast to smart zoom, the diagram of one hierarchy level remains quite compact when zooming in. This makes reading and navigating large diagrams a lot easier.

Figure 8 shows the effect of zooming in a top-down layout. The main advantage of top-down layout is that entire hierarchy levels can be read at once. The layout stays consistent and does not depend on the layout of the inner graph in a deeper hierarchy level. Here, zooming is the main control interaction to view details.

We previously examined the effect of top-down layout on the readability of diagrams [12]. The key takeaway from that evaluation was that top-down layout does indeed increase readability of labels in large diagrams across different zoom levels. GLSP currently does not support top-down layout since it is a relatively new feature of ELK, but there are no technical limitations to supporting it.

The raw data used for the smart zoom and top-down layout analysis is available for download[4].

5 Conclusion

We presented the KIELER framework that utilizes both the textual and the graphical model and provides general solutions for common problems of languages that aim to have a graphical view.

A text-only approach for developing complex systems misses the advantages that filtered, graphical representations provide for some languages. Furthermore, compared to a diagram-first approach on the example of GLSP, the text-first approach, as used by KIELER framework, still provides all the advantages of textual editing, while being able to utilize the best of a graphical view. Filtering is a first-class citizen in KIELER and it is common to hide parts of the model that are better edited textually or that are irrelevant for the current use case. In a diagram-first approach, however, filtering does not integrate well with editing. A well implemented diagram-first approach should not sacrifice the benefits of textual editing but rather emulate and enhance them. A text-first approach naturally does this by building on top of text-only tooling. Linking text and diagram lets users utilize the representation that helps with their current problem.

We presented the KIELER synthesis, options sidebar, proxies, smart-zoom, and top-down layout approach to efficiently work with diagrams of complex systems and help to increase the readability of large human-made hierarchical models and compared each to its GLSP pendant if it exists.

Some features presented in this paper are experimental and have not yet found their way into the official releases of the open-source projects. Nevertheless, we provide a tool demo (see footnote 4) that is a pair of Visual Studio Code extensions that can be installed locally to test smart zoom, top-down layout, and proxies. The download also includes several example models and a usage guide.

Future Work and Outlook

The KLighD component currently implements many features of the KIELER framework. In the long run, they would serve better as modules within the

---

[4] https://figshare.com/s/51c31e4f7910d53df3a7.

Sprotty framework for languages that fit the KIELER approach to modeling. Hence, we plan to migrate part of the KLighD micro layout capabilities as well as proxies and the synthesis option sidebar into Sprotty itself. This would not only allow more users to easily configure diagrams for their textual language, but specifically allow using them together with GLSP. Furthermore, we plan to make more of our framework available on the client-side such as the micro and macro layout. Currently, our synthesis API is quite restrictive as a synthesis must be written in Java—this could be expanded with a more open API, e.g., a JSON schema, allowing syntheses in other languages.

References

1. Burigat, S., Chittaro, L., Gabrielli, S.: Visualizing locations of off-screen objects on mobile devices: a comparative evaluation of three approaches. In: Proceedings of the 8th Conference on Human-Computer Interaction with Mobile Devices and Services, pp. 239–246. ACM (2006). https://doi.org/10.1145/1152215.1152266
2. De Carlo, G., Langer, P., Bork, D.: Advanced visualization and interaction in GLSP-based web modeling: realizing semantic zoom and off-screen elements. In: Proceedings of the 25th International Conference on Model Driven Engineering Languages and Systems, pp. 221–231. Association for Computing Machinery, New York (2022). https://doi.org/10.1145/3550355.3552412
3. Domrös., S., Riepe., M., von Hanxleden., R.: Model order in Sugiyama layouts. In: Proceedings of the 18th International Joint Conference on Computer Vision, Imaging and Computer Graphics Theory and Applications - Volume 3: IVAPP, pp. 77–88. INSTICC, SciTePress (2023). 10.5220/0011656700003417
4. Domrös, S., von Hanxleden, R., Spönemann, M., Rüegg, U., Schulze, C.D.: The eclipse layout kernel. arXiv preprint arXiv:2311.00533 [cs.DS] (2023). https://doi.org/10.48550/arXiv.2311.00533
5. Eker, J., et al.: Taming heterogeneity-the Ptolemy approach. Proc. IEEE **91**(1), 127–144 (2003). https://doi.org/10.1109/JPROC.2002.805829
6. Eumann, P., Wechselberg, N.: Application of SCCharts in the railway domain (2023). https://rtsys.informatik.uni-kiel.de/~biblio/downloads/Synchron23/Day1/Day1-0900-Eumann-SCChartInRailway.pdf. International Open Workshop on Synchronous Programming
7. Frisch, M., Dachselt, R.: Visualizing offscreen elements of node-link diagrams. Inf. Vis. **12**(2), 133–162 (2013). https://doi.org/10.1177/1473871612473589
8. Frisch, M., Dachselt, R., Brückmann, T.: Towards seamless semantic zooming techniques for UML diagrams. In: Proceedings of the 4th ACM Symposium on Software Visualization, pp. 207–208 (2008)
9. Glaser, P.L., Bork, D.: The bigER tool - hybrid textual and graphical modeling of entity relationships in VS Code. In: 25th International Enterprise Distributed Object Computing Workshop, pp. 337–340. IEEE (2021)
10. von Hanxleden, R., et al.: Pragmatics twelve years later: a report on Lingua Franca. In: Margaria, T., Steffen, B. (eds.) ISoLA 2022. LNCS, vol. 13702, pp. 60–89. Springer, Cham (2022). https://doi.org/10.1007/978-3-031-19756-7_5

11. Jöhnk, F.: Structure-based editing for SCCharts. Master thesis, Kiel University, Department of Computer Science (2022). https://rtsys.informatik.uni-kiel.de/~biblio/downloads/theses/fej-mt.pdf
12. Kasperowski, M., von Hanxleden, R.: Top-down drawings of compound graphs. arXiv preprint arXiv:2312.07319 [cs.DS] (2023). https://doi.org/10.48550/arXiv.2312.07319
13. Lohstroh, M., Menard, C., Bateni, S., Lee, E.A.: Toward a Lingua Franca for deterministic concurrent systems. ACM Trans. Embed. Comput. Syst. (TECS) **20**(4) (2021). https://doi.org/10.1145/3448128
14. Metin, H., Bork, D.: On developing and operating GLSP-based web modeling tools: lessons learned from bigUML. In: 26th International Conference on Model Driven Engineering Languages and Systems, pp. 129–139. IEEE (2023). https://doi.org/10.1109/MODELS58315.2023.00031
15. Perlin, K., Fox, D.: Pad: an alternative approach to the computer interface. In: Proceedings of the 20th annual Conference on Computer Graphics and Interactive Techniques, pp. 57–64. ACM, New York (1993). https://doi.org/10.1145/166117.166125
16. Petre, M.: Why looking isn't always seeing: readership skills and graphical programming. Commun. ACM **38**(6), 33–44 (1995). https://doi.org/10.1145/203241.203251
17. Petzold, J., Domrös, S., Schönberner, C., von Hanxleden, R.: An interactive graph layout constraint framework. In: Proceedings of the 18th International Joint Conference on Computer Vision, Imaging and Computer Graphics Theory and Applications - Volume 3: IVAPP, pp. 240–247. INSTICC, SciTePress (2023). https://doi.org/10.5220/0011803000003417
18. Petzold, J., Kreiß, J., von Hanxleden, R.: PASTA: pragmatic automated system-theoretic process analysis. In: 53rd International Conference on Dependable Systems and Network, pp. 559–567. IEEE (2023). https://doi.org/10.1109/DSN58367.2023.00058
19. Prochnow, S., von Hanxleden, R.: Statechart development beyond WYSIWYG. In: Engels, G., Opdyke, B., Schmidt, D.C., Weil, F. (eds.) MODELS 2007. LNCS, vol. 4735, pp. 635–649. Springer, Heidelberg (2007). https://doi.org/10.1007/978-3-540-75209-7_43
20. Rentz, N., von Hanxleden, R.: SPViz: a DSL-driven approach for software project visualization tooling. arXiv preprint arXiv:2401.17063 [cs.SE] (2024). https://doi.org/10.48550/arXiv.2401.17063
21. Rodriguez-Echeverria, R., Izquierdo, J.L.C., Wimmer, M., Cabot, J.: Towards a language server protocol infrastructure for graphical modeling. In: Proceedings of the 21th International Conference on Model Driven Engineering Languages and Systems, pp. 370–380. ACM, New York (2018). https://doi.org/10.1145/3239372.3239383
22. Schneider, C., Spönemann, M., von Hanxleden, R.: Transient view generation in Eclipse. In: Proceedings of the First Workshop on Academics Modeling with Eclipse, Kgs. Lyngby, Denmark (2012)
23. Schneider, C., Spönemann, M., von Hanxleden, R.: Just model! – Putting automatic synthesis of node-link diagrams into practice. In: Proceedings of the Symposium on Visual Languages and Human-Centric Computing, pp. 75–82. IEEE, San Jose (2013). https://doi.org/10.1109/VLHCC.2013.6645246
24. Smyth, S., Motika, C., von Hanxleden, R.: A data-flow approach for compiling the sequentially constructive language (SCL). In: 18. Kolloquium Programmiersprachen und Grundlagen der Programmierung, Pörtschach, Austria (2015)

25. Vujović, V., Maksimović, M., Perišić, B.: Sirius: a rapid development of DSM graphical editor. In: 18th International Conference on Intelligent Engineering Systems, pp. 233–238 (2014). https://doi.org/10.1109/INES.2014.6909375
26. Wechselberg, N., Schulz-Rosengarten, A., Smyth, S., von Hanxleden, R.: Augmenting state models with data flow. In: Lohstroh, M., Derler, P., Sirjani, M. (eds.) Principles of Modeling. LNCS, vol. 10760, pp. 504–523. Springer, Cham (2018). https://doi.org/10.1007/978-3-319-95246-8_28

Open Access This chapter is licensed under the terms of the Creative Commons Attribution 4.0 International License (http://creativecommons.org/licenses/by/4.0/), which permits use, sharing, adaptation, distribution and reproduction in any medium or format, as long as you give appropriate credit to the original author(s) and the source, provide a link to the Creative Commons license and indicate if changes were made.

The images or other third party material in this chapter are included in the chapter's Creative Commons license, unless indicated otherwise in a credit line to the material. If material is not included in the chapter's Creative Commons license and your intended use is not permitted by statutory regulation or exceeds the permitted use, you will need to obtain permission directly from the copyright holder.

Historical Aspects of Diagrams

Drawing Technology: Sketches of Isambard Kingdom Brunel

Guy Clarke Marshall[1,2,3]

[1] Fuza Ltd., Manchester, UK
guy@fuza.co.uk
[2] PorthouseDean Structural Engineers Ltd., Manchester, UK
[3] University of Manchester, Manchester, UK

Abstract. This paper examines the sketchbooks of the British Engineer, Isambard Kingdom Brunel, famed for his work on the Great Western Railway and the first purpose-built transatlantic steam ship. His sketching methods are described, through exploratory analysis of the 57-volume Brunel sketchbook archive (1830–1866). Through two in-depth case-studies of diagram series enumerated by Brunel, we capture the role of sketching as an aid to Brunel's reasoning about engineering design, and explore diagramming as externalised cognition, central to the process of Brunel's engineering.

Keywords: Brunel · Engineering · Cognition · Diagram · Technical Drawing

1 Introduction

1.1 Brunel in Brief

Much has been written, biographically and otherwise, about Brunel. The present work will outline this briefly, and draw only on this research when it directly impacts our subject matter. Isambard Kingdom Brunel was born in 1806. His father, Sir Marc Isambard Brunel, was an engineer noteworthy in his own right for his drawing skill and inventions, including the tunneling shield used in the creation of the Thames Tunnel, the first tunnel under a navigable river.

Isambard Kingdom Brunel (henceforth "Brunel") was "one of the most ingenious and prolific figures in engineering history" [11]. He was famed for his drawing, particularly for a lifelike horse (drawn age 6) and reputed ability to hand-draw a perfect circle. [7] note the influence of his French father and French schooling, themselves influenced by Gaspard Monge's descriptive geometry, on the precision and accuracy of Brunel's engineering sketches.

Brunel's engineering contributions include the Great Western Railway (headquartered at Paddington Station), three steamships, and many docks, bridges and tunnels. Many of these works continue to be used in the present day.

© The Author(s), under exclusive license to Springer Nature Switzerland AG 2024
J. Lemanski et al. (Eds.): Diagrams 2024, LNAI 14981, pp. 421–428, 2024.
https://doi.org/10.1007/978-3-031-71291-3_34

1.2 Examining Engineering Sketches

The Middle Ages: Popplow [10] notes that "any attempt to explain when and why drawings came to be employed in mechanical engineering in the Middle Ages must take into consideration the tradition of late medieval architectural drawings. As the different roles of machine builder, architect, and fortification engineer emerged more clearly only in the course of the sixteenth century, it can be assumed that the employment of such a crucial medium as drawing in earlier periods still showed similar characteristics in all three of these fields." Further, "The materiality of technology was often ignored in these presentational treatises. Machines in these books should be understood as a product of the engineer's brain, his ingenium; their material realization was not the topic of these books. The organizational activities of the engineer on the building site were mentioned as scarcely as materials, measurements or gear ratios". Rubio et al. [12] examine Leonardo Da Vinci's Madrid Codex I, a major collection of Da Vinci's workbooks (1490–1499). The project collects 100 diagrams from this corpus of almost 1000 drawings, to capture the main mechanisms and machine element designs. The focus of the work is on the content of the drawings, rather than the drawings themselves. Sketching styles or methods are not analysed.

The Industrial Revolution: Founded in 1824, Manchester Mechanics' Institution was the largest outside London. Through much of its history the "mechanical and architectural drawing" evening class was the best attended of all classes at the Manchester Mechanics' Institution, as recorded in their archived Minutes [6], highlighting the recognised importance of drawing to the engineering of the time. Kelly and Kelly [7] note that the foundation, in 1837, of the Royal College of Art was "to forge and maintain links with industry and to encourage an element of design within engineering by absorbing it into the process as a whole". Dutz and Schlimm [4] note that in the 1860's, mathematician and computer pioneer Charles Babbage advocated the use of diagrammatic notation to "rigorously express complex ideas and to facilitate the discovery of new ones", something Babbage himself expressed as "immense power of signs" [2]. These contemporaneous resources further highlight the centrality of drawing to the practice of engineering.

Corpus Analysis: McGee [9] cautions against two problems of historical drawing analysis: that the corpus is biased and that we lack the conceptual tools to properly analyse them. However, in the case of Brunel, his fame during and after his lifetime was such that even poor quality drawings have been preserved. The problem of conceptual tooling remains relevant, and it is perhaps only through attempts at analysis that we will refine the required conceptual tooling.

Throughout this exploratory data analysis paper, we will use the word "sketch" to describe the epistemic images of Brunel, as they are contained within "sketchbooks". Some of the subjects could be described as "drawings" or "diagrams", though as noted by Marr [8] the definitions surrounding technical images are neither plentiful nor yet consistent.

2 Background

2.1 Brunel's Sketchbooks

The overall process of sketching was for Brunel to create a conceptual drawing or sketch. These were sent to a "drawing office" for the project, comprised of a team of engineers and draughtsman to draw scale drawings of parts and details, replicated by tracers, to allow unambiguous creation of the subject by fabricators. Brunel was involved in "signing off" (literally) the designs. Depending on the subject, watercolours or lithographs may also have been created, to visualise the finished product. Thanks to Brunel's popularity, during his life and afterwards, many of his artefacts are preserved. The Brunel Institute archive consists of 57 volumes of sketchbooks, including thousands of sketches of engineering projects.

2.2 Brunel's Other Drawings and Paintings

Brunel carried pocket notebooks. These appear to primarily contain text, shorthand, and financial notes. Some drawings occur, but more as a drawing and a record than for reasoning. Brunel's diaries also contain some drawings. Brunel's watercolours sometimes served as working drawings, such as those for the Great Western Railway, evidenced by pinholes in the corners from being used in drawing offices. Other watercolours were used as promotional materials, such as those for the Thames Tunnel (credited to Isambard Kingdom Brunel and acquired by The Brunel Museum Thames Tunnel).

Brunel used his inability to draw a perfect circle whilst travelling by train to motivate his broad gauge for the Great Western Railway, which he believed would result in a smoother journey. This drawing experience is recreated on an iPad as part of the SS Great Britain museum.

2.3 Drawing to Reason

Heiser [5] examines production and comprehension of text and diagrams in instructions through five experiments with students, concluding "Effective diagrams depends on knowing how people mentally represent the information to be communicated and how people perceive and interpret diagrams". Through examination of series of science diagrams, including unfinished works, Burnston et al. [3] demonstrate the process of ideas being distilled diagrammatically, and identify two uses of diagrams in science: facilitating human reasoning; and as explanatory. Abrahamsen and Bechtel [1] identify "distinctive functions served by diagrams in three aspects of mechanistic research: (1) delineating the phenomenon to be explained; (2) identifying explanatory relations (relations between variables that are relevant to explaining the phenomenon); (3) constructing and revising a mechanistic explanation of the phenomenon". These works suggest that diagrams are a reflection of how the diagram creators are mentally representing the information. As such, for Brunel, the diagrams may also be a window into his cognitive processes.

3 Brunel's Sketching Techniques

This section is based on an examination of Brunel's sketchbooks, held by the University of Bristol and the Brunel Institute.

3.1 Summary of the Sketchbook's Content

The sketchbooks are made of squared paper, mounted. The majority have been rebound, presumably as part of the acquisition by University of Bristol in 1990 [13], and the original condition was not recorded. They contain sketches, engineering calculations, some accompanying text, and some financial calculations. Most of the sketches are not titled and have not been linked to Brunel's projects. Some pages are dated, and each of the sketchbooks often span several years.

Throughout the sketches, as a *drawing media*, Brunel uses pencil, with the heaviness of pressure varying greatly. Black pen is used sometimes, especially where a series of sketches progresses. In these, Brunel appears to use pen for the parts he is satisfied with, though these may also change in subsequent sketches. *Red or brown pen* is very occasionally used, not for any obvious specialised purpose beyond "this line is of a different type to black". Tools other than pens and pencil are used to create the sketches. A *compass* is often used, despite Brunel's reputation for being able to draw a perfect circle. This is evidenced by the hole markings. His set of drawing equipment, preserved at the SS Great Britain, contains rulers, compasses, pens, and dividers. The set contains mostly compasses, confirming their importance. Occasionally, rubbings can be seen, where an error occurs and the paper is damaged. In the archive, some paper is remounted, and the infrequent occurrences of scratching away the ink are not discussed. A *ruler* is seldom used in the sketchbooks. It can be seen that most lines are not perfectly straight, with some of the natural variation one would expect drawing a line freehand. Where rulers are used, this is sometimes alongside the existing grid. Freehand straight lines are also done for the diagonals, in short sections of around 5cm before replacing the pen or pencil to continue the line.

The *squared paper* appears crucial to Brunel's sketching method. He appears to rely on this for drawing straight lines freehand, as well as for ensuring consistent relative scale. *Scale* is very occasionally indicated explicitly (as in one radial measurement in Fig. 2). Most of the sketches are conceptual.

The *details* included vary. Sometimes, threads on screws are drawn completely, sometimes not. Sometimes human figures are included, usually a gentleman wearing a top hat, presumably for scale. Sometimes trees or bushes are included. Despite often involving moving mechanisms, *movement* is seldom indicated. Arrows are not used, nor other sense of movement beyond occasionally repositioned objects (see Fig. 3). *Annotations* (such as "A goes here") are sometimes found. Labels (such as "A") are also used. Very occasionally, areas are shaded using the side of the pencil to give texture. More commonly, additional lines are used for 3D effect (occasionally crossed, usually parallel).

The sketches are almost all *2-dimensional* (2D), but some use projection to make 3D, though usually only at the conclusion of a set of 2D plans. Figure 1

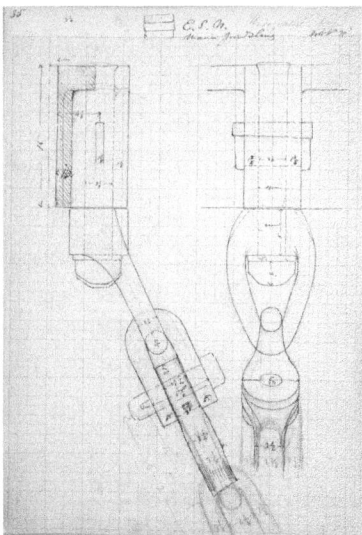

Fig. 1. DM162/8/1/1/Large Sketchbook 5, folio 35. Detail of rigging, Great Eastern steamship, showing a transition from 2D to 3D. By Courtesy of the Brunel Institute – a collaboration of the ss Great Britain Trust and the University of Bristol

shows a transition within the sketch from a 2D plan to 3D, with pencil sketchlines and pencil dimensions contrasting with those lines overlaid in pen, as Brunel appears to work through some of the details. Dashed lines are used to indicate "hidden" information in the 2D sketches, where this information is seem as important. The sketches were usually done top-to-bottom, left-to-right, as space permitted. This is evident from the occasional use of numbering a series of sketches.

The majority of sketches are evidently deemed "good enough" *first time*, and do not require revisions. In the subsequent case studies, two examples are described which involve a series of sketches, evidencing reasoning through sketching.

3.2 Crane Series

Figure 2 shows a series of numbered sketches of a crane. In these, Brunel appears to be initially deciding about the shape of the crane and its supporting wheel (sketches 1–4), and then reasoning about the position of the cable (sketches 5–7). There is a faint un-numbered outline of the shape of the crane can also be seen above sketch 4. Brunel appears to have thought sketch 5 would work out, and the radius of the arc of the crane is labelled. However, the cable attachment evidently required more work, crammed into the page with sketches 6 and 7. In sketch 7, he decides it will be necessary to add a bolt-reinforced outrigger. An eighth, unnumbered crane in faint pencil can be seen to the left of sketch 7, with

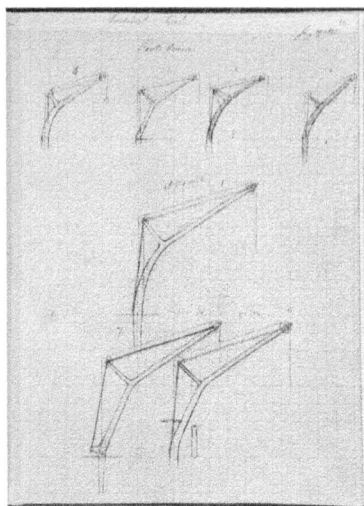

Fig. 2. DM162/8/1/1/Large Sketchbook 4, folio 16. Seven numbered sketches of port cranes at Paddington Station. July 29, 1851. By Courtesy of the Brunel Institute – a collaboration of the ss Great Britain Trust and the University of Bristol

the same cable profile as sketch 7. This may have contributed to the overlap of sketch 7 with sketches 6 and 5, as he wanted to keep them on the same page. In this example, red pen is used for the cable itself, and pencil overlaid with black pen for the crane structure. This includes the first four sketches in which the cable is not the focus. Only pencil is used on the two, unnumbered sketches above sketch 4, and to the left of sketch 7.

3.3 Spring Series

Figure 3 shows a series of three sketches of a spring mechanism. It is notable for the additional straight lines, which indicate the range of possible positions of the right-hand side of the mechanism, as their method of attachment is reasoned about. The middle sketch is conceptually between the top and bottom sketches, as it is evident Brunel was not satisfied with the range of motion afforded by the original static positioning on the left-hand side. In the bottom sketch, a sketch from the other profile is included, to the same scale, using the squared paper lines. When combined with information from the other sketches it shows how the central overlapping portion would be arranged, with what is the right-hand side on the other sketches being a thinner piece than the left-hand side. Note also the careful sketch of the threading on the left-hand side, and the careful-but-partial details of the spring on the right-hand side. In general in his sketches, it is not clear at what stage Brunel decides to stop his precise, minute, repetitive sketch (such as threading, or shading bricks, or architectural ornamentation), but he often continues diligently to the end (even when this does not provide

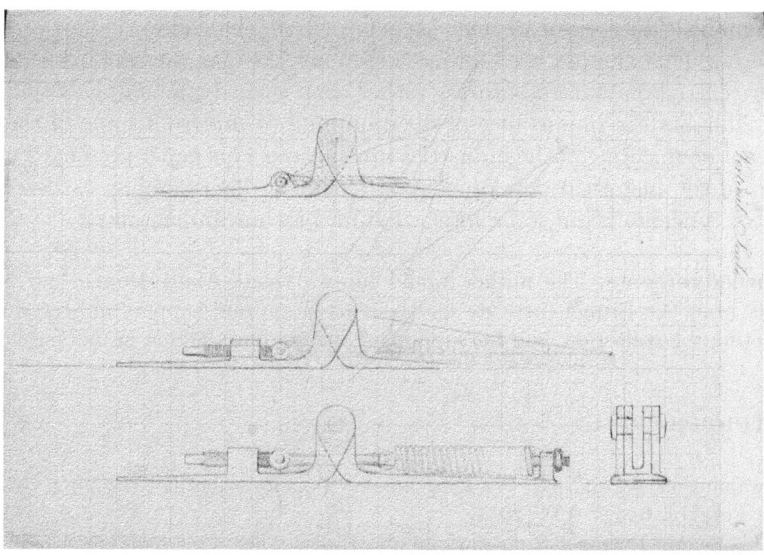

Fig. 3. DM162/8/1/1/Large Sketchbook 2, folio 6. Three sketches of spring loaded levers. By Courtesy of the Brunel Institute – a collaboration of the ss Great Britain Trust and the University of Bristol

additional information). No example of Brunel being casual in his attention to visual details (e.g. threading) was found in this partial (13 of 57 volumes) examination of the archive, only that he stopped the sketching process (as seen in this spring mechanism sketch).

3.4 Summary of Insights from These Series of Sketches

- Sketching was central to Brunel's engineering design process, as demonstrated by the quantity and type of surviving architects.
- Brunel was not using sketches solely to communicate with others, they also served an externalised cognition function.
- Brunel was decisive, as demonstrated by mostly doing a single sketch, and committing to pen.
- Brunel may have thought in terms of *objects* more than *motion*, even on dynamic objects.

4 Conclusion

This paper outlines the uses of sketches in Brunel's engineering work, ranging from promotional watercolours through to engineering designs. Through two case studies selected from his sketchbooks, Brunel's use of sketching to reason about engineering problems is described. Finally, it is postulated that in his diagramming choices we can infer aspects of Brunel's conceptualisation about engineering. For example, even for moving items, the diagrams are presented in a manner centred on material objects rather than their dynamics or motion. This work highlights the importance of diagramming to underpin some of the most significant engineering of the Industrial Revolution. This paper presents a critical portion of the historical development of technical diagramming, as descriptive geometry concepts begin to be integrated into technical diagrams.

Acknowledgements. The author would like to thank Molly Bowen and Joanna Mathers from the Brunel Institute for their attention and support in accessing and interpreting the collection, and for supporting image reproduction in this publication.

References

1. Abrahamsen, A., Bechtel, W.: Diagrams as tools for scientific reasoning. Rev. Philos. Psychol. **6**, 117–131 (2015)
2. Babbage, C.: Passages from the Life of a Philosopher. Longman green (1864)
3. Burnston, D.C., Sheredos, B., Abrahamsen, A., Bechtel, W.: Scientists' use of diagrams in developing mechanistic explanations: a case study from chronobiology. Pragmat. Cogn. **22**(2), 224–243 (2014)
4. Dutz, J., Schlimm, D.: Babbage's guidelines for the design of mathematical notations. Stud. Hist. Philos. Sci. Part A **88**, 92–101 (2021). https://doi.org/10.1016/j.shpsa.2021.03.001
5. Heiser, J.L.: External representations as insights to cognition: production and comprehension of text and diagrams in instructions. Ph.D. thesis, Stanford University (2004)
6. Institution, M.M.: Archive of the Manchester mechanics' institution. GB 133 MMI (1824–1891)
7. Kelly, A., Kelly, M.: Brunel, in Love with the Impossible: A Celebration of the Life, Work, and Legacy of Isambard Kingdom Brunel. Bristol Cultural Development Partnership (2006)
8. Marr, A.: Knowing images. Renaiss. Q. **69**(3), 1000–1013 (2016)
9. McGee, D.: The origins of early modern machine design. In: Picturing Machines 1400–1700. The MIT Press (2004)
10. Popplow, M.: Why draw pictures of machines? The social contexts of early modern machine drawings. In: Picturing Machines 1400–1700, pp. 17–48. The MIT Press (2004)
11. Rolt, L.T.C.: Isambard Kingdom Brunel: A Biography. Longmans, Green (1957)
12. Rubio, H., Bustos, A., Castejon, C., Meneses, J.: Analysis of the first treatise on machine elements: Codex Madrid I. Found. Sci. 1–22 (2023)
13. University of Bristol: The Brunel collection (2010). https://www.bristol.ac.uk/media-library/sites/library/documents/special-collections/brunel-collection.pdf

On the Expressivity of Byzantine Diagrams in Logic

Jens Lemanski[1,2](✉) and Reetu Bhattacharjee[2](✉)

[1] Philosophisches Seminar, University of Münster, Münster, Germany
[2] Faculty of Humanities and Social Sciences, FernUniversität in Hagen, Hagen, Germany
{jenslemanski,reetu.bhattacharjee}@uni-muenster.de

Abstract. 'Byzantine logic diagrams' have been used since at least late antiquity, and became popular in Europe in the 16th century. However, since the criticism of W. Hamilton and J. Venn in the 19th century, Byzantine diagrams have been largely dismissed as obsolete. This paper challenges this prevailing view. Initially, we provide a comprehensive overview of the current state of research pertaining to these diagrams, illustrating their applicability in analyzing assertoric syllogisms. Subsequently, we propose that the expressive capacity of these diagrams extends far beyond their traditional use. In particular, we demonstrate that the expressivity of these diagrams is not limited to checking the validity of syllogisms. Rather, we show that there are various extensions of Byzantine diagrams, including opposition relations, modal logic, particular instances, n-term diagrams and logical connectives. These extensions are compositional and can therefore be combined to create very complex Byzantine diagrams that go far beyond what has been commonly believed since Hamilton and Venn.

Keywords: Byzantine Logic · Syllogistic · Diagrammatic Reasoning · Early Modern Philosophy · History of Logic · Aristotle

1 Introduction

In recent years, there has been a growing scholarly interest in Byzantine logic diagrams. These diagrams, introduced by Byzantine scholars into the European logic tradition during the early modern period, reached their zenith in the 16th century. By the 17th century, however, they were eclipsed in Europe by alternative diagrammatic forms in logic, which subsequently became more predominant. In the 19th century, scholarly interpretations diminished the perceived significance of these diagrams, leading to their relegation as antiquated and trivial curiosities until the 21st century. Since 1999, however, numerous studies have challenged some prejudices of 19th and 20th century logicians. Nevertheless, the exact workings and potential of these diagrams remain unclear.

The present paper seeks to contribute to this reevaluation by focusing first on the basic function and then on the extensions of the Byzantine diagrams. Our objective is to elucidate the logical expressiveness inherent in these diagrams through some examples. Due to space limits, we restrict ourselves from delving deeply into the technical details of the usage of these diagrams[1]. To this end, the paper is structured as follows. Section 2 provides a comprehensive review of the existing research. Section 3 explicates the process of diagrammatic reasoning using Byzantine diagrams. The main aim is to demonstrate how Byzantine diagrams were used to prove the validity of a syllogism directly or indirectly. Unlike some other diagram types of the modern era, however, this does not exhaust the function and expressiveness of these diagrams. Section 4 uses six examples to show that there were also other diagrammatic techniques that transcended representation and reasoning via direct/indirect methods in syllogistic. Finally, Sect. 5 uses a few examples to show that the extended diagrammatic techniques can be combined and that both the combination and the application of Byzantine diagrams with other types of diagrams have resulted in very complex diagrams whose exact function is still unknown today.

2 Research Review

Contemporary scholarship has established that Byzantine scholars from the late Middle Ages onwards utilized various types of logic diagrams, not limited to a single form. It is recognized that Byzantine scholars sometimes employed circle diagrams akin to those of Euler and Venn. However, it is the triangular and crescent diagrams, particularly prevalent in Central Europe during the 16th century, that have garnered significant attention. This review will primarily focus on these latter diagrams, especially those known in Europe, and their logical expressiveness.

The digital repository *History of Euler-Venn Diagrams*[2], as of February 2024, lists 34 manuscripts under the tag 'Byzantine diagrams'. Among these manuscripts, twenty-one are dated between 1501 and 1586, three from the 17th and 18th centuries each, and the remainder from previous or subsequent periods. Notably, post-16th century manuscripts exhibit a reduced frequency of diagrams compared to the earlier period before. In particular, writings from the 19th century onward indicate a diminished understanding of the functional utility of these diagrams. Yet, the perspective of the 19th century has long influenced the perception of Byzantine logic diagrams.

Significant in the contemporary comprehension of Byzantine logic diagrams, particularly those featuring crescents and triangles, are the commentaries of William Stirling Hamilton of Preston and, importantly, John Venn and his successors. Hamilton follows Immanuel Kant's idea that Byzantine diagrams go

---

[1] We will encourage interested readers to check [4], which presents a part of the current ongoing research with detailed technical aspect. More papers on various applications of the Byzantine diagrams are on their way toward publication.

[2] https://www.zotero.org/groups/319026/history_of_euler-venn-diagrams.

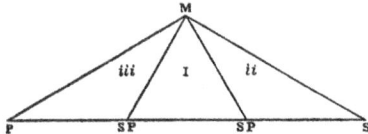

Fig. 1. Hamilton's isosceles triangle [10, 668] showing $\mathcal{F}_3$(iii), $\mathcal{F}_1$(I) and $\mathcal{F}_2$(ii)

back to Aristotle, who used combinatorics to identify the three syllogistic figures [11, XVI, 726]. A number of authors in the history of logic, including Hamilton, have proposed that Aristotle's term 'figure' (or $\mathcal{F}$) was not merely an empty metaphor used to characterise the position of the middle term. Rather, they have argued that it referred to diagrams, although these have unfortunately been lost in the surviving writings from antiquity. Hamilton sought to reconstruct Aristotle's original schemes using an isosceles triangle (Fig. 1), aligning with Aristotle's method of categorizing the three $\mathcal{F}_{1-3}$ of syllogism based on the position (subject or predicate position in a sentence) of the middle term (M) [10, 666ff.]. Here P and S denote the major and minor terms, and the first two lines or sentences are the premises and the last one the conclusion.

1st Figure ($\mathcal{F}_1$) $\begin{bmatrix} M & P \\ S & M \\ S & P \end{bmatrix}$ 2nd Figure ($\mathcal{F}_2$) $\begin{bmatrix} P & M \\ S & M \\ S & P \end{bmatrix}$ 3rd Figure ($\mathcal{F}_3$) $\begin{bmatrix} M & P \\ M & S \\ S & P \end{bmatrix}$

In Hamilton's Fig. 1 three triangles are given, each comprising the letters S, P and M. If one of these letters is positioned to the right of another, it is in the place of the subject in the sentence; if it is positioned to the left, it is in the place of the predicate. For example, in the triangle (iii) M is always in the subject position (to the right of P and S) as in $\mathcal{F}_3$, but in the triangle (i) it is once to the right of P and once to the left of S, i.e. as in $\mathcal{F}_1$. Hamilton perceived a similar endeavour in the three basic Byzantine diagrams, as given in Fig. 2. The Byzantine diagrams also show whether M occupies a central position ($\mathcal{F}_1$ which is represented by a crescent and M is the vertex in the centre), or whether S and P lead towards M ($\mathcal{F}_2$ which is represented by a triangle and M is the top vertex) or away ($\mathcal{F}_3$ which is represented by the inverted triangle and M the bottom vertex). However, Hamilton raised both general and specific objections against Byzantine diagrams.

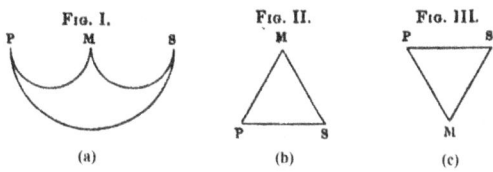

Fig. 2. Diagrams for (a) $\mathcal{F}_1$, (b) $\mathcal{F}_2$, (c) $\mathcal{F}_3$

In the 2nd edition of *Symbolic Logic*, ch. XX, Venn simplifies Hamilton's critiques and asserts that Byzantine diagrams predominantly represent syllogisms with an emphasis on the middle term, as given in Fig. 2. However, he argues that they are unsuitable for diagrammatic reasoning and distinctly separates them from so-called 'analytical diagrams', such as those by Euler or his own creations (today known as 'Venn diagrams'). His critique goes as follows.

> It is obvious that in diagrams of this description [sc. Byzantine diagrams such as given in Fig. 2] no kind of analysis of the proposition is attempted, and it cannot be claimed for them that they afford any real aid to the mind when dealing with trains of reasoning. For the last two or three centuries they have been entirely abandoned. [20, p. 506]

This perspective has predominated throughout the 20th century, and it is against this backdrop that Sect. 3 and 4 of this paper present a counterargument. References to 20th-century critics who followed Venn are found in the studies that will be discussed subsequently. Only Bocheński [6, p. 141], presents a relatively complex Byzantine diagram, but its function is discernible only through contextual analysis. The Kneales also acknowledge such diagrams, particularly referencing the so-called 'pons asinorum', which both Venn and Bocheński associate closely with Byzantine diagrams [13, pp. 72, 185].

Since the 2000s, there has been a renewed interest in Byzantine diagrams. In 1999, Panizza critically examined the historical genesis and dissemination of the diagrams as described by Venn [16]. Her work not only contributes significantly to historical understanding but also implicitly counters Venn's assertion that these diagrams lack expressive function and merely represent the differentiation of $\mathcal{F}_{1-3}$ according to the middle term. Although Panizza does not explicitly state that these diagrams extend beyond the diagrammatic representation of syllogisms, she demonstrates that the crescent and triangle vertices indicate a reading direction for analyzing syllogistic propositions. She elucidates this direction using arrows (see Fig. 3), absent in the original Byzantine diagrams but intuitively understood by logicians until the 16th century. In Fig. 3, the arrow tail represents the position of the subject and the arrowhead represents the position of the predicate. Panniza also introduces numbers to denote premises and conclusions with ease. Here 1, 2 and 3 represent the first premise, second premise and conclusion respectively.

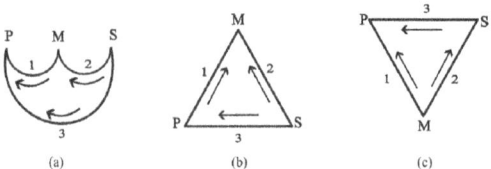

Fig. 3. Reading direction for (a) $\mathcal{F}_1$, (b) $\mathcal{F}_2$, (c) $\mathcal{F}_3$.

In 2001, Cacouros, in a brief section on "figures (diagrammes ou schémas) syllogistiques)" [7, p. 30], reveals the existence of complex Byzantine diagrams in ancient manuscripts, noting that $\mathcal{F}_{1-3}$ Venn displayed are sometimes combined [7, p. 31]. He does not elaborate on the rationale for this combination but observes that at least one triangle shape, as given in Fig. 2b and c, is often present when diagrams are amalgamated [7, p. 32].

Wesoły, in 2012, aligns with Kant's and Hamilton's project (though not explicitly) and posits that these diagrams originate from antiquity. He argues that Byzantine diagrams correlate with Aristotelian terminology, thus suitable for reconstructing diagrams Aristotle presumably employed [21]. His evidence, however, does not concentrate on diagrammatic reasoning; instead, he aims to demonstrate that Byzantine diagrams reveal combinatorics of $\mathcal{F}_{1-3}$ identifiable in Aristotle's verbal metaphors ('figures') which refer to imaginative diagrams.

A recent article by Triantafyllou [19] resembles the projects of Kant, Hamilton and Wesoły, suggesting that Aristotle's language is imbued with geometric metaphors indicative of diagrams. Triantafyllou, like Hamilton, inadvertently approaches Byzantine diagrams, showing the feasibility of diagrammatic thinking with Aristotelian diagrams, yet his reconstructions differ markedly in form and function from Byzantine ones.

Other recent studies have primarily analyzed Byzantine diagrams through the lenses of church history, culture studies, art history, music theory and theology. Noteworthy contributions by Safran [18], Roberts [17], Hamburger [9], Lourié [14], and Wilson [22] indicate a significant phase of Byzantine diagrams in medieval manuscripts. These studies reveal their inclusion in syllogistic writings, the existence of forms beyond those discussed by Hamilton and Venn, and their theological and aesthetic functions. However, these works do not primarily focus on logic, leaving many diagrammatic functions and their precise logical expressiveness unexplored.

In summary, since the works of Panizza and Cacouros, it has been evident that Byzantine diagrams encompass more than previously thought by Hamilton, Venn, and their 20th-century successors. Both the reading direction rediscovered by Panizza and the diagram combinations highlighted by Cacouros suggest that diagrammatic reasoning with Byzantine diagrams is feasible. Nonetheless, the historical diagrams in logic remained enigmatic until Bhattacharjee's breakthrough in 2024 [4]. She built upon Panizza's approach, combined diagrams in the vein of Cacouros, and devised an algorithm for direct syllogism testing using these diagrams. The following section presents a simplified version of this methodology and extends it to include indirect proofs.

3 Diagrammatic Reasoning with Byzantine Diagrams

In this section, we aim to delineate the fundamental functionality of Byzantine diagrams, employing straightforward examples of both direct and indirect proofs. Cacouros' observation (see Sect. 2) is correct, as the diagram types mentioned in Fig. 2 are combined with each other. Each crescent or triangle represents a syllogism and shows a categorical proposition at each arc or edge, viz: universal affirmative or 'a' (All A are B), universal negative or 'e' (No A is B), particular affirmative or 'i' (Some A are B) and particular negative or 'o' (Some A are not B). Together with four categorical propositions $\mathcal{F}_{1-3}$ yields 256 syllogistic combinations and only 24 are considered valid. Each of these combinations in each of $\mathcal{F}_{1-3}$ are called 'moods'. The $\mathcal{F}_1$ moods 'Barbara', 'Celarent', 'Darii' and 'Ferio' are considered 'perfect moods'. The rest of the 20 moods are considered to be 'imperfect moods'. In order to prove the validity of an imperfect mood directly with Byzantine diagrams, let us consider the following example from Pacius' 1584-edition of Aristotle [3, p. 142].

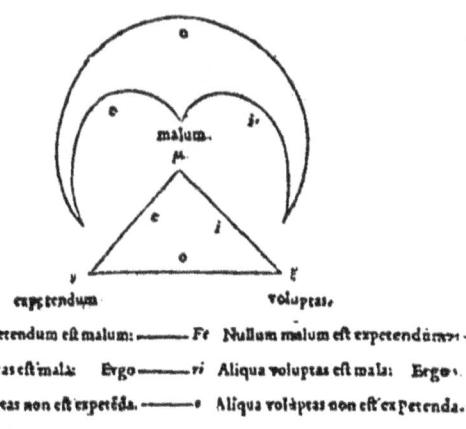

Fig. 4. Validity of Festino (μ: bad; ν: pleasures; ξ: desired)

In Fig. 4, the triangle represents the $\mathcal{F}_2$ mood Festino and the crescent represents the $\mathcal{F}_1$ perfect mood Ferio. The ν denotes the major term, ξ denotes the minor term, and μ denotes the middle term in both $\mathcal{F}_{1-2}$. 'e', 'i' and 'o' respectively denote the categorical propositions 'universal negative', 'particular affirmative' and 'particular negative' respectively. The translation of the premises and conclusion of both Festino and Ferio is given below.

Nothing to be desired is bad.	Nothing bad is to be desired.
Some pleasures are bad.	Some pleasures are bad.
∴ Some pleasures are not to be desired.	∴ Some pleasures are not to be desired.
(Festino)	(Ferio)

Even though a text representation of Festino and Ferio was presented in [3], we do not need it to check the validity of Festino. The text was just an example for which the diagram in Fig. 4 was drawn. The point to be noted is that we check the validity of any imperfect syllogistic mood by comparing it to one of the $\mathcal{F}_1$ perfect moods, i.e. Barbara or Celarent or Darii or Ferio which are considered to be valid. For more information check [4]. We first introduce the arrows representing the subject-predicate relationship. Hence, Fig. 4 now looks like the diagram in Fig. 5a. Here, the arrows work similarly as shown in Fig. 3. However we have replaced the numbers of Fig. 3 with the categorical propositions for each of the premises and conclusion. The next step is to compare the direction of the arrows of Festino with that of Ferio. We find that the direction of the arrow in the first premise of Festino is in the reverse direction of that of Ferio, denoted by wavy arrows in Fig. 5b[3]. As the categorical proposition for the first premise is e, we can reverse the direction of the arrow in Festino (Fig. 5c). That is, we reverse the subject-predicate relationship of the e proposition. This is permitted as saying 'No ν is μ' (i.e. $\nu \cap \mu = \emptyset$) is the same as saying 'No μ is ν' (i.e. $\mu \cap \nu = \emptyset$). This rule is called '*conversio simplex*' rule in the traditional syllogism. Now in Fig. 5c, we find that all the arrows in Festino are in the same direction as that of Ferio. Furthermore, the categorical propositions that take place in the premises and conclusion are the same as Ferio. Hence, we say that Festino has been reduced to Ferio through direct reduction, and therefore, Festino is a valid mood.

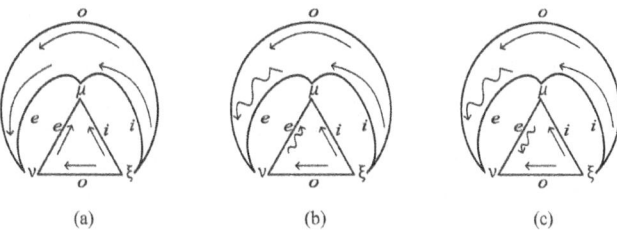

Fig. 5. *Conversio simplex*

As shown in Fig. 2, the inverted triangle and the crescent represent the $\mathcal{F}_3$ and $\mathcal{F}_1$ respectively. Similarly, in Fig. 6 [3, p. 148], the inverted triangle represents the $\mathcal{F}_3$ mood Darapti. Here also the crescent represents the $\mathcal{F}_1$ mood Darii. However, the crescent is inverted here. This is because we combine these diagrams to compare their subject-predicate relationships and categorical propositions. For this reason, we place the diagrams together in such a way that the middle term of the crescent should always be in the same position as the middle term of the triangle/inverted triangle. Similarly to Fig. 4, the π, ρ and σ denote the major term, the minor term, and the middle term respectively. The letter 'a' denotes the categorical proposition 'universal affirmative'. We are not going to translate the text representation as the text is not needed for checking the validity of the mood Darapti.

Again, like Fig. 5a, we first introduce the arrows (see Fig. 7a). By comparing the direction of the arrows, we find that the second premise in Darapti has the arrow in the reverse direction of that of Darii (see Fig. 7b). Since the second premise of Darapti has an a proposition, we cannot reverse the direction of the

---

[3] Only the straight arrows are part of the syntax for this kind of diagrammatic reasoning. Here we have introduced the wavy arrows just for explanation purposes.

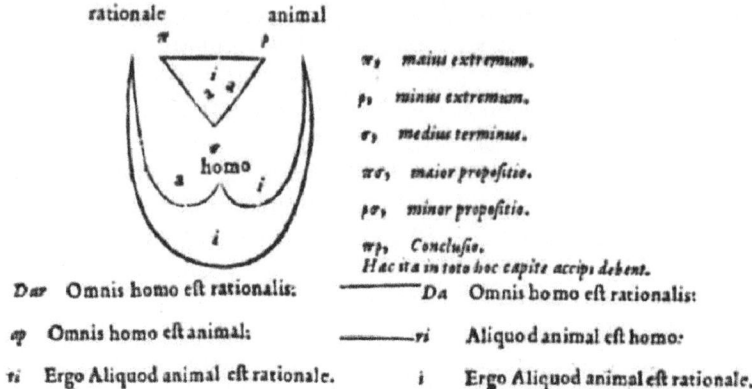

Fig. 6. Validity of Darapti (π: rational; ρ: animal; σ: human)

arrow as we did in Fig. 5c because 'All σ is ρ' (i.e. $\sigma \subseteq \rho$) is not same as the 'All ρ is σ' (i.e. $\rho \subseteq \sigma$). However, we can convert the a proposition to its subalternation, i.e. the i proposition (see Fig. 7c) (as $\sigma \cap \rho \neq \emptyset$ follows from $\sigma \subseteq \rho$)[4]. This is similar to the so-called '*conversio per accidens*' rule in traditional syllogistic. We can now reverse the direction of the arrow through *conversio simplex* rule (see Fig. 7c) which is also applicable to i proposition. Thus, Darapti is valid based on the same argument presented for the case of Festino.

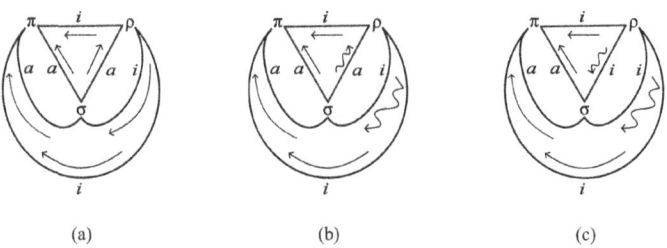

Fig. 7. *Conversio per accidens*

Now we check the validity of the $\mathcal{F}_3$ mood Disamis by reducing it to the $\mathcal{F}_1$ mood Darii through the example shown in Fig. 8 [3, p. 150]. Following the previous convention, Disamis and Darii are represented by the inverted triangle and crescent respectively. However, in this particular example, π and ρ denote different kinds of terms for different diagrams. π denotes the major term for Disamis whereas for Darii, π denotes the minor term. For Disamis, ρ denotes the minor term whereas ρ denotes the major term for the mood Darii. σ denotes the middle term for both the moods.

---

[4] Although it is a debatable issue whether existential import was part of Aristotelian syllogism, we assume the existential import here without losing generality.

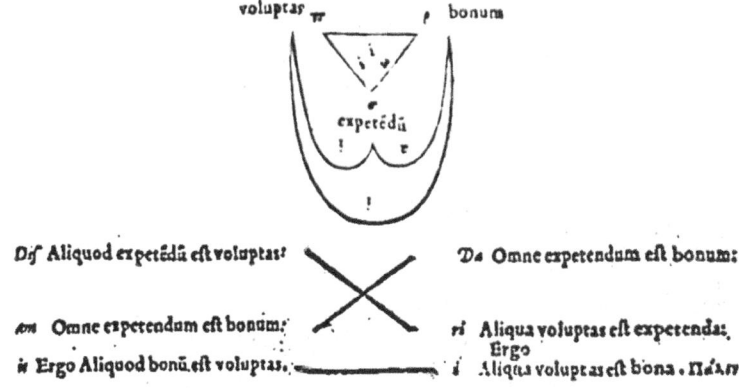

Fig. 8. Validity of Disamis (π: desired; ρ: good; σ: pleasures)

We can again proceed with a similar validity-checking method as shown in the previously mentioned examples. However, in [4] a little different method has been taken. According to [4], instead of having a Darii where the major and the minor terms are different than that of Disamis, we choose the same terms for both moods (see Fig. 9a). By comparing the categorical proposition for all the premises for both moods we find that they are in reverse order. We reverse the order of the premises in Disamis (see Fig. 9b). This reversal of premises is known as the *mutatio* rule in traditional syllogistic. In Fig. 9b, ρ is now the major term and π is the minor term. So, we rename ρ and π as π' and ρ' (see Fig. 9c). Next, we compare the diagram for Disamis in Fig. 9c with the mood Darii which has π' and ρ' as the major and minor terms respectively (see Fig. 9d). Finally, we compare the direction of the arrows and find the difference as shown in Fig. 9d. Since both the arrows of Disamis with different directions denote the subject-predicate relationship of the i propositions, we can reverse the direction by the *conversio simplex* rule. Hence, Disamis is also valid.

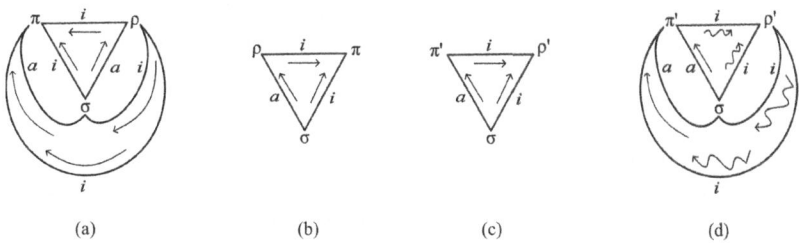

Fig. 9. *Mutatio*

In [4], only the direct method for proving the validity using the Byzantine diagrams has been shown. That is why, the methods described [4] fail to show the validity of moods like Baroco and Bocardo which can only be proven to be valid through indirect methods. In this paper, we introduce the indirect proof method for showing validity. We choose the example in Fig. 10 [3, p. 158]. In this example, the inverted triangle represents the $\mathcal{F}_3$ mood Darapti. For Darapti, α, β and γ denote the major, minor and middle terms respectively. The crescent represents the mood Celarent, where α also represents the major term. However, β and γ denote the middle and minor terms respectively.

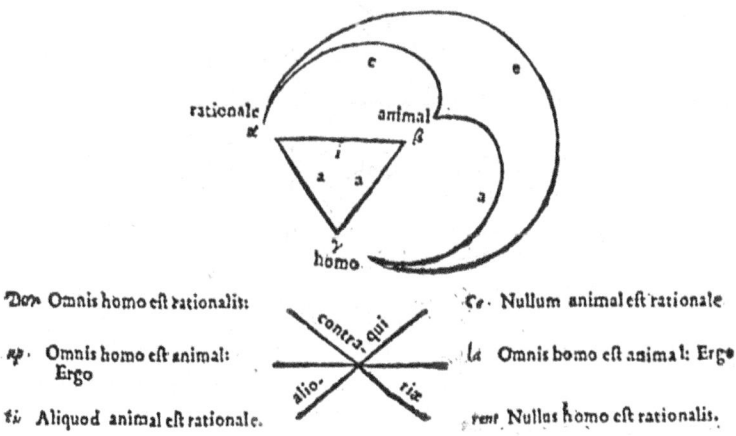

Fig. 10. Checking validity of Darapti using the indirect method (α: rational; β: animal; γ: humans)

The method of using this indirect method is as follows. First, draw the diagram for the mood whose validity needs to be checked. In this case, it is Darapti whose diagram is in Fig. 11a. Next, we draw a crescent by considering the following conditions.

1. The conclusion of Darapti is the major premise of the crescent.
2. The major terms remain the same for both moods.
3. The minor term of Darapti is now the middle term of the crescent.

The resulting diagram is shown in Fig. 11b. This is the basic difference between the direct and indirect methods. In the direct method, all the terms of the crescent lie in the same position as the inverted triangle. Due to this reason, any arc of the crescent represents the same premise as the neighbouring edge of the inverted triangle. However, in the indirect method, the position of the minor and middle terms for the crescent are different from that of the inverted triangle. Also, the arc of the crescent, which represents the major premise, is a neighbour to the edge of the inverted triangle that represents the conclusion. As we now know the terms and we know the direction of the arrows between major, middle and minor terms for the Fig. 3a, we also draw the respected arrows inside this crescent (see Fig. 11c).

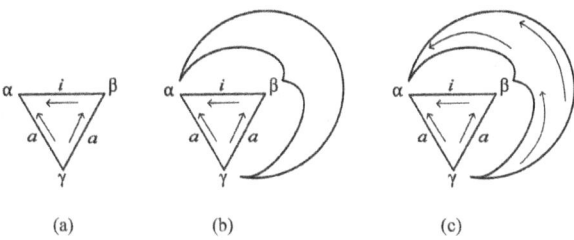

Fig. 11. Conditions for drawing combined diagrams in indirect method

Next, we add the categorical premises for the crescent. The condition is that the major premise of the crescent is the contradiction of the conclusion of Darapti. Hence, we put e in Fig. 12a in order to contradict i. The categorical proposition for the minor premises remains the same as Darapti (see Fig. 12a). We know that the only perfect mood that has e and a as the premise is Celarent. Hence, we put the categorical proposition for the conclusion of the crescent which is e (see Fig. 12b). Finally, we check whether this conclusion of the Celarent mood is a contradiction or contrary to that of the first premise of Darapti. The basic idea is that since at the beginning, we have started with taking the contradicting proposition for the conclusion of Darapti as the major premise of the Celarent, thus the major premise of Darapti should also have a contradictory/contrary proposition to the conclusion of the Celarent. If it is indeed contradictory or contrary then we will say that Darapti is valid. In this case, the conclusion of Celarent, i.e. e, is contrary to the first premise of Darapti which is a (see Fig. 12b). Hence, Darapti is valid.

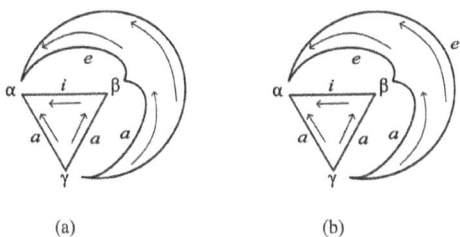

Fig. 12. Validity checking in indirect method

4 Enhancing the Expressivity of Byzantine Diagrams

As explained in Sect. 3, Byzantine diagrams serve the purpose of checking the validity of syllogisms. The information of this diagrammatic technique will aid readers in interpreting Byzantine diagrams in a wider context. Contrary to the beliefs of Hamilton and Venn, the utility of these diagrams extends beyond the mere representation of $\mathcal{F}_{1-3}$; they are instrumental in the analysis of syllogisms. In this section, our objective is to explore the additional applications of Byzantine diagrams, particularly focusing on their use in the 16th century. It is important to acknowledge that this exploration is not exhaustive, given the extensive array of Byzantine diagrams encountered across numerous texts, amounting to thousands of pages. However, through six selected examples, we intend to demonstrate that the logical expressivity of Byzantine diagrams is significantly more profound than what was recognized after the 19th century.

Invalidity. A notable aspect in the examination of syllogisms using Byzantine diagrams is the representation of invalid syllogisms. In some diagrams, invalid syllogisms are depicted in a manner that precludes the formation of a coherent connection between the subject and predicate in the conclusion. For example, in Fig. 13a, the upper arcs show: 'All humans (homo) are animals (animal)', 'No γ is a human'. But, the two lower arcs (equus, lapis) do not meet γ at the end of the upper right arc. Consequently, the crescent diagram is 'open' (invalid) and not 'closed' as e.g. in Fig. 13c or 14a (both valid). That is from the above-mentioned premises neither the conclusion 'All γ (e.g. horses/ equus) are animals' nor 'No γ (e.g. stone/ lapis) is an animal' follows. This is explained in detail in Aristotle's *Analytica Priora*, I, 7 (26a2–9). Of course, these diagrams are only drawn if a proof of invalidity as in Sect. 3 has been provided beforehand, i.e. if no reduction method has been found.

Opposition. Byzantine diagrams were also employed for what are known as Aristotelian diagrams, specifically to visualize opposition relations. A notable technique involves expressing the contrariety between three terms. An illustrative example of this can be found in Stapulensis [8, p. 43], where the term 'contraria' is inscribed along the arcs of a crescent diagram such as Fig. 2a. At the vertices, terms from Aristotle's *Nicomachean Ethics* III and IV are positioned, such as 'excess, moderation, deficiency', 'daring, bravery, fearfulness', and 'flattery, kindness, quarrelsomeness'.

Thus the arcs of the Byzantine diagram of $\mathcal{F}_1$ are used in a manner analogous to the edges of a square of opposition. However, the terms do not delineate the corners of the square, but are arranged in a linear sequence. Figure 13b is more complex and shows four judgments by refering and commenting to Aristotle's *De interpretatione* 14 (23a27–24b6): (1) Being good is good. (2) Not being good is not good. (3) Being good is not good. (4) Not being good is good. The arcs between the judgements show four contrary relations ('contraria'), some in which the quality of the predicate changes from affirmation to negation (1 and 3 as well as 2 and 4), some in which the subject and the predicate are different ("differunt subiecto et figura") (1 and 2 as well as 3 and 4).

Connectives. In very rare cases, Byzantine diagrams were also used for inferences including non-simple assertibles as given in propositional logic in the Aristotelian or Stoic vein [5]. In some of these cases, assertoric syllogisms including connexives are represented by Byzantine diagrams. One example is Fig. 13c in which the following syllogism is depicted: "All animals are either mortal or immortal. All humans are animals. Thus, all humans are either mortal or immortal".

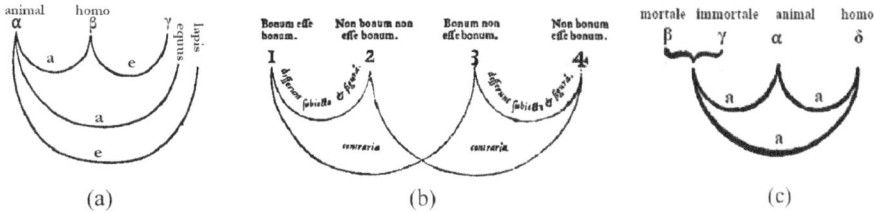

Fig. 13. (a) Invalid diagram [3, p. 134]; (b) Oppositional diagram [3, p. 121]; (c) Connective diagram [3, p. 252]

Modality. Byzantine diagrams are not only used in assertoric syllogistics, but also in modal logic. Several authors use the diagrams to check modal syllogisms following Aristoteles' *Analytica Priora* I.3 and 8–22 [15]. In Fig. 14a, the modal operators are indicated by diacritical marks above the letters for the categorical judgements. For example, a trema indicates contingency and a macron indicates necessity. The validity of a modal syllogism is proven by transforming it into a perfect modal syllogism. Diagrammatically, this procedure is similar to the checking of assertoric syllogisms described in Sect. 3, but with adapted rules. The diagram in Fig. 14a shows the modus cëlārënt (modern: 'Celarent QNQ') as an example for *Analytica Priora* I.15, 36a17-31: "(ë) It is possible that no animal (β) moves (α). (ā) It is necessary for every human (γ) being to be an animal (β). (ë) Thus: It is possible that no human being (γ) moves (α)".

***n*-Terms.** Diagrams with more than three terms are not uncommon. At least in Sect. 3, we already saw that three terms of the imperfect mode are already combined with the three terms of the perfect mode when proving validity. However, Byzantine diagrams offer numerous other techniques that combine more than three terms, such as sorites [12]. Sorites are inferences that have more than three terms and can be broken down into several syllogisms, each with three terms. In this way, the conclusion of the first decomposed syllogism becomes a premise of the next syllogism contained in the sorites. The example in Fig. 14b represents a sorites in which four terms and six arcs showing *a*-judgements. These arcs build four Byzantine diagrams for $\mathcal{F}_1$ as given in Fig. 2a, all represent a modus Barbara: (1) δ, γ, β, (2) γ, β, α, (3) δ, γ, α, and (4) δ, β, α. The final conclusion is then drawn by the most lower arc from δ to α.

Instances. In some cases, diagrammatic techniques can be found where instances and examples are given, which often have an explanatory function. These instances are usually attached to a complete Byzantine diagram as appended arcs. The term on the attached arc supplements the argument, in the same way that the 'backing' in modern Toulmin argument maps can optionally be attached to the premises or to the conclusion. However, the respective use and the exact technique differs slightly from textbook to textbook. The instance is best explained by the diagram in Fig. 14c, which reads roughly as follows: "All c are b, no b is a, thus no c is a, such as d".

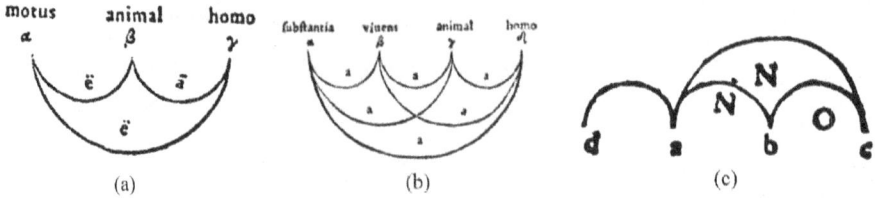

Fig. 14. (a) Modal diagram [3, p. 196]; (b) n-term diagrams [3, p. 235]; (c) Instance diagram [8, p. 158]

5 Complex Byzantine Diagrams

In this section, we would like to present four examples of complex Byzantine diagrams in logic. We will not explain the logical functions of the respective diagrams, as a brief explanation would raise many questions and a detailed explanation of some of these diagrams would fill a whole paper. The purpose of the four examples is twofold. Firstly, to demonstrate that the extensions presented in Sect. 4 can be combined with each other, in a manner analogous to that which was done in Sect. 3 with the basic diagrams given in Fig. 2. Secondly, to illustrate that Byzantine diagrams can also be combined with other logic diagrams. Let us examine the two diagrams that combine two extensions.

Example 1. The diagram in Fig. 15a does not provide a proof as defined in Sect. 3. Instead, it illustrates the interaction of two Byzantine shapes and demonstrates a complex inference over the terms presented in the centre. Specifically, the 'connective' and 'instance' expansions, as shown in Sect. 4, are combined. The connective extension is depicted in the bifurcation of the two terms in the centre of the diagram, while the instance is added at the bottom of the left-hand Byzantine diagram.

Example 2. Figure 15b illustrates the combination of the two expansions 'invalidity' and 'opposition'. Invalidity is identified by the absence of an arc between terms a and c, giving the diagram a similar form to the representation of the invalid syllogism in Fig. 13a. The semicircle connecting the quantity of the two lower arcs indicates opposition between many (*multa*) and one (*alicui*).

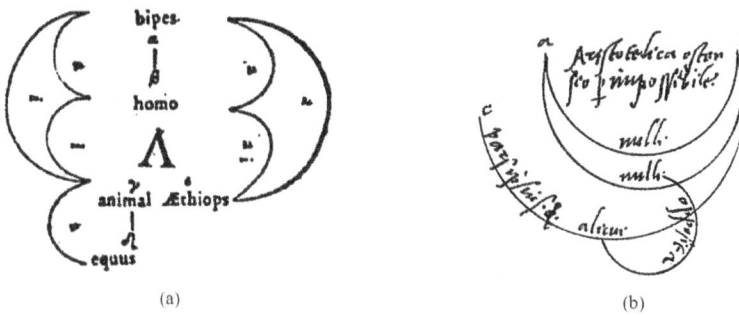

Fig. 15. (a) Connective and instance [3, 286]; (b) Oppositions and invalidity [1, fol. 73]

Byzantine diagrams are compositional, meaning their forms can be combined with each other or with other types of diagrams. Two examples will be provided to demonstrate the amount of information that can be integrated into one diagram.

Example 3 is taken from the extensive collection of manuscripts by the Venetian mathematician Bartolomeo Zamberti. He analysed the logical and rhetorical writings of Aristotle and Porphyrius using diagrams. Figure 16 shows three crescents gathered around a triangle. The mode, represented by the central triangle, is to be reduced to three rules, each represented by a crescent. For example, reductions 'per impossibile' can be observed on the left and right crescents.

Example 4 is taken from a 15th-century edition of Aristotle with various commentaries. Figure 16b is announced as illustrating an example by Alexander of Aphrodisias. The edges show relations such as oppositions ("asystaton") or conversions between the capital letters at the corners. The respective relations are explained in the sentences along the edges and the Byzantine diagrams serve as an explanation of the relations.

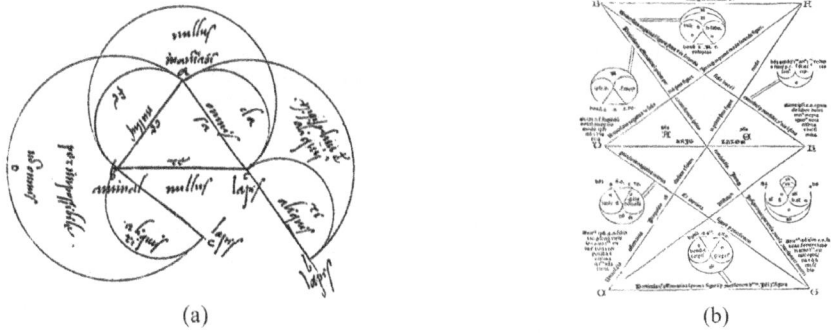

Fig. 16. (a) Many byzantine diagrams [1, fol. 34r]; (b) Combining relations and Byzantine diagrams [2]

6 Conclusion

Since Hamilton and Venn, most logicians have believed that the primary function of Byzantine diagrams is to emphasize the middle term of a syllogism. However, it has been recently discovered that there are even more complex Byzantine diagrams in logic. In the previous sections, we have provided examples to illustrate how Byzantine diagrams can be used for checking validity. Although the basic diagrams in the form of a crescent and a triangle may be modelled on $\mathcal{F}_{1-3}$, the function of the diagrams is far more intricate.

Byzantine diagrams were commonly used in early modern works, particularly in the 16th century, to test the validity of syllogisms, either directly or indirectly. This feature alone gives them a similar level of expressiveness to verbal scholastic mnemonics or diagrammatic Euler diagrams. However, due to their connective capacity, Byzantine diagrams have a wider range of applications beyond syllogistic. They are not thematically limited to syllogisms and can be combined with many extensions. The paper demonstrates that Byzantine diagrams have greater potential than previously believed. However, further research is required to fully understand their capabilities.

Acknowledgements. The authors of this paper are supported by the Fritz Thyssen Foundation (project: *History of Logic Diagrams in Kantianism*) and from the ViCom-project *Gestures and Diagrams in Visual-Spatial Communication* funded by the German Research Foundation (DFG – RE 2929/3-1).

References

1. Aphrodisias, A., Philoponus, J., Zamberti, B.: Alexandri Aphrodisias in Aristotelis priorum resolutivorum ad Eudemum primum commentarius B. Zamberto interprete – BSB Clm 117. Venice (1500)
2. Aristotle: Opera: Organon. Bernardinus Stagninus, Venice (1489)
3. Aristotle, Pacius, J.: Organon: Libri Omnes ad Logicam Pertinentes. Laimarius, Morges (1584)
4. Bhattacharjee, R.: Direct reduction of syllogisms with Byzantine diagrams. Hist. Philos. Log. (forthcoming)
5. Bobzien, S.: Stoic logic. In: Inwood, B. (ed.) The Cambridge Companion to Stoic Philosophy. Cambridge University Press, Cambridge (2003)
6. Bochenski, J.M.: A History of Formal Logic. University of Notre Dame Press, Notre Dame (1961)
7. Cacouros, M.: Les schémas dans les manuscrits grecs de contenu logique. Gazette du livre médiéval **39**(1), 21–33 (2001)
8. d'Etaples, J.L.: Libri Logicorum ad Archetypos Recogniti. Regnault, Paris (1525)
9. Hamburger, J.F.: Drawing conclusions: logic diagrams as a matrix for the making and meaning of Christian images in the middle ages. In: Hamburger, J.F., Roxburgh, D.J., Safran, L. (eds.) The Diagram as Paradigm: Cross-cultural Approaches, pp. 419–455. Dumbarton Oaks, Washington, DC (2022)
10. Hamilton, W.: Discussions on Philosophy and Literature, Education and University Reform, 2nd edn. Longman, Brown, Green and Longmans, London (1853)

11. Kant, I.: Kant's Gesammelte Schriften. de Gruyter et al., Leipzig et al. (1928)
12. Kittsteiner, C.: Schopenhauer's sorites diagram. In: Lemanski, J., Johansen, M.W., Manalo, E., Viana, P., Bhattacharjee, R., Burns, R. (eds.) Diagrams 2024, LNAI, vol. 14981, pp. 145–152. Springer, Cham (2024)
13. Kneale, W., Kneale, M.: The Development of Logic. Clarendon Press, Oxford (1962)
14. Lourié, B.: Diagram reasoning and paraconsistent thinking: Hieromonk Hierotheos, his ancestry, and legacy. SUBBTO **LXVII**(2), 61–125 (2022)
15. Malink, M.: Aristotle's Modal Syllogistic. Harvard University Press, Cambridge and London (2013)
16. Panizza, L.: Learning the syllogisms: Byzantine visual aids in Renaissance Italy - Ermolao Barbaro (1454–93) and others. In: Blackwell, C., Kusukawa, S. (eds.) Philosophy in the Sixteenth and Seventeenth Centuries, pp. 22–47. Routledge, Brookfield (1999)
17. Roberts, A.: Byzantine-Islamic scientific culture in the astronomical diagrams of Chioniades on John of Damascus. In: Hamburger, J.F., Roxburgh, D.J., Safran, L. (eds.) The Diagram as Paradigm: Cross-cultural Approaches, pp. 113–148. Dumbarton Oaks, Washington, DC (2022)
18. Safran, L.: Byzantine diagrams. In: Hamburger, J.F., Roxburgh, D.J., Safran, L. (eds.) The Diagram as Paradigm: Cross-cultural Approaches, pp. 13–32. Dumbarton Oaks, Washington, DC (2022)
19. Triantafyllou, V.: Aristotle's syllogistic as a form of geometry. Hist. Philos. Log. Anal. 1–49 (forthcoming)
20. Venn, J.: Symbolic Logic, 2nd ed. Macmillan and Co. (1894)
21. Wesoły, M.: Analysis peri ta schemata: Restoring Aristotle's lost diagrams of the syllogistic figures. Peitho **3**(1), 83–114 (2012)
22. Willson, J.: On the aesthetic of diagrams in Byzantine art. Speculum **98**(3), 763–801 (2023)

Posters

An Innovative Approach to Diagrams Representation: The Marlo Diagrams Web Page

Fernando Soler Toscano[1](✉) and Marcos Bautista López Aznar[2]

[1] Logic, Language, and Information Research Group (HUM 609), Seville University, Seville, Spain
fsoler@us.es
[2] Logic, Language, and Information Research Group (HUM 609), Huelva University, Huelva, Spain

Abstract. In this work we present a web version of Marlo diagrams, which are an innovative way to visualize basic principles of logical reasoning, returning to the tradition of the Quantification of the Predicate. The page contains concise instructions, multiple examples of classical reasoning, and more than one hundred proposed exercises. Thus, it is a teaching resource that can be executed on computers, mobile phones, and tablets as an ICT tool for the development of critical thinking. We present the essential definitions of Marlo diagrams and their operations and provide JavaScript code examples to illustrate their implementation. We also present the line of research that we are currently working on to expand the scope of Marlo diagrams by adding disjunctions to their regions.

Keywords: Logic diagrams · Critical Thinking · First order logic · Teaching of logic

1 Introduction and Essential Definitions

The graphical representation of logical relationships is essential across diverse fields. The Marlo Diagrams web page [4] offers a distinctive perspective [2, 3], enabling users to craft diagrams with an intuitive and expressive approach. These diagrams, recognized as innovative tools for teaching logic [1], earned their creator a Margarita Salas scholarship in 2021, leading to a year-long collaboration with the University of Seville. During this period, in partnership with Professor Fernando Soler Toscano, the theoretical foundations of the diagrams were refined, and a website was developed as a dedicated tool for logic education. The primary objective was to translate the principles governing the creation and manipulation of Marlo diagrams into JavaScript.

A Marlo Diagram (hereafter MD) is a formal structure that represents a proposition such as *Every Andalusian (A) is European (E),* or *Some philosopher (F) is not bearded (¬B)*. It can also represent more complex expressions like *All philosophers (F) are either Andalusian (A) or bearded (B)* or *There are philosophers (F) who are both Andalusian and bearded (B)*.

Let us delve into the structure. The examples provided serve as illustrations of formalism and should not be taken as expressions of truth. For instance, the proposition *All philosophers (F) are either Andalusian (A) or bearded (B)* is not necessarily true; We are interested in its formal nature only.

The MD consists of the following elements:

- Subject, SUJ: A. Corresponds to the subject of the represented proposition. In *Every Andalusian (A) is European (E)*, the subject is *A*, which represents Andalusian.
- All Part, ALL: Contains properties shared by all individuals that fulfill the subject. In the case of *Every Andalusian (A) is European (E)*, the ALL part contains the proposition *E*, which represents European.
- In Parts, IN: Contains different sets expressing options held by some individuals fulfilling the subject but not all. Thus, *Some philosophers (F) are Andalusian (A)* would have two IN parts, one with *A* and the other blank, indicating that *F* can be *A* or ¬*A*.
- Out Parts, OUT: Propositions that individuals not belonging to the subject can have. Therefore, *Every Andalusian (A) is European (E)* allows for Europeans not being Andalusian, placing *E* in the OUT part. However, if we state *Every Andalusian is European*, and there are no Europeans who are not Andalusian, *E* will not be in the OUT part.

We represent a Marlo diagram like *Every Andalusian (A) is European (E)* as follows: {SUJ: A, ALL: [E], IN: [], OUT: [E]} i.e.:

- The subject SUJ is *A*.
- The ALL part contains *E*, affirmed for all *A*.
- There are no IN parts as there is no particular information.
- In the OUT part, *E* is present because non-Andalusian Europeans are allowed.

2 Current Operations and Work in Progress

There are six operations in the current version of Marlo Diagrams. First, you must use the operation Creation. Select the toto-partial relations established between the subject and the predicate of the proposition. Once the premises are represented, use the conversion and Transformation operations to match the subject of their diagrams. Then, use the Inference operation to synthesize Diagrams with the same subject and draw conclusions. Convert the diagram obtained by inference to present the conclusion from the term you prefer. Finally, use the Extraction operation to eliminate the middle term and present a clear conclusion (see Fig. 1).

The next objective is to create a program that allows us to convert, transform, and synthesize diagrams whose subjects and/or predicates contain conjunctions or disjunctions, without losing sight of the fact that a teaching tool must be clear and intuitive. By adding disjunctions to Marlo diagrams, we lose some clarity, but we believe that it is worth developing this route to work on increasingly complex arguments in the classroom. Figure 2 shows the two ways in which twelve-year-old students represented the premise *All M is A or S*. In this Figure, we have numbered the regions of the diagrams for ease of understanding, and we must remember that letters outside of *M* (regions 3,

6, and 9) are merely possible. But although children in a pilot study intuitively applied disjunction elimination to obtain B when in the same region of the diagram they found $(A \lor B) \land \neg A$, it will be necessary to establish the semantics of our diagrams based on Predicate Logic to represent disjunctions of type $\exists x\, (Px \lor Qx)$ and $\forall x\, (Px \lor Qx)$. There are examples of the semantics we are developing in diagrams D5 to D8 in Fig. 2.

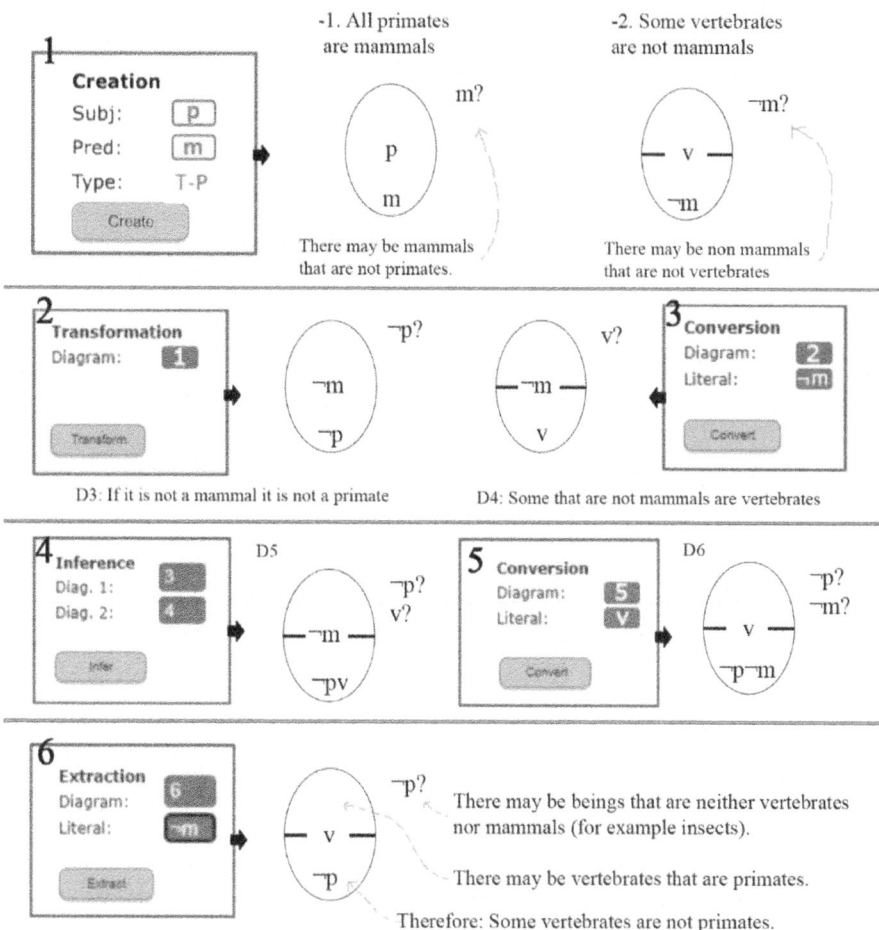

Fig. 1. Steps to Solve Syllogisms on Marlo Diagram Website.

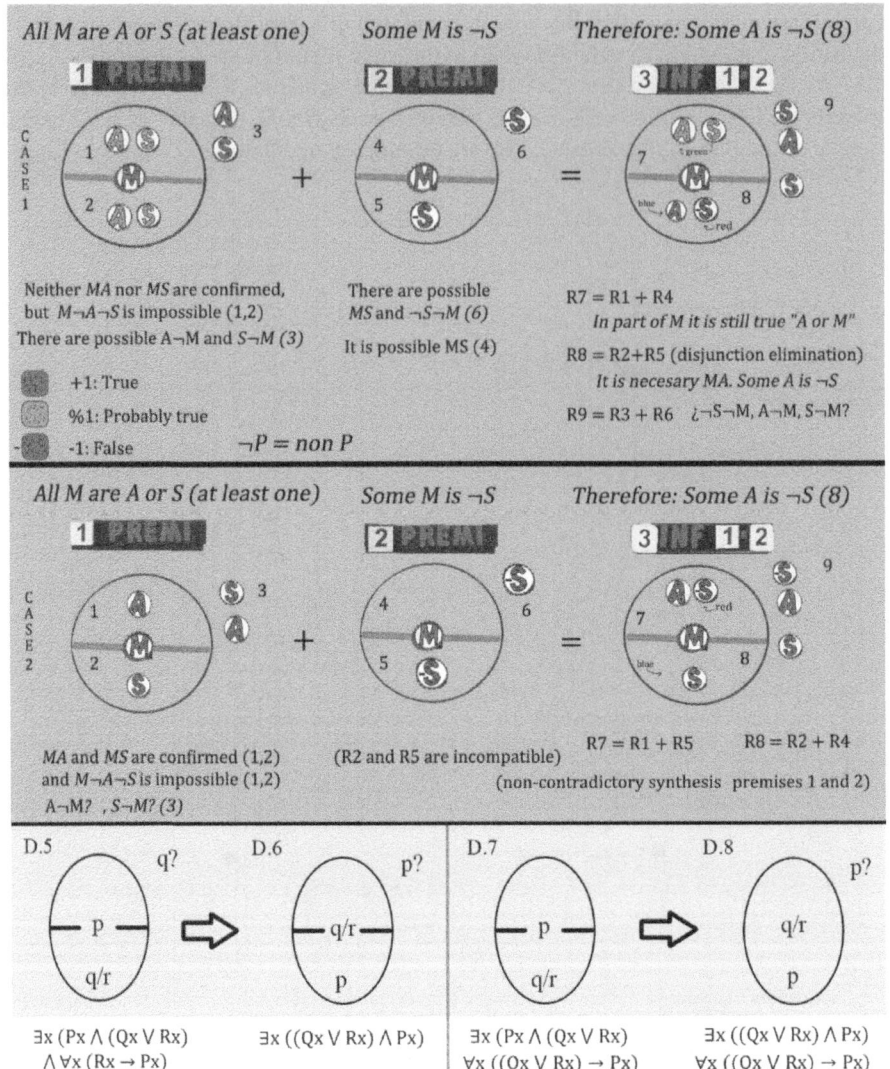

Fig. 2. Inferences with universal and particular disjunctions in Marlo diagrams.

3 Conclusions

The Marlo Diagrams Website offers a unique and powerful approach to representing elementary logical relationships. We hope to be able to implement it by adding diagram negation and disjunctions to the predicates. It will be necessary to interpret the diagrams as a whole from the semantics of predicate logic, at the same time that we operate with propositional calculus to convert the formulas contained in each of the regions. This is a significant computer programming challenge, but we believe it is worth continuing to explore and develop the didactic potential of our proposal.

References

1. Aznar, M.B.L., Acosta, G.C., Gadea, W.F.: Significance in Marlo diagrams versus thoroughness of venn diagrams. In: Arai, K. (ed.) SAI 2022. LNNS, vol. 506, pp. 207–227. Springer, Cham (2022). https://doi.org/10.1007/978-3-031-10461-9_14
2. Boole, G.: An Investigation of the Laws of Thought. Dover, New York (1854)
3. Mays, W., Henry, D.P.: Jevons and logic. Mind **62**, 484–505 (1953)
4. Soler-Toscano, F., Aznar, M.B.L.: fersoler/MarloDiagrams (2022). https://fersoler.github.io/MarloDiagrams/. Accessed 31 Jan 2024

Codifying Visual Representations

Wode Ni[1(✉)], Sam Estep[1], Hwei-Shin Harriman[1], Jiří Minarčík[2], and Joshua Sunshine[1]

[1] Carnegie Mellon University, Pittsburgh, PA, USA
nimo@cmu.edu
[2] Pittsburgh, USA

Abstract. Making visually appealing and meaningful diagrams involves craftsmanship in designing the visual representation, drawing shapes, and laying them out. Can the effort spent on diagrams by an expert be reused by others, especially those without the expertise in design and drawing? In this paper, we outline our prior work on PENROSE, a diagramming tool with first-class support for reusing visual representations. The nature of our approach to reusability necessitates a domain-agnostic method to automatically lay out a diagram. We highlight our existing approach for general diagram layout and styling, and propose a new composable approach for codifying visual representations to reuse expertise that cuts across domains.

Keywords: Diagram Authoring Tools · Automatic Diagram Layout · Natural Diagramming Interface

1 Defining Visual Representations

Authoring good diagrams requires both knowledge of the domain being illustrated and graphic design sense. An effective diagram author develops *visual representations* by illustrating domain concepts with appropriate visual elements and spatial layout. In prior work, we interviewed diagram authors and found that defining visual representations is an important part of their diagramming process and authors cared about reusing these representations to improve productivity [2]. However, existing tools limited reuse to either duplicating prior diagrams and manually tweaking them. Authors found these methods to be tedious and time-consuming.

This result suggests that diagramming tools should be *representationally salient*: tools should allow authors to define visual representations for domain-specific concepts in a manageable, scalable, and composable way. With representation salience as an explicit goal, we build an open-source tool called PENROSE [4]. Authors make diagrams in PENROSE using two languages: SUBSTANCE describes the conceptual content of a diagram while STYLE describes the visual

---

W. Ni, S. Estep and H.-S. Harriman—Authors contributed equally.
J. Minarčík—Independent Researcher.

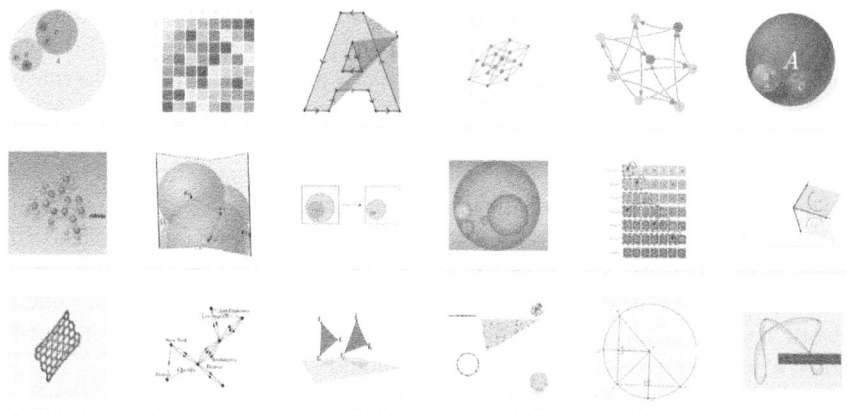

Fig. 1. Selected PENROSE-authored diagram examples. Each diagram is generated from a different visual representation, codified as its own STYLE program.

representation. To date, PENROSE has been used to define visual representations across a wide range of domains,[1] including diagrams shown in Fig. 1, and PENROSE generates diagrams for external tools [1,3].

2 Reusing Visual Representations

In PENROSE, the STYLE language encodes visual representations by mapping conceptual objects from SUBSTANCE to primitive visual shapes and descriptions of the visual layout. For instance, in Fig. 2, the STYLE program describes how to map sets and relationships in the domain to visual shapes (e.g., Circle) and constraints, and this STYLE is reused to generate multiple Euler diagrams.

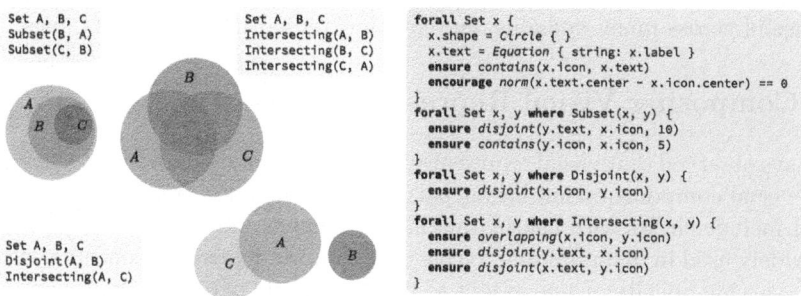

Fig. 2. Three Euler diagrams generated from different SUBSTANCE programs (left) using the same visual representation from a single STYLE program (right).

---

[1] https://penrose.cs.cmu.edu/examples.

We say that STYLE achieves representation salience because it encodes the visual representation of Euler diagrams for all possible SUBSTANCE programs. Importantly, representation salience *necessitates* automatic layout. Alternatives could include specifying layout in the SUBSTANCE program or requiring manual tweaks, both of which imply that important aspects of the visual representation were not sufficiently encoded in STYLE. That is, reusability is a necessary consequence of representation salience.

While STYLE serves to abstract visual layout from the perspective of the SUBSTANCE writer, it is itself built on top of a set of PENROSE primitives for describing a visual representation as a numerical optimization problem. The `ensure` keyword denotes a hard constraint which must be satisfied, while `encourage` denotes a soft objective which the system should strive to achieve.

The `disjoint`, `contains`, and `overlapping` constraints used in Fig. 2 are examples of visual layout primitives provided by PENROSE. Many different STYLE programs need to talk about shapes that must not overlap or must be nested, for instance, so we've developed a mathematical framework to describe these for arbitrary shapes. Using the signed distance function

$$\phi_A(x) = \begin{cases} -d(x, \partial A) & x \in A, \\ d(x, \partial A) & x \notin A. \end{cases} \quad \text{where} \quad d(x, \partial A) = \min_{y \in \partial A} |x - y|$$

and the Minkowski difference $A - B = \{a - b : a \in A, b \in B\}$, we can perform layout by composing together these two operations in various ways:

$$\begin{aligned}
\texttt{disjoint(A, B)} &\iff \text{minimize } \max(0, -\phi_{A-B}(0)) \\
\texttt{contains(A, B)} &\iff \text{minimize } \max(0, -\phi_{A^c - B}(0)) \\
\texttt{overlapping(A, B)} &\iff \text{minimize } \max(0, \phi_{A-B}(0))
\end{aligned}$$

We find these Minkowski penalties particularly useful for label placement, which is often a very tedious subtask of diagramming. To support STYLE construction, PENROSE provides a library of over 200 built-in functions and over 50 pre-defined layout constraints and objectives. These functions and primitives are useful across many domains, and are thus reused in many STYLE files.

3 Composing Visual Representations

We have observed that visual representations in different domains often share common visual components and layout patterns. For instance, multiple examples in Fig. 1 include circles with nearby text labels. Further, common visual techniques are widely used in diagramming to convey domain-independent concepts, such as using varying opacity or line weight to highlight parts of a diagram, maintaining layout consistency across multiple diagrams to form a visual narrative, or using sliders or other widgets to drive real-time physical simulations in interactive webpages. It seems natural to separate out these common patterns into their own components, suggesting that PENROSE's existing reusability of visual representations in STYLE does not provide sufficient flexibility for the needs of digital diagrammers.

In the current version of PENROSE, authors can reuse geometric and layout primitives (Sect. 2) to create new STYLE programs, and users consume these programs by writing different SUBSTANCE programs with them. Each STYLE program is standalone and self-contained, meaning that everything from the styling of points to the color palettes must be defined within that program. In practice, this means that common visual design patterns are copied and pasted between STYLE files. Additionally, it is common for individual diagrams within a domain to have customized visual elements to draw focus or illustrate a concept. Currently, the only way to override the domain-wide visual style in PENROSE is by using workarounds that involve more copying/pasting code in STYLE. These two limitations result in repetitive and lengthy programs that require high effort to edit and maintain, even for expert PENROSE users.

While code duplication and multiple versions of STYLE may be manageable on a small scale (e.g., Fig. 1), we plan to build a broader ecosystem of diagrams and this requires more flexible reuse mechanisms. We propose **composability** as the main design goal for improving PENROSE. The existing layout primitives are an example of composability: authors can reuse and *combine* multiple primitives to form new layout problems. Looking forward, we plan to allow diagrammers to create *modules* of visual components and layout patterns. Through this mechanism, an author can draw together multiple different modules they need for their own diagram. And these modules can themselves be composed from other modules: for instance, a module for visualizing complex analysis might make use of lower-level modules for visualizing a coordinate plane and plotting curves, but build on top of that with domain-specific visuals for singularities in holomorphic functions. In addition to user-defined modules, there are also opportunities to build domain-independent visual techniques, such as individual object-level highlighting or annotations, into our languages or as standard library modules.

We believe this composable approach will open up new possibilities for diagrammers to collaborate and create more flexible, reusable, and expressive visual representations. Going forward, we plan to leverage research on common building blocks of and layout patterns in specific domains of diagramming, to construct a substrate for composable visual representations.

References

1. Clark, C., Bohrer, R.: Homotopy type theory for sewn quilts. In: Proceedings of the 11th ACM SIGPLAN International Workshop on Functional Art, Music, Modelling, and Design, FARM 2023 (2023)
2. Ma'ayan, D., Ni, W., Ye, K., Kulkarni, C., Sunshine, J.: How domain experts create conceptual diagrams and implications for tool design. In: Proceedings of the 2020 CHI Conference on Human Factors in Computing Systems, CHI 2020 (2020)
3. Nawrocki, W., Ayers, E.W., Ebner, G.: An extensible user interface for lean 4. In: Proceedings of the 14th International Conference on Interactive Theorem Proving, ITP 2023 (2023)
4. Ye, K., et al.: Penrose: from mathematical notation to beautiful diagrams. ACM Trans. Graph (2020)

A Diagram Helping the Mathematical Problem Solving Procedure

Tullio Aebischer

Department of Mathematics, University of Rome Tor Vergata, Rome, Italy
aebischer@axp.mat.uniroma2.it

Abstract. The Author verified in the middle school that its setting up a diagram for solving problems in arithmetic or in geometry was considered completely different from that developed in Primary school (grade 7). For this reason, he has decided to verify, since 2017, with pupils and the collaboration of a Primary's teacher, the possibility of sharing, as possible, the same solution diagram both to make the aforementioned passage as less traumatic as possible, and to better understand the question of the mathematization of reality and the metacognitive processes underlying it. The proposed diagram, analyzed only in middle school, must not give the false certainty of solving the problem correctly, i.e. it must not be seen as a solution algorithm but only an helping for better and ordered writing the resolution procedure for error checking.

Keywords: Mathematics · Problem solving · Metacognitive strategies · Student error

1 Introduction

Various research highlights that many students consider mathematical problems difficult hence their aversion to all mathematics. At the same time, however, the use of problem solving is useful by highlighting the visual aspect of the resolution [1] for which a diagram is certainly helpful.

The proposed diagram, the object of daily classroom practice in the Author's mathematics lessons in the Middle school (the Primary experience will not be considered here), has the aim of helping the planning of the solution to a problem but also the clarity of communication thereof. This last aspect is fundamental for identifying any errors even if the proposed diagram must not give the false certainty of solving the problem correctly, i.e. it must not be seen as a solution algorithm [2]. However, as reported in [3], research indicates that the use of diagrams improves the ability to correctly solve problems and reduces the associated cognitive stress. To reach this goal, students must learn the details of the construction and use of the diagrams. In our case the class activity had this purpose and the proposed diagram became a shared tool as well as analyzed in detail with examples.

In the mathematical field, Polya [4] defined the steps to solve the problem: 1) understanding the problem, 2) planning solution, 3) doing the plan, and 4) correcting back.

Based on point 4), the Author's professional experience reports that in problem solving at school the educational usefulness of error analysis is not always highlighted. The error at school is seen as a nightmare but for the teacher, on the contrary, it is an important discussion point for the students' improvement [5, 6] allowing, thus, to verify their level of knowledge and skills, their metacognitive approach to reality. The student who does not argue, who does not know how to communicate, is a black box who does not help anyone.

2 The Problem Solution Diagram (PSD)

The resolution of a problem can be divided into two macrophases: a) translation of the proposed text from natural to formal language [7] and b) the correct writing of the formal language to perform the calculations. The PSD for problem solving (Fig. 1) has three purposes: 1) clear and logical writing of the problem resolution; 2) allow the student to identify any errors; 3) make the corrector understand the solution reasoning.

The idea of a diagram as an aid to problem solving goes back to points 3) and 4) of Polya. For this reason the PSD is only one component of problem solving and must serve as a guide to the solution [8]. In particular, the delicate topic of errors analysis must be supported by the possibility of identifying them along the resolution path. For this reason, errors can be of three types: 1) transcription (Data), 2) conceptual planning (Solution scheme) and 3) calculation (Procedure). In this, the proposed diagram allows for immediate identification both during the course and later in the reanalysis phase.

Furthermore, while school textbooks generally provide a textual description of the solution method, the PSD allows the connections, formalism and preparatory features of the individual parts to be highlighted.

In [9] the conclusion is reported that the use of diagrams is useful both for error analysis and in the transition towards a more abstract approach to resolution. At the same time it was seen, and the Author confirms, that the use of the diagrams adopted in Primary, valid for that context, could not be easily transferred to the transition to grades 6–8 when the problems become more complex. Our experimentation, in fact, considered fifth grade students who entered middle school the following year (grade 6).

Ultimately, the PSD can be seen as an educational contract between students and teacher to understand each other better. The Author's experience has even verified its usefulness for students with difficulties, especially dyslexia, and there is evidence of effective use even in secondary schools.

The PSD is divided into three sections: a) reading and understanding the text of the problem, b) writing the data and questions, c) writing the solution procedure with formulas according to the Solution Scheme (Fig. 2). The text of the above simple problem for which the PSD was used is:

A triangle with a perimeter of 48 cm is given, and two sides, one 20 cm long and the other 17 cm long. Calculate the length of the third side.

At the beginning of solving the problem the first difficulty is that of understanding the text of the problem. For this reason, solving a mathematics problem requires both a good command of natural language and knowledge of technical terminology. Collaboration with a mother tongue teacher it would be very useful.

Writing data and questions can be made difficult by the presence of implicit data (paraphrases) or expressed in a non-numerical manner.

The writing of the solution procedure introduces the novelty of a solution scheme in which the symbols of data must be connected with the unknowns like as a tree graph. From a procedural point of view, the aforementioned connections materialize in the solution procedure formulas.

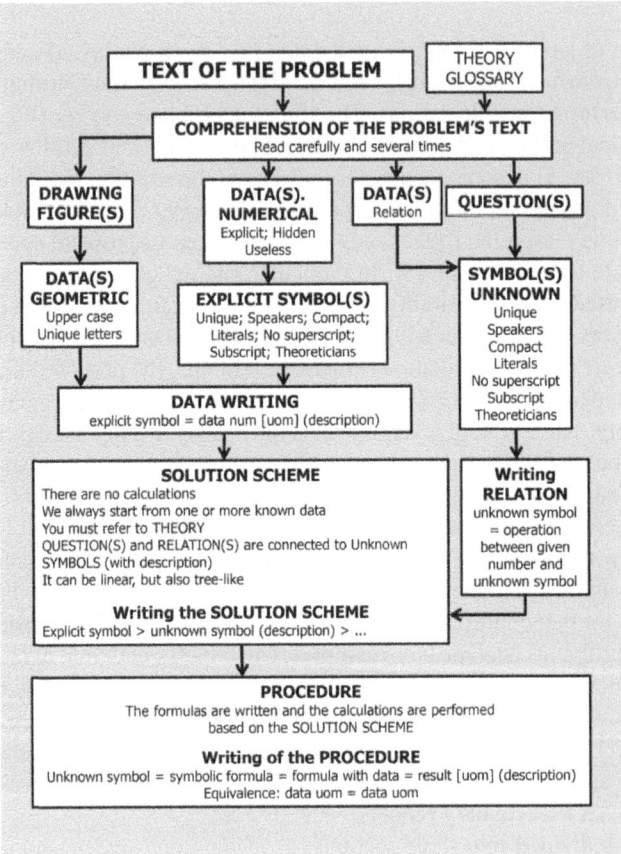

Fig. 1. The Problem Solution Diagram (PSD) with its steps (*uom* means unit of measure). Last version: October 2022.

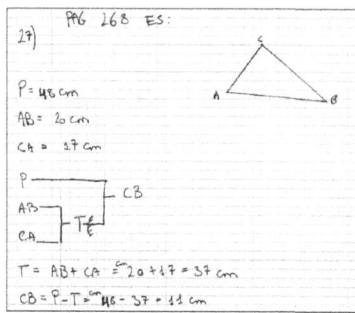

Fig. 2. A simple problem solved with PSD. See the Solution Scheme written after data and the geometric figure (school's grade 8). See the error in writing the Procedure where the dimension (*cm*) is written both before the numerical values and during the calculations.

Acknowledgments. The research activity was carried out at the IC Orsa Maggiore (Roma, Italy) in collaboration with the teacher M. Menna of the Primary school. This poster, with updates, is based on the article Aebischer, T., Menna, M.: Mappa per la risoluzione del problema in aritmetica o geometria (MRP): esperienza laboratoriale tra linguaggio naturale e formale. Formazione & Insegnamento, **XVIII**(3), 198–209 (2020) (doi: https://doi.org/10.7346/-fei-XVIII-03-20_16).

Disclosure of Interests. The author have no competing interests to declare that are relevant to the content of this poster.

References

1. Mudaly, V.: Constructing mental diagrams during problem-solving in mathematics. Pythagoras – J. Assoc. Math. Educ. South Afr. **42**(1), 1–8 (2021). https://doi.org/10.4102/pythagoras.v42i1.633)
2. D'Amore, B., Fandiño Pinilla, M.I.: Sugli scivolamenti metadidattici. Alcuni esempi. L'Insegnamento Della Mat. Delle Sci. Integrate **43A**(2), 108–136 (2020)
3. Ayabe, H., Manalo, E., de Vries, E.: Problem-appropriate diagram instruction for improving mathematical word problem solving. Front. Psychol. **13**.992625 (2022). https://doi.org/10.3389/fpsyg.2022.992625
4. Polya, G.: How to Solve It. Princeton University Press (1945)
5. Jihe, C., Pereira, J., Zhouli, H., Mingli, Z., Wiranota, H., Tamur, M.: The analysis of students' error in solving mathematics problem. Int. J. Latest Res. Humanit. Soc. Sci. (IJLRHSS) **4**(7), 40–49 (2021)
6. Lai, C.: Error analysis in mathematics, Technical report #1012, pp. 1–7. University of Oregon (2012)
7. Demartini, S., Sbaragli, S.: La porta di entrata per la comprensione di un problema: la lettura del testo. Didattica della Mat. **5**, 9–43 (2019)
8. Jitendra, A.K., Hoff, K., Beck, M.M.: L'uso degli schemi visivi per la risoluzione dei problemi matematici. Difficoltà apprendimento **8**(1), 9–20 (2002)
9. Booth, J.L., Koedinger, K.R.: Are diagrams always helpful tools? Developmental and individual differences in the effect of presentation format on student problem solving. Br. J. Educ. Psychol. **82**(3), 1–20 (2011). https://doi.org/10.1111/j.2044-8279.2011.02041.x

Collaborative Graph-Document Composition is Efficient and Enhances Critical-Thinking Skills Without Extra Cost

Kôiti Hasida[1,2(✉)], Zilian Zhang[2], Zifan Yao[2], Vili Valtteri Karilas[2], Shitao Fang[2], Kuanghuan Tan[2], Kenichi Shibata[1], and Yusuke Matsubara[2]

[1] RIKEN AIP, 15th floor, Nihonbashi 1-chome Mitsui Building, 1-4-1 Nihonbashi, Chuo-ku, Tokyo 103-0027, Japan
koiti.hasida@riken.jp
[2] The University of Tokyo, 7-3-1, Hongo, Bunkyo-ku, Tokyo 113-8656, Japan

Abstract. Graph documents (GDs) can express the content of any nonartistic text documents (TDs). Experiments have shown that the collaborative GD composition is more efficient than collaborative TD composition. Another experiment demonstrated that regular high school classrooms can not only seamlessly integrate collaborative GD composition without incurring additional costs, but also improve students' CT skills. Opting for GDs over TDs hence promises an enhanced efficiency not only in collaborative document composition but also in intellectual works in general.

Keywords: graph document · collaborative composition · critical thinking

1 Introduction

Graphs such as concept maps and argument maps have served as supplementary resources for education, brainstorming, idea generation, etc. [1–3]. However, graphs could be used as formal nonartistic documents better than text documents (TDs), if graphs communicate logical content more straightforwardly than TDs.

Graph documents (GDs) are labeled directed graphs with nodes representing semantic entities, which encompass eventualities and objects, expressed in brief texts or other data such as images and videos. The links among those nodes represent semantic/pragmatic relations among those semantic entities. Figure 1 shows a GD outlining part of this paper in the screen of Semantic Editor (SE), which is an authoring tool for collaboratively composing GDs.

Partially supported by NEDO (New Energy and Industrial Technology Development Organization) under grant No. 20001217-0.

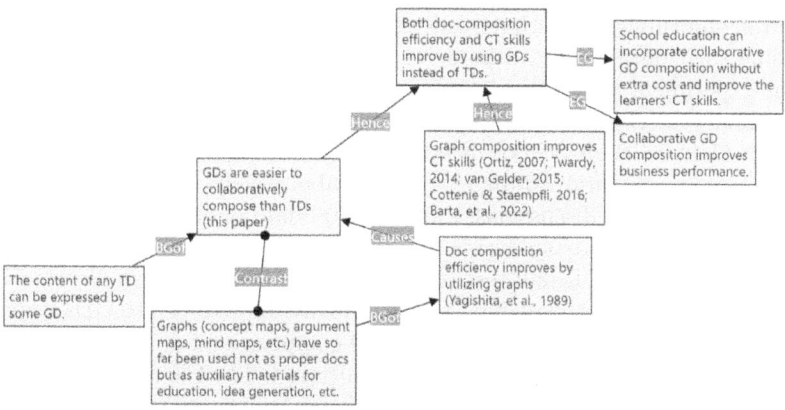

Fig. 1. A GD partially summarizing this paper.

To effectively express the content of arbitrary nonartistic text documents (TDs) by GDs, ontologies for GDs specify semantic/pragmatic relations including discourse relations (such as EG and Hence in Fig. 1, typically expressed in TDs by conjunctions such as "for instance" and "hence", respectively), because TDs are sets of sentences/phrases connected through such conjunctions (relations).

The authors conducted two experiments to compare GDs and TDs regarding the efficiency of collaborative composition. Both experiments confirmed that collaborative composition of GDs is significantly more efficient than collaborative composition of TDs. Readers are referred to Zhang [4] for details.

2 GD Composition Improves Critical-Thinking Skills Without Extra Cost

Numerous studies [5,6] have demonstrated that graph composition enhances critical-thinking (CT) skills. Despite such proven benefits, widespread adoption of graphs has been hindered by the perception of significant costs associated with training of graph composition [7].

It is hence essential to incorporate graph composition into regular curricula, without imposing an additional workload on teachers, while improving the students' CT skills. Since group discussions are popular in recent school education in Japan, the authors tested hypothesis [H] below by an experiment.

[H] High-school students can collaboratively compose GDs to keep records of their group discussions in regular lessons without incurring extra cost, and thereby improve their CT skills.

2.1 Method

The experiment was approved by the research ethics board at RIKEN. The authors obtained consents from both the participating students and their parents, after informing them that students can participate in the lessons without disclosing their data to the authors.

The experiment took place between October 2022 and January 2023, involving five first-year classes at Kawaguchi City High School and one first-year class at Miura Gakuen High School. Students in each group (involving two to five students) used Semantic Editor (SE) to collaboratively compose GDs to keep records of their group discussions. Experimental data of exactly 100 of them was available for the authors' analysis.

At each school, a CT-skill test (CT test) was administered in October 2022, after which the teachers explained to the students how to create GDs using SE. Then five lessons involving collaborative GD composition followed until January 2023, and finally a second CT test was administered. The CT tests were based on the WGCTA (Watson-Glaser Critical Thinking Appraisal) method of multiple choice questions. Each test consisted of about 40 questions. There was no overlap between the first and second CT test questions at either high school. The five classes at Kawaguchi City High School took the same test each time.

For each school, the authors normalized the data in two steps. The first step is to reduce the diminishing-return effect: i.e., it's hard for good students to become better. $I = S_2 + CS_1^2$ is the corrected increase of the CT-test score for each student, where S_1 and S_2 are her scores of the first and the second CT test, respectively, and C is a positive constant associated with the school, such that the average of I is about the same for any S_1. A simple quadratic function is used for this non-linear correction. Many other nonlinear functions may be better, but we need not find the best one. The second step is to normalize I and the number of each type of GD manipulation so that the average be 0.0 and the standard deviation be 1.0 for each school, because the four CT tests (the first and second ones at the two schools) are all different and the two schools are also different in the teaching method, etc., which may affect frequencies of GD manipulation types.

2.2 Result

The experiment proved the first half of [H]: high school students can collaborate on creating GDs to keep records of their group discussions in regular lessons without incurring extra cost. The teachers did not prepare extra teaching materials. They did not make any special efforts in conducting these lessons, either. They had to prepare for the students' GD composition, including the instruction on how to set up and use SE, but that workload could be ignored if followed by much more lessons involving GD composition. Also, the lessons involved no hindrance, except for inconveniences due to bugs of SE in October 2022. The teachers were able to immediately comment on the GDs created by the students, which means that the students were able to create good enough GDs.

Proven by data analytics was the second half of [H]: the collaborative composition of GD to keep records of group discussions improves students' CT skills. The authors analyzed the CT-skill improvement and the amount of GD manipulation (including link creations, link edits, node creations, node edits, and node movements) of exactly 100 students, excluding those who were absent from the first or second CT test.

Table 1 shows a multiple regression analysis of the normalized improvement of CT skills in relation to the normalized amount of GD manipulation. This analysis ignores node creations, because its contribution to the CT-score improvement was much smaller than that of the others. The analysis results are summarized as below.

- It is 99.73% probable (i.e., the two-sided p-value is 0.0027) that the amount of GD manipulation and the improvement of CT skills are positively correlated.
- The correlation coefficient was 0.30, and accordingly the determination coefficient was 9%.
- Among the five types of GD manipulation, link creation, link edit, and node edit were the major factors that improved CT skills.
- The average improvement of CT skills at the two schools was 3.1%, which is a rough estimate without any normalization as mentioned above.

Table 1. Multiple regression analysis of the normalized improvement of CT skills as to the normalized amount of GD manipulation, where the sample size is 100.

	link creations	link edits	node edits	node moves	multiple regression
total number	1,372	562	445	2,541	
correlation	0.2307	0.2290	0.2472	0.1497	0.297
partial regression coefficient	0.1029	0.0651	0.1023	0.0278	
t score	2.3466	2.3286	2.5253	1.4993	3.082
two-sided p value	0.0210	0.0219	0.0132	0.1370	0.00267

2.3 Discussion

GD manipulation contributes to 9% (determination coefficient) of the factors affecting CT skills. Other factors could include studying various subjects, health condition during CT tests, cognitive development, etc. The correlation and determination coefficients are relatively low because each student participated only in five graph lessons in this experiment. Over the course of their three-year high school education, they could complete more than 50 similar lessons, potentially increasing the determination coefficient to 50%.

3 Conclusion

Collaborative GD composition is more efficient than collaborative TD composition and enhances CT skills without extra cost. Therefore, replacing TDs with GDs will enhance efficiency in not only collaborative document creation but also intellectual works in general.

References

1. Novak, J.D.: Learning, Creating, and Using Knowledge: Concept Maps as Facilitative Tools in Schools and Corporations. Lawrence Erlbaum Associates, Mahweh (1998)
2. Shi, Y., Yang, H., Dou, Y., Zeng, Y.: Effects of mind mapping-based instruction on student cognitive learning outcomes: a meta-analysis. Asia Pac. Educ. Rev. **24**, 303–317 (2023). https://doi.org/10.1007/s12564-022-09746-9
3. De Simone, C., Schmid, R.F., Mcewen, L.: Supporting the learning process with collaborative concept mapping using computer-based communication tools and processes. Educ. Res. Eval. **7**(2–3), 263–283 (2001)
4. Zhang, Z.: Collaborative graph composition is more productive than collaborative text composition. Master thesis, the University of Tokyo (2020)
5. Cottenie, K., Staempfli, M.: Concept mapping as means to critical thinking. Teach. Learn. Innov. **18** (2016)
6. van Gelder, T.: Using argument mapping to improve critical thinking skills. In: The Palgrave Handbook of Critical Thinking in Higher Education, pp. 183-192. Palgrave Macmillan, New York (2015). https://doi.org/10.1057/9781137378057_12
7. Nussbaum, E.M.: Argumentation and student-centered learning environments. In: Jonassen, D.H., Land, S.M. (eds.) Theoretical Foundations of Learning Environments, 2nd edn., pp. 114–141. Routledge, New York (2012)

An Eye-Tracking Study on the Effects of Using Highlighted Multi-attribute Tables: A Preliminary Report

Masahiro Morii[1(✉)], Takashi Ideno[2(✉)], Yuki Tamari[3], Kazuhisa Takemura[4], and Mitsuhiro Okada[5(✉)]

[1] Tokai University, 4-1-1 Kitakaname, Hiratsuka-Shi, Kanagawa, Japan
masa.morii@gmail.com
[2] Tokyo University of Science, 1-11-2 Fujimi, Chiyoda-Ku, Tokyo, Japan
idenodei@gmail.com
[3] University of Shizuoka, 52-1 Yada, Suruga-Ku, Shizuoka, Japan
[4] Waseda University, 1-24-1 Toyama, Shinjuku-Ku, Tokyo, Japan
[5] Keio University, 2-15-42 Mita, Minato-Ku, Tokyo, Japan
mitsu@abelard.flet.keio.ac.jp

Abstract. Our previous research, aimed at improving the design of multi-attribute tables (such as product catalogs), found that black-and-white graphics aided decision-making more than numerical or textual formats (Ideno *et al.*, Diagrams 2020). We extended these findings by examining how combining graphic and numerical presentation (highlighting the cells with superior values) influences decision-making. Twenty-one participants chose the most desirable digital cameras from five alternatives. Results showed faster decision-making with highlighted cells compared to non-highlighted ones. A two-stage decision strategy emerged, with more attribute-based gaze shifts initially and equal vertical and horizontal shifts later. We discussed the implications and prospects of highlighted multi-attribute tables.

Keywords: Highlighted multi-attribute table · Decision-making · Eye-movements

1 Introduction

This is a preliminary report of our new results in studying the design of diagrammatic or graphic aided multi-attribute tables to facilitate decision-making. Multi-attribute tables for decision-makers are widely used in many areas of our lives; typical examples include, among others, product catalogs, insurance plan catalogs, and so on.

We initiated our study on the diagrammatic and graphic effects of multi-attribute tables by referencing Shimojima & Katagiri [1]. We examined the cognitive effects of Black-and-white tables using eye-tracking measures and compared them with traditional Numerical tables (where the value levels are limited to two), as shown in Fig. 1 (left and center). Our previous research [2, 3] found that participants' decision-making using the

Black-and-white table was more effective than using with the Numerical table. The main findings indicated that: (i) the response latencies for decisions were faster with Black-and-white table than with Numerical table; (ii) for Numerical table, participants adopted a typical two-stage decision-making strategy: initially comparing alternatives widely, followed by confirming the utility measures of those alternatives. In contrast with the Black-and-white tables, participants adopted the single decision-making strategy. In the current study, we used Highlighted table, which is made by superposing a numerical table and the corresponding black-and-white table. Figure 1 shows an example of Highlighted table (*e.g.* Fig. 1-3 table is made of superposing Fig. 1-1 table and Fig. 1-2 table.)

The effects of combining text and highlighting were examined in the field of multimedia learning (*c.f.* [4] and its references); however, no fundamental research has been conducted from the standpoint of decision-making studies. Our research question was to examine how Highlighted table reflect the characteristics of Numerical table and Black-and-white table.

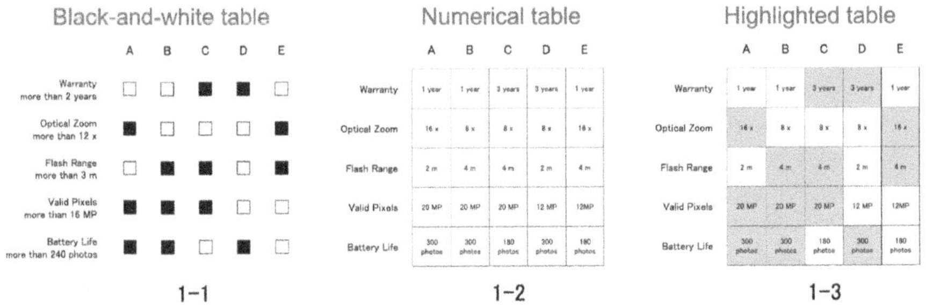

Fig. 1. Multi-attribute tables used in Ideno *et al.*, 2020 [3] (Black-and-white table and Numerical table) and the present study (Highlighted table).

2 Methods

Twenty-one Asians (13 males, 7 female, 1 not-answered, 19–51 years old, $M = 24.4$) participated in the experiment and we omitted the two participants data whose fixations were not measured. Participants were individually tested and were remunerated with JPY 1,100 (approximately USD 7.33) for their participation. Information search process was recorded by the eye-tracking system, Tobii Pro spark (Tobii Technology AB).

To compare the results with our previous study, we used the same stimuli and the same procedure as in the numerical conditions of the previous study [3], differing only in that the cells of the superior values for each attribute were highlighted with yellow color (Fig. 2-2). 5-alternative with 5-attribute (valid pixels, optical zoom, warranty, battery life, flash range) of digital cameras were presented on the 23-inch display. The attribute values were restricted to two levels. Participants were asked to choose the most desirable alternative by pressing the key. Participants was exposed to 32 trials. The half of 32 tables had the dominant alternative (Dominance tables), in which one alternative having four better attributes which was dominant over three other alternatives except one alternative which had two better attributes (Fig. 2-2). And the other half didn't have any dominant alternatives (Non-dominance tables, Fig. 1-3).

3 Results

The response latency was defined as that the time from multi-attribute table was presented on the screen to the decision by key pressing. Figure 3 shows the result of response latencies. Response latency for dominance table was shorter than for non-dominance table ($t\,[18] = 3.81, p < .01$).

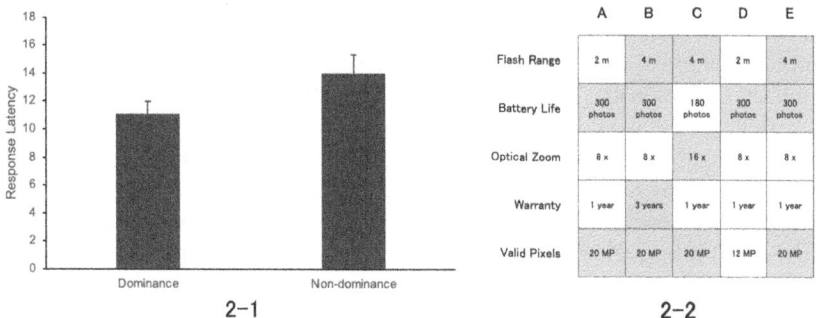

Fig. 2. Results of response latencies (2-1) and an example of Dominance table (2-2).

We calculated the number of fixation shifts for each participant. The area of interest was set to correspond to each cell of the multi-attribute table. A vertical shift was defined as a fixation shift within the same row, and a horizontal shift was defined as a fixation shift within the same column.

As a results, two-factor ANOVA showed that the interaction of shift direction and first or second half was significant ($F\,[18, 1] = 11.05, p < .01$). For further analysis, we calculated the transition score defined as follows [5]:

$$Transition\ Score = \frac{Vertical\ Shifts - Horizontal\ Shifts}{Vertical\ Shifts + Horizontal\ Shifts}$$

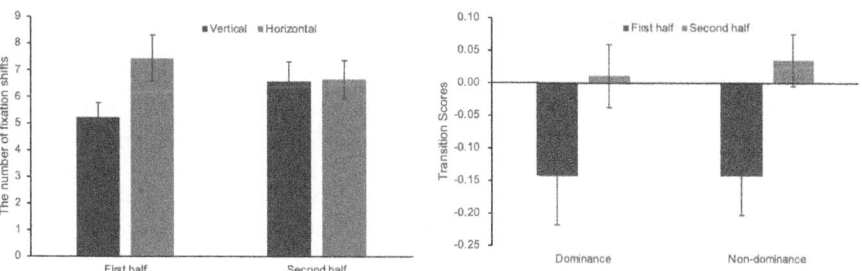

Fig. 3. The number of fixations shifts and the transition scores.

We removed the data of two participants because there were trials where neither vertical nor horizontal shifts were present. The transition scores were lower in the first half than in the second half ($F\,[16, 1] = 12.67$, $p < .01$). There was no main effect between Dominance and Non-dominance table ($F\,[16, 1] = 0.23$, n.s.).

4 Discussions

The main findings are as follows: (i) concerning response latencies, the mean response latency for Highlighted table (12.53 s) was shorter than that for Numerical table (14.00 s) but longer than that for Black-and-white table (11.41s) in the previous study [3]. Furthermore, response latency for Dominance table is significantly shorter than that for Non-Dominance table (Fig. 2-1). A similar result was observed for Black-and-white table but not for Numerical tables in the previous study [3], suggesting that the Highlighted table reflects characteristics more akin to Black-and-white table. (ii) concerning decision-making strategy, the fixation shift patterns varied between the first and second halves of the trials (Fig. 3), indicating that participants employed a two-stage decision-making strategy akin to Numerical table [3]. On the other hand, overall horizontal shifts decreased compared to Numerical table. These results suggest that the Highlighted table may facilitate easier information search within specific alternative(s) during the second stage of the two-stage strategy.

These indicate that the Highlighted table potentially facilitates decision-makers' decisions. Due to differences in experimental settings, the comparisons between the current study and our previous study [3] were limited. We will conduct further experiments in the same setting in our future work.

Acknowledgments. This work was supported by JSPS KAKENHI Grant Numbers 19KK0006, JP21H00467, JP21K18339, JP22K13506, JP22K20170, JP23K20416, and JP24K16466.

Disclosure of Interests. The authors declared no potential conflicts of interest with respect to the research, authorship, and/or publication of this article.

References

1. Shimojima, A., Katagiri, Y.: An eye-tracking study of integrative spatial cognition over diagrammatic representations. In: Hölscher, C., Shipley, T.F., Olivetti Belardinelli, M., Bateman, J., Newcombe, N.S. (eds.) Spatial Cognition 2010. LNCS, vol. 6222, pp. 262–278. Springer, Heidelberg (2010). https://doi.org/10.1007/978-3-642-14749-4_23
2. Morii, M., Ideno, T., Takemura, K., Okada, M.: Qualitatively coherent representation makes decision-making easier with binary-colored multi-attribute tables: an eye-tracking study. Front. Psychol. **8**, 1388 (2017). https://doi.org/10.3389/fpsyg.2017.01388
3. Ideno, T., Mori, M., Takemura, K., Okada, M.: On effects of changing multi-attribute table design on decision making: an eye-tracking study. In: Pietarinen, A.V., Chapman, P., Bosveld-de Smet, L., Giardino, V., Corter, J., Linker, S. (eds.) Diagrams 2020. LNCS, vol. 12169, pp. 365–381. Springer, Cham (2020). https://doi.org/10.1007/978-3-030-54249-8_29

4. Richter, J., Scheiter, K., Eitel, A.: Signaling text-picture relations in multimedia learning: a comprehensive meta-analysis. Educ. Res. Rev. **17**, 19–36 (2016). https://doi.org/10.1016/j.edurev.2015.12.003
5. Payne, J.W.: Task complexity and contingent processing in decision making: an information search and protocol analysis. Organ. Behav. Hum. Perference **16**, 366–387 (1976). https://doi.org/10.1016/0030-5073(76)90022-2

On the Formal Cause of Diagrams: Mimesis and Phenomenology

Noah Greenstein

The Bronx, New York, USA
me@noahgreenstein.com
https://www.noahgreenstein.com

Abstract. We investigate the formal cause of diagrams, initially realizing that diagrams have no obvious form. It is argued their form is to mimic expert perspectives. This perspective provides a organizational structure that represents the relations important in understanding the worldly situation. We then shift to a study of how we are to understand an expert perspective. Using the distinction between intuitive and formal logic, *logica utens* versus *logica docens*, we identify games of habituation: games of focus and distraction. The skills required for games of focus and distraction are phenomenological in regulating the way we see, *i.e.*, how we approach and analyze different situations. This unique phenomenological skill is characteristic of diagrammatic reasoning.

Keywords: Diagrams · Utens/Docens · Mimesis · Phenomenology · Games · Formal Cause

We inquire into the metaphysics of diagrams: what are their four causes? The material and efficient causes seem straight-forward. Diagrams are materially constituted by the writing systems of the day, be it ink on paper, digital pixels, *etc.* and are 'efficiently' drafted by expert agents. The final cause, too, seems uncontroversial: diagrams are tools that assist us to complete tasks. But what of the form of diagrams? Diagrams cannot be the form of thing they represent, as then the diagram would be no different from the object itself. Nor can diagrams have a totally unique form, as then they could not represent anything else. Hence there is no obvious form to a diagram.

So, consider having to leave a building in an emergency. We could follow someone tasked with public safety who already knows where all the exits are and the fastest routes to get to them. They also have to know which routes lead to dead ends in order to avoid those passages if their current way is blocked and need to reroute. These personnel need this expert understanding and the ability to act appropriately in emergency situations... or, we could follow a map. An exit map attempts to give a lay-person this expert perspective without having to learn the building's layout: if we have a map that indicates the shortest routes, and we can read the map, it is as if we were the expert.

N. Greenstein—Independent Scholar.

Diagrams, then, obviate having to train our *own ability* to react to stimuli in the appropriate manner, to interpret phenomena correctly. That is, a diagram replaces the *subjective experience of expertise*: it is an expert agent that can tell the difference between an easy and difficult route. But, having a guide or diagram means we gain an 'expert perspective' without having to do the work of becoming an expert. To be explicit: a diagram does not actually give a layperson the subjective experience or expertise of an expert. But it does act as a replacement for the expert's ability in that it *converts* a difficult situation into a tractable one for someone of non-expert ability. While the expert understands how to read the *situation*, the lay-person just has to understand the *diagram*.

We can interpret this as a kind of mimicry. What a diagram does is mimic something of the subjective experience of an expert, enabling an interpreter to take on the *perspective* of the creator. This perspective is a kind of organization: to organize the world in a certain structured way, prioritizing certain features as more important than others. Merely taking on such a perspective will not grant a person expertise, but it will reveal certain relations between worldly content. If a diagram is useful, these relations will be relevant to solving worldly problems. Interpreting such a diagram will orient the agent to see certain relations, and, in doing so, imitate how an expert *might* see the world. This suggests a diagram's form is a simulacrum of an expert perspective.

This raises the question: if diagrams are simulacra of expert perspectives *that we do not have*, how do they enable us to have expert thoughts?

Note, the distinction between the expert's subjective experience and the diagrammatic perspective recalls an old Scholastic distinction, championed by Peirce, between *logica utens* and *logica docens*: the way the logic is used, versus the way the logic is described in a 'theoretical,' 'scientific,' or 'educational' way [6, p. 357]. While practitioners *use* logic, just as experts know how to do things, this is not necessarily equivalent to how the logic is represented, e.g., diagrammatically. And, just as many diagrams can be seen as representing formal systems [2], then they can be interpreted as revealing formal logical truth.

The problem is that the truth of *docens*, formal truth, is usually taken to be trivial. That is, truth tells us nothing we already didn't know: asserting some proposition P to be True, "'The sky is blue' is true," is no different than just asserting P, 'The sky is blue'. What we want is something of *utens*, to be able to *expertly* figure out the truth ourselves.

This triviality does not mean there are no further interpretations of logical truth: let us consider the tradition of logic and games, which have a long history together [5]. Importantly, a game-theoretic semantics (GTS) has been given for logic [4]. GTS interprets each of the logical symbols as different features of a game, where they present the players with different choices, such as instantiation. The strategy of choosing different ways to instantiate, conjoin, *etc.*, describes the 'game moves' that a player makes to ensure the winning condition obtains, that is, the truth of the proof. Truth, then, is defined as the *existence of a winning strategy* for one of the players and logic is a game of 'seeking and finding' such a strategy [4, pp. 415-416].

Is there, then, a corresponding concept to the formal *docens* 'existence of a winning strategy' for *utens*? As Pietarinen [6, p. 361] notes, for Peirce: "This is not to say that the *logica utens* is in some sense utterly pretheoretical, or not subject to its own laws and intrinsic structure of how it comes to be constituted... the camouflage of *utens* as a form of instinct is not only consistent with, but also paramount to, Peirce's desire to implant instinctive aspects of reasoning in the conception of reasoning guided by habits." It is this notion of 'habit' that is crucial: habit instinctively guides the reason of the expert. Pietarinen [6, p. 364] continues along this line, saying: "the concept of habit... came to be replaced by the institutionalized concept of a strategy". We therefore have the instinctive 'habits' of *utens* in parallel to the formal 'strategy' of *docens*.

What concerns us here, then, are our *habits of interpretation*. Each of us interprets the world uniquely and has skills in different areas. But many will have significant overlap due to historical accident or by training. A diagram is effective because it exploits how we, for the most part, habitually interpret the world. By harnessing our extant habits of interpretation, it can orient our perspective as desired.

How well any diagram, notation, *etc.*, actually accomplishes orienting our perspective will depend on the skill of the drafter and the abilities of the viewer. On this point De Toffoli [1, p. 163] defines 'Transparency': "Transparency quantifies... to what extent [a notation] can be interpreted and used directly, by exploiting our cognitive abilities and our training. Therefore, the transparency criterion cannot be fixed once and for all, but it is indexed to the practitioner's background." Hence transparency is a contingent notion, as it is measured relative to particular practitioners. This contingency could signal an end to the philosophical investigation: there could be further research into, say, the psychological effects of different media, but this is an empirical question. If all that is left is empirical, then philosophical reflection is at its end.

However, this would be too quick. On the *docens* side, we described logic as a game of 'seeking and finding' [4, p. 415], kind of like a game of 'Hide and Seek' or 'Hot and Cold' [3, p. 14]. If there is a corresponding game for *utens*, then we could investigate it along the same game-inspired lines.

In this vein, consider games of 'focus and distraction', such as Musical Statues or a Staring Contest. Musical Statues, or Freeze, is a game where one player controls the music and the others dance to it. The DJ makes the music suddenly stop every so often and the Dancers need to freeze in place, like a statue, when the music cuts out. Anyone who gets too wrapped up in their dancing and doesn't freeze, loses. Similarly, in a Staring Contest, two people have to stare into each other's eyes and not break contact. Easier said than done. Due to psychological and biological factors, maintaining eye contact becomes increasingly difficult, almost a battle of the wills, until one person has to look away.

What is common to these games is that the participants need to focus their attention in spite of distractions. In Musical Statues, we become habituated to dancing, and then become distracted from noticing the music stopping. In a Staring Contest, we need to focus our attention on keeping our eyes fixed on the other without being distracted by fatigue. Importantly, we can get *better* at these games, *i.e.* we can be trained to focus on arbitrary targets and block out distractions. This shows that there is a skill to being focused and implies that there is more than just our disposition and familiarity with a subject matter: there is more than just transparency. There is also a skill in combining the two, in maximizing how our natural disposition and training interact. Let us call this skill our 'meditative' ability.

The better we are at meditating on a topic, the more we can fix our focus upon it and hold everything else irrelevant. Given a diagram to analyze, it is our meditative ability that allows us to re-orient our perspective to make it most salient, while simultaneously blocking everything extraneous from our thought. Only by 'starting fresh' can we begin to search for how the diagram relevantly matches up with the world and provides an expert perspective.

That we 'start fresh' or 'with fresh eyes' is critical. These phrases refer to the *phenomenological* skill of bracketing our default way of 'seeing' a problem so that we do not become misled by our presuppositions. Nor is this just a blank slate: meditating on a topic is not to ignore our disposition and training, but to re-orient both our disposition and training with respect to it. In this way, by exercising our phenomenological skills, we maximize our chances of appropriating whatever the diagram has to offer. Note, this makes the meditative ability a skill governing phenomenology itself, a phenomenological meta-skill regulating how we approach seeing problems.

Thus, by inquiring into the formal cause of diagrams, we've shown them to have a distinctive metaphysics: they mimic expert perspectives and, while we may sometimes use our meditative ability in other contexts, diagrammatic reasoning is characterized by it.

Disclosure of Interests. No competing interests exist.

References

1. De Toffoli, S.: Chasing the diagram-the use of visualizations in algebraic reasoning. Rev. Symb. Log. **10**(1), 158–186 (2017). https://doi.org/10.1017/s1755020316000277
2. De Toffoli, S.: Who's afraid of mathematical diagrams? Philos. Impr. **23**(1) (2023). https://doi.org/10.3998/phimp.1348
3. Dreyfus, H.: A phenomenology of skill acquisition as the basis for a Merleau-Pontian nonrepresentational cognitive science (manuscript). https://philpapers.org/rec/dreapo. Accessed 11 June 2024
4. Hintikka, J., Sandu, G.: Game-Theoretical Semantics. In: ter Meulen, A., van Benthem, J. (eds.) Handbook of Logic and Language, 2nd edn., pp. 361–410. Elsevier, Amsterdam (2011)
5. Hodges, W., Väänänen, J.: Logic and Games. In: Zalta, E.N. (ed.) The Stanford Encyclopedia of Philosophy. Fall 2019. Metaphysics Research Lab, Stanford University (2019). https://plato.stanford.edu/archives/fall2019/entries/logic-games/. Accessed 2 Aug 2024
6. Pietarinen, A.-V.: Cultivating habits of reason: peirce and the logica utens versus logica docens distinction. Hist. Philos. Q. **22**(4), 357–372 (2005)

The Region Connection Calculus, Euler Diagrams and Aristotelian Diagrams

Claudia Anger[1] and Lorenz Demey[2](✉)

[1] Institut für Philosophie, FernUniversität Hagen, Universitätsstraße 33, 58097 Hagen, Germany
`claudia.anger@fernuni-hagen.de`

[2] Center for Logic and Philosophy of Science/Leuven.AI, KU Leuven, Kardinaal Mercierplein 2, 3000 Leuven, Belgium
`lorenz.demey@kuleuven.be`

Abstract. The Region Connection Calculus (RCC) is a qualitative spatial reasoning formalism, developed in knowledge representation and geographical information systems. We argue that RCC can be viewed as a more fine-grained approach to the use of Euler diagrams to visualize categorical statements like 'all A are B'. We present RCC using the syntax of first-order modal logic and a topological semantics. We compare the Gergonne relations (a well-known set of 5 jointly exhaustive and pairwise disjoint relations between two non-empty sets, visualized using Euler-type diagrams) with RCC8 (a similar set of 8 relations for regions).

Keywords: Region Connection Calculus · Euler diagram · Aristotelian diagram · RCC8 · Gergonne · logical geometry

Euler Diagrams and Boundary Issues. Euler diagrams are widely used across science and philosophy to visualize sets/concepts and the logical relations among them. By using concrete circles to visualize abstract sets/concepts, they offer definite cognitive advantages, often called 'free rides' [5]. For example, Fig. 1(a) shows how the transitivity of physical enclosure can be used to represent the transitivity of set inclusion (from $A \subseteq B$ and $B \subseteq C$ to $A \subseteq C$).

In certain cases, however, the use of concrete/physical circles and curves to represent abstract sets can become problematic. For example, sets do not have boundaries (every element in the domain either belongs to the set or does not belong to the set), whereas physical circles do have boundaries (an element can be inside the circle or outside the circle, but also on the boundary). This is relevant for the visualization of certain set relationships; e.g., the Euler diagrams in Fig. 1(b–c) both seem to represent $A \subseteq B$, and those in Fig. 1(d–e) both seem to represent $A \cap B = \emptyset$. Focusing on Fig. 1(e), if we disregard the boundaries, we indeed find that $A \cap B = \emptyset$, but if we do take the boundaries into account, then we actually find that $A \cap B$ contains a (boundary) point, and is thus non-empty after all. Several authors have proposed *no brushing points* as a well-formedness property of Euler diagrams, and cognitive research suggests that the violation of this property, as in Fig. 1(c,e), adversely impacts diagram comprehension [5].

The second author holds a Research Professorship (BOFZAP) from KU Leuven.

© The Author(s), under exclusive license to Springer Nature Switzerland AG 2024
J. Lemanski et al. (Eds.): Diagrams 2024, LNAI 14981, pp. 476–479, 2024.
https://doi.org/10.1007/978-3-031-71291-3_42

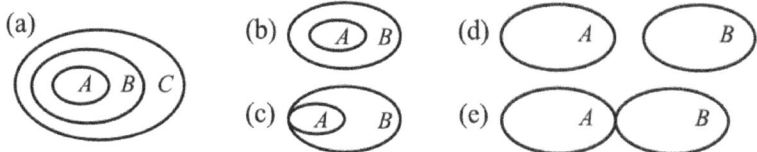

Fig. 1. Several examples of Euler diagrams.

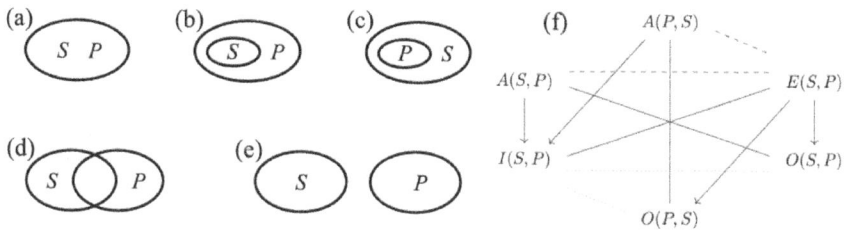

Fig. 2. The five Gergonne relations (a) $\mathbb{I}(S,P)$, (b) $\mathbb{C}(S,P)$, (c) $\mathbb{O}(S,P)$, (d) $\mathbb{X}(S,P)$, (e) $\mathbb{H}(S,P)$; (f) Kraszewski's hexagon. (Contradiction, contrariety, subcontrariety and subalternation are shown using resp. solid, dashed, dotted lines and arrows.)

Euler Diagrams and the Gergonne Relations. In the early 19th century, Gergonne showed that two non-empty sets S and P always stand in exactly one of five relations to each other [3]. These jointly exhaustive and pairwise disjoint (JEPD) relations, which Gergonne called $\mathbb{I}$, $\mathbb{C}$, $\mathbb{O}$, $\mathbb{X}$ and $\mathbb{H}$, are shown as Euler diagrams in Fig. 2(a–e). The five Gergonne relations can be used to provide a semantics for the categorical statements; for example, $A(S,P)$ gets interpreted as $\mathbb{I}(S,P) \vee \mathbb{C}(S,P)$, and $I(S,P)$ as $\mathbb{I}(S,P) \vee \mathbb{C}(S,P) \vee \mathbb{O}(S,P) \vee \mathbb{X}(S,P)$.[1]

The classical square of opposition is not sufficiently expressive to generate all five Gergonne relations; in particular, it does not differentiate between $\mathbb{I}$ and $\mathbb{C}$, nor between $\mathbb{X}$ and $\mathbb{O}$. However, in the 1950s, Kraszewski considered the hexagon in Fig. 2(f), which does generate all Gergonne relations [3]. In particular, applying the techniques from [4], we find that the partition induced by this hexagon is $\{A(S,P) \wedge A(P,S), A(S,P) \wedge O(P,S), O(S,P) \wedge A(P,S), I(S,P) \wedge O(S,P) \wedge O(P,S), E(S,P)\}$, which is precisely $\{\mathbb{I}(S,P), \mathbb{C}(S,P), \mathbb{O}(S,P), \mathbb{X}(S,P), \mathbb{H}(S,P)\}$.

The Region Connection Calculus. The Region Connection Calculus (RCC) is a qualitative spatial reasoning formalism, which was developed in the 1990s in the fields of knowledge representation and geographical information systems [2, 6]. This formalism is standardly used to study relations among (physical) regions,

---

[1] We use the standard abbreviations A, I, E, O for the categorical statements from syllogistics. (Note that the assumption of existential import in this logical system corresponds to Gergonne's decision to only consider non-empty sets.) Gergonne wrote I, C, etc. instead of $\mathbb{I}$, $\mathbb{C}$, etc. for his relations, but we have adapted this in order to avoid confusion between the Gergonne relation $\mathbb{I}$ and the categorical statement I.

but we will argue that RCC can also be viewed as a more fine-grained approach to visualizing relations among sets, which allows us to avoid the aforementioned 'boundary issues' of Euler diagrams. To achieve this shift in perspective, we need to reinterpret the physical regions of RCC (incl. the distinction between the 'boundary' and the 'interior' of a region) in set-theoretical/logical terms.

Pure RCC is a first-order theory, with quantifiers that range over *regions* and a binary predicate symbol C that expresses that two regions are *connected* to each other. This relation is axiomatized as being reflexive ($\forall x Cxx$) and symmetric ($\forall x \forall y (Cxy \to Cyx)$). Several other relations over regions can be defined in terms of C. For example, given regions x and y, we say (i) that x and y are *disconnected*, $\mathsf{DC}(x,y)$, iff $\neg \mathsf{C}(x,y)$, (ii) that x is a *part* of y, $\mathsf{P}(x,y)$, iff $\forall z (\mathsf{C}(z,x) \to \mathsf{C}(z,y))$, (iii) that x *overlaps* with y, $\mathsf{O}(x,y)$, iff $\exists z (\mathsf{P}(z,x) \wedge \mathsf{P}(z,y))$, and (iv) that x and y are *externally connected*, $\mathsf{EC}(x,y)$, iff $\mathsf{C}(x,y) \wedge \neg \mathsf{O}(x,y)$ [2]. The cases of $\mathsf{DC}(S,P)$ and $\mathsf{EC}(S,P)$ are shown in Fig. 3(a–b); note the similarity with Fig. 1(d–e).

In pure RCC, the notions of 'region' and 'connected' are taken as primitive, but there exist several topological interpretations.[2] For example, we can interpret *regions* as non-empty, regular closed sets in the Euclidean plane $\mathbb{R}^2$, and say that two regions are *connected* iff they have a non-empty intersection [6]. Under this topological interpretation, $\mathsf{DC}(S,P)$ means that $S \cap P = \emptyset$, while $\mathsf{EC}(S,P)$ means that $S \cap P \neq \emptyset$ and $int(S) \cap int(P) = \emptyset$; again compare with Fig. 3(a–b).

This topological semantics also inspires a translation from RCC into a *modal* language. After all, it is well-known that the modal logic S4 has a topological semantics, with the $\Box$- and $\Diamond$-modalities corresponding to the topological interior and closure operators, respectively; i.e., given a topological model $\mathcal{M}$, we have $[\![\Box \varphi]\!]^{\mathcal{M}} = int([\![\varphi]\!]^{\mathcal{M}})$ and $[\![\Diamond \varphi]\!]^{\mathcal{M}} = cl([\![\varphi]\!]^{\mathcal{M}})$ [1,6].[3] These ideas were already used to translate RCC into *propositional* modal logic [6], but in order to maintain the connection with the categorical statements, it is more natural to translate RCC into *first-order* modal logic [1]. We therefore start from the system of first-order S4 [1], and add the axioms $\forall x (\Phi x \leftrightarrow \Diamond \Box \Phi x)$ (since regions are regular closed sets) and $\exists x \Box \Phi x$ (since regions are non-empty sets), for all unary predicate symbols Φ.[4] Under this modal translation, $\mathsf{DC}(S,P)$ becomes $\forall x (Sx \to \neg Px)$, while $\mathsf{EC}(S,P)$ becomes $\exists x (Sx \wedge Px) \wedge \forall x (\Box Sx \to \neg \Box Px)$; again cf. Fig. 3(a–b).

The RCC diagrams in Fig. 3 thus express categorical statements, just like the Euler diagrams in Fig. 2. The only difference is that, while the Euler diagrams from Fig. 2 are restricted to expressing *assertoric* statements, the RCC diagrams from Fig. 3 are capable of expressing *modal* statements. The 'boundary issues' from Fig. 1(b–e) can thus be leveraged to express logically significant differences.

---

[2] For reasons of space, we have to presuppose a basic understanding of topology. In particular, recall that given a topological space (X, τ) and a set $S \subseteq X$, the *interior* of S is $int(S) := \bigcup \{O \in \tau \mid O \subseteq S\}$ and the *closure* of S is $cl(S) := \overline{int(S)} = \bigcap \{\overline{O} \mid O \in \tau, S \subseteq \overline{O}\}$. Finally, we say that S is *regular closed* iff $S = cl(int(S))$.
[3] Also note that $[\![\varphi \wedge \neg \Box \varphi]\!]^{\mathcal{M}} = [\![\varphi]\!]^{\mathcal{M}} \setminus int([\![\varphi]\!]^{\mathcal{M}})$, so $\varphi \wedge \neg \Box \varphi$ corresponds to the boundary of φ, i.e., the part of φ that remains after 'subtracting' its interior.
[4] If S is regular closed, we have $S \neq \emptyset$ iff $int(S) \neq \emptyset$, so we need to add the axioms $\exists x \Box \Phi x$ instead of 'merely' $\exists x \Phi x$. Also note that $\forall x (\Box \Phi x \leftrightarrow \Phi x)$ is *not* a theorem.

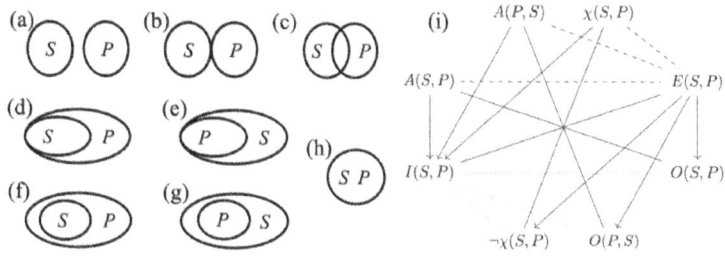

Fig. 3. (a–h) The RCC8 relations; (i) an octagon of opposition that generates RCC8.

RCC8. A well-known result in RCC is that two regions always stand in exactly one of eight relations [2,6]. These jointly exhaustive and pairwise disjoint (JEPD) RCC8 relations are shown in Fig. 3(a–h). They can be defined directly in terms of C and also in topological terms, but for reasons of space, we only give their modal characterizations: (a) *disconnectedness*/DC(S,P) and (b) *external connectedness*/EC(S,P) were already defined above, (c) *partial overlap*/PO(S,P) is $\exists x(\Box Sx \wedge \Box Px) \wedge \exists x(\Box Sx \wedge \neg Px) \wedge \exists x(\neg Sx \wedge \Box Px)$, (d) *tangential proper parthood*/TPP(S,P) is $\forall x(Sx \rightarrow Px) \wedge \exists x(Sx \wedge \neg\Box Px) \wedge \exists x(\neg Sx \wedge Px)$ and (e) its *inverse* is TPPi$(S,P) = $ TPP(P,S), (f) *non-tangential proper parthood*/NTPP(S,P) is $\forall x(Sx \rightarrow \Box Px) \wedge \exists x(\neg Sx \wedge Px)$ and (g) its *inverse* is NTPPi$(S,P) = $ NTPP(P,S); finally, (h) *equality*/EQ(S,P) is $\forall x(Sx \leftrightarrow Px)$.

The (modal) RCC8 is thus a refinement of the (assertoric) Gergonne relations: $\mathbb{I}(S,P) \equiv $ EQ(S,P), $\mathbb{C}(S,P) \equiv $ TPP$(S,P) \vee $ NTPP(S,P), $\mathbb{D}(S,P) \equiv $ TPPi$(S,P) \vee$ NTPPi(S,P), $\mathbb{X}(S,P) \equiv $ PO$(S,P) \vee $ EC(S,P) and $\mathbb{H}(S,P) \equiv $ DC(S,P). Finally, if we put $\chi(S,P) := $ EC$(S,P) \vee $ TPP$(S,P) \vee $ TPPi(S,P) and add $\chi(S,P)$ and its negation to Kraszewski's hexagon from Fig. 2(f), then we obtain the octagon in Fig. 3(i). The partition induced by this octagon is precisely RCC8 [4].

References

1. Awodey, S., Kishida, K.: Topology and modality: the topological interpretation of first-order modal logic. Rev. Symb. Log. **1**, 146–166 (2008)
2. Cohn, A., et al.: Qualitative spatial representation and reasoning with the Region Connection Calculus. GeoInformatica **1**, 275–316 (1997)
3. Demey, L., Erbas, A.: Boolean subtypes of the U4 hexagon of opposition. Axioms **13**(76), 1–20 (2024)
4. Demey, L., Smessaert, H.: Combinatorial bitstring semantics for arbitrary logical fragments. J. Philos. Log. **47**, 325–363 (2018)
5. Rodgers, P.: A survey of Euler diagrams. J. Vis. Lang. Comput. **25**, 134–155 (2014)
6. Wolter, F., Zakharyachev, M.: Qualitative spatio-temporal representation and reasoning: a computational perspective. In: Lakemeyer, G., Nebel, B. (eds.) Exploring Artificial Intelligence in the New Millennium, pp. 175–216. Morgan Kauffman (2002)

Between Pro/Con-Lists and Argument Graphs Finding the Right Level of Complexity in Argumentation Representation

Joannes B. Campell[1(✉)] and Michael A. Müller[2]

[1] Universität von Bern, Bern, Switzerland
j_campell@gmx.net
[2] Universiteit van Amsterdam, Amsterdam, Netherlands
mam258@cantab.ac.uk

Abstract. Pro/Con-lists are one of the most common ways of summarising the main points of debates surrounding practical decisions. However, their simplicity has attracted criticism and argument graphs have been proposed as an alternative. Since graphs can easily become very complex, they tend to be difficult to use in everyday life. Our proposal, the Concern-Oriented REasoning (CORE) table, lies in-between the two. It improves on the drawbacks of pro/con-lists without having to go to graph-level complexity. We start by identifying and organising the main reasons for and against an action into a pro/con-list. We then add structure to the CORE-table by organising the main reasons according to their subject matter. Additional arguments surrounding the main reasons are then placed in designated spots based on their function.

Keywords: Pro/Con-Lists · Argumentation Graphs · Reasons · Practical Reasoning · Debate Reconstruction · Voter Information

1 Introduction

Sometimes in practical reasoning, we are directly presented with some candidate action and have to decide whether or not to do this action (cf. [7]). For instance, in a national referendum, the voters have to decide whether or not to accept it. In such situations, it is crucial that voters are well-informed about the arguments in favour of and against the candidate action. This raises the question of which method to use for representing arguments.

Pro/con-lists are a very popular tool to inform people about such debates (cf. [5,9,11]). The main points in favour of and against a certain action are listed in an easily accessible way. Based on this, we can then weigh both sides and come to a decision. Of course, their simplicity is also a source of limitations: Common criticisms are that they can't depict different relations between the pro/con-arguments, that they usually reduce whole arguments to single statements, and that they suggest in some sense that counting pros and cons is the proper way to make a decision [3].

In contrast, argument graphs depict all the complexities of argumentation by using nodes to represent arguments and edges to represent various relations between the arguments such as supports and attacks (cf. [4,6]). While graphs seem ideal for detailed argument analysis, it is difficult and time-intensive to process all the information presented in them. Though the field of computational argumentation deals with automatic evaluation of such graphs, see [1], we aim at providing information in a form that is geared towards typical voters. Further, graphs don't seem to capture much of the experience of making practical decisions: Namely that of having to balance various reasons in favour of and against some specific course of action.

2 The CORE-Table

The Concern-Oriented REasoning (CORE) table is a tool for representing debates surrounding practical decisions. It combines the strengths of argument graphs and pro/con-lists by being accessible while keeping enough of the relevant argumentative structure.

To achieve this goal, we make use of the argument scheme for proposing reasons in [8], which informs the structure of the table. The main idea behind the scheme is that a reason in favour of or against an action has two essential components: a change and a concern. Without a change—ranging from an obvious material change to a subtle change in moral properties—there can be no reason. A concern, on the other hand, is what makes us care about such changes: a raise in taxes, for instance, is only relevant if one cares about the amount of taxes owed. As debates don't usually provide reasons in these terms, creating a CORE-table based on this idea requires substantive reconstructive work.

The CORE-table (Fig. 1) uses this view of reasons to distinguish them from other argumentation. The idea and simplicity of pro/con-lists is preserved in the *main points* section for each concern. These main points are the reasons that need to be weighed in order to assess whether a concern benefits from the action

	Concern		
Support	**Main Points**		**Support**
Facts and other supports for the main pro-points.	+ Reasons that speak in favour of the action via the concern.	− Reasons that speak against the action via the concern.	Facts and other supports for the main con-points.
	Discussion		
	+ Replies as well as appeasements and discussion of the main points of the other side.	− Replies as well as appeasements and discussion of the main points of the other side.	
	Overall Points		
+ Positive considerations between and about different concerns.		− Negative considerations between and about different concerns.	

Fig. 1. An illustration of the structure of a minimal CORE-table with one concern.

Support	Main Points	Support
\multicolumn{3}{c}{**Fairness towards and Dignity of Pensioners**}		
The raise in cost of living is about one month's AHV-pension.	+ A 13th AHV-pension balances out the risen cost of living for all.	− A lot of pensioners would get but not need a 13th AHV-pension.
The AHV-pension is at the moment no longer sufficient.	+ A 13th AHV-pension allows struggling pensioners to get by.	
Only the AHV recognises unpaid carework—overwhelmingly done by women.	+ Only the AHV alleviates the problem of lower pension for women.	
	Discussion	
	− The social benefits are low: most people have other pensions, income or assets.	
	− Pensioners who can not meet their subsistence needs can apply for supplementary benefits.	

	Cost of living for the Youths	
	Main Points	**Support**
	− Labour or consumption would get much more expensive.	Either VAT or salary contributions would have to be increased.
	− The government would have to raise taxes or cut expenses.	It would cost the government CHF 800m a year.
	Discussion	
	+ Augmenting the contribution from the salary by 0.4% suffices to cover the costs in the long run.	

	Sustainability of AHV	
	Main Points	
	− The financing of the 13th pension would place too much financial burden on the AHV	
	Discussion	
	+ The AHV has reserves of CHF 50bn and expects a surplus of CHF 3.5bn.	

\multicolumn{3}{c}{**Overall Points**}		
+ The AHV should give the pensioners the dignified pension they deserve.		− Instead of extending the expenses of the AHV, we need to secure its pensions.
+ The AHV is the best solution for decent pensions.		

Fig. 2. An example of a CORE-table for the debate around raising pensions in Switzerland [10].

or not. Other arguments, not involved in the weighing, then either *support* main points or (critically) *discuss* and compare them. *Overall points* relate the different concerns and argue for or against their importance. An extended example is given in Fig. 2, depitcting a debate about introducing a 13th monthly pension in the Swiss pension system (AHV) [10].

The structure of the CORE-table suggests the following decision procedure: (1) Since the arguments are sorted according to concern, the decision-maker can cross out all arguments pertaining to a concern irrelevant to them. (2) Left with the concerns that matter, the decision-maker can then look at the main points, keeping the ones judged to be well supported. (3) The remaining main points per concern then allow the decision-maker to judge the influence of the proposed action on each concern. (4) Finally, the decision-maker can make an overall judgement on whether the affected concerns taken together favour accepting or rejecting the proposed action. The CORE-table doesn't prescribe any specific mechanism for each of these steps. Rather, it is compatible with a wide range of methods (see e.g. [2]).

All arguments that occur in a CORE-table have a clear function and are placed such that their relations are immediate. This structure allows for integrating CORE-tables and argument graphs. For instance, the table can be integrated in an argument graph spanning a whole debate: The main points, together with their respective concern, form direct arguments in favour of accepting or rejecting the proposed state action under discussion. Support- and discussion-arguments then build up the rest of the graph in arbitrary complexity.

In conclusion, the CORE-table provides clarity about the functions and concerns of the arguments in a debate surrounding a practical decision. It can serve as a valuable tool for argument analysis, especially for voter information.

References

1. Baroni, P., Gabbay, D., Giacomin, M., van der Torre, L.: Handbook of Formal Argumentation. College Publications, UK (2018)
2. Blair, J.A., Johnson, R.H.: Conductive Argument. An Overlooked Type of Defeasible Reasoning. College Publications, UK (2011)
3. Brun, G., Betz, G.: Analysing practical argumentation. In: Hansson, S., Hirsch Hadorn, G. (eds.) The Argumentative Turn in Policy Analysis. Logic, Argumentation and Reasoning, vol. 10, pp. 39–77. Springer, Cham (2016). https://doi.org/10.1007/978-3-319-30549-3_3
4. DebateGraph. https://debategraph.org. Accessed 03 Mar 2024
5. Easyvote. https://www.easyvote.ch. Accessed 03 Mar 2024
6. Kialo. https://www.kialo.com. Accessed 03 Mar 2024
7. Macagno, F., Walton, D.: Practical reasoning arguments: a modular approach. Argumentation **32**, 519–547 (2018)
8. Müller, M.A., Campell, J.B.: Arguing in direct democracy: an argument scheme for proposing reasons in debates surrounding public votes. Topoi **42**(2), 593–607 (2023)

9. ProCon.org. https://www.procon.org. Accessed 03 Mar 2024
10. Schweizerische Bundeskanzlei: Volksabstimmung vom 3. März 2024. Erläuterungen des Bundersrates. Schweizerische Bundeskanzlei (2024)
11. Tsikerdekis, M.: Pro/Con lists and their use in group decision support systems for reducing groupthink. INFOCOMP **11**(2), 10–20 (2012)

Diagrammatic Analogical Reasoning

Henri Prade[✉] and Gilles Richard

IRIT – CNRS, Université Paul Sabatier, 118, Route de Narbonne, Toulouse, France
{henri.prade,gilles.richard}@irit.fr

Abstract. Analogical proportions are statements of the form "a is to b as c is to d". Their logical modeling is at the basis of an inference mechanism used in classification and reasoning tasks. This note proposes an original view of this inference based on a structure of opposition between a, b, c, d, which may be feature-based representations of items, or just logical formulas.

Keywords: Analogy · Analogical Proportions · Square of Opposition

1 Introduction

Analogical reasoning has stood apart from logic for a long time. About fifteen years ago, a logical modeling of the notion of analogical proportion, i.e., statements of the form "a is to b as c is to d", has been proposed [10,12], where the items a, b, c, d are represented by vectors of Boolean attributes values. Based on analogical proportions, a mechanism of analogical inference can be defined and applied to classification and reasoning tasks, e.g., [2,3]. This view of analogical reasoning departs from the classical psychological view, known as structured mapping theory [6], which amounts to look for similarity in the relationships between objects in a domain with the relationships between objects in another different domain [6]. Analogical proportions also acknowledge that analogies are not just a matter of similarity, but also a matter of dissimilarity, which is more on line with a view defended in philosophy by [7]. In this short paper we propose a diagrammatic view of analogical inference either for assessing that an analogical proportion holds, or for inferring d from a, b, c when possible. The approach relies on a particular structure of opposition between items represented by bitstrings. It enables us to handle both features-based representations and logical formulas. The paper starts with a brief refresher on the logical modeling of analogical proportions.

2 Refresher on Boolean Analogical Proportions

Following the behavior of numerical proportions, analogical proportions "a is to b as c is to d" are usually supposed to obey three postulates: i) $a:b::a:b$ (*reflexivity*); ii) $a:b::c:d \Rightarrow c:d::a:b$ (*symmetry*); iii) $a:b::c:d \Rightarrow a:c::b:d$ (*central permutation*) [5]. As a consequence, some other noticeable properties hold, such as $a:a::b:b$ (*identity*); $a:b::c:d \Rightarrow d:b::c:a$ (*extreme permutation*).

Viewing a, b, c, d as Boolean variables with value in $\mathbb{B} = \{0,1\}$, various equivalent Boolean formulas satisfy the postulates of an analogical proportion. One of them makes explicit that "a differs from b as c differs from d (and vice-versa)" [10]. This is logically expressed as a quaternary connective [10] by:

$$a : b :: c : d = ((a \wedge \neg b) \equiv (c \wedge \neg d)) \wedge ((\neg a \wedge b) \equiv (\neg c \wedge d))$$

which is true only for the 6 valuations given in the opposite table. It offers the minimal Boolean model assuming reflexivity and stability under central permutation. This model is symmetrical. As can be seen, 0 and 1 play exchangeable roles.

a	b	c	d
0	0	0	0
1	1	1	1
0	0	1	1
1	1	0	0
0	1	0	1
1	0	1	0

Analogical proportions can be extended componentwise to items represented by vectors, such as $\mathbf{a} = (a_1, \ldots, a_n)$, defined on the same set of Boolean attributes, namely: $\mathbf{a} : \mathbf{b} :: \mathbf{c} : \mathbf{d}$ iff $\forall i \in [1, n]$, $a_i : b_i :: c_i : d_i$.

Analogical inference relies on the solving of analogical equations $a_i : b_i :: c_i : x_i$ for some unknown value x_i of some attribute i. The solution is unique when it exists (as can be seen in the above table), but it may not exist. Indeed $1 : 0 :: 0 : x$ and $0 : 1 :: 1 : x$ have no solution x in $\mathbb{B}$ w. r. t. the logical view of analogical proportion.

3 A Particular Square of Opposition

An analogical proportion can be obtained by taking pairs of mutually exclusive Boolean properties, $(p, p'), (q, q'), (r, r')$ (i.e., such that $p \wedge p' = \bot, q \wedge q' = \bot, r \wedge r' = \bot$), and then by considering the four items $\mathbf{a}, \mathbf{b}, \mathbf{c}, \mathbf{d}$ respectively described on the six properties (p, q, r, r', q', p') according to Table 1 below:

Table 1. Analogical proportion induced by mutually exclusive properties

	p	p'	q	q'	r	r'
a	1	0	1	0	1	0
b	1	0	1	0	0	1
c	1	0	0	1	1	0
d	1	0	0	1	0	1

It can be seen that for any vector component, $(a_i \wedge \neg b_i \equiv c_i \wedge \neg d_i) \wedge (\neg a_i \wedge b_i \equiv \neg c_i \wedge d_i)$ holds true, where $\mathbf{a} = (a_1, a_2, a_3, a_4, a_5, a_6)$, and each component a_i ($i = 1, \cdots, 6$) refers to the truth-value of $\mathbf{a}$ for properties p, p', q, q', r, r' respectively, $\mathbf{b}, \mathbf{c}, \mathbf{d}$ being similarly defined. We can observe that (a_i, b_i, c_i, d_i) for $i = 1, \cdots, 6$ takes respectively the values $(1,1,1,1), (0,0,0,0), (1,1,0,0), (0,0,1,1), (1,0,1,0)$, and $(0,1,0,1)$ which are the six 4-tuples that make an analogical proportion true. Thus the proportion $\mathbf{a} : \mathbf{b} :: \mathbf{c} : \mathbf{d}$ holds true.

It is worth noticing that $\mathbf{a}, \mathbf{b}, \mathbf{c}, \mathbf{d}$ make a kind of square of opposition (not to be confused with the traditional one [1, 11]), as pictured below:

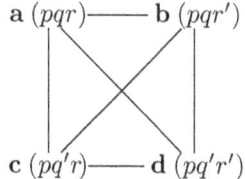

Indeed the edges of the above square reflect oppositions since the two items they link differ on one property (either q vs. q', or r vs. r') and then they cannot be true together, while for the diagonals, the opposition concerns two of the three pairs ((q, q') and (r, r')) of mutually exclusive properties. Note also the four items share the property p (and are false on p'). Indeed (a_i, b_i, c_i, d_i) is equal to $(1, 1, 1, 1)$, and to $(0, 0, 0, 0)$ respectively for $i = 1, 2$. In order a, b, c, d be four *distinct* items, we need *at least two attributes* i exhibiting both a pattern (s, s, t, t) as here for $i = 3, 4$, and a pattern (s, t, s, t) as here for $i = 5, 6$.

We now assume that the items are associated with bitstrings (this type of encoding has been recently considered in relation with (classical) structures of opposition [4], but the idea of binary coding dates back to [9]). Now if we consider for **a** a bitstring of a given length (at least 2), say, e.g., 10100 and for **b** a bitstring of the same length, but different from **a** say, e.g., 11101, then define **c** as the complement of **b**, which gives 00010, and **d** as the complement of **a**, which gives 01011. Possibly complete the four bitsprings with the same chain of 1 or 0, say 011, then we get

$a : \mathbf{b} :: \mathbf{c} : \mathbf{d} = 10100011 : 11101011 :: 00010011 : 01011011$,

which is obviously an analogical proportion, as can be checked, on each digit of the four items. This procedure is general and always yields an analogical proportion between four distinct items. Moreover, it guarantees that on the digits where **a** and **b** differ, **a** and **b**, as well **c** and **d** are in opposition. Similarly, **a** and **c**, and **b** and **d** are in opposition on the digits where they differ. So we always obtain a square of opposition of the same kind as the one previously introduced. This is exemplified on the opposite figure. A simple visual check is enough for concluding if the analogical proportion holds or not.

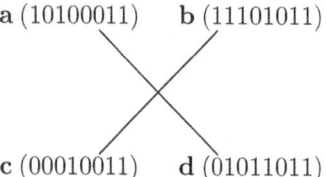

4 Analogical Proportions Between Formulas

Any propositional formula with n variables can be described by a bitstring of size 2^n. Thus, if we consider two Boolean variables x and y, we have the interpretations xy, $x\neg y$, $\neg xy$, $\neg x\neg y$. Taking in this order, x is encoded by 1100, $x \wedge y$ by 1000, and so on. On the diagram below we can check that a known universal analogical proportion [12], namely $x : x \wedge y :: x \vee y : y$, since we have indeed 1100 : 1000 :: 1110 : 1010, digit by digit. We can see that the first and the fourth digit are equal for the four items, while the second and third digit are in opposition between x and y and between $x \wedge y$ and $x \vee y$.

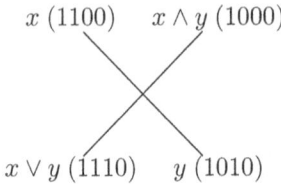

This approach enables us not only to check if an analogical proportion holds or not between four formulas, but also to perform analogical inference, as explained on the following example. Let us consider 3 variables x, y and z. We take the interpretations in the following order: xyz, $x\neg yz$, $xy\neg z$, $x\neg y\neg z$, $\neg xyz$, $\neg x\neg yz$, $\neg xy\neg z$, $\neg x\neg y\neg z$. Let us consider the 3 formulas $\mathbf{a} = (x \vee y) \wedge z$, $\mathbf{b} = (x \wedge y) \vee (\neg x \wedge y \wedge z)$, $\mathbf{c} = x \wedge z$. Is there a formula $\mathbf{d}$ making an analogical proportion with the 3 others ? This corresponds to the following diagram below where the bitsprings representing the formulas are given.

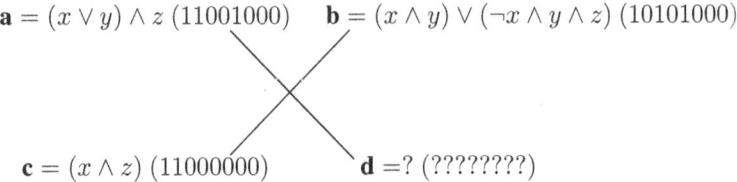

We can see that the first, the seventh, and the eigth digits are equal for $\mathbf{a}$, $\mathbf{b}$, $\mathbf{c}$. So their values should be the same for $\mathbf{d}$. The analogical equations are solvable on the other digits, as can be seen. We obtain for $\mathbf{d}$, (10100000), so $\mathbf{d} = (x \wedge y)$. If there is no solution $\mathbf{d}$, given 3 formulas $\mathbf{a}$, $\mathbf{b}$, $\mathbf{c}$, we can discover it as well by simple inspection of the digits.

5 Concluding Remarks

We have presented a simple diagrammatic machinery for checking analogical proportions, or for solving analogical equations when possible. This approach could be extended to partial models, replacing unknown digits by variables and taking track of the oppositions that should hold. Interestingly enough S. Klein [8] has shown long time ago that analogical proportions could bring some light on the Chinese I-ching hexagrams, which inspired [9]!

Acknowledgements. This research was supported by the ANR project "Analogies: from theory to tools and applications" (AT2TA), ANR-22-CE23-0023.

References

1. Béziau, J.Y.: New light on the square of oppositions and its nameless corner. Log. Invest. **10**, 218–233 (2003)
2. Bounhas, M., Prade, H., Richard, G.: Analogy-based classifiers for nominal or numerical data. Int. J. Approx. Reason. **91**, 36–55 (2017)
3. Correa Beltran, W., Prade, H., Richard, G.: Constructive solving of Raven's IQ tests with analogical proportions. Int. J. Intell. Syst. **31**(11), 1072–1103 (2016)

4. Demey, L., Smessaert, H.: Combinatorial bitstring semantics for arbitrary logical fragments. J. Philos. Logic **47**, 325–363 (2018)
5. Dorolle, M.: Le Raisonnement par Analogie. Bibliothèque de Philosophie Contemporaine, Presses Universitaires de France, Paris (1949)
6. Gentner, D.: Structure-mapping: a theoretical framework for analogy. Cogn. Sci. **7**(2), 155–170 (1983)
7. Hesse, M.B.: On defining analogy. Proc. Aristot. Soc. **60**, 79–100 (1959)
8. Klein, S.: Culture, mysticism & social structure and the calculation of behavior. Technical report, Computer Science Department, University of Wisconsin-Madison, C. S. Technical report #462 (1981)
9. Leibniz, G.W.: Explication de l'arithmétique binaire, qui se sert des seuls caractères 0 & 1; avec des Remarques sur son utilité, & sur ce qu'elle donne le sens des anciennes figures Chinoises de Fohy. Compte Rendu de l'Acad. des Sci. (Paris), Mémoires, pp. 85–89 (1703)
10. Miclet, L., Prade, H.: Handling analogical proportions in classical logic and fuzzy logics settings. In: Sossai, C., Chemello, G. (eds.) ECSQARU 2009. LNCS (LNAI), vol. 5590, pp. 638–650. Springer, Heidelberg (2009). https://doi.org/10.1007/978-3-642-02906-6_55
11. Parsons, T.: The traditional square of opposition. In: Zalta, E.N. (ed.) The Stanford Encyclopedia of Philosophy. Metaphysics Research Lab, Stanford University (2017)
12. Prade, H., Richard, G.: From analogical proportion to logical proportions. Log. Univers. **7**(4), 441–505 (2013)

Correction to: A Building-Block Approach to the Diversity of Visualization Types – Each Type Expressed Visually, and as a Systematically Generated Sentence

Yuri Engelhardt and Clive Richards

Correction to:
Chapter 3 in: J. Lemanski et al. (Eds.): *Diagrammatic Representation and Inference*, **LNAI 14981**,
https://doi.org/10.1007/978-3-031-71291-3_3

In the originally published version of this chapter, figure 4 was displayed incorrectly. This now has been corrected.

The updated version of this chapter can be found at
https://doi.org/10.1007/978-3-031-71291-3_3

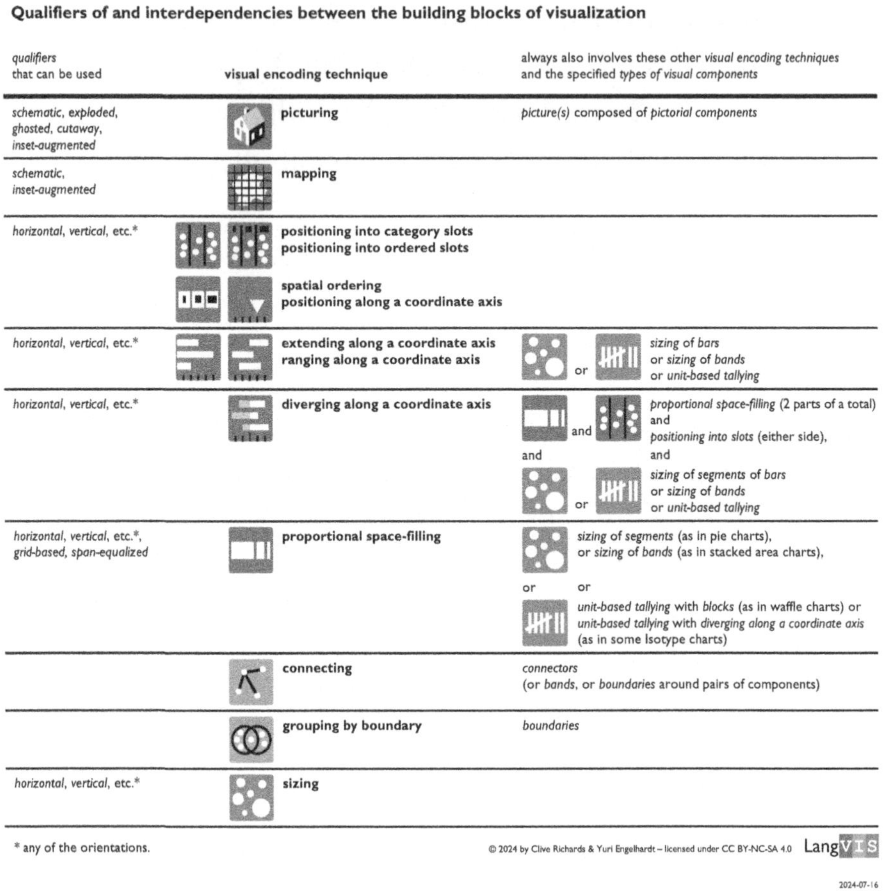

Qualifiers of and interdependencies between the building blocks of visualization

qualifiers that can be used	visual encoding technique	always also involves these other visual encoding techniques and the specified types of visual components
schematic, exploded, ghosted, cutaway, inset-augmented	picturing	picture(s) composed of pictorial components
schematic, inset-augmented	mapping	
horizontal, vertical, etc.*	positioning into category slots / positioning into ordered slots / spatial ordering / positioning along a coordinate axis	
horizontal, vertical, etc.*	extending along a coordinate axis / ranging along a coordinate axis	sizing of bars or sizing of bands or unit-based tallying
horizontal, vertical, etc.*	diverging along a coordinate axis	proportional space-filling (2 parts of a total) and positioning into slots (either side), and sizing of segments of bars or sizing of bands or unit-based tallying
horizontal, vertical, etc.*, grid-based, span-equalized	proportional space-filling	sizing of segments (as in pie charts), or sizing of bands (as in stacked area charts), or or unit-based tallying with blocks (as in waffle charts) or unit-based tallying with diverging along a coordinate axis (as in some Isotype charts)
	connecting	connectors (or bands, or boundaries around pairs of components)
	grouping by boundary	boundaries
horizontal, vertical, etc.*	sizing	

* any of the orientations.

© 2024 by Clive Richards & Yuri Engelhardt – licensed under CC BY-NC-SA 4.0

Fig. 4. Some visual encoding techniques (middle column in bold type) can be specified by *qualifiers*, such as *orientations* (left column). In addition, interdependencies exist between building blocks – some visual encoding techniques are always used in combination with one or more other visual encoding techniques and/or with a certain type of visual component (right column).

Author Index

A
Abe, Hirohiko 232
Aebischer, Tullio 458
Ando, Risako 232
Anger, Claudia 476
Archambault, Daniel 190
Aznar, Marcos Bautista López 449

B
Ballout, Mohamad 61
Barker-Plummer, Dave 101
Başkent, Can 300
Beisecker, Dave 137
Bellucci, Francesco 182
Bhattacharjee, Reetu 207, 429
Blake, Andrew 377
Boehm, Timon G. 11

C
Campell, Joannes B. 480
Chakrobarty, Mihir Kumar 284
Chapman, Peter 377
Chen, Binfeng 350
Cheng, Peter C-H. 11

D
De Klerck, Alexander 153
de Vries, Erica 335
Demey, Lorenz 111, 153, 476
Dobler, Alexander 377
Doğan, Kerem 44
Domrös, Sören 76, 402
Dürrschnabel, Dominik 165

E
Engelhardt, Yuri 28
Erhas, Atahan 275
Estep, Sam 454

F
Fang, Shitao 462
Frijters, Stef 275
Fukuda, Mari 358

G
Galilée, Martin 335
Gottfried, Björn 215
Greenstein, Noah 472

H
Harriman, Hwei-Shin 454
Hasida, Kôiti 462
Haunert, Jan-Henrik 190
He, Lin 350
Helmke, Stefan 44
Hossain, Sohail 284

I
Ideno, Takashi 467

K
Karilas, Vili Valtteri 462
Kasperowski, Maximilian 402
Kittsteiner, Christina 145
Kozak, Piotr 129
Kühnberger, Kai-Uwe 61

L
Lemanski, Jens 207, 429
Lu, Ran 366

M
Manalo, Emmanuel 358, 366
Marshall, Guy Clarke 421
Matsubara, Yusuke 462
Mayer, Richard E. 327
Minarčík, Jiří 454
Mineshima, Koji 232, 393
Miotti, Giulia 3

Moktefi, Amirouche 207
Moretti, Alessio 84
Morii, Masahiro 467
Morishita, Takanobu 232
Müller, Michael A. 480

N
Ni, Wode 454
Nöllenburg, Martin 377

O
Okada, Mitsuhiro 232, 467
Oostra, Arnold 267
Ozeki, Kentaro 232

P
Pietarinen, Ahti-Veikko 316
Prade, Henri 485
Priss, Uta 165

R
Rentz, Niklas 402
Richard, Gilles 485
Richards, Clive 28
Rodgers, Peter 190, 377
Rottmann, Peter 190

S
Sato, Yuri 393
Scheffler, Robert 44
Schoenherr, Johanna 327
Schorlemmer, Marco 61
Schumann, Andrew 251

Schwartz, Neil 335
Shibata, Kenichi 462
Shimojima, Atsushi 101
Smessaert, Hans 111
Sunshine, Joshua 454
Suzuki, Ayaka 393

T
Takemura, Kazuhisa 467
Tamari, Yuki 467
Tan, Kuanghuan 462
Toscano, Fernando Soler 449

V
Vignero, Leander 153
von Hanxleden, Reinhard 76, 402

W
Wallinger, Markus 377
Wang, Bei 190
Wrobel, Gregor 44

X
Xu, Fangzhou 316

Y
Yan, Xinyuan 190
Yao, Zifan 462

Z
Zhang, Zilian 462
Zhao, Jing 350

SPRINGER NATURE

GPSR Compliance

The European Union's (EU) General Product Safety Regulation (GPSR) is a set of rules that requires consumer products to be safe and our obligations to ensure this.

If you have any concerns about our products, you can contact us on ProductSafety@springernature.com

In case Publisher is established outside the EU, the EU authorized representative is:

Springer Nature Customer Service Center GmbH
Europaplatz 3
69115 Heidelberg, Germany

The manufacturer's authorised representative in the EU is Springer Nature Customer Service Centre GmbH, Europaplatz 3, 69115 Heidelberg, Germany. If you have any concerns regarding our products, please contact ProductSafety@springernature.com

Printed and bound by CPI Group (UK) Ltd, Croydon, CR0 4YY

25/03/2026

02078187-0019